Phthalocyanine Research and Applications

Author

Arthur L. Thomas
Chemical Engineer
Greenwich, Connecticut

CRC Press
Boca Raton Ann Arbor Boston

Library of Congress Cataloging-in-Publication Data

Thomas, Arthur Louis.
Phthalocyanine research and applications / Arthur L. Thomas.
p. cm.
Includes bibliographical references.
ISBN 0-8493-4624-X
1. Phthalocyanines. I. Title.
QD441.T48 1990
547.8′6--dc20 89-25290
CIP

This book represents information obtained from authentic and highly regarded sources. Reprinted material is quoted with permission, and sources are indicated. A wide variety of references are listed. Every reasonable effort has been made to give reliable data and information, but the authors and the publisher cannot assume responsibility for the validity of all materials or for the consequences of their use.

Direct all inquiries to CRC Press, Inc., 2000 Corporate Blvd., N.W., Boca Raton, Florida 33431.

International Standard Book Number 0-8493-4624-X

Library of Congress Card Number 89-25290
Printed in the United States

INTRODUCTION

The new information in the field of the phthalocyanines which has been generated from 1984 through 1987 has been brought together, under one cover, with the exception of the topics of dyes and naphthalocyanines.

Phthalocyanine compounds have, since their final discovery and elucidation of structure in 1934, attained a signal success as an object of research and of practical application. Indeed, they have been honored as one of the first class of compounds the crystals of which have been grown in the microgravity of space aboard a space shuttle. Citations covered in this volume number 3000. In a sense, like a scythe cutting a swath across a field of grain, each stalk and each grain have been harvested and identified from the fertile field of creativity. As much discussion or description to each citation that can reasonably be given *non troppo* in one volume has been presented.

In the field of the phthalocyanines, science and technology have become connate. The ease of preparation of phthalocyanine compounds from inexpensive and abundant raw materials, their ideality as a molecule for theoretical study, and their ready acceptance in the marketplace for a variety of uses have earned them a prominent and respected place in chemical science and technology.

Another indicator of the interest in and the health of the phthalocyanines is the estimated 1987 U.S. production of about 16 million lb of phthalocyanine blue pigment or toner at a unit value of about $5 to 12/lb and production of phthalocyanine green pigment of about 2 million lb at a unit value of about $8/lb. In terms of tons produced in 1987, phthalocyanine blue (pigment blue 15:3 beta form copper phthalocyanine) is one of the most popular toners or pigments in the U.S., in second place only to pigment yellow 12. In addition, about 758,000 lb of copper phthalocyanine sodium disulfonate (C. I. direct blue 86) were also manufactured in 1987.

Phthalocyanine blues and greens are used as pigments because of their outstanding stability to light, heat, acids, and alkalies and their insolubility in organic solvents and water. They are used extensively in printing inks, coatings, paints, and plastics. They yield outstanding, durable automobile colors and they can be used in the matching of colors in plastics for car interiors.

The phthalocyanines are also used as catalysts in sulfur oxidation in the petroleum industry and they find use in lasers, lubricants, medicine, photography, and xerography and as indicators.

About three research articles and patents relating to phthalocyanine compounds are published every working day of the week. They represent an expanding base of knowledge from which new uses can be found. Over 5000 phthalocyanines are now known, and new phthalocyanines are being created every year.

The author's wish is that this monograph, like its predecessor volumes of which he was co-author, published in 1963 and 1983, will respond to the requirements of its readers for a ready access to the world literature in the field of phthalocyanine science and technology.

ACKNOWLEDGMENT

The author especially acknowledges with many thanks the courtesy extended to him by the library staff at The University of Connecticut at Stamford to make use of reference materials and services, under the guidance of the Head Librarian, Mrs. Carmel O'Neill.

The author also thanks Mrs. Mildred Hunt of The Chemists' Club Library in New York City and Mrs. Judith A. Douville, Information Scientist, of NdS Information Consultants, Middletown, CT, who promptly provided essential information when it was required.

I also express my very great appreciation to my wife, Charlotte, who tacitly encouraged my efforts as time well spent in a constructive endeavor.

Arthur L. Thomas
March 1989

DEDICATION

This monograph is dedicated to the memory of Jacob Poorman (1742—1784), his son William Poorman (1780—1852), and his grandson Samuel Poorman (1824—1914).

Jacob Poorman, blacksmith and yeoman, served as a private in the Lancaster County, PA, militia; he was captured at Fort Washington, Manhattan Island, November 16, 1776, by the British and Hessians. In 1778 he supplied a bushel of grain and two bushels of forage to the Continental Army. William Poorman served in the Lancaster County militia during the War of 1812. His son Samuel was a ship's carpenter in the U.S. Navy from 1842 to 1847 and later devoted his career to the construction of church organs. The author is a direct descendant of Jacob Poorman.

TABLE OF CONTENTS

Chapter 1

INTRODUCTION

I. HISTORY

Phthalocyanine is derived from the Greek term for naphtha (rock oil) and cyanine (dark blue).

The word "phthalocyanine" was first used by Professor Reginald P. Linstead of the Imperial College of Science and Technology in 1933 to describe a class of organic compounds, which consists of metal-free or di hydrogen phthalocyanine, and metal phthalocyanines, and their derivatives. The phthalocyanine molecule is shown in the structure on the next page.

The two hydrogen atoms in the center of the molecule have been replaced by metals from every group of the periodic table to form the metal phthalocyanines. At least 70 metal phthalocyanines have been prepared. Table 1 enumerates these metals in the form of the Periodic Table.

As a result of the 16 reactive sites on the four benzene units that are part of the phthalocyanine molecule, an astonishing number, over 5000, of phthalocyanine compounds have been prepared.

The phthalocyanines were discovered by happenstance. In 1907, at the South Metropolitan Gas Company in London, Braun and Tcherniac found a blue substance after heating *o*-cyanobenzamide. This substance probably was phthalocyanine. In 1927 de Diesbach and von der Weid at the University of Fribourg attempted to make nitriles of benzene by reacting *o*-dibromobenzene with cuprous cyanide. In doing so, they obtained a blue product, which was probably copper phthalocyanine.

In 1928, at the Grangemouth works of Scottish Dyes Ltd., during the preparation of phthalimide from phthalic anhydride and ammonia, a blue impurity was observed in the reaction mass. This blue substance was iron phthalocyanine, formed by the reaction of phthalimide with the iron wall of the reaction vessel across a flaw in the glass lining.

The first patent issued to compounds that we now know are phthalocyanines was British Patent 322,169, to Dandridge, Drescher, and Thomas of Scottish Dyes Ltd., in 1929.

Also beginning in 1929, Professor Linstead and his students, at the University of London, supported by grants from Imperial Chemical Industries, in a series of investigations determined the structure of the phthalocyanine molecule and announced this achievement in 1933 and 1934.

X-ray studies of phthalocyanine and of its metal derivatives including nickel, copper, and platinum, by Professor J. Monteath Robertson and co-workers, from 1933 to 1940, demonstrated that phthalocyanine, and the nickel, copper, and platinum phthalocyanines form monoclinic crystals, space groups P2/a, with two molecules per unit cell, each with center symmetry. The entire molecule lies in one plane, with the possible exception of the two central hydrogen atoms in di hydrogen phthalocyanine.

The neat fit of the central metal atom in phthalocyanine, the coordination and covalent coordination of the central metal atom, and the high degree of aromaticity of the surrounding nitrogen and benzene structures which form the molecule may explain the unusual stability of the phthalocyanines. They are exceptionally stable to acids, alkalies, moisture, heat, light, and solvents of all kinds. Generally, they decompose by pyrolysis at or above 500°C. Copper phthalocyanine, among others, is soluble in concentrated sulfuric acid and precipitates from dilute sulfuric acid. The phthalocyanines are only slightly soluble in organic solvents and they are insoluble in water.

Because of their excellent tinctorial qualities as well as of the properties of stability and

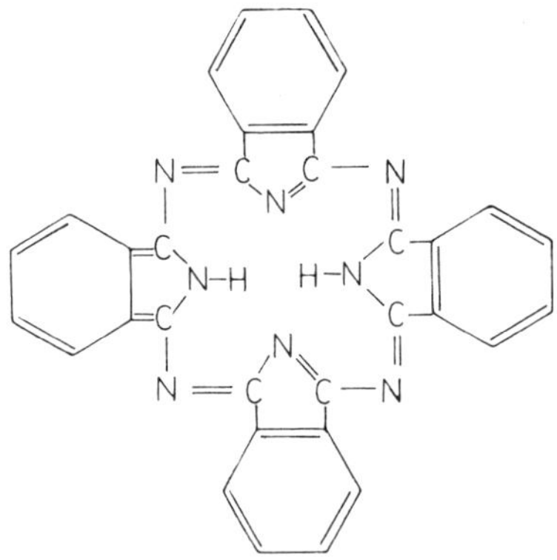

TABLE 1
Metals that Have Been Used to Make Metal Phthalocyanines

H																	
Li	Be																
Na	Mg											Al	Si	P			
K	Ca	Sc	Ti	V	Cr	Mn	Fe	Co	Ni	Cu	Zn	Ga	Ge	As			
		Y	Zr	Nb	Mo	Tc	Ru	Rh	Pd	Ag	Cd	In	Sn	Sb			
	Ba	La	Hf	Ta	W	Re	Os	Ir	Pt	Au	Hg	Tl	Pb				
		Ac															
Lanthanides			Ce	Pr	Nd	Pm	Sm	Eu	Gd	Tb	Dy	Ho	Er	Tm	Yb	Lu	
Actinides			Th	Pa	U	Np	Pu	Am									

insolubility of the phthalocyanines, and of copper phthalocyanine in particular, copper phthalocyanine was readily accepted as an outstanding blue pigment. *The New York Times* related, in 1935, "BLUE PIGMENT DISCOVERY IS THE FIRST IN 100 YEARS. London, Nov. 25 — Discovery of a new blue dye possessing all the qualifications of a pigment was announced today by the Imperial Chemical Industries, Ltd. It is named Monastral fast blue and is the first discovery of a blue pigment for over a century." Monastral fast blue was copper phthalocyanine. In 1936 I. G. Farbenindustrie began to produce copper phthalocyanine at Ludwigshafen, and the Du Pont Company followed suit at Deepwater Point, NJ, in the late 1930s. It is estimated that during 1985 production of phthalocyanine pigments and dyes had reached 14 million pounds annually in the U.S. alone, and was still growing, achieved by about 12 producers.

In the 1930s through the 1950s, X-ray spectra, polymorphism, absorption spectra, magnetic properties, catalytic properties, oxidation and reduction, photoconductivity, and physical properties such as solubility, and photochemical, dielectric, and semiconductor properties were studied. Also, 45 metal phthalocyanines in all groups of the periodic table were synthesized.

Commercial manufacturing methods of phthalocyanine and copper phthalocyanine as well as methods for eliminating crystallization and flocculation of the pigment form evolved. In addition, partially chlorinated and fully chlorinated copper phthalocyanines and methods for their manufacture were developed.

The first phthalocyanine dye was a polysulfonate, as described in British Patent 322,169. Since then, in the 1930s through the 1960s, phthalocyanine sulfonic acid dyes, phthalocyanine sulfonic chlorides, other water soluble phthalocyanine dyes, sulfur dyes, azo dyes, vat dyes, leuko dyes, chrome dyes, triazine dyes, and dye precursors were demonstrated and patented.

Also, the polymorphism was recognized by the discovery of α, β, γ, δ, π ρ, R, and ϵ-crystal forms of Cu and other metal phthalocyanines. The ρ form was reported in 1977. In 1967 the x form of metal free phthalocyanine was reported, with application to electrophotography. η- and τ-metal-free and metal phthalocyanines were reported in the early 1980s, with application as photoreceptors in photoconductive layers in electrophotography. Also, in the 1930s through the 1960s the use of phthalocyanines as colorants for paints, plastics, printing inks, and textiles, as well as their use in ball point pen inks, roofing granules, leather, gasoline, oil, photographic color prints, and even food coloring were announced. It was also found possible to polymerize the phthalocyanines.

From the time of their discovery to 1962, about 1200 literature references and patents signaled the novel developments and progress in phthalocyanine science and technology.[1] From 1963 to 1975 about 3500 journal and patent citations marked the advances in this field.[2] It is estimated that from 1975 to 1987 about 9000 journal articles and patents that related in some way to the phthalocyanines were published.

II. CU PHTHALOCYANINE

Copper phthalocyanine is the most sought after phthalocyanine pigment or product in terms of tonnage. It is extremely simple to prepare from such starting materials as urea, phthalic acid anhydride, cuprous chloride, and molybdenum trioxide catalyst, either in or with no solvent, with reaction at 200°C at atmospheric pressure. The crude copper phthalocyanine thus prepared in yields of over 90% is then purified by sulfuric acid treatments.

Copper phthalocyanine is the Process Blue of four color process printing. The important five commercially available forms of copper phthalocyanine pigment blue 15 are

Color index name	Color index number
Pigment blue 15	74160
α, crystallizing	
Pigment blue 15:1	74160
α, NC	74250 (copper monochloro phthalocyanine)
Pigment blue 15:2	74160
α, NCNF	74250 (copper monochloro phthalocyanine)
Pigment blue 15:3	74160
β, NC	
Pigment blue 15:4	74160
β, NCNF	

NC = noncrystallizing and NCNF = noncrystallizing and nonflocculating. α-Copper phthalocyanine is a desirable red shade blue pigment but it crystallizes to the less desirable green shade but more stable β-copper phthalocyanine. The α-form can be stabilized, however, and as such becomes pigments 15:1 and 15:2.

III. NOMENCLATURE

In this monograph, ''pc'' is used to stand for phthalocyanine. In addition, the metals in metal pcs are abbreviated by their standard periodic table abbreviations. Copper phthalocyanine becomes Cu pc, for example. Metal-free phthalocyanine may also be called free base pc, metal-free pc, di hydrogen pc, H_2 pc, or simply pc. When conversing about the pcs, one may be relating to both metal, M, and nonmetal pcs; however, when conversing about pc, it almost invariably means that one is relating to metal-free pc.

IV. RECENT ADVANCES

M pcs are gaining attention as electrocatalysts, with focus on Co and Fe pcs. Pcs are notable contributors to inks because of their intense dark blue color. Pcs are contenders as components in the development of high energy density batteries. When M pcs form crystals in which the molecules are packed parallel to each other, and with suitable doping, such as with iodine, high conductive ''molecular metals'' result. Pcs are candidates for use in photodynamic therapy, a cancer treatment that employs one of several dyes that are injected and selectively adsorbed by cancer cells. The cancerous tissue interacts with the dye to form a toxin under the influence of lasers, and the toxin kills the cancer cells. The Beckman Laser Institute, and many other centers, are active in this research. Pcs offer the possibility of use as photovoltaic devices and as photocatalysts for the reduction of water to hydrogen. Pcs can also photocatalytically reduce methyl viologen. The photoconductive properties of the pcs make them obvious candidates for use in electrophotographic processes and this interest has spawned an extensive patent literature, particularly Japanese. Electrochromism is a property that places the Pcs under scrutiny as candidates in display devices. pcs are also under study as sensors for minute or trace amounts of gases in the atmosphere. The review ''The Other Periodic Chart'' with 49 references summarizes well the current thrust of interest in the development of pcs for useful purposes.[3]

REFERENCES

1. **Moser, F. H. and Thomas, A. L.,** *Phthalocyanine Compounds,* Reinhold Publishing, New York, 1963.
2. **Moser, F. H. and Thomas, A. L.,** *The Phthalocyanines,* CRC Press, Boca Raton, FL, 1983.
3. **Lever, A. B. P.,** The other periodic chart, *CHEMTECH,* August, 506, 1987.

Chapter 2

SYNTHESIS

I. DIMERS AND BRIDGED BINUCLEAR PCS

μ-Cyano M phthalocyanines (pcs) $(M\ pc\ CN)_n$, Co, Cr, Fe, Mn, and $(H\ pc\ Fe\ CN)_n$ are prepared. $(Co\ pc\ CN)_n$ and $(Fe\ pc\ CN)_n$ have ~0.01 1/Ω-cm conductivity at room temperature without doping, in the same range as iodine doped $(Si\ pc\ O)_n$.[1]

The reaction of CrO tetraphenylporphyrin with Fe(II) pc forms an μ-oxo complex.[2]

μ-Thiocyanato Co, Fe, and Mn pcs and μ-N_3 Co, Fe, and Mn pcs are prepared.[3]

The μ-nitrido-bridged Fe pc dimer pc-Fe-N-Fe-pc is prepared by reacting Fe pc and sodium azide in refluxing α-$ClC_{10}\ H_7$.[4,5]

Di hydrogen, Co, and Cu pc dimers, in which the two pc nuclei are covalently linked through five atom bridges, are prepared and characterized. Some new di hydrogen, Co, Cu, and Zn 2,9,16,23 tetraalkoxy pcs, some of which are quite soluble in organic solvents, are also prepared.[6,7]

In the preparation of Lu di pc, in addition to the green Lu octa-4-*tert*-butyl di pc, a stable neutral radical of a sandwich-type structure is formed as a blue diamagnetic complex.[8]

The bis(tetra-4-*tert*-butyl pcs) of Er, Gd, La, Lu, Nd, Pr, and Sm are made and their spectral and electrochemical properties are studied.[9] Heating a rare earth acetate with 4-*tert*-butylphthalonitrile to 290°C leads to the green Er, Gd, and Lu and to the blue Nd and Sm complexes.

$Ln_2\ pcs_3$ (Ln = lanthanide) are prepared and characterized.[10] $(Fe\ pc\ pyridine)_2\ O$ is prepared by the interaction of $(Fe\ pc)_2$ (μ-O) with gaseous or liquid pyridine.[11]

The crystalline isomer of Fe(III) (μ-oxo) pc_2, earlier identified as μ-oxo(2), is further studied by magnetic susceptibility measurements, ESR, and Moessbauer spectra.[12]

Monomeric and bridged Fe(II) pc and Ru(II) pc compounds are formed with 2,3,5,6-tetramethyl-1,4-diisocyanobenzene and 2,3,5,6-tetrachloro-1,4-diisocyanobenzene as axial ligands.[13]

Face-to-face (1:1) complexes are formed between Mn(III) tetrakis(4-methylpyridyl) porphyrin and Mn(III) tetrakis(sulfonatophenyl) porphyrin or Mn(II) tetrasulfo pc and between Cu(II) tetrakis (4-methylpyridyl)porphyrin and Cu(II) tetrasulfo pc.[14]

There is no evidence that visible light from a spectrophotometer alters the dimerization of Co(II) tetrasulfonated pc.[15,16]

II. POLYNUCLEAR PCS

In a study of binuclear "clamshell" M pcs, the following are prepared: Co, Cu, and Zn tetraneopentoxy pc; Co_2 or Cu_2 1,3-bis(trineopentoxyphthalocyaninoxy)-2,2-dimethylpropane; and Co_2 or Cu_2 1,3-bis(trineopentoxyphthalocyaninoxy)-2-ethyl-2-methylpropane.[17,18]

A 1,8-naphthalene bis pc is prepared from isoindolines of 1,2-$(CN)_2$-3-$[(CH_3)_3CCH_2O]C_6H_3$ and 1,8-tris(3′,4′-dicyanophenyl)naphthalene and its Co, Cu, and Zn complexes are studied.[19]

A binuclear pc-type compound, as shown in Figure 1, is made by cyclocondensation of diamino-β-isoindigo in the presence of cobalt acetate in ethylene glycol.[20]

The preparation, characterization, and intermolecular interactions of a planar metal-free binuclear pc and its cobalt derivatives are described.[21]

With pentaerythritol as framework, a tetranuclear Co pc tetrakis[(9,16,23-trineopentoxy

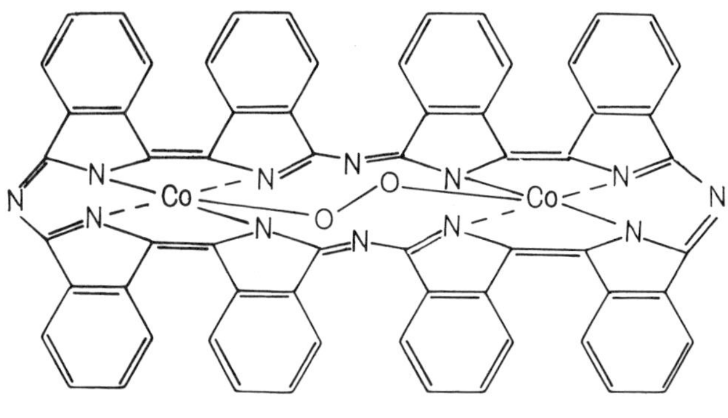

FIGURE 1. A binuclear pc-type compound.

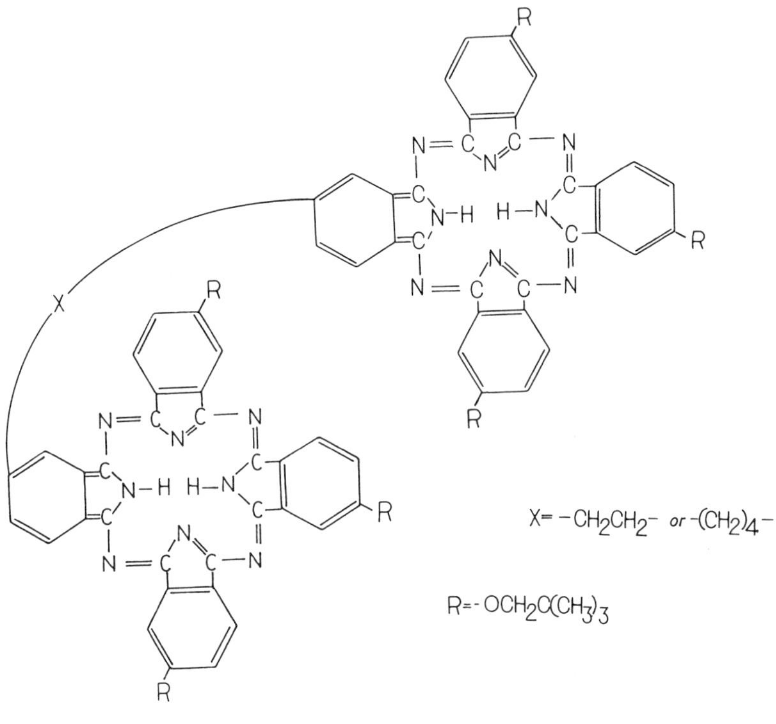

FIGURE 2. Binuclear pcs covalently linked through two- and four-atom bridges.

pc-2yl)oxy)methyl]methane with a spiro-type linkage is prepared. This cobalt dimer aggregates very strongly, the monolayers of which on a graphite electrode electrocatalytically reduce oxygen more efficiently than previously described analogous mononuclear and binuclear pcs.[22,23]

The diiminoisoindolines from 4-neopentoxyphthalonitrile and bis(3,4-dicyanophenyl) ether give, by mixed condensation, binuclear and trinuclear pcs covalently linked by one atom oxygen bridges with the dimer predominant.[24]

Binuclear and polynuclear pcs are prepared and oxygen reduction using these pcs as monolayers on graphite electrodes are studied.[25]

Binuclear pcs are prepared. They are covalently linked through two- and four-atom bridges, such as -CH_2CH_2- and -$(CH_2)_4$-, as shown in Figure 2.[26]

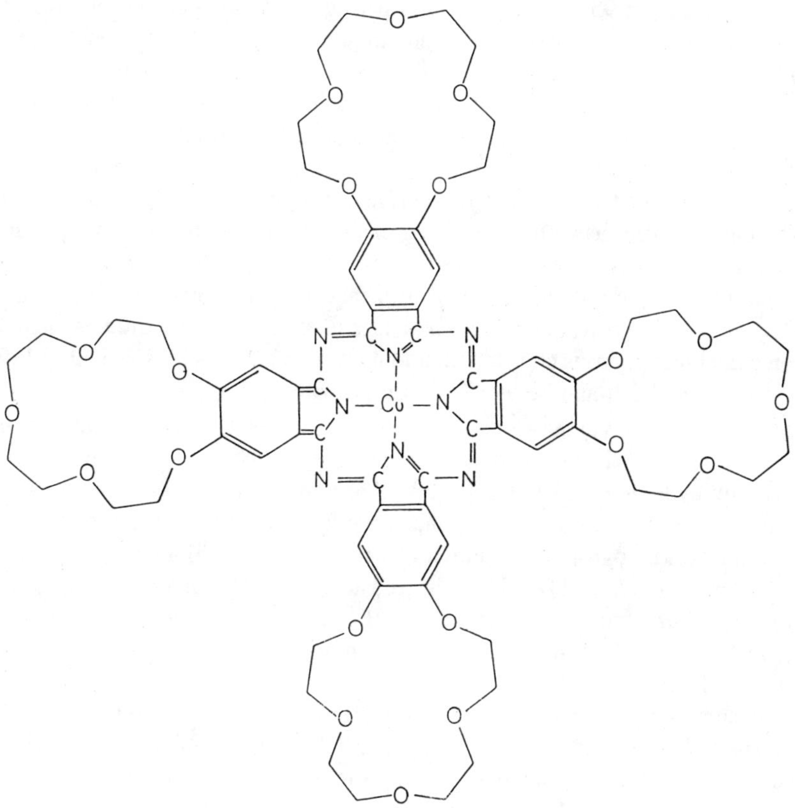

FIGURE 3. A crowned Cu pc.

III. CROWN PCS

Refluxing CuCN and 4,5-dibromobenzene-15-crown-5 in DMF in the presence of pyridine gives (Figure 3) a crowned Cu pc. Addition of potassium bromide leads to dimerization.[27]

Monoclinic crystalline Kryptofix 222 potassium pc ferrate, (KC222) (Fe pc) and (18-crown-6)Li(18-crown-6)(tetrahydrofuran) Li pc ferrate, [LiC18c.6][Li(THF)C18c.6][Fe pc] are prepared. The crown ether is made by adding a THF solution to $LiAlH_4$ and 18-crown-6 to a THF suspension containing Fe pc.[28]

IV. CO, CU, AND ZN PCS

The catalyzed ammonium molybdate condensation of phthalic acid derivatives, urea, and copper salts (cuprous chloride) to prepare Cu pc is effected with a small amount of organic solvents such as trichlorobenzene to obtain a pasty reaction mixture. The method avoids disadvantages of both the solution process (solvent recovery and higher energy use) and of the dry process (mediocre mixing and heat transfer).[29]

Phthalic anhydride, urea, cuporus chloride, and ammonium molybdate are heated in 4-$CH_3C_6H_4COOCH_3$ to 200°C to give 88.3% crude Cu pc of 95.1% purity. The ester solvent replaces toxic and carcinogenic aromatic solvents.[30]

Space time yield and product quality in the preparation of Cu pc by the urea bake process are improved by 0.1 to 5.0 wt % of sodium stearate antifoaming agent.[31]

Improved space time yields and less foaming are obtained to make Cu pc by the urea bake process if the source of copper is CuO.[32]

Cu pc is prepared in 93.4% yield from $K_2pc(diglyme)_2$ and cuprous chloride.[33]

Phthalimide, urea, cuprous chloride, and ammonium molybdate are heated in nontoxic *tert*-hexylbenzene at 190 to 195°C for 3 h to give a bright blue pigment.[34]

Phthalic anhydride, urea, cupric sulfate, ammonium molybdate, and calcined soda are mixed and heated to ~225°C; 5% sulfuric acid is added, forming Cu pc of 95% purity in 95% yield.[35]

In the preparation of Cu pc, cuprous chloride containing 2.0% cupric chloride dihydrate mixed with phthalonitrile has a heating rate greater than that with cuprous chloride without cupric chloride dihydrate.[36]

A phthalimide-urea mixture in trichlorobenzene, phthalic anhydride, and urea is heated to 170°C with water removal for 1 h; urea is added, and the mixture is mixed with a suspension containing cuprous chloride, ammonium molybdate, and is heated to 180°C. The preparation requires 4.5 h and gives 87.4% Cu pc with 96.6% purity.[37]

M pcs are prepared in yields >90% and with purity >95% by electrochemical conversion of aromatic o-dinitriles with metal salts. For example, Cu pc is prepared by electrolysis in the presence of hydrogen at 108°C and 280 A/m^2 from a solution of *o*-phthalodinitrile, triethylamine, anhydrous cupric sulfate, and tetraethylammonium sulfate.[38]

Phthalic anhydride, pyromellitic anhydride, urea, cuprous chloride, ammonium molybdate, *p*-phenylenediamine, and Hisol P are stirred at 180°C for 2.5 h, concentrated in vacuum, taken up in 1% hydrochloric acid, stirred at 80°C for 1 h, filtered, and washed with hot water to give pigment with 73-m^2/g surface area compared with 25 m^2/g for a control prepared without *p*-phenylenediamine.[39]

Phthalic anhydride, pyromellitic anhydride, urea, cuprous chloride, and ammonium molybdate are heated in Hisol P to give Cu pc, which is then ground with sodium chloride, diethylene glycol, and xylene to give pigment. Inks prepared from this pigment show a relative coloring strength of 112 compared with 100 using Cu pc made without pyromellitic anhydride.[40]

Phthalic anhydride, pyromellitic anhydride, urea, cuprous chloride, and ammonium molybdate are heated in sulfolane at 200°C under pressure to give π-Cu pc which is dispersed in a resin to give a clear blue coating.[41]

From 0.1 to 20% Cu pc Cl (m = 1 to 8) are added to crude Cu pc, dry pulverized, and immersed in crystallizing solvents such as xylene, filtered, and dried to give a pigment of specific surface area of 70 m^2/g and 100% β-crystal structure.[42]

Rate and selectivity of condensation of 1-amino-3-iminoisoindolenine with Co, Cu, or Ni cations to form M pcs are increased by adding alcoholic sugars, or polyhydroxy aldehydes, ketones, and acids. The condensation is catalyzed with urea in 1,2-propanediol to give pure M pcs in high yields.[43]

A novel method to prepare M pcs in inorganic semiconductive powders is described. Coverage roughly equivalent to a monolayer is achieved for Cd pc on CdS and Zn pc on ZnO.[44]

Co pc is prepared from phthalic anhydride, urea, and cobalt chloride in kerosine, with yields >98%.[45]

Cu pcs, the α form of which is more stable in aromatic solvents than α-Cu pc, are prepared in good yield by adding 4-nitrophthalic acid and/or its derivatives to the customary phthalic anhydride, cuprous chloride, urea, and ammonium molybdate in Hisol P alkylbenzene solvent.[46]

In the preparation of β-Cu pc by dispersing Cu pc in ethylene glycol in the presence of sodium chloride by use of spent perchloroethylene and distilling, pigment resistance to flocculation is improved by carrying out the dispersion in the presence of 0.06 to 0.15 part Cu pc $-(CH_2NHCH_2CH_2OH)_2$ and 0.1 to 0.3 part perchloroethylene per 1 part technical grade Cu pc.[47]

A combination of 95 parts crude Cu pc and 5 parts Cu pc -CH_2R (R = phthalimido) is dry milled to 31% β-Cu pc and then milled with xylene to 100% β-Cu pc.[48]

Crude Cu pc in concentrated sulfuric acid is dispersed in a nozzle by nitrogen at 20 l/min above the surface of water. After filtering, washing, and adding rosin, α-Cu pc is obtained with high color strength.[49]

The effect of reaction conditions on selectivity in crude Cu pc synthesis from phthalic anhydride, urea, and cuprous chloride in trichlorobenzene in the presence of ammonium molybdate is studied.[50]

Cu pc is purified by preparing an aqueous slurry of crude Cu pc of 65 to 85%, heating it to >75°C, and filtering at 200 to 300 psi at that temperature.[51]

A 1:10 weight ratio suspension of crude Cu pc in N-methylpyrrolidone is refluxed and filtered hot, and the residue washed with N-methylpyrrolidone and water to give pure Cu pc.[52]

Cu pc pigments are produced by treating them with a 1 to 4% Cu pc derivative such as Cu pc -$[CH_2NH(C_4H_9)_2]$. Their flocculation resistance is increased by this treatment.[53]

An apparatus for the fluidized bed drying of dispersions of Cu pc is described.[54]

R-Cu pc, having a redder color than α-Cu pc with no crystal conversion to β-Cu pc in aromatic solvents, is prepared by milling mixtures of α-Cu pc and with practically colorless compounds obtained from organic isocyanates and aliphatic amines together with grinding aids.[55]

The α- and β-Cu pc transformations in acid pasting and in organic solvents are described. A crystal seeding method to make β-Cu pc from α-Cu pc requires less energy than conventional kneading.[56]

The crystal phase transformation of α-Cu pc in toluene is first order with activation energy 60.2 kJ/mol.[57]

In drying α-Cu pc, organic solvent flushing and salt coprecipitation do not cause any crystal phase transformations when the proper solvent is used.[58]

Cu pc powder is mixed with butyl cellosolve and water, heated, and stirred. The mixture is treated with 2% aqueous sulfuric acid at 80 to 90°C, filtered, water washed, and dried to give a clear β-Cu pc.[59]

The course of micronization of Cu pc pigment in a ball mill is evaluated by the color strength development according to the Pigenot equation.[60]

Green Cu pc is prepared by heating a mixture of Cu pc, SO_2Cl_2, S_2Cl_2, $AlCl_3$, and CuCl.[61]

Color-strengthened Cu polyhalo pcs are prepared by recrystallization of finely agglomerated, unpurified Cu polyhalo pcs warmed in an organic-aqueous phase with carboxylic acid among other acids. The green pigment obtained has a high luster and powder softness.[62]

Cu pc pigment is synthesized in the laboratory in a masters degree program in fine arts.[63]

The Cu pc derivatives 4,11,18,25-tetrakis(chlorosulfonyl) Cu pc and its 3,10,17,24-isomer are used to prepare organic-soluble 4,11,18,25-tetrakis(2-pyridylmethylaminosulfuryl) Cu pc and its 3,10,17,24 isomer. These compounds are then used to prepare the water-soluble tetrakis(*N*-methyl-2-pyridiniomethylaminosulfonyl) Cu pc tetrakis(methyl sulfate) isomers. These complexes are thought to form dimers.[64]

The mechanism of Cu pc formation by the phthalic anhydride-urea process is elucidated by using nitrogen-14, nitrogen-15, and copper-63 in the preparation of Cu pc.[65]

Cu pc reacts with CH_2Cl_2 and nitric acid to give Cu pc $(NO_3) \cdot HNO_3$. In contrast to Cu pc, Cu pc $(NO_3) \cdot HNO_3$ is very soluble in organic solvents in the presence of formic acid.[66]

Co, Cu, Ni, and Zn 4,9,16,23-pc tetraamines are prepared.[67]

β-Co pc is treated with pyrazine (among other pyrazines and pyridines) to give Co pc pyrazine and Co pc $(\text{pyrazine})_2$, and Co pc-μ-pyrazine-Co pc. The stable polymer [Co pc $(\text{pyrazine})]_n$ can also be made.[68]

Co(III) pc thiocyanate and isothiocyanate compounds are synthesized. The crystal and molecular structure of Co(III)thiocyanato pc-pyridine is studied.[69]

Dimercaptide complexes of Co(III) tetrasulfo pc, Co tetrasulfo pc $(SC_6F_5)_2$, are made by reacting Co tetrasulfo pc with KSC_6H_5 in DMF.[70]

Peripherally substituted Co pcs with bisaxially coordinated ligands are prepared, such as $Co[(CH_3)_8$ pc] Cl_2 and $Co[(CH_3)_8$ pc]$(CN)_2$.[71]

Co pc·pyridine (NO_2) is formed by suspending Co pc (NO) in ethylene dichloride containing pyridine in an oxygen atmosphere at 50 to 60 atm at room temperature for 1 to 2 d.[72]

[Co X_2 pc] (1-) (X = Cl,Br) and [Co X_2(pc)](2^-)(X = OH, 7, Cl, Br) are prepared and characterized spectroscopically.[73]

Faujasite supported Co, Cu, and Ni pcs are prepared by reaction of ion exchanged zeolites with 1,2-dicyanobenzene.[74]

Solid photoactivator preparations include sulfonated Zn pc purified by ultrafiltration.[75]

Co(III) catena-μ-ethynylene pc is synthesized. This polymer is a novel one-dimensional polymer with 0.0003 1/Ω–cm conductivity at room temperature without doping, explained by an increase of carriers induced by the radical pc rings.[76]

Dimethylaminomethylated Cu pc is prepared by the methylation of Cu pc with formic acid and formaldehyde.[77]

Cu pc or Ni pc is prepared in high yield and purity with a reduced amount of chlorobenzene solvents by condensing phthalic anhydride or phthalimide with urea, and CuCl or nickel chloride catalyzed with ammonium molybdate in the presence of ammonium nitrate or sodium nitrate.[78]

Room light promotes the formation of di hydrogen pc in the reaction of phthalonitrile with sodium alkoxide at elevated temperature. Yields of Co pc and of Ni pc are only slightly higher with light than in the dark.[79]

The effectiveness of the use of feedstocks in Azerbaidzhan to manufacture aromatic nitriles is determined. The processes considered include the manufacture of Cu pc.[80]

Cu pc $(OC_2H_5)_8Cl_8$ is prepared from Cu pc Cl_{16}, ethanol, and potassium hydroxide in xylene. This compound absorbs in the IR and near IR.[81]

Zn 2,9,16,23-tetrakis(3-pyridyloxy) pc is prepared.[82]

H_2 5,5′,5″,5‴-tetrabromo-7,7′,7″,7‴-tetra-*tert*-butyl-2,3-naphthalocyanine and its Al, Cu, Lu, V, and Zn complexes are prepared. The Lu complex is a sandwich compound.[83]

Naphthalocyanines are prepared (Figure 4) where R-R^4 = $C_{5\text{-}12}$ alkyl.[84] Cu and Ni *tert*-butyl pcs are prepared.[85]

Cu benzaza pcs prepared from phthalic anhydride and quinolinic acid are fractionated into 15 fractions.[86]

Co(I) inner complex compounds are formed by reduction of Co pc by azobis (isobutyronitrile).[87]

A water-soluble Cu pc derivative (Figure 5) is prepared that is useful to color optical glass bright green.[88]

3,6-Bis(dodecyloxy)phthalonitrile, cuprous chloride and 1,8-diazabicyclo[5.4.0]-7-undecane in amyl alcohol are heated under reflux for 6 h to obtain a Cu alkoxy pc (Figure 6).[89]

Iodine containing pcs are obtained by placing di hydrogen, Cu, Fe, Mg, or Zn pc and iodine in two different areas of a closed vessel and heating in a reduced inert atmosphere or in a vacuum with the di hydrogen pc or M pc at 540 to 580°C and iodine at 70 to 100°C at an I_2/H_2 pc or M pc weight ratio of 2 to 10:1.[90]

Cu pc is purified by treating it with isobutyl phosphates.[91]

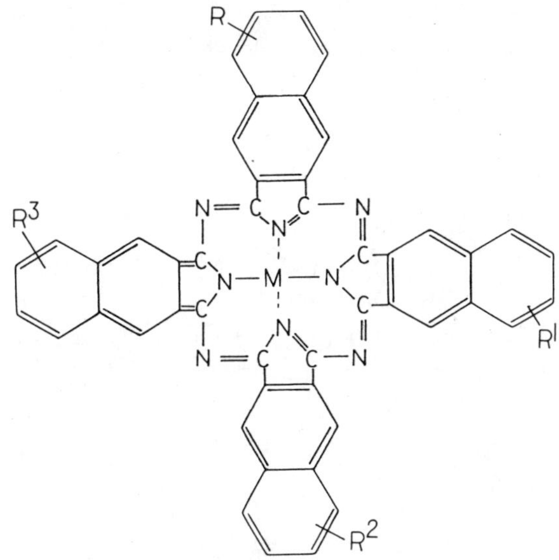

FIGURE 4. A naphthalocyanine.

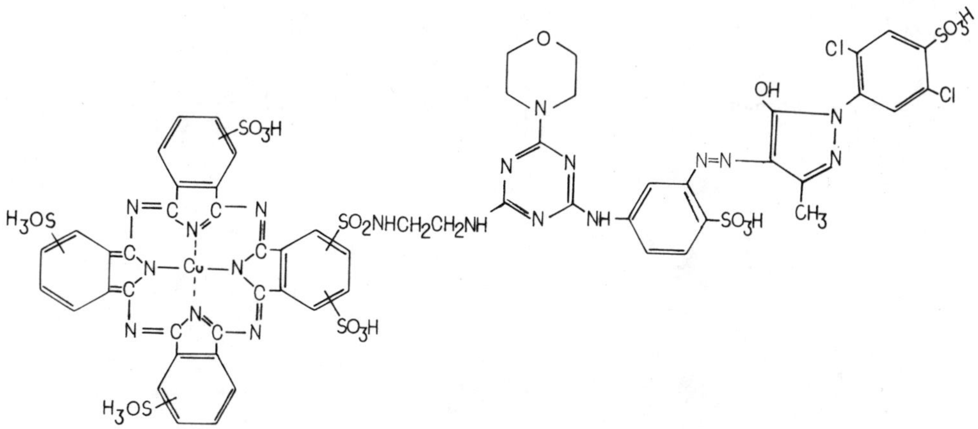

FIGURE 5. A water-soluble pc compound useful to color optical glass.

V. CR(II) AND MN(II) PC COMPOUNDS

Cr and Mn pc (pyrazine)$_2$ and Cr and Mn pc (piperidine)$_2$ are prepared from Cr(II) pc and Mn(II) pc and pyrazine and piperidine, respectively.[92]

Mn pc is conveniently prepared from a dinitrile and Mn powder in the presence of iodine and $(NH_4)_6Mo_7O_{24}\cdot 4H_2O$ at about 300°C.[93]

Interaction of a thin layer of sublimated α-Mn pc with bromine or iodine vapor oxidizes Mn(II) to Mn(III), forming Mn(III) pc Br or Mn(III) pc I. Reaction of α-Mn pc or a pyridine solution of Mn pc Br or Mn pc I with atmospheric oxygen gives Mn pc-O-Mn pc, whereas β-Mn pc is unreactive toward oxygen.[94]

VI. GE, SI, AND SN PCS

The preparation and electrochemical behavior of Si pc-[$OSi(C_6H_{13})_3$] and Si nc-[$OSi(C_6H_{13})_3$] (nc = naphthalocyanine) are described.[95,96]

A new method to prepare $(C_6H_{13})_3SiO(Si\ pc\ O)_2Si(C_6H_{13})_3$ is described.[97]

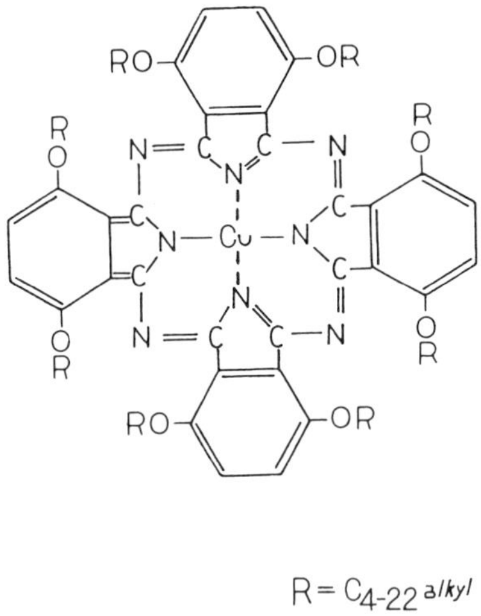

FIGURE 6. A Cu alkoxy pc.

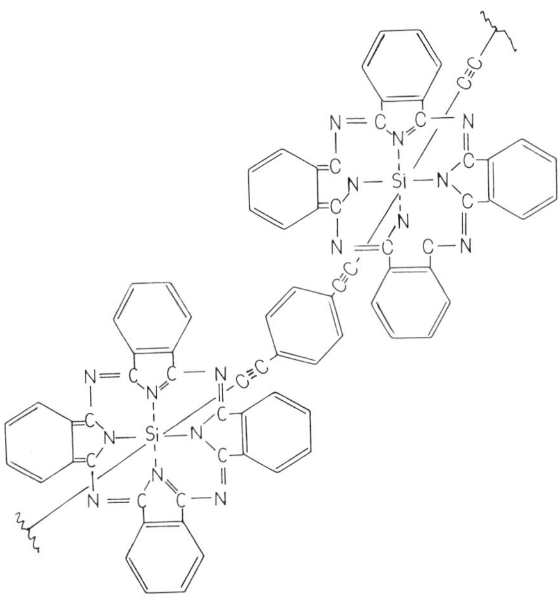

FIGURE 7. Low-dimensional polyyne polymers containing Si pc in the polymer backbone.

Low-dimensional polyyne polymers containing pc silicon and dimethyl silicon in the polymer backbone are synthesized (Figure 7).[98]

The polymers are stable in air and have good stability.[99]

Silicon pcs are prepared with strategies to simplify the route to pcs with β,β′-fused linearly annellated polycyclic aromatic rings.[100]

Transmission electron microscopy shows that Ge pc $(OH)_2$ and (Ge pc $O)_n$ prepared from solid-state topotactic polymerization of Ge pc $(OH)_2$ consist of nearly identical, lath-like single crystals.[101]

(Ge pc $O)_n$ is made from single crystals of Ge pc $(OH)_2$ by thermal solid-state condensation polymerization.[102]

A germanium pc exhibiting unusual electronic and mass spectra is identified as Ge (TBC) (OH) where H_3TBC = α,β,γ-triazatetrabenzcorrole.[103]

trans-Dichloro Ge pc in THF is treated with CH MgI in ether and refluxed 3 h to give *trans*-dimethyl Ge pc, which is then heated, losing its methyl groups at 200°C in a vacuum to obtain reddish-black Ge pc.[104]

Soluble silicon pcs Si pc $[OSi(CH_3)_2\ OCH_2ethynyl]_2$ are prepared from Si pc $(OH)_2$.[105]

Sn(II) pc is prepared in one step in good yield by condensing *o*-dicyanobenzene and stannous chloride in the presence of 4-(dimethylamino)pyridine in an alcohol.[106]

VII. LI PCS

LiH pc, lithium hydrogen pc, is prepared by the thermal decomposition of $(Alkyl\ N)_4$ Li pc, yielding monoclinic blue needles.[107]

Li pc (1-) and the halogen adducts Li pc (1-) X (X = Cl, Br, I) are prepared and characterized spectroscopically.[108]

Refluxing phthalonitrile, lithium chloride, and $CH_3OCH_2CH_2OH$ over 2 h, then refluxing with DBU for 6 h gives reddish-purple Li pc · DBU.[109]

VIII. FE PCS

An optimum yield from various Fe compound candidates is obtained from $FeCl_2{\cdot}4H_2O$ or Fe $C_2O_4{\cdot}2H_2O$ with the catalysts $TiCl_4$ or ammonium molybdate by the phthalic anhydride urea process.[110]

After thermally decomposing $Fe(CO)_5$-adsorbed zeolite NaY, the product is mixed with di hydrogen pc and heated slowly to give Fe pc on the zeolite surface.[111]

Fe $(CO)_5$ adsorbed on zeolite NaY containing phthalonitrile is exposed to phthalonitrile vapors. Initial phthalonitrile adsorption gives Fe pc on the upper surface of the crystals.[112]

Zeolite-included Fe pc is prepared by treating Fe- or $Fe(CO)_5$-containing NaY zeolite with phthalonitrile. The pc guest molecule is encapsulated within the zeolite cavity.[113]

Substituted Co and Fe pcs and their pyridine adducts, such as $(CH_3O)_4$ pc Fe $(pyridine)_2$, are prepared.[114]

Dicyano ferrates (III) are prepared and characterized.[115]

Bis[bis(triphenylphosphine)iminium]dicyano pc ferrate (II) dichloromethane is synthesized. The Fe(II) ion is coordinated by six nearly equidistant ligands that form an almost regular octahedron.[116]

Sodium thioanisolyl pc ferrate (II) · 4THF is prepared.[117]

Low-spin Fe pc $(trialkylphosphine)_2$ or Fe pc $(trialkyl\ phosphite)_2$ is prepared and characterized by Moessbauer spectroscopy.[118]

A high-spin Fe(III) pc is prepared by esterification of the iron pc (Figure 8), where R = Cl, R^1 and R^2 absent, by $CH_3(CH_2)_9$ OH gives the iron pc in Figure 8 where R = $CH_3(CH_2)_9$ O which on imidazole addition gives R = $CH_3(CH_2)_9$ O, R^1 = imidazole, R^2 = absent, and R = $CH_3(CH_2)_9$ O, and R^1 = R^2 = imidazole. The two imidazole derivatives are low-spin Fe(III) and Fe(II) complexes. Without imidazole, the complex is a high-spin Fe(III) complex.[119]

A novel class of metal [bis- and tetrakis(decyloxycarbonyl)] pc is made. The metals are Fe(III) (Figure 9), Co, Cu, and Ni. The synthesis involves condensation of the corresponding acid chloride derivatives with 1-decanol.[120]

IX. SEMICONDUCTING AND CONDUCTING PCS

There is a variety of macrocyclic metal complexes, including metal tetra-*tert*-butyl pc

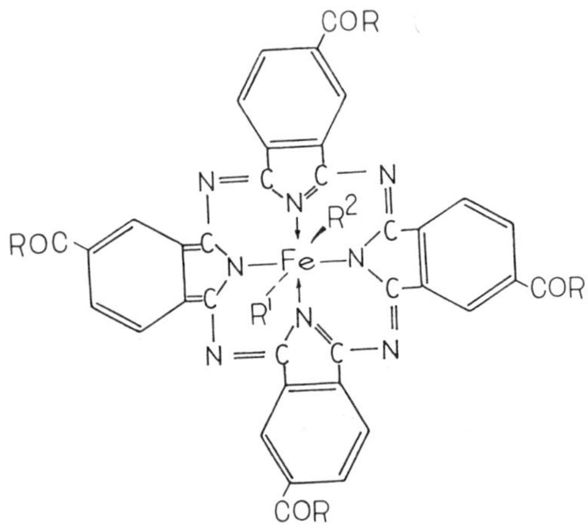

FIGURE 8. A substituted iron pc.

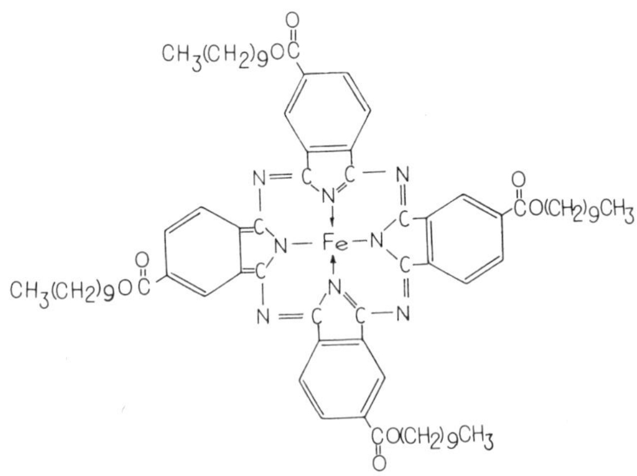

FIGURE 9. A Fe tetrakis (decyloxycarbonyl) pc.

of Group IVA with and without iodine (I) doping. Polymerization increases the electrical conductivity of some of the macrocycles by as much as sevenfold.[121]

β-Fe pc reacts with aqueous potassium cyanide under nitrogen or oxygen to give [Fe (H pc) CN]$_n$ or (Fe pc CN)$_n$. The electrical conductivities are measured.[122]

Fe pc [1,4-diazabicyclo(2.2.2)octane)$_2$ and Fe pc [1,4-diazabicyclo(2.2.2)octane]$_n$ are prepared. The electrical conductivity of the latter is fourfold less than that of the former.[123]

[Ru pc (pyrazine)]$_n$ is prepared and has an electrical conductivity of 0.00001 1/Ω-cm less than that of [Fe pc (pyrazine)]$_n$.[123]

Poly (Cu 2,3,9,10,16,17,23,24-octacyano pc is prepared with <5 to 8 1/Ω-cm conductivity.[124]

Fe-bridged polymer pcs, both doped and nondoped, are prepared. Their conductivities are given.[125]

The DC dark conductivity of [(H_4 pc) Ge (NCN)]$_n$ polymer is determined with and without iodine doping.[126]

(Fe pc Cl)$_2$ I_2 and the molecular conductor Fe pc I are synthesized. The single crystal conductivity of Fe pc I is 20 1/Ω-cm at room temperature and increases with decreasing temperature.[127]

Stacked-bridged [Fe pc(μ-pyrazine)]$_n$ is doped with iodine. The room temperature conductivity is ~0.01 1/Ω-cm without doping.[128]

Co, Cu, Fe, Mg, Ni, Zn, and di hydrogen pcs are doped with iodine vapor at >480°C to give the corresponding metal pc I_x (x = 0.2 to 1.7) which exhibit high electrical conductivity.[129]

Treating Cu pc with iodine yields Cu pc I which stacks in the form of partially oxidized Cu pc surrounded by linear chains of I_3^-. Single Cu pc I crystals exhibit a room temperature conductivity of about 1000 1/Ω-cm along the stacking direction. Cu pc I may be the most highly conductive porphyrinic molecular metal studied so far and one of the most conductive molecular crystals known.[130]

A number of I-doped bridged Fe(II) pcs and Ru(II) pcs such as (Fe pc-1,4-diisocyanobenzene-$I_y)_n$ are formed. An increasing I-dopant level increases the electric conductivity of these pcs, which exhibit room temperature conductivities of 0.01—0.1 1/Ω-cm.[131]

Fe and Ru pc form highly semiconducting *s*-tetrazine bridged compounds without additional doping. Their conductivities compare with cyanide bridge metallomacrocycles. The preparation of semiconducting fibers containing a bridged cyanide species $(\text{Co pc CN})_n$ is described for the first time.[132]

X. PCS FROM GROUPS IIIA (AL, GA), IVA (GE), AND IVB-VIIIB (TI, VO, MO, RE, TC, AND RU)

Fluorine bridged polymeric $(\text{Al pc F})_n$ and $(\text{Ga pc F})_n$ react smoothly with $NOBF_4$ or $NOPF_6$ in dry methyl nitrite or in methylene chloride to give partially oxidized Al pc F $(BF_4)_x$, Ga pc $(BF_4)_x$, Al pc F $(PF_6)_x$, and Ga pc $(PF_6)_x$, where $0 < x < 0.9$. Their pressed pellets have conductivities ≤10 1/Ω-cm.[133]

Reacting a suspension of Al pc X (X = F, Cl, Br, I) in methylene chloride with $(\text{butyl})_4NF \cdot 3H_2O$ gives $(\text{butyl})_4N(\text{Al pc } F_2) \cdot 2H_2O \cdot 0.3$ HF. Hydrolysis of this compound yields $[(\text{Al pc F})_2\ O]^{2-}$.[134]

Stirring 4-sulfophthalic anhydride, urea, and aluminum trichloride in sulfolane in the presence of ammonium molybdate at 170 to 270°C for 8 h gives 91 mol % Al pc Cl disulfonic acid in 98% purity.[135]

α-Al pc OH is transferred into a novel polymorph by treatment with aqueous acid solution.[136]

Li_2 pc $(SC_2H_5)_{16}$ is treated with indium trichloride in amyl alcohol at 130°C for 2 h to give 18.5% In Cl pc $(SC_2H_5)_{16}$.[137]

Tetrachlorophthalonitrile is treated with ethanol to yield tetraethoxyphthalonitrile, which is treated with $LiOC_5H_{11}$ to yield Li_2pc $(OC_2H_5)_{16}$. This product is treated with indium trichloride in amyl alcohol to give (InBr) pc $(OC_2H_5)_{16}$ with maximum light absorption at 830 nm.[138]

Ti pc Cl_2 is made from phthalonitrile and $TiCl_4$ in α-chloronaphthalene. Substitution reactions of the chloride ions in Ti pc Cl_2 lead to Ti pc catecholate (2-), oxalate (2-), and O_2 (2-).[139]

The sodium salt of tetrasulfo pc complexes of fourth period transition metal complexes are prepared by heating a mixture of the sodium salt of 4-sulfophthalic acid, urea, and a fourth period transition metal in the presence of ammonium molybdate.[140]

VO tetra-2,9,16,23-*tert*-butyl pc is prepared and its electronic absorption spectrum is drawn.[141]

Soluble VO pcs with 1.4, 1.6, 2.0, 2.5, and 4 *tert*-butyl groups per molecule are prepared by reacting vanadium trichloride with phthalonitrile and 4-*tert*-butylphthalonitrile.[142]

Reaction of 4-aminophthalonitrile or 4-aminophthalimide with urea and cobalt dichloride, cuprous chloride, or vanadium trichloride in quinoline in the presence of ammonium molybdate yields the respective M tetra-4-ureido pcs in 50 to 98% yield. They are stable in dilute alkaline solution.[143]

VO pc suitable for photoresponsive, e.g., electrophotographic, purposes is prepared from vanadium pentoxide and phthalonitrile in a polyhydric alcohol boiling at >180°C.[144] 6-*tert*-Amyl-2,3-dicyanonaphthalene, urea, and $VOCl_3$ are reacted to give VO tetra-6-*tert*-amyl nc.[145]

6-Butoxy-2,3-dicyanonaphthalene, urea, and $VOCl_3$ are heated at 195 to 200°C for 2 h to give a yellowish-green VO tetra-6-butoxy-nc.[146]

Violet crystals of Mo pc O are synthesized by reacting Mo $(CO)_6$ with phthalodinitrile in toluene at 180°C. Partially oxidizing Mo pc O with iodine gives Mo pc O I_x ($x = 1.4$ to 2.6) with 0.00001 to 0.0001 1/Ω-cm electrical conductivity.[147]

Reacting $(NH_4)_6Mo_7O_{24} \cdot 4H_2O$ with molten phthalodinitrile gives blue Mo pc O, which sublimes at 400°C in a vacuum to give violet crystals.[148]

NH_4ReO_4 (Re = rhenium) reacts with phthalodinitrile to give Re pc N, rhenium nitrido pc, the electrical resistance of 800-nm thick layers of which is 1 to 6 × 10^{11} Ω-cm.[149]

Tc N pc (Tc = technetium) is prepared from NH_4TcO_4 and phthalodinitrile.[150]

$ReOCl_3[P(C_6H_5)_3]_2$, ammonium chloride, and phthalonitrile mixture is melted at 280 to 300°C and held at that temperature 0.5 h, and extracted with ethanol to give >90% dark blue crystalline Re nitrido pc.[151]

Ru pc is prepared by thermally decomposing Ru pc $[(CH_3)_2SO]_2 \cdot 2(CH_3)_2$ SO and reacted with pyrazines, pyridines, pyridazines, or pyrimidines to yield bis axially coordinated Ru pc derivatives such as Ru pc $(pyrazine)_2$, which can lead to [Ru pc $(pyrazine)_2]_n$. The polymers have much higher electrical conductivities than the monomers.[152]

Axial ligand substitution reactions of Ru(II) pc are the subject of a thesis.[153]

Fe pc or Ru pc react with *tert*-butyl NC forming trans-Fe pc (*tert*-butyl $NC)_2$ or trans-Ru pc (*tert*-butyl $NC)_2$. Differences between these two compounds indicate the existence of pentacoordinate Ru(II) pc compounds.[154]

XI. LANTHANIDE PCS

Ce (tetra-4-*tert*-butyl $pc)_2$ (Ce = cerium) is prepared and characterized.[155]

$KNd(pc)_2$ or $HNd(pc)_2$ is refluxed in THF to give a Nd(III) (Nd = neodymium) pc derivative with an unusual absorption spectrum at 650 to 750 μ.[156]

Pc Nd pc H, (pc Nd pc)· radical, and pc Nd O (acetyl) are separated from a crude Nd(III) pc mixture by column chromatography.[157]

Dy, Gd, Lu, Nd, Tb, and Yb (Dy = dysprosium, Gd = gadolinium, Lu = lutetium, Nd = neodymium, Tb = terbium, and Yb = ytterbium) pc are prepared from their respective M (O $acetyl)_3$ (M = lanthanide) and phthalonitrile. The difficulty in preparation increases from the heaviest, lutetium, to the lightest, lanthanum.[158]

(μ-pc) bis[di(2,2,6,6-tetramethylheptane-3,5-dionato)lanthanide(III)] compounds are prepared. Treating ML_3(M = Sm, Eu, Gd, Tb, Dy, Er, Tm, Yb, Y: samarium, europium, gadolinium, terbium, dysprosium, erbium, thulium, ytterbium, yttrium) with Li_2 pc gives M_2 (μ-pc)L_4.[159]

Lu $(pc)_2$ (CH_2Cl_2) and [Lu pc O (Acetyl) $(H_2O)_2](H_2O)(2CH_3OH)$ are prepared in boiling hexanol from $Lu(Oacetyl)_3$ and 1,2-$C_6H_4(CN)_2$ in the presence of 1.8-diaza-bicyclo[5.4.0]undec-7-ene and purified chromatographically.[160]

A triple decker-like sandwich structure is prepared where two metal atoms are sandwiched between one porphyrin and two macrocycles. They are porphyrin bis pc dilanthanide (III)s.[161]

Ln pc (O $acetyl)_2$ (Ln = Sm, Eu, Gd, Tb, Dy, Ho, Er, Tm, Yb) (samarium, europium, gadolinium, terbium, dysprosium, holmium, erbium, thulium, ytterbium) are prepared by the cyclocondensation of Ln (O $acetyl)_3$ with diiminoisoindole in DMF.[162]

The di pc Lu pc tetra-6-*tert*-butyl-2,3-nc is prepared by fusion of Lu (tetra-4-*tert*-butyl pc) with phthalic acid dinitrile.[163]

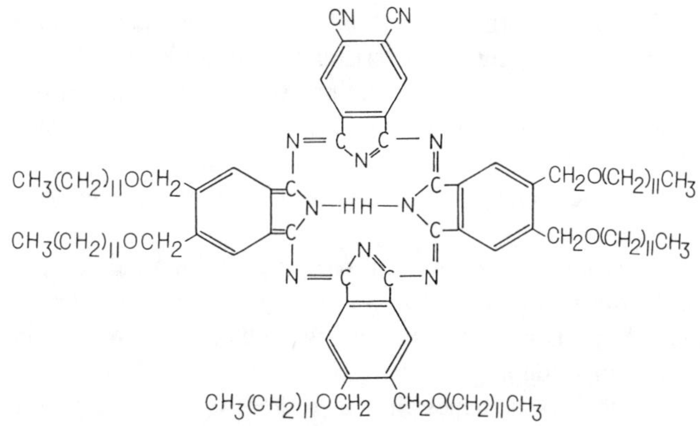

FIGURE 10. A polar discogen.

Ln pc Cl (Ln = Pr, Eu, Dy, Ho, Er, Yb) and Ln pc (O acetyl) (Ln = Nd, Lu) are prepared and their spectral and electrochemical properties are studied.[164]

"Super complex" Ln_2pc_3 (Ln-La, Nd, Eu, Gd, Dy, Yb, and Lu) are prepared. The super complex is progressively formed when approaching the lighter lanthanides.[165]

Lanthanide salts are heated gradually with tetra-4-*tert*-butyl pc forming, in progression: $LnL^+ \rightarrow LnL_2$ (green) $\rightarrow LnL_2^-$ (blue).[166]

The solid reaction of Ln(O acetyl)$_3$ lanthanide acetates and phthalonitrile is studied by differential thermal analysis. The reaction products are Ln pc and Ln pc_2 complexes; the ratio of these two compounds depends on the heating rate and the particular lanthanide used.[167]

Refluxing Ln (β-diketone)$_3$ with Li_2 pc in organic solvents yields mixed ligand complexes. Ln_2pc (dpm)$_4$ and Ln pc (dpm)$_2$ are obtained with dipivaloylmethane (dpm).[168]

New transition metal compounds which have the potential of high electrical conductivities are prepared. Pc sandwich compounds, H pc Ln pc, are discussed.[169]

XII. ACTINIDE PCS

Th(IV) pc (2,4-pentanedione)$_2$ and U(IV) pc (2,4-pentanedione)$_2$ are prepared. They are the first examples of 1:1 actinide pc complexes.[170]

XIII. DI HYDROGEN PCS

A polar discogen, 9,10,16,17,23,24-hexakis[(dodecyloxy)methyl]-H_2 pc-2,3-dicarbonitrile (Figure 10) is synthesized. The high permanent electric dipole moment of the discogen leads to a new type of discotic mesophase.[171]

Metalloporphyrins are made by insertion of metal cations into the H_2 pc ring in the presence of an electron acceptor capable of complexing with the porphyrin.[172]

The synthesis and properties of potential pc coupling agents are the subject of a thesis.[173]

A study is made of the lithium surface-initiated conversion of phthalodinitrile to H_2 pc, including the kinetics of formation.[174]

α-Pc is dissolved with Cu pc-$(COCH_2NHC_2H_5)_{1.1}$, Cu pc-(CH_2 pyrrolidino)$_{2.2}$, Cu pc-$(CH_2CH_2\ NHCH_3)_{1.61}$, and Cu pc ($CH_2$piperidino)$_{2.2}$ in 98% sulfuric acid, precipitated by diluting with water, filtered, washed, and dried. The mixture is added to sodium chloride and polyethylene glycol, and milled at 60 to 130°C for 8 h to give bright green blue η-H_2 pc.[175]

Modified η-H_2 pc comprises 100 parts H_2 pc and <50 parts H_2 pc with substituents on the benzene rings, pc nitrogen homolog with or without substituents on the benzene rings, and/or metal containing pcs, giving characteristic absorption bands.[176]

η- or modified η-H_2 pc is prepared by heating 100 parts H pc precursors and <50 parts substituted pc precursors or pc in alcohol containing bases and seed crystals of M pc or H_2 pc with crystal shapes similar to those of η- or modified η-pc.[177]

α-H_2pc, sodium chloride, and polyethylene glycol are milled at 60 to 120°C for 7 to 15 h to give a bright green-blue τ-H_2 pc.[178]

Modified τ-H_2 pc is prepared by milling α-H_2 pc, sodium chloride, and polyethylene glycol at 60 to 120°C for 7 to 15 h to give a bright green blue-modified τ-H pc.[179]

γ- or modified γ-H pc is prepared by heating phthalonitrile with isoindoline or isoindolenine pc precursors in alcohols containing bases and seed crystals of γ- or modified γ-H pc to give H_2 pcs with crystal shapes similar to those of the γ- or modified γ- form.[180]

x-H_2 pc is prepared by reacting phthalonitrile in 1,8-diazabicyclo[4.3.0]nonene-5, solubilizing and purifying the crude with solvents, regenerating to give H_2pc and ball milling to change to the x-H_2 pc structure.[181]

Phthalic anhydride chlorination reaction product mixtures are used in the preparation of stabilized pc pigments.[182]

Cross-condensations of 1,3-diiminoisoindoline with 5-neopentoxy-1H-isoindole-1,3(2H)-dithione gives mixtures of pcs from which 2-neopentoxy pc and 2,16-dineopentoxy pc can be purified separately from other pcs.[183]

XIV. REVIEW

A review with 112 references relates to the preparation, structure, and coordination chemistry of low molecular weight pcs. Ways to combine pcs with polymers are surveyed.[184]

A review with 43 references discusses the synthesis and properties of pc pigments, with reference to Cu pc.[185]

A recent review with 33 references of progress on organic pigments that includes pc includes their preparation, properties, and applications.[186]

A review with 174 references relates to the formation, structure, and properties of pcs and their use with respect to functional materials.[187]

REFERENCES

1. **Datz, A., Metz, J., Schneider, O., and Hanack, M.,** Synthesis and conductivities of μ-cyano) phthalocyaninatometal compounds, *Synth. Met.*, 9(1), 31, 1984.
2. **Liston, D. J., Kennedy, B. J., Murray, K. S., and West, B. O.,** Oxochromium compounds. I. Synthesis and properties of μ-oxo chromium-iron porphyrin and phthalocyanine compounds, *Inorg. Chem.*, 24(10), 1561, 1985.
3. **Hanack, M., Hedtmann-Rein, C., Datz, A., Keppeler, U., and Muenz, X.,** Synthesis and properties of (μ-thiocyanato) and (μ-azido) phthalocyaninato-metal-complexes, *Synth. Met.*, 19(1—3), 787, 1987.
4. **Goedken, V. L. and Ercolani, C.,** Nitrido-bridged iron phthalocyanine dimers: synthesis and characterization, *J. Chem. Soc., Chem. Commun.*, 6, 378, 1984.
5. **Bottomley, L. A., Gorce, J. N., Goedken, V. L., and Ercolani, C.,** Spectroelectrochemistry of a μ-nitrido-bridged iron phthalocyanine dimer, *Inorg. Chem.*, 24(23), 3733, 1985.
6. **Leznoff, C. C., Marcuccio, S. M., Greenberg, S., Lever, A. B. P., and Tomer, K. B.,** Metallophthalocyanine dimers incorporating five-atom covalent bridges, *Can. J. Chem.*, 63(3), 623, 1985.
7. **Leznoff, C. C., Marcuccio, S. M., Greenberg, S., Lever, A. B. P., and Tomer, K. B.,** Metallophthalocyanine Dimers Incorporating Five Atom Covalent Bridges, Report 1985, TR-1, Order No. AD-A154365, National Technical Information Service, Springfield, VA, 35 pp.
8. **Gavrilov, V. I., Luk'yanets, E. A., and Shelepin, I. V.,** New phthalocyanine complex of lutetium — eight single-electron redox steps, *Elektrokhimiya,* 21(11), 1581, 1985.

9. **Tomilova, L. G., Chernykh, E. V., Ioffe, N. T., and Luk'yanets, E. A.,** Phthalocyanines and related compounds. XXII. Synthesis and spectroscopic-electrochemical study of rare earth element diphthalocyanines, *Zh. Obshch. Khim.*, 53(11), 2594, 1983.
10. **Kasuga, K., Ando, M., Morimoto, H., and Isa, M.,** Preparation of new phthalocyanine complexes of yttrium(III) and some lanthanoid(III) ions, *Chem. Lett.*, 7, 1095, 1986.
11. **Ercolani, C., Gardini, M., Murray, K. S., Pennesi, G., Rossi, G., and Zwack, P. R.,** Molecular and electronic structure of six-coordinate low-spin oxo-bridged dimers formed by the interaction of (μ-oxo)bis(phthalocyaninato)iron(III) with nitrogen bases, *Inorg. Chem.*, 26(21), 3539, 1987.
12. **Ercolani, C., Gardini, M., Murray, K. S., Pennesi, G., and Rossi, G.,** Crystalline isomerism in (μ-oxo)bis[(phthalocyaninato)iron(III)]: further characterization of the isomer having a linear or quasi-linear Fe-O-Fe bond system (μ-oxo(2)), *Inorg. Chem.*, 25(22), 3972, 1986.
13. **Keppeler, U. and Hanack, M.,** Monomeric and bridged (phthalocyaninato)iron(II) and -ruthenium(II) compounds with 2,3,5,6-tetramethyl-1,4-diisocyanobenzene and 2,3,5,6-tetrachloro-1,4-diisocyanobenzene as axial ligands, *Chem. Ber.*, 119(11), 3363, 1986.
14. **Radzki, S., Gaspard, S., and Giannotti, C.,** Optical study of the face-to-face complexation of water-soluble metalloporphyrins and metallophthalocyanines, *J. Chem. Res. Synop.*, 10, 360, 1986.
15. **Seiders, R. P. and Ward, J. R.,** Effect of visible light on the dimerization of cobalt(II) tetrasulfonated phthalocyanine in water, *Anal. Lett.*, 17(A15), 1763, 1984.
16. **Seiders, R. P. and Ward, J. R.,** Effect of Visible Light on the Dimerization of Cobalt(II) Tetrasulfonated Phthalocyanine in Water, Report 1983, ARCSL-TR-83058, Order No. AD-134847, National Technical Information Service, Springfield, VA, 18 pp.
17. **Leznoff, C. C., Greenberg, S., Marcuccio, S. M., Minor, P. C., Seymour, P., Lever, A. B. P., and Tomer, K. B.,** Binuclear 'clamshell' metallophthalocyanines, *Inorg. Chim. Acta,* 89(3), L35, 1984.
18. **Leznoff, C. C., Greenberg, S., Marcuccio, S. M., Minor, P. C., and Seymour, P.,** Binuclear 'Clamshell' Metallophthalocyanines, Report 1984, TR-27, Order No. AD-A143509, National Technical Information Service, Springfield, VA, 15 pp.
19. **Leznoff, C. C., Lam, H., Nevin, W. A., Kobayashi, N., Janda, P., and Lever, A. B. P.,** 1,8-Naphthalenebis(phthalocyanines), *Angew. Chem.*, 99(10), 1065, 1987.
20. **Smirnov, R. P., Andreyanov, V. V., and Vorob'ev, Y. G.,** Synthesis and electrical properties of a binuclear cobalt-containing macroheterocyclic compound, *Izv. Vyssh. Uchebn. Zaved., Khim. Khim. Tekhnol.*, 27(10), 1239, 1984.
21. **Leznoff, C. C., Ham, H., Marcuccio, S. M., Nevin, W. A., and Janda, P.,** Planar Binuclear Phthalocyanine and its Di-Cobalt Derivatives, Report 1987, TR-12, Order No. AD-A177316/7/GAR, National Technical Information Service, Springfield, VA, 19 pp.
22. **Nevin, W. A., Liu, W., Greenberg, S., Hempstead, M. R., Marcuccio, S. M., Melnik, M., Leznoff, C. C., and Lever, A. B. P.,** Synthesis, aggregation, electrocatalytic activity, and redox properties of a tetranuclear cobalt phthalocyanine, *Inorg. Chem.*, 26(6), 891, 1987.
23. **Nevins, W. A., Liu, W., Greenberg, S., Hempstead, M. R., and Marcuccio, S. M.,** Synthesis, Aggregation, Electrocatalytic Activity, and Redox Properties of Tetranuclear Cobalt Phthalocyanine, Report 1986, Order No. AD-A174376/4/GAR, National Technical Information Service, Springfield, VA, 55 pp.
24. **Greenberg, S., Marcuccio, S. M., and Leznoff, C. C.,** Selective synthesis of binuclear and trinuclear phthalocyanines covalently linked by a one-atom oxygen bridge, *Synthesis,* 5, 406, 1986.
25. **Lever, A. B. P., Hempstead, M. R., Leznoff, C. C., Liu, W., and Melnik, M.,** Recent Studies in Phthalocyanine Chemistry, Report 1986, TR-7. Order No. AD-A172939/1/GAR, National Technical Information Service, Springfield, VA, 12 pp.
26. **Marcuccio, S. M., Svirskaya, P. I., Greenberg, S., Lever, A. B. P., Leznoff, C. C., and Tomer, K. B.,** Binuclear pcs covalently linked through 2-#1 and 4- atom bridges; *Can. J. Chem.*, 63(11), 3057, 1985.
27. **Hendriks, R., Sieleken, O. E., Drenth, W., and Nolte, R. J. M.,** Polytopic ligand systems: synthesis and complexation properties of a crowned phthalocyanine, *J. Chem. Soc., Chem. Commun.*, 19, 1464, 1986.
28. **Tahiri, M., Doppelt, P., Fischer, J., and Weiss, R.,** Synthesis and molecular structure of kryptofix 222 potassium phthalocyaninatoferrate, [K⊂ 222][FePc], and (18-crown-6)lithium (18-crown-6)(tetrahydrofuran)lithium phthalocyaninatoferrate, [Li⊂ 18c.6][Li(THF)⊂ 18c.6][FePc], *Inorg. Chim. Acta,* 127(1), L1, 1987.
29. **Gutwirth, K.,** Czech Patent 202,486, 1983.
30. **Gutwirth, K., Holeci, I., and Vrana, M.,** Czech Patent 202,609, 1983.
31. **Pieschel, F., Lange, E., Kleine, F., Deissner, A., Kasper, J., Kochmann, W., Schnabel, S., Chojnacki, K., Reinholz, S., and Wetzel, K. (to VEB Chemiekombinat Bitterfeld),** East German Patent 221,189, 1985.
32. **Pieschel, F., Kleine, F., Lange, E., Kochmann, W., Schmidt, F., Dost, H., Kasper, J., Kluger, K. H., Koennecke, W., et al. (to VEB Chemiekombinat Bitterfeld),** East German Patent 221,190, 1985.

33. **Kinoshita, S. and Yonehara, Y. (to Dainippon Ink and Chemicals, Inc. and Kawamura Physical and Chemical Research Institute)**, Japan Patent 86,190,562, 1986.
34. **Nippon Petrochemicals Co., Ltd., Toyo Ink Manufacturing Co., Ltd.**, Japan Patent 84,89,361, 1984.
35. **Kleine, F., Pieschel, F., Dost, H., Kluger, K. H., Kochmann, W., Koennecke, W., Lange, E., Schmidt, F., Mewes, H., et al. (VEB Chemiekombinat Bitterfeld)**, East German Patent, 225,700, 1985.
36. **Mack, E. and Witzke, L. (to Th. Goldschmidt AG)**, German Patent 3,513,273, 1986.
37. **Vrana, M., Loukota, O., Pridal, J., Roman, J., and Stepanek, J.**, Czech Patent 223,337, 1986.
38. **Habermann, W., Kranz, J., and Thoma, P. (to BASF AG)**, German Patent 3,538,104, 1986.
39. **Tanaka, M. and Sato, T. (to Dainippon Printing Co., Ltd.)**, Japan Patent 86,203,175, 1986.
40. **Tanaka, M. and Sato, T. (to Dainippon Ink and Chemicals, Inc.)**, Japan Patent 86,195,166, 1986.
41. **Tanaka, M. and Sato, T. (to Dainippon Ink and Chemicals, Inc.)**, Japan Patent 87,48,769, 1987.
42. **Shirao, M., Sagara, H., and Muramatsu, S. (to Toyo Ink Manufacturing Co., Ltd.)**, Japan Patent 87,25,175, 1987.
43. **Thorovsky, Z., Smisek, J., Vichova, I., and Jenista, P.**, Czech Patent 204,072, 1983.
44. **Harbour, J. R., Dietelbach, B., and Duff, J.**, Phthalocyanization of cadmium sulfide and zinc oxide. Effects on the photochemistry in aqueous dispersion, *J. Phys. Chem.*, 87(26), 5456, 1983.
45. **Agrawal, B. B. and Fish, S. F.**, Synthesis of metal-phthalocyanines, *Res. Ind.*, 31(4), 319, 1986.
46. **Tanaka, M. and Honda, Y. (to Dainippon Ink and Chemicals, Inc.)**, Japan Patent 85,199,066, 1985.
47. **Smrchek, V. A., Astaf'eva, M. V., Smirnova, S. V., Shchepkina, O. S., Ladyzhenskii, Y. B., and Kogan, E. M.**, U.S.S.R. Patent 790,750, 1984.
48. **Tokyo Ink Manufacturing Co., Ltd.**, Japan Patent 84,155,467, 1984.
49. **Vojtisek, V., Vrana, M., Janda, F., Vorel, M., Stepanek, J., Dittrich, F., and Vagner, M.**, Czech Patent 219,416, 1985.
50. **Pridal, J.**, Objective color methods in the technology of phthalocyanine pigments, *Chem. Prum.*, 35(12), 644, 1985.
51. **Wong, M. K. L., Schroeder, R. G., and Donegan, T. E. (to BASF Wyandotte Corp.)**, Canada Patent 1,211,431, 1986.
52. **Pascalide, R., Dianconu, G., Stefan, L., Nedelcu, M., Moraru, G., and Todiresei, A. (to Intreprinderea de Coloranti "Colorom")**, Romania Patent 88,282, 1985.
53. **Smrchek, V. A., Sinitsyna, I. N., Vil'ner, S. G., Klimova, T. S., Aristov, B. G., Poznyakevich, A. L., Ladyzhenskii, Y. B., Vorob'eva, I. I., Pudilov, M. P., et al.**, U.S.S.R. Patent 1,219,623, 1986.
54. **Holeci, I., Kotlan, J., and Navratil, F.**, Czech Patent 224,418, 1985.
55. **Mori, A., Kotaki, M., and Wada, N. (to Dainichiseika Color and Chemicals Manufacturing Co., Ltd.)**, Japan Patent 86,266,471, 1986.
56. **Zhao, Y., Chen, P., and Zhuang, P.**, Copper phthalocyanine pigments. I. Properties and crystal phase adjustment of copper phthalocyanine, *Huagong Xuebao*, 4, 435, 1986.
57. **Zhuang, P., Cheng, P., and Zhao, Y.**, Copper phthalocyanine pigments. II. Rate and mechanism of crystal phase change of copper phthalocyanine in organic solvents, *Huagong Xuebao*, 4, 445, 1986.
58. **Chen, P., Zhao, Y., and Zhuang, P.**, Copper phthalocyanine pigments. III. Pigment particle size control and surface treatment, *Huagong Xuebao*, 4, 455, 1986.
59. **Mori, A. and Kibe, S. (to Dainichiseika Color and Chemicals Manufacturing Co., Ltd.)**, Japan Patent 87,131,072, 1987.
60. **Necas, M., Plechacek, V., and Buchtova, J.**, Micronization and modification of copper phthalocyanine, *Chem. Prum.*, 37(9), 472, 1987.
61. **Pepoy, L. J. (to BASF Wyandotte Corp.)**, U.S. Patent 4,452,740, 1984.
62. **Kranz, J. (to BASF AG)**, German Patent 3,442,118, 1986.
63. **Butler, I. S. and Furbacher, R. J.**, Chemistry and artists' pigments, *J. Chem. Educ.*, 62(4), 334, 1985.
64. **Skorobogaty, A., Smith, T. D., Dougherty, G., and Pilbrow, J. R.**, Synthesis and physicochemical properties of cationic derivatives of phthalocyaninatocopper(II), *J. Chem. Soc., Dalton Trans.*, 4, 651, 1985.
65. **Doescher, B., Priess, R., and Gunsser, W.**, Study of the formation of copper phthalocyanine by EPR spectroscopy, *Org. Magn. Reson.*, 22(10), 658, 1984.
66. **Homborg, H.**, Preparation and characterization of phthalocyanine (π-cation radicals of hydrogen ion, magnesium(2+) and copper (2+), *Z. Anorg. Allg. Chem.*, 507, 35, 1983.
67. **Achar, B. N., Fohlen, G. M., Parker, J. A., and Kesavayya, J.**, Synthesis and structural studies of metal(II) 4,9,16,23-phthalocyanine tetraamines, *Polyhedron*, 6(6), 1463, 1987.
68. **Metz, J. and Hanack, M.**, Synthesis and properties of (phthalocyaninato)cobalt(II) adducts with nitrogen bases, *Chem. Ber.*, 120(8), 1307, 1987.
69. **Hedtmann-Rein, C., Hanack, M., Peters, K., Peters, E. M., and Von Schnering, H. G.**, Synthesis and properties of (phthalocyaninato)cobalt(III)thiocyanate and isothiocyanate compounds. Crystal and molecular structure of (phthalocyaninato)(pyridine)(thiocyanato)cobalt (III), *Inorg. Chem.*, 26(16), 2647, 1987.

70. **Berezin, I. V., Kozlyak, E. I., Erokhin, A. S., and Yatsimirskii, A. K.,** Dimercaptide complexes of cobalt(III) tetrasulfophthalocyanine, *Izv. Akad. Nauk SSSR, Ser. Khim.*, 12, 2830, 1986.
71. **Hanack, M. and Fay, R.,** Synthesis and properties of peripherally substituted phthalocyaninatocobalt complexes with bisaxially coordinated ligands, *Recl. Trav. Chim. Pays-Bas,* 105(10), 427, 1986.
72. **Ercolani, C. and Pennesi, G.,** Nitrosyl and nitro derivatives of phthalocyaninatocobalt(II): reactivity with dioxygen and oxygen atom transfer, *Inorg. Chim. Acta,* 101(3), L41, 1985.
73. **Homborg, H. and Kalz, W.,** Phthalocyanines of cobalt(III): preparation and resonance-Raman, infrared, and electronic absorption spectra of $[CoX_2Pc(1-)]$ (X = Cl, Br) and $[CoX_2Pc(2-)]^-$ (X = OH, F, Cl, Br), *Z. Naturforsch., B,* 39(11), 1490, 1984.
74. **Meyer, G., Woehrle, D., Mohl, M., and Schulz-Ekloff, G.,** Synthesis of faujasite supported phthalocyanines of cobalt, nickel and copper, *Zeolites,* 4(1), 30, 1984.
75. **Preiswerk, W. and Bruttel, B. (to Ciba-Geigy AG),** German Patent 3,410,611, 1984.
76. **Orihashi, Y., Kobayashi, N., Tsuchida, E., Matsuda, H., Nakanishi, H., and Kato, M.,** Synthesis of catena-μ-ethynylene-phthalocyaninatocobalt(III) and its conductivity, *Nippon Kagaku Kaishi,* 3, 410, 1986.
77. **Takei, T. and Tanaka, T. (to Dainippon Ink Chemical Industry Co.),** German Patent 3,641,891, 1987.
78. **Vrana, M., Loukota, O., and Pridal, J.,** Czech Patent 215,594, 1984.
79. **Tomoda, H., Saito, S., and Shiraishi, S.,** The participation of light in the formation of phthalocyanines from phthalonitrile, *Nippon Kagaku Kaishi,* 1, 125, 1984.
80. **Pryanikov, E. I., Khadyrov, B. S., Rizaev, R. G., Ismikhanova, M. A., Belau, R. M., Arzumanova, I. A., Alizade, G. A., and Khalilova, T. V.,** Effectiveness of organization of the manufacture of aromatic nitriles from local feedstock in Azerbaidzhan, *Sb. Tr.-Akad. Nauk Az. SSR, Inst. Neftekhim. Protsessov im. Akad. Yu. G. Mamedalieva,* 15, 158, 1986.
81. **Kuroiwa, A., Nanba, N., and Kamijo, T. (to TDK Corp.),** Japan Patent 86,152,769, 1986.
82. **Oliver, S. W. and Smith, T. D.,** Synthesis and characterization of 2,9,16,23-tetrakis(3-pyridyloxy)phthalocyanine in the form of its zinc(II) chelate, *Heterocycles,* 22(9), 2047, 1984.
83. **Gal'pern, M. G., Talismanova, T. D., Tomilova, L. G., and Luk'yanets, E. A.,** Phthalocyanines and related compounds. XXV. 5,5′,5″,5‴-Tetrabromo-7,7′,7″,7‴-tetra-*tert*-butyl-2,3-naphthalocyanines, *Zh. Obshch. Khim.,* 55(5), 1099, 1985.
84. **Eda, T. (to Yamamoto Synthetic Chemical Co., Ltd.),** Europe Patent Application 134,518, 1985.
85. **Fujiki, M. and Tabei, H. (to Nippon Telegraph and Telephone Public Corp.),** Japan Patent 86,287,964, 1986.
86. **Sakamoto, K., Yoshioka, M., and Shibamiya, F.,** Position isomers constituting copper phthalocyanine-*N*-isologue and its separation, *Shikizai Kyokaishi,* 58(3), 121, 1985.
87. **Gridnev, A. A., Belgovskii, I. M., and Enikolopyan, N. S.,** Formation of cobalt(I) inner complex compounds in the presence of free radicals, *Dokl. Akad. Nauk SSSR,* 289(3), 616, 1986.
88. **Yamamura, S. and Hirasawa, Y. (to Hippon Kayaku Co., Ltd.),** Japan Patent 86,87,759, 1986.
89. **Ichimura, K., Ono, O., and Sasaki, T. (to Agency of Industrial Sciences and Technology),** Japan Patent 86,223,056, 1986.
90. **Grigoryan, L. S., Simonyan, M. V., and Sharoyan, E. G.,** Iodine-containing phthalocyanines, U.S.S.R. Patent 1,120,686, 1986.
91. **Spietschka, E. and Urban, M. (to Hoechst AG),** German Patent 3,432,319, 1986.
92. **Hanack, M. and Datz, A.,** (Phthalocyaninato)chromium(II)- and -manganese(II) compounds with pyridine, pyrazine, and piperidine as axial ligands, *Chem. Ber.,* 119(4), 1281, 1986.
93. **Camenzind, M. J. and Hill, C. L.,** Reactions of dinitriles with manganese powder: first synthesis of manganese(II) octaphenylporphyrazine and a convenient synthesis of manganese(II) phthalocyanine, *Inorg. Chim. Acta,* 99(1), 63—67, 1985.
94. **Vinogradskii, A. G.,** Oxidized forms of manganese phthalocyanine, *Zh. Fiz. Khim.,* 57(11), 2883, 1983.
95. **Wheeler, B. L., Nagasubramanian, G., Bard, A. J., Schechtman, L. A., and Kenney, M. E.,** A silicon phthalocyanine and a silicon naphthalocyanine: synthesis, electrochemistry, and electrogenerated chemiluminescence, *J. Am. Chem. Soc.,* 106(24), 7404, 1984.
96. **Wheeler, B. L., Nagasubramanian, G., Bard, A. J., Schechtman, L. A., and Kenney, M. E.,** Silicon Phthalocyanine and a Silicon Naphthalocyanine: Synthesis, Electrochemistry, and Electrogenerated Chemiluminescence, Report 1984, TR-32, Order No. AD-A146441/1/GAR, National Technical Information Service, Springfield, VA, 28 pp.
97. **DeWulf, D. W., Leland, J. K., Wheeler, B. L., Bard, A. J., Batzel, D. A., Dininny, D. R., and Kenney, M. E.,** Isolation, spectroscopic properties, and electrochemical properties of two oligomeric silicon phthalocyanines, *Inorg. Chem.,* 26(2), 266, 1987.
98. **Shim. I. W. and Risen, W. M., Jr.,** Synthesis and spectral study of low dimensional polyyne polymers containing phthalocyanine silicon and dimethylsilicon in the polymer backbone, *J. Organomet. Chem.,* 260(2), 171, 1984.
99. **Shim. I. W. and Risen, W. M., Jr.,** Synthesis and Spectral Study of Low Dimensional Poly-yne Polymers Containing Phthalocyanine Silicon and Dimethyl Silicon in the Polymer Backbone, Report 1983, TR-83-05, Order No. AD-A130580, National Technical Information Service, Springfield, VA, 21 pp.

100. **Dirk, C. W.**, Preparation and Properties of Silicon Hemiporphyrazines. Strategies to Simplify the Route to Phthalocyanines with β,β′-Fused Linearly Annellated Polycyclic Aromatic Rings, Ph.D. thesis, Northwestern University, Evanston, IL, 1983.
101. **Zhou, X., Marks, T. J., and Carr, S. H.**, Topotactic polymerization mechanism of phthalocyaninatogermanium dihydroxide in the solid state, *Polym. Mater. Sci. Eng.*, 51, 651, 1984.
102. **Zhou, X., Marks, T. J., and Carr, S. H.**, Direct imaging of the lattice in poly(phthalocyaninatogermoxane) single crystals, *Mol. Cryst. Liq. Cryst.*, 118(1—4), 357, 1985.
103. **Fujiki, M., Tabei, H., and Isa, K.**, New tetrapyrrolic macrocycle: α,β,γ-triazatetrabenzcorrole, *J. Am. Chem. Soc.*, 108(7), 1532, 1986.
104. **Nippon Telegraph and Telephone Public Corp.**, Japan Patent 84,179,664, 1984.
105. **BernalCastillo, J., Jaime, G. A., and Johanna Sanabria, O.**, New soluble phthalocyanines, *Rev. Ion*, 9(1), 25, 1986.
106. **Nippon Telegraph and Telephone Public Corp.**, Japan Patent 83,127,763, 1983.
107. **Homborg, H. and Kalz, W.**, Preparation and characterization of lithium hydrogenphthalocyanine, *Z. Anorg. Allg. Chem.*, 514, 115, 1984.
108. **Homborg, H. and Teske, C. L.**, Lithium phthalocyanines: preparation and characterization of the monoclinic and tetragonal modifications of LiPc(1-) and the halogen adducts LiPc(1-)X (X = Cl, Br, I), *Z. Anorg. Allg. Chem.*, 527, 45, 1985.
109. **Kinoshita, S. and Yonehara, Y. (to Dainippon Ink and Chemicals, Inc., and Kawamura Physical and Research Institute)**, Japan Patent 86,186,386, 1986.
110. **Komarov, R. D. and Maizlish, V. E.**, Synthesis of Iron Phthalocyanine from Phthalic Anhydride, Deposited Doc. 1982, SPSTL 585 Khp-D82, 5 pp. Avail. SPSTL.
111. **Korol'kova, T. V., Zakharov, A. N., and Romanovskii, B. V.**, Complexes of transition metals in zeolites. Preparation of iron(II) phthalocyanine, *Vestn. Mosk. Univ., Ser. 2: Khim.*, 25(4), 362, 1984.
112. **Zakharov, A. N., Korol'kova, T. V., and Romanovskii, B. V.**, Matrix preparation of phthalocyaninatoiron(II) by ligand substitution with iron pentacarbonyl adsorbed on zeolite, *Koord. Khim.*, 12(1), 94, 1986.
113. **Zakharov, A. N. and Romanovskii, B. V.**, Zeolite included iron(II) phthalocyanine, *J. Inclusion Phenom.*, 3(4), 389, 1985.
114. **Metz, J., Schneider, O., and Hanack, M.**, Synthesis and properties of substituted(phthalocyaninato)-iron and -cobalt compounds and their pyridine adducts, *Inorg. Chem.*, 23(8), 1065, 1984.
115. **Kalz, W., Homborg, H., Kueppers, H., Kennedy, B. J., and Murray, K. S.**, Preparation and characterization of dicyano(phthalocyaninato)ferrates(III), *Z. Naturforsch., B: Anorg. Chem., Org. Chem.*, 39B(11), 1478, 1984.
116. **Kueppers, H., Eulert, H. H., Hesse, K. F., Kalz, W., and Homborg, H.**, Synthesis and crystal structure of bis[bis(triphenylphosphine)iminium] dicyanophthalocyaninatoferrate(II) dichloromethane, *Z. Naturforsch., B Anorg. Chem., Org., Chem.*, 41B(1), 44, 1986.
117. **Steinborn, D. and Taube, R.**, Nature of transition metal-carbon σ-bond. X. Preparation, characterization and thermal decomposition of bis (η^5-cyclopentadienyl)titaniumbis(thioanisolyl) and sodium thioanisolylphthalocyaninatoferrate(II).4THF, *J. Organomet. Chem.*, 284(3), 395, 1985.
118. **Ohya, T., Morohoshi, H., and Sato, M.**, Preparation and characterization of low-spin iron(II) porphyrin complexes with bis(phosphine) or bis(phosphite) axial ligands, *Inorg. Chem.*, 23(9), 1303, 1984.
119. **Kobayashi, N., Koshiyama, M., Funayama, K., Osa, T., Shirai, H., and Hanabusa, K.**, A mononuclear high-spin iron(III) phthalocyanine, *J. Chem. Soc., Chem. Commun.*, 16, 913, 1983.
120. **Shirai, H., Hanabusa, K., Kitamura, M., Masuda, E., Hirabaru, O., and Hojo, N.**, Functional metal porphyrazine derivatives and their polymers. XIV. Synthesis and properties of [bis- or tetrakis(decyloxycarbonyl)phthalocyaninato]metal complexes, *Makromol. Chem.*, 185(12), 2537, 1984.
121. **Hanack, M., Datz, A., Fischer, K., Kobel, W., Koch, J., Metz, J., Mezger, M., and Schneider, O.**, Synthesis and properties of conducting bridged macrocyclic metal complexes, *Polym. Prepr. (Am. Chem. Soc., Div. Polym. Chem.)*, 23(1), 126, 1982.
122. **Schneider, O. and Hanack, M.**, Cyanide-containing phthalocyaninatoiron compounds: synthesis and electrical conductivities, *Z. Naturforsch., B: Anorg. Chem., Org. Chem.*, 39B(2), 265, 1984.
123. **Hanack, M., Datz, A., Kobel, W., Koch, J., Metz, J., Mezger, M., Schneider, O., and Schulze, H. J.**, Synthesis and properties of conducting bridged macrocyclic metal complexes, *J. Phys. Colloq.*, 633, 1983.
124. **Lin, J. W. P. and Dudek, L. P.**, Thermal treatment, spectroscopic analyses, and electrical properties of poly(copper 2,3,9,10,16,17,23,24-octacyanophthalocyanines, *J. Polym. Sci., Polym. Chem. Ed.*, 23(5), 1589, 1985.
125. **Hedtmann-Rein, C., Keppeler, U., Muenz, X., and Hanack, M.**, Doped and nondoped polymeric bridged macrocyclic transition metal complexes, *Mol. Cryst. Liq. Cryst.*, 118(1—4), 361, 1985.
126. **Hanack, M. and Fischer, K.**, Synthesis and conductivities of phthalocyaninato-germanium-carbodiimido complexes, *Synth. Met.*, 10(5), 347, 1985.

127. **Palmer, S. M., Stanton, J. L., Jaggi, N. K., Hoffman, B. M., Ibers, J. A., and Schwartz, L. H.,** Preparation, structures, and physical properties of two products from the iodination of (phthalocyaninato)iron(II), *Inorg. Chem.*, 24(13), 2040, 1985.
128. **Hanack, M.,** Synthesis and properties of conducting bridged macrocyclic metal complexes, *Isr. J. Chem.*, 25(3—4), 205, 1985.
129. **Grigoryan, L. S. and Sharoyan, E. G.,** Synthesis of a new class of conductive organic compounds based on phthalocyanines and iodine, *J. Chem. Soc., Chem. Commun.*, 19, 1302, 1985.
130. **Ogawa, M. Y., Martinsen, J., Palmer, S. M., Stanton, J. L., Tanaka, J., Greene, R. L., Hoffman, B. M., and Ibers, J. A.,** The (phthalocyaninato)copper iodide complex Cu(pc)I: a molecular metal with a one-dimensional array of local moments embedded in a "Fermi sea" of charge carriers, *J. Am. Chem. Soc.*, 109(4), 1115, 1987.
131. **Hanack, M., Keppeler, U., and Schulze, H. J.,** Iodine-doped bridged phthalocyaninatoiron(II) and -ruthenium(II) compounds, *Synth. Met.*, 20(3), 347, 1987.
132. **Hanack, M., Deger, S., Keppeler, U., Lange, A., Leverenz, A., and Rein, M.,** Synthesis and properties of new semiconducting bridged phthalo- and naphthalocyaninatometal compounds, *Synth. Met.*, 19(1—3), 739, 1987.
133. **Brant, P., Weber, D. C., Haupt, S. G., Nohr, R. S., and Wynne, K. J.,** Partially oxidized Group 3A fluorometallophthalocyanines, *J. Chem. Soc., Dalton Trans.*, 2, 269, 1985.
134. **Homborg, H. and Murray, K. S.,** Preparation and properties of anionic fluoro complexes of magnesium, zinc, and aluminum phthalocyanine(2-), *Z. Anorg. Allg. Chem.*, 517, 149, 1984.
135. **Segawa, T. and Maruyama, K. (to Kawasaki Kasei Chemicals, Ltd.),** Japan Patent 85,193,992, 1985.
136. **Dao, L. H. and Perrier, G.,** A new polymorph of hydroxyaluminum phthalocyanine. Characterization and photoactivity, *Chem. Lett.*, 8, 1259, 1986.
137. **Kuroiwa, A., Nanba, N., and Kamijo, T. (to TDK Corp.),** Japan Patent 86,152,685, 1986.
138. **Nanba, N. and Kamijo, T. (to TDK Corp.),** Japan Patent 86,152,689, 1986.
139. **Goedken, V. L., Dessy, G., Ercolani, C., and Fares, V.,** Synthesis, reactivity, and x-ray crystal structure of dichloro(phthalocyaninato)titanium(IV), *Inorg. Chem.*, 24(7), 991, 1985.
140. **Ivasenko, V. L. and Velichko, A. V.,** U.S.S.R. Patent 1,320,216, 1987.
141. **Freyer, W. and Minh, L. Q.,** Synthesis of metal complexes of tetra-(2,2-anthra)tetraazaporphines and comparison of their electronic absorption spectra with other annellated tetraazaporphine systems, *Monatsch. Chem.*, 117(4), 475, 1986.
142. **Law, K. Y.,** Synthesis and properties of *tert*-butyl-substituted vanadylphthalocyanine dyes, *Inorg. Chem.*, 24(12), 1778, 1985.
143. **Mikhalenko, S. A., Solov'eva, L. I., Ivanova, T. M., and Luk'yanets, E. A.,** Phthalocyanines and related compounds. XXVI. Ureido-substituted phthalocyanines, *Zh. Obshch. Khim.*, 55(5), 1106, 1985.
144. **Page, G. A., Tokoli, E. G., Cosgrove, R. T., and Spiewak, J. W. (to Xerox Corp.),** U.S. Patent 4,557,868, 1985.
145. **Yamamoto Synthetic Chemical Co., Ltd.,** Japan Patent 85,23,451, 1985.
146. **Eda, T. (to Yamamoto Chemical Industry Co., Ltd.),** Japan Patent 86,215,663, 1986.
147. **Boerschel, V. and Straehle, J.,** Synthesis and structure of oxophthalocyaninatomolybdenum(IV), *Z. Naturforsch., B,* 39(12), 1664, 1984.
148. **Edmondson, S. J. and Mitchell, P. C. H.,** Molybdenum phthalocyanine, *Polyhedron,* 5(1—2), 315, 1986.
149. **Mrwa, A., Rummel, S., and Starke, M.,** Preparation and electrical properties of nitridophthalocyaninerhenium(V), *Z. Chem.*, 25(5), 186, 1985.
150. **Rummel, S., Hermann, M., and Schmidt, K.,** Preparation and characterization of nitridophthalocyanine technetium(V), *Z. Chem.*, 25(4), 152, 1985.
151. **Rummel, S., Herrmann, M., and Schmidt, K.,** East German Patent 219,486, 1985.
152. **Kobel, W. and Hanack, M.,** Bis axially coordinated (phthalocyaninato)ruthenium(II) compounds, *Inorg. Chem.*, 25(1), 103, 1986.
153. **Doeff, M. M.,** Axial Ligand Substitution Reactions of Ruthenium(II) Phthalocyanine. Hydrogen Bonding in Metalloporphyrin Reactions. Electrocatalytic Ligand Substitution in Tricarbonyl(arene)chromium Complexes, Ph.D. thesis, Brown University, Providence, RI, 1983.
154. **Keppeler, U., Kobel, W., Siehl, H. U., and Hanack, M.,** (Phthalocyaninato)iron(II)- and -ruthenium(II) compounds with isocyanides as axial ligands, *Chem. Ber.*, 118(5), 2095, 1985.
155. **Tomilova, L. G., Chernykh, E. V., Nikolaeva, T. B., Zelentsov, V. V., and Luk'yanets, E. A.,** Cerium(IV) bis(phthalocyanine), *Zh. Obshch. Khim.*, 54(7), 1678, 1984.
156. **Kasuga, K. and Morimoto, H.,** A phthalocyanine complex of neodymium(III) showing an unusual absorption spectrum, *Inorg. Chim. Acta,* 119(1), L17, 1986.
157. **Kasuga, K., Ando, M., and Morimoto, H.,** The radical formation of diphthalocyanine complexes of lanthanum(III), neodymium(III) and yttrium(III) with *p*-benzoquinone, *Inorg. Chim. Acta,* 112(1), 99, 1986.

158. **Clarisse, C. and Riou, M. T.**, Synthesis and characterization of some lanthanide phthalocyanines, *Inorg. Chim. Acta,* 130(1), 139, 1987.
159. **Sugimoto, H., Higashi, T., Maeda, A., Mori, M., Masuda, H., and Taga, T.**, Preparation of (μ-phthalocyaninato)bis[di(2,2,6,6-tetramethylheptane-3,5-dionato)lanthanide(III)] complexes and the crystal structure of the samarium complex, *J. Chem. Soc., Chem. Commun.*, 21, 1234, 1983.
160. **De Cian, A., Moussavi, M., Fischer, J., and Weiss, R.**, Synthesis, structure, and spectroscopic and magnetic properties of lutetium(III)phthalocyanine derivatives: $LuPc_2.CH_2Cl_2$ and $[LuPc(OAc)(H_2O)_2].H_2O.2CH_3OH$, *Inorg. Chem.*, 24(20), 3162, 1985.
161. **Moussavi, M., De Cian, A., Fischer, J., and Weiss, R.**, (Porphyrinato)bis(phthalocyananinato)dilanthanide(III) complexes presenting a sandwich triple-decker-like structure, *Inorg. Chem.*, 25(13), 2107, 1986.
162. **Zhuang, W., Ni, J., and Xu, H.**, Rare earth phthalocyanine compounds. I. Synthesis and properties of rare earth monophthalocyanine, *Huaxue Xuebao,* 44(5), 509, 1986.
163. **Subbotin, N. B., Tomilova, L. G., Chernykh, E. V., Kostromino, N. A., and Luk'yanets, E. A.**, Synthesis of unsymmetrical rare-earth-element diphthalocyanines, *Zh. Obshch. Khim.*, 56(1), 232, 1986.
164. **Subbotin, N. B., Tomilova, L. G., Kostromina, N. A., and Luk'yanets, E. A.**, Phthalocyanines and related compounds. XXVIII. Synthesis and spectral-electrochemical study of rare earth monophthalocyanines, *Zh. Obshch. Khim.*, 56(2), 397, 1986.
165. **M'Sadak, M., Roncali, J., and Garnier, F.**, Lanthanide-phthalocyanine complexes: from diphthalocyanine Pc_2 Ln to a super complex Pc_3 Ln_2, *J. Chim. Phys. Phys.-Chim. Biol.*, 83(3), 211, 1986.
166. **Tomilova, L. G., Cherynkh, E. V., and Luk'yanets, E. A.**, Synthesis of blue forms of rare earth diphthalocyanines, *Zh. Obshch. Khim.*, 55(11), 2631, 1985.
167. **Zhuang, W., Fan, G., and Ni, J.**, Rare earth phthalocyanines. III. DTA study of the solid reaction of rare earth acetates with phthalonitrile, *Yingyong Huaxue,* 2(4), 26, 1985.
168. **Sugimoto, H., Higashi, T., Maeda, A., Hirai, Y., Teraoka, J., Mori, M., Masuda, H., and Taga, T.**, Mixed-ligand rare earth complexes of phthalocyanine and β-diketones, *J. Less-Common Met.*, 112, 387, 1985.
169. **Hatfield, W. E.**, Synthesis and Characterization of New Low-Dimensional Metal Complex Conductors, Report 1986, Order No. AD-A172659/5/GAR, National Technical Information Service, Springfield, VA, 12 pp.
170. **Guilard, R., Dormond, A., Belkalem, M., Anderson, J. E., Liu, Y. H., and Kadish, K. M.**, First example of 1:1 actinide-phthalocyanine complexes: synthesis, electrochemical, and spectral characterization of bis(diketonato)thorium(IV) and -uranium(IV) phthalocyaninates, *Inorg. Chem.*, 26(9), 1410, 1987.
171. **Piechocki, C. and Simon, J.**, Synthesis of a polar discogen. A new type of discotic mesophase, *J. Chem. Soc., Chem. Commun.*, 5, 259, 1985.
172. **Guilleux, L., Krausz, P., Nadjo, L., and Uzan, R.**, French Patent, 2,566,776, 1986.
173. **Sounik, D. F.**, Synthesis and Properties of Potential Phthalocyanine Coupling Agents, Ph.D. thesis, Case Western Reserve University, Cleveland, 1986.
174. **Cicek, A., Guthrie, J. T., and Squires, L. J.**, Phthalocyanine ring systems from phthalodinitrile via radical-anionic cyclization, *Dyes Pigm.*, 5(3), 225, 1984.
175. **Toyo Ink Manufacturing Co., Ltd.**, Japan Patent 83,183,758, 1983.
176. **Toyo Ink Manufacturing Co., Ltd. and Hitachi, Ltd.**, Japan Patent 85,20,970, 1985.
177. **Toyo Ink Manufacturing Co., Ltd. and Hitachi, Ltd.**, Japan Patent 85,87,289, 1985.
178. **Toyo Ink Manufacturing Co., Ltd.**, Japan Patent 83,183,757, 1983.
179. **Toyo Ink Manufacturing Co., Ltd. and Hitachi, Ltd.**, Japan Patent 85,20,969, 1985.
180. **Toyo Ink Manufacturing Co., Ltd. and Hitachi, Ltd.**, Japan Patent 85,87,288, 1985.
181. **Kinoshita, S. and Yonehara, Y. (to Dainippon Ink and Chemicals, Inc. and Kawamura Physical and Chemical Research Institute)**, Japan Patent 85,243,089, 1985.
182. **Rejlek, A. and Kroupa, J.**, Czech Patent 212,686, 1984.
183. **Leznoff, C. C., Greenberg, S., Khouw, B., and Lever, A. B. P.**, The syntheses of mono- and disubstituted phthalocyanines using a dithioimide, *Can. J. Chem.*, 65(8), 1705, 1987.
184. **Woehrle, D. and Meyer, G.**, Phthalocyanines — a system with an unusual structure and properties. I. *Kontakte (Darmstadt),* 3, 38, 1985.
185. **Kumano, I.**, Synthesis and physical properties of phthalocyanine pigments, *Denshi Shashin Gakkaishi,* 22(2), 111, 1984.
186. **Nishi, H.**, Chemistry of high-performance organic pigments, *Kagaku Kogyo,* 37(5), 353, 1986.
187. **Woerhle, D. and Kaneko, M.**, Characteristics of phthalocyanines and their application to functional materials, *Yuki Gosei Kagaku Kyokaishi,* 45(9), 837, 1987.

Chapter 3

ANALYSIS

I. DETERMINATION OF PCS IN OTHER MATERIALS

Traces of tetrasulfonated Cu phthalocyanines (pcs) can be detected by surface enhanced Raman spectroscopy (SERS), since silver island films are stable and are easy to use as optical sensors for detection of nanogram amounts.[1]

High-resolution/higher order derivative spectrophotometry is used to identify and to estimate nanogram amounts of Cu pc in artists' paints.[2]

Cu pc 4,4′,4″,4‴-tetrasulfonic acid tetrakis(1-isopropyl-2,5-dihydroxyphenylamide) is detected by thin layer chromatography on silica gel.[3]

II. DETERMINATION OF OTHER MATERIALS IN PCS

Cyanuric acid is determined in the manufacture of Cu pc in the presence of acids by sodium hydroxide titration for total acidity and treating another sample with melamine, converting cyanuric acid to an addition compound, and titrating.[4]

Neutron activation analysis and X-ray diffraction are used to determine the amount and nature of impurities in semiconductor pcs: di hydrogen, Cu, Fe, Ni, and Zn.[5]

Contaminants are determined in Cu pc by decomposing the Cu pc sample in boiling 65% nitric acid, diluting with distilled water, and comparing with a standard or drying, dissolving in sulfuric acid, and analyzing spectrophotometrically.[6]

III. DETECTION OF CU PC PIGMENTS AND DYES

Identification of organic pigments including Cu pc is accomplished by Raman microprobe spectrometric techniques.[7]

Copper is identified in a mixture containing pigment yellow and Cu pc by simultaneous atomic and molecular spectrometry with electrothermal atomization and diode array detection.[8]

IV. USE OF PCS IN ANALYSIS

Cu pc is used to detect organic or inorganic solvents by colorimetry.[9]

A blue silica gel containing a pc reactive dye adsorbs and desorbs polycyclic organic substances in solution and can be used for the separation or removal of mutagenic substances from the environment or foods, or as a column filler for HPLC.[10]

Electrocatalytic chemically modified electrodes are prepared by adding weighed amounts of Co pc to a carbon paste mixture; they are sensing electrodes for the electrochemical detection of hydrazine in a typical liquid chromatography flow cell.[11]

REFERENCES

1. **Aroca, R. and Martin, F.,** Trace analysis of tetrasulfonated copper phthalocyanine by surface enhanced Raman spectroscopy, *J. Raman Spectrosc.*, 17(3), 243, 1986.
2. **Talsky, G. and Ristic-Solajic, M.,** High resolution/higher order derivative spectrophotometry for identification and estimation of synthetic organic pigments in artists' paints, *Anal. Chim. Acta,* 196, 123, 1987.

3. **Nikonova, L. Z. and Kirillova, K. M.,** U.S.S.R. Patent 1,154,614, 1985.
4. **Lakomy, J.,** Czech Patent 206,300, 1983.
5. **Marsan, B., Crine, J. P., and Belanger, G.,** Impurities in phthalocyanines and structural modifications induced during thin-film deposition, *Mater. Chem. Phys.,* 14(3), 269, 1986.
6. **Pikhart, J.,** Czech Patent 218,405, 1985.
7. **Boyer, H.,** Possible applications for Raman microanalysis, *Spectrochim. Acta,* Part B, 39(12), 1527, 1984.
8. **Tittarelli, P., Lancia, R., and Zerlia, T.,** Simultaneous molecular and atomic spectrometry with electrothermal atomization and diode array detection, *Anal. Chem.,* 57(9), 2002, 1985.
9. **Yanagi, K.,** Japan Patent 85,91,243, 1985.
10. **Hayatsu, H., Nakano, M., and Matsuo, Y.,** Europe Patent Application 157,549, 1985.
11. **Korfhage, K. M., Ravichandran, K., and Baldwin, R. P.,** Phthalocyanine-containing chemically modified electrodes for electrochemical detection in liquid chromatography/flow injection systems, *Anal. Chem.,* 56(8), 1514, 1984.

Chapter 4

CRYSTALS

I. STUDIES WITH THE ELECTRON MICROSCOPE

A. HIGH-RESOLUTION ELECTRON MICROSCOPY (HREM)

The nonstoichiometric compound Ni pc · I_x ($x < 1.0$) is studied by HREM. The mechanism of the topotactic formation is revealed by using two kinds of epitaxial films taking different orientations.[1]

Structural defects or crystal imperfections in pc metal derivatives are studied by HREM. The existence of a structural isomer of Th pc generates some defects.[2]

Molecular arrangement is studied at the green boundaries in fully chlorinated Cu pc Cl_{16} by HREM.[3]

Practical high-resolution observations such as for pc green are made for direct imaging of organic molecules in crystals by HREM at the atomic level.[4]

HREM imaging of a surface complex of pc 2,3,9,10,16,17,23,24-octacarbonitrile is made on potassium iodide.[5]

Two new methods of nonlinear image processing are applied to high-resolution micrographs of chlorinated Cu pc taken on a Kyoto 500 kV electron microscope. Point-to-point resolution is about 1.2 to 1.4 Å.[6]

Molecular images of chlorinated Cu pc show the arrangement of molecules at the interface between five crystals.[7]

A discussion with 18 references on HREM includes pc polymers at the molecular level.[8]

B. RADIATION DAMAGE

Epitaxial films of Pt pc formed on a mica surface are studied by HREM.[9]

Radiation damage of pc by HREM is studied.[10] Radiation damage to pc in potassium chloride encapsulated by carbon in an electron microscope is examined.[11]

Epitaxially grown molecular crystals of fully brominated and fully chlorinated pcs are isomorphous and have only the α-type polymorph structure. These crystals are sensitive to beam damage.[12]

Electron irradiation of the (001) orientation of Cu pc induces deformation and produces dislocations.[13]

Radiation damage in the electron microscope proceeds by loss of peripheral atoms in Cu pc.[14]

C. TRANSMISSION (TEM) AND SCANNING (SEM) ELECTRON MICROSCOPY

TEM directly reveals defects in thin single crystals of x-pc polymorph.[15]

The morphology of fully and partially polymerized poly(phthalocyaninato germoxane), $(Ge\ pc\ O)_n$ crystals is studied by SEM and TEM. The crystal laths are 1000 to 2000 Å thick, 2000 to 10,000 Å wide, and 1 to 5 μm long.[16]

The crystallographic behavior of Zn pc is studied by TEM during the $\alpha \rightarrow \beta$ transition in alcoholic vapor.[17] The phase behavior of Co, Ni, and Zn pc is studied by DSC, IR absorption spectroscopy, and TEM.[18]

D. STRUCTURAL IMAGES

EM structural images of Si/O pc polymer and Al/F pc polymer are obtained. These polymers have sufficient space within their lattice for high concentrations of halogen dopant.[19] Cu pc Cl is sublimed onto potassium chloride cleavage faces and the films are examined by EM and diffraction.[20]

II. MOLECULAR BEAM

It is the incident angle of the molecular beam to the Cu pc substrate that decides the crystallite orientation of the Cu pc evaporated film.[21]

Ultrathin Cu pc films are deposited on amorphous quartz by the molecular beam method.[22]

III. FILM DEPOSITION

A method for film deposition involves use of Cu pc, the transmittance of which changes according to the changes in the film thickness, and carrying out the deposition while monitoring the transmittance.[23]

Oriented thin films of Cu pc are formed by evaporation onto scratched substrates.[24]

Cu pc films are deposited as α-Cu pc at 0.01 to 0.000001 torr. At 0.1 torr, the film is X-ray amorphous. A transition is observed from α-Cu pc → x-Cu pc at 0.01 to 0.1 torr.[25]

IV. FILMS GROWN IN SPACE

An analysis is made of the results of the growth of films of Cu pc on space shuttle flight STS-511. Preliminary effects due to gravitation are discussed.[26]

V. UNIT CELL CONSTANTS

Unit cell constants are measured of the charge transfer complex of Zn pc with piperidine. The lattice constants are (a) 24.5 and (c) 17.5 Å.[27]

A new crystalline Zn pc Cl is prepared by electrolysis in an organic solvent. Zn pc Cl is tetragonal, space group P4/nnc, with a 13.85(1), c 13.09(1) Å; d_m = 1.58 and d_c = 1.62. The final R = 0.098 and R_w = 0.11.[28]

β-Neodymium di pc is monoclinic, space group C2/c, a 19.010, b 19.066, c 15.538 Å, and β = 116.1°; d = 1.54 for Z = 4; R = 0.052.[29]

The crystal and molecular structure is studied of Ga pc Cl and Al pc Cl. Ga pc Cl is triclinic, space group $P\bar{1}$, a 9.301(3), b 11.272(1), c 13.143(3) Å, α 105.46(2), β 105.61(2), and γ 96.80(2)°; Z = 2 for d_c = 1.64. Final R = 0.036 and R_w = 0.042.[30]

The unit cell constants are given for fluorophthalocyaninatogallium(III) $(\text{Ga pc F})_n$ and (μ-oxo)bis(phthalocyaninato)aluminum(III) $(\text{Al pc})_2$ 0.[31]

The unit cell constants are given for bis(triphenylphosphine)iminiumbis-(cyano)(phthalocyaninato)ferrate (III), $[N\{P(C_6H_5)_3\}_2][Fe(CN)_2(C_{32}H_{16}N_8)]$ at 180 K.[32]

The unit cell constants are given for Th pc_2. The thorium atom is in a distorted anti-prismatic coordination.[33]

Unit cell constants are given a lithium pc complex.[34]

Unit cell constants are also given for the β-forms of [phthalocyaninato(2-)]Fe, -Co, and -Ni.[35]

Unit cell constants are given for α-neodymium di pc.[36]

Calculated d-spacings for Cu pc and Zn pc are tabulated.[37]

VI. A VARIETY OF TOPICS

IR absorption spectroscopy and differential scanning calorimetry are used to study the phase transitions of Cu pc and Zn pc. Both compounds undergo polymorphic transition in two stages.[38]

A model is developed to simulate the kinetic data for the α → β polymorphic transformation in Cu pc.[39]

The adsorption of polar and nonpolar low molecular weight compounds such as water and hexane on α-Cu pc is independent of the α-Cu pc pigment particle size but dependent on pigment shape.[40]

The thermotropic polymorphism of bis(octaalkyloxyphthalocyaninato)lutetium complexes is studied by polarizing microscopy, DSC (differential scanning calorimetry), and X-ray diffraction.[41]

Octasubstituted pc derivatives forming discotic mesophases are prepared.[42] Octakis(dodecyloxy)methyl M pcs forming discotic mesophases are studied by EPR.[43]

Octasubstituted pc derivatives are studied using optical microscopy and X-ray diffraction. They exhibit columnar mesophases with hexagonal column packing.[44]

Evidence is given for the columnar mesophase of the octakis(octyloxymethyl) derivative of Pb pc at $<<$ room temperature, obtained with flat discotic molecules showing an antiferroelectric stacking.[45,46]

The epitaxial growth of chlorinated Cu pc on the (001) potassium chloride surface are described by molecular energetics computations.[47]

The Madelung energy of the one-dimensional molecular metal Ni pc I is calculated as a function of charge transfer and stacking motif along the unique axis. The Ni pc I stacking axis has alternate rings rotated by 39.5°.[48]

The crystal growth of β-pc single crystals from the vapor phase in closed ampoules is studied.[49]

IR, FTIR and visible absorption, and X-ray diffraction peaks of τ-H pc are drawn.[50]

Dye-in-polymer films consisting of 20 wt % *tert*-butylsubstituted VO pc in vinyl chloride-vinyl acetate copolymer are prepared by a spin-coating technique. The pc precipitates as a glassy solid within the polymer. Upon exposure of the films to solvent vapors, the pc crystallizes to a more stable form.[51]

Time-of-flight experiments are made for vacuum-evaporated H_2pc layers. Transport phenomena in them are more complex vs. those in simple organic crystals.[52]

VII. LIQUID CRYSTALS

Polymer octasubstituted μ-oxo-(phthalocyaninato)tin(IV) forms columnar liquid crystals.[53]

H and Cu 1,4,8,11,15,18,22,25-octa-alkyl pcs with C_6-C_{10} alkyl chain lengths show thermotropic liquid crystalline behavior with one or more mesophases.[54]

Color liquid crystal display devices have color images containing Cu pc.[55]

Alkoxy pcs as shown in Figure 1 ($R^{1\text{-}8}$ = $C_{1\text{-}6}$ alkyl, R^9,R^{10} together form an optionally substituted benzo group or are each halo) have strong absorption bands at about 250 to 360 nm and at about 700 to 900 nm. They are useful in smectic liquid crystal materials for laser-addressed storage devices.[56]

A H_2 pc octasubstituted by paraffinic chains has an increased tendency to aggregate compared with H_2 pc, leading to a red-shifted fluroescence in concentrated solutions. In benzene there is an E-type delayed fluorescence, which is partially quenched by octasubstituted Cu pc. Solid crystals can clearly be differentiated from thermotropic liquid crystals by emission spectroscopy.[57]

VIII. REVIEW

A review with 39 references discusses the morphology of organic pigments with emphasis on Cu pc.[58]

A review with 42 references discusses the crystal structure of pc and its metal complexes.[59]

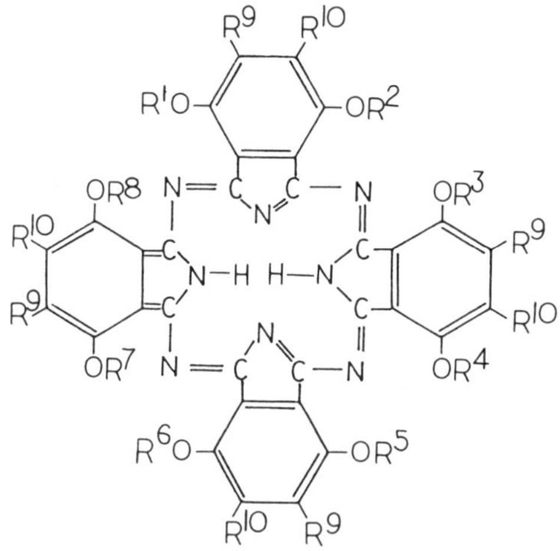

FIGURE 1. Alkoxy pcs.

A review with eight references includes crystal growth, preparation, pigmentation, and flocculation of Cu pc.[60]

A review with 20 references compares the lattice parameters of polymorphic forms of Cu pc complexes by different authors.[61]

REFERENCES

1. **Kobayashi, T., Yase, K., and Uyeda, N.,** Direct observation of structure change in nickel-phthalocyanine caused by iodine doping, *Acta Crystallogr., Sect. B,* 40(3), 263, 1984.
2. **Kobayashi, T., Fujiyoshi, Y., and Uyeda, N.,** High-resolution electron microscopy of structural defects in organic crystals, *J. Cryst. Growth,* 65(1—3), 511, 1983.
3. **Fryer, J. R., Smith, D. J., and Camps, R. A.,** High resolution study of localized molecular arrangements in phthalocyanines, *Conf. Ser.-Inst. Phys.,* 68, 15, 1984.
4. **Uyeda, N.,** Direct imaging of organic molecules in crystals by high resolution electron microscopy at atomic level, *Dyes Pigm.,* 6(2), 115, 1985.
5. **Uyeda, N., Kobayashi, T., Ishizuka, K., Fujiyoshi, Y., Inokuchi, H., and Saito, G.,** Direct molecular imaging of low dimensional solids by high resolution electron microscopy, *Mol. Cryst. Liq. Cryst.,* 125(1—4), 103, 1985.
6. **Kirkland, E. J., Sigel, B. M., Uyeda, N., and Fujiyoshi, Y.,** Improved high resolution image processing of bright field electron micrographs, *Ultramicroscopy,* 17(2), 87, 1985.
7. **Fryer, J. R. and Smith, D. J.,** High resolution electron microscopy of interfaces in chlorinated phthalocyanine molecular crystals, *J. Microsc. (Oxford),* 141(1), 3, 1986.
8. **Fryer, J. R.,** Intermolecular structure in organic crystals shown by high resolution electron microscopy, *Mol. Cryst. Liq. Cryst.,* 137(1—4), 49, 1986.
9. **Yase, K., Kobayashi, T., and Uyeda, N.,** Direct observation of disorders of molecular stacking in platinum-phthalocyanine epitaxial film, *Bull. Inst. Chem. Res., Kyoto Univ.,* 63(5—6), 383, 1985 (Publ. 1986).
10. **Fryer, J. R. and Holland, F.,** High resolution electron microscopy of molecular crystals. III. Radiation processes at room temperature, *Proc. R. Soc. London, Ser. A,* 393(1805), 353, 1984.
11. **Holland, F., Fryer, J. R., and Baird, T.,** Studies of radiation damage in the electron microscope, *Conf. Ser.-Inst. Phys.,* 68, 19, 1984.
12. **Smith, D. J., Fryer, J. R., and Camps, R. A.,** Radiation damage and structural studies: halogenated phthalocyanines, *Ultramicroscopy,* 19(3), 279, 1986.
13. **Natsvlishvili, G. I.,** Radiation-induced lattice disturbances in copper phthalocyanine crystals, *Izv. Akad. Nauk SSSR, Ser. Fiz.,* 48(9), 1735, 1984.
14. **Fryer, J. R., McNee, C., and Holland, F. M.,** The further reduction of radiation damage in the electron microscope, *Ultramicroscopy,* 14(4), 357, 1984.

15. **Perovic, A.,** Transmission electron microscopic observation of dislocations in x-phthalocyanine crystals, *J. Mater. Sci.,* 22(3), 835, 1987.
16. **Zhou, X., Marks, T. J., and Carr, S. H.,** Morphology of poly(phthalocyaninatogermoxane) crystals, *J. Polym. Sci., Polym. Phys. Ed.,* 23(2), 305, 1985.
17. **Iwatsu, F.,** Crystal behavior of zinc phthalocyanine films in alcohols, *J. Cryst. Growth.,* 71(3), 629, 1985.
18. **Collins, R. A. and Mohammed, K. A.,** Phase behavior of cobalt, nickel and zinc phthalocyanines, *Thermochim. Acta,* 109(2), 397, 1987.
19. **Fryer, J. R. and Kenney, M. E.,** Structural arrangements in polymeric phthalocyanines, *Inst. Phys. Conf. Ser.,* 78, 441, 1985 (Publ. 1986).
20. **Saito, Y.,** Epitaxial growth mechanism of chlorinated copper phthalocyanine on potassium chloride surfaces, *Appl. Surf. Sci.,* 22—23(2), 574, 1985.
21. **Shi, Z. and Cai, Z.,** Effect of incident molecule beam direction on the crystallite orientation of the evaporated film of copper phthalocyanine, *Kexue Tongbao (Foreign Lang. Ed.),* 29(2), 280, 1984.
22. **Komiyama, M., Sakakibara, Y., and Hirai, H.,** Preparation of highly ordered ultrathin films of copper(II) phthalocyanine on amorphous substrates by molecular beam deposition, *Thin Solid Films,* 151(1), L109, 1987.
23. **Canon, K. K.,** Japan Patent 85,46,371, 1985.
24. **Sakamoto, M. and Mori, Y. (to Toshiba Corp.),** Japan Patent 86,77,322, 1986.
25. **Shi, Z., Yang, Q., and Chang, L.,** Effect of vacuum on the crystal form and morphology of the evaporated film of copper phthalocyanine, *Kexue Tongbao (For. Lang. Ed.),* 31(16), 1108, 1986.
26. **Debe, M. K., Cook, E. L., Poirier, R. J., Miller, L. R., Follett, G. J., and Spiering, S. M.,** Closed cell vapor transport growth of organic solids in microgravity and unit gravity: some comparisons and results to date, *Polym. Prepr. (Am. Chem. Soc., Div. Polym. Chem.),* 28(2), 453, 1987.
27. **Kobayashi, T., Yase, K., and Uyeda, N.,** Unit cell constants and crystal habit of charge transfer complex of zinc phthalocyanine with piperidine, *J. Cryst. Growth,* 66(3), 553, 1984.
28. **Mossoyan-Deneux, M., Benlian, D., Pierrot, M., Fournel, A., and Sorbier, J. P.,** Structure and conductivity of chloro(phthalocyaninato) zinc single crystals, *Inorg. Chem.,* 24(12), 1878, 1985.
29. **Darovskikh, A. N., Tsytsenko, A. K., Frank-Kamenetskaya, O. V., Fundamenskii, V. S., and Moskalev, P. N.,** Polymorphism of neodymium di phthalocyanine. Molecular and crystal structure of the β-phase, *Kristallografiya,* 29(3), 455, 1984.
30. **Wynne, K. J.,** Crystal and molecular structure of chloro (phthalocyaninato) gallium (III), Ga(Pc)Cl, and chloro (phthalocyaninato) aluminum (III), Al(Pc)Cl, *Inorg. Chem.,* 23(26), 4658, 1984.
31. **Wynne, K. J.,** Two ligand-bridged phthalocyanines: crystal and molecular structure of fluoro(phthalocyaninato)gallium(III), $[Ga(Pc)F]_n$, and (μ-oxo)bis[(phthalocyaninato)aluminum(III)], $[Al(Pc)]_2O$, *Inorg. Chem.,* 24(9), 1339, 1985.
32. **Kueppers, H., Kalz, W., and Homberg, H.,** Structure of bis(triphenylphosphine)iminium bis(cyano)(phthalocyaninato)ferrate(III), $[N\{P(C_6H_5)_3\}_2]$ $[Fe(CN)_2$ $(C_{32}H_{16}N_8)]$, at 180 K, *Acta Crystallogr., Sect. C,* 41(10), 1420, 1985.
33. **Darovskikh, A. N., Frank-Kamenetskaya, O. V., Fundamenskii, V. S., and Golynskaya, O. A.,** Refinement of the molecular and crystal structure of thorium diphthalocyanine. Crystal chemical characteristics of a monoclinic modification of diphthalocyanines of the actinides and rare earth elements, *Kristallografiya,* 30(6), 1085, 1985.
34. **Sugimoto, H., Mori, M., Masuda, H., and Taga, T.,** Synthesis and molecular structure of a lithium complex of the phthalocyanine radical, *J. Chem. Soc., Chem. Commun.,* 12, 962, 1986.
35. **Ruiz-Ramirez, L., Martinez, A., Javier Sosa, J., Luis Brianso, J., Estop, E., Alcobe, X., and Chinchon, J. S.,** Obtaining the high resolution powder diffraction diagrams of phthalocyanines of iron, cobalt, and nickel, *Afinidad,* 43(404), 337, 1986.
36. **Darovskikh, A. N., Frank-Kamenetskaya, O. V., and Fundamenskii, V. S.,** Molecular and crystal structure of tetragonal α-phase of neodymium diphthalocyanine, *Kristallografiya,* 31(5), 901, 1986.
37. **Ruiz-Ramirez, L., Martinez, A., Mendieta, J., Brianso, J. L., Estop, E., and Chinchon, J. S.,** Production of the powder diffraction diagrams with high resolution power, of copper and zinc β-phthalocyanines, *Afinidad,* 44(407), 45, 1987.
38. **Mohammed, K. A. and Collins, R. A.,** Phase behavior of copper and zinc phthalocyanines, *Thermochim. Acta,* 104, 377, 1986.
39. **Cardew, P. T. and Davey, R. J.,** The kinetics of solvent-mediated phase transformations, *Proc. R. Soc. London, A,* 398(1815), 415, 1985.
40. **Aristove, B. G., Kuznetsov, B. V., and Mishchenko, V. N.,** Effect of morphology of phthalocyanine pigments on the properties of their surfaces, *Lakokras. Mater. Ikh Primen.,* 2, 19, 1986.
41. **Piechocki, C., Simon, J., Andre, J. J., Guillon, D., Petit, P., Skoulios, A., and Weber, P.,** Synthesis and physiochemical studies of neutral and chemically oxidized forms of bis(octaalkyloxyphthalocyaninato)lutetium, *Chem. Phys. Lett.,* 122(1—2), 124, 1985.
42. **Piechocki, C. and Simon, J.,** Annelides. XI. Elaboration of molecular materials. Synthesis of octasubstituted phthalocyanine derivatives forming discotic mesophases, *Nouv. J. Chim.,* 9(3), 159, 1985.

43. **Andre, J. J., Bernard, M., Piechocki, C., and Simon, J.,** Octakis[(dodecyloxy)methyl]-metallophthalocyanines forming discotic mesophases studied by electron paramagnetic resonance Annelides. XVII, *J. Phys. Chem.*, 90(7), 1327, 1986.
44. **Guillon, D., Weber, P., Skoulios, A., Piechocki, C., and Simon, J.,** Columnar mesophases from metal and metal-free derivatives of phthalocyanine, *Mol. Cryst. Liq. Cryst.*, 130(3—4), 223, 1985.
45. **Weber, P., Guillon, D., and Skoulios, A.,** Antiferroelectric stacking in lead phthalocyanine columnar mesophases, *J. Phys. Chem.*, 91(9), 2242, 1987.
47. **Saito, Y. and Shiojiri, M.,** Molecular energetics of the epitaxial growth of chlorinated copper phthalocyanine on potassium chloride surfaces, *J. Cryst. Growth*, 67(1), 91, 1984.
48. **Euler, W. B.,** Madelung energy calculations on the highly conducting molecular metal nickel phthalocyanine iodide, *Inorg. Chem.*, 23(17), 2645, 1984.
49. **Simonyan, M. V., Grigoryan, L. S., Gasparyan, A. G., and Sharoyan, E. G.,** Crystal growth of β-phthalocyanines from the vapor phase in closed ampuls, *Cryst. Res. Technol.*, 19(4), 441, 1984.
50. **Takano, S., Enokida, T., Kakuta, A., and Mori, Y.,** A new polymorph of metal-free phthalocyanine, *Chem. Lett.*, 12, 2037, 1984.
51. **Law, K. Y.,** An investigation of solvent-vapor-induced crystallization soluble vanadyl phthalocyanine dyes in polymer matrixes, *J. Phys. Chem.*, 89(12), 2652, 1985.
52. **Ziolkowska-Pawlak, B. and Mycielski, W.,** Anomalous transport in disordered organic layers, *J. Non-Cryst. Solids*, 90(1—3), 645, 1987.
53. **Sirlin, C., Bosio, L., and Simon, J.,** Annelides. XIX. Spinal column liquid crystals. Polymeric octasubstituted μ-oxo-(phthalocyaninato)tin(IV), *J. Chem. Soc., Chem. Commun.*, 5, 379, 1987.
54. **Cook, M. J., Daniel, M. F., Harrison, K. J., McKeown, N. B., and Thomson, A. J.,** 1,4,8,11,15,18,22,25-Octa-alkyl phthalocyanines: new discotic liquid crystal materials, *J. Chem. Soc., Chem. Commun.*, 14, 1086, 1987.
55. **Dainippon Printing Co., Ltd.,** Japan Patent 84,59,491, 1984.
56. **Harrison, K. J., Cook, M. J., Howe, S. D., and Thomson, A. J.,** British Patent 2,168,372, 1986.
57. **Knoesel, R., Piechocki, C., and Simon, J.,** Annelides. XII. Photochemical association properties of phthalocyanine-based discogens, *J. Photochem.*, 29(3—4), 445, 1985.
58. **Schroeder, J.,** Morphology of organic pigments with special reference to copper phthalocyanine, *Prog. Org. Coat.*, 12(2), 181, 1984.
59. **Luis Brianso, J., Regull, P., and Victori, L.,** Study of the correlation between crystal structure and photoelectrochemical behavior of phthalocyanine and its metallic derivatives. I. Classification of the crystal structures in different models, *Afinidad*, 41(393), 442, 1984.
60. **Wanser, C. C.,** Physical and chemical considerations on copper phthalocyanine, *J. Coat. Technol.*, 57(729), 57, 1985.
61. **Shi, Z. and Qian, R.,** Polymorphy of copper phthalocyanine, *Huaxue Tongbao*, 9, 42-5 and 69, 1983.

Chapter 5

REACTIONS: THEIR MECHANISMS AND RATES

I. TERTIARY BUTYL PC

The kinetics of complexation of tetra-*tert*-butyl phthalocyanine (pc) with cupric acetate are studied in proton-donor solvents such as acetic acid and pentanol.[1]

O_2^- is formed from O_2 in alkaline solutions of *tert*-butyl pc cobaltate (1-) in dimethyl formamide (DMF) solutions.[2]

The kinetics are studied of the coordination of 4,4′,4″,4‴-tetra-*tert*-butyl pc (pc H_2) to Cd, Co, Cu, Ni, and Zn by using the corresponding metal salt in pyridine solutions.[3]

II. A VARIETY OF TOPICS

A review covers intramolecular electron transfer reactions including oxide-bridged M pc complexes.[4]

The thermal oxidative degradation of Co pc is studied in alkali solutions.[5]

The kinetics and equilibria of the reaction between pyridine and Fe pc $(DMSO)_2$ are studied at 20°C. In excess pyridine, Fe pc $(pyridine)_2$ is formed by pseudo first-order kinetics.[6]

The equilibria and kinetics between NO and Fe pc in DMSO are studied at 20°C. The dissociation rate constant is estimated.[7]

The one-electron oxidations of Co(II), Co(III), Cu(II), and Ni(II) tetrasulfo pc are studied by using stop-flow, pulse radiolysis, and photolysis techniques.[8]

Adduct–ion formation by Co pc is studied with a methane plasma under methane negative chemical ionization conditions.[9]

The lability of ligands *trans* to methylimidazole in the Fe pc complex is studied.[10]

Equilibrium constants for $P(C_4H_9)_3$ and $P(OC_4H_9)_3$ binding to methylimidazole complexes of Fe protoporphyrin IX dimethyl ester are compared with new and published values for Fe pc and Fe(II) dimethyl glyoxime analogs.[11]

Isothermal thermogravimetric measurements are used to study the thermal stability of H_2, Cu, Ni, Pb, and Zn pc in air. A probable reaction mechanism is proposed.[12]

The irradiation of water containing pcs as photosensitizers gives hydrogen. A mechanism is proposed.[13]

The dissociation of Ta pcs in sulfuric acid is 3.5 order in $[H_3O^+]$ indicating a mechanism more complex than in the case of divalent M pcs.[14]

The dissociation mechanism of a Co pc complex in aqueous sulfuric acid is studied.[15]

Sn–CH_3COOH succeeds to reduce a Zn pc type compound nitro derivative to the acetamide derivative[16] (Figure 1).

The effect of solvent on some excited state processes of Mg and Zn pcs is studied. The triplet state of these compounds decays with mixed first and second order kinetics or mainly second order kinetics depending on the solvent. Excited state quenching by methyl viologen or *p*-benzoquinone is discussed.[17]

The formation of dimeric and polymeric forms of the sulfo pc derivatives and their disproportionation have a significant effect on the oxidation of water-soluble Co or Zn tetrasulfo pc in dilute acidic solutions. Oxidation of the metal ion in sulfo pc derivatives proceeds at a greater rate than oxidation of the macrocycles.[18]

When Fe tetra-4-*tert*-butyl pc $(pyridine)_2$ is heated in oxygen at 215°C for 3 h, pyridine evolves and pc^{2-} decomposes to tetra-4-*tert*-butylphthalimide.[19]

Heating the two crystalline modifications of the μ-oxo dimer of Fe(III) pc in 1-chlo-

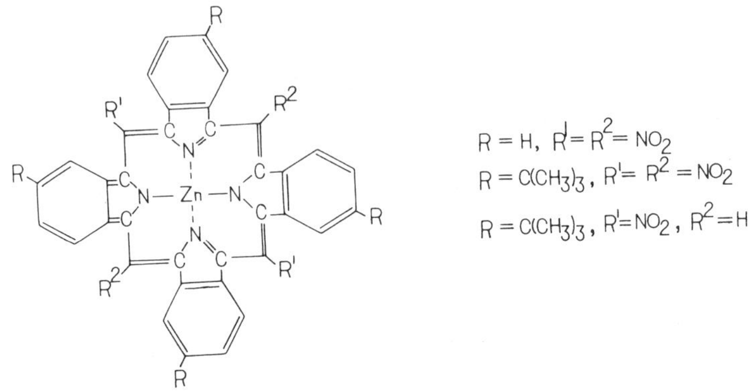

FIGURE 1. Zn pc-type compound nitro derivative.

ronaphthalene or DMF, in vacuum-sealed tubes, gives as a final product β-Fe(II) pc. The same procedure in chlorobenzene or nitrobenzene yields the pure oxygen-containing species isomorphous with α-Fe(II) pc. Until now this product was obtained only fortuitously.[20]

Li pc is oxidized by chloranil, bromanil, TCNQ, and so on to form a stable radical which is easily reduced to H_2pc by inorganic acids. Partial oxidation of the radical leads to a charge transfer complex with increased electrical conductivity.[21]

REFERENCES

1. **Borovkov, N. Y. and Akopov, A. S.,** Solvent effect on the kinetics of complexation tetra-tert-butyl-phthalocyanine with cupric acetate, *Zh. Obschch. Khim.*, 55(11), 2549, 1985.
2. **Dubrovina, A. S., Malkova, A. I., and Tupikov, V. I.,** Formation of oxygen complexes of cobalt tertiary-butylphthalocyanine in *N,N*-dimethylformamide solutions, *Koord. Khim.*, 10(9), 1207, 1984.
3. **Borovkov, N. Y. and Akópov, A. S.,** Kinetics of the complex formation of 4,4′,4″,4‴-tetra-*tert*-butyl-phthalocyanine with metal salts in pyridine, *Koord. Khim.*, 10(4), 455, 1984.
4. **Hale, P. D. and Ratner, M. A.,** Intramolecular electron transfer, with bridge assistance and without, in molecules and models, *Int. J. Quant. Chem., Quant. Chem. Symp.*, 18, 195, 1984.
5. **Sapozhnikova, I. N., Kovtunenko, S. V., Kiva, E. A., and Sharipov, A. K.,** Thermal oxidative degradation of cobalt polyphthalocyanine in alkali solutions, *Khimiyai Tekhnol. Neftekhim. SintezaM*, 137, 1984.
6. **Pennesi, G., Ercolani, C., Ascenzi, P., Brunori, M., and Monacelli, F.,** Equilibrium and kinetic study of pyridine binding to phthalocyaninato(II) in dimethyl sulfoxide, *J. Chem. Soc., Dalton Trans.*, 6, 1107, 1985.
7. **Ascenzi, P., Brunori, M., Pennesi, G., Ercolani, C., and Monacelli, F.,** Equilibrium and kinetic study of nitric oxide binding to phthalocyaninato(II) in dimethyl sulfoxide, *J. Chem. Soc., Dalton Trans.*, 2, 369, 1987.
8. **Geiger, D. K., Ferraudi, G., Madden, K., Granifo, J., and Rillema, D. P.,** Redox reactivity of transition-metal phthalocyanines: ligand radical formation vs. metal center oxidation, *J. Phys. Chem.*, 89(18), 3890, 1985.
9. **Dillow, G. W., Gregor, I. K., and Guilhaus, M.,** Adduct-ion formation by metal complexes under methane negative chemical-ionization conditions, *Org. Mass Spectrom.*, 21(3), 151, 1986.
10. **Siddiqui, N. and Stynes, D. V.,** Comparative spectra and kinetics of ferrous bis(dioxime) complexes of π-acid ligands, *Inorg. Chem.*, 25(12), 1982, 1986.
11. **Stynes, D. V., Fletcher, D., and Chen, X.,** Equilibria for phosphine ligation to ferrous protoporphyrin IX dimethyl ester and related systems in toluene, *Inorg. Chem.*, 25(19), 3483, 1986.
12. **Starke, M.,** Isothermal TG studies on phthalocyanines and polytetrafluoroethylene and determination of the kinetic parameters for the thermal oxidation mechanism, *Z. Phys. Chem. (Leipzig)*, 267(2), 305, 1986.
13. **Barmasov, A. V., Korotkov, V. I., and Kholmogorov, V. E.,** Mechanism of the photodecomposition of water, sensitized by phthalocyanine, *Khim. Fiz.*, 5(3), 414, 1986.

14. **Akopov, A. S. and Borovkov, N. Y.,** Acid-base properties and kinetic stability of tantalum phthalocyanine in proton-donor media, *Zh. Obshch. Khim.*, 56(4), 931, 1986.
15. **Berezin, B. D., Khelevina, O. G., Stuzhin, P. A., and Glazunov, A. V.,** Stability of cobalt complexes of tetraazaporphyrines and their dissociation mechanism in sulfuric acid, *Zh. Fiz. Khim.*, 60(9), 2189, 1986.
16. **Kopranenkov, V. N., Makarova, E. A., and Luk'yanets, E. A.,** Nitration of tetrabenzoporphines, *Khim. Geterotsikl Soedin.*, 9, 1189, 1986.
17. **Kim, D.,** Effect of solvent on some excited state processes of magnesium and zinc phthalocyanines, *Bull. Korean Chem. Soc.*, 7(6), 416, 1986.
18. **Parmon, V. N., Elizarova, G. L., Matvienko, L. G., Lozhkina, N. V., and Maizlish, V. E.,** Oxidation of cobalt, copper, and iron sulfophthalocyanines in dilute acid aqueous solutions, *Izv. Akad. Nauk SSSR, Ser. Khim.*, 8, 1735, 1984.
19. **Borisenkova, S. A., Posii, V. S., Gulina, T. Y., Derkacheva, V. M., Mekhryakova, N. G., Kaliya, O. L., and Luk'yanets, E. A.,** Low-temperature high autooxidation of iron phthalocyanine, *Vestn. Mosk. Univ., Ser. 2*, 28(3), 307, 1987.
20. **Collamati, I.,** Additional characterization of the two oxygen-containing compounds formed by the interaction of phthalocyaninato(II) and dioxygen. Reproducible synthesis of the species isomorphic with the α form of Fe(II)Pc, *Inorg. Chim. Acta,* 124(2), 61, 1986.
21. **Zhukhovitskii, V. B., Ezhova, N. N., Krinichnyi, V. I., Letuchii, Y. A., Ton'shin, A. M., Dyumaev, K. M., and Khidekel, M. L.,** Reaction and products of lithium phthalocyanine with various oxidants, *Izv. Akad. Nauk SSSR, Ser. Khim.*, 12, 2694, 1985.

Chapter 6

ORGANIC CHEMISTRY

Ether and ester functions, such as in high molecular weight dyes, are cleaved by treatment with bromobenzodioxaborole in methylene chloride at room temperature. Thus, Cu N,N′,N″,N‴-tetrakis[2-(2,5-dimethoxyphenyl)-1-methylethyl] pc tetrasulfonamide is deblocked with bromobenzodioxaborole.[1]

Di hydrogen pc is referenced to highly reactive magnesium and its application to organic synthesis. A method is given for preparing highly reactive magnesium slurries.[2]

REFERENCES

1. **King, P. F. and Stroud, S. G.,** Mild homogeneous deblocking employing 2-bromo-1,3,2-benzodioxaborole, *Tetrahedron Lett.*, 26(11), 1415, 1985.
2. **Burns, T. P. and Rieke, R. D.,** Highly reactive magnesium and its application to organic syntheses, *J. Org. Chem.*, 52(16), 3674, 1987.

Chapter 7

SOLUTION CHEMISTRY

I. DIMERIZATION

The dimerization constant of Co(II) tetrasulfonated phthalocyanine (pc) in methanol-water solution is measured as a function of temperature at 15 to 45°C. The heat capacity of dimerization varies linearly with temperature. In pure water, the dimerization heat capacity is zero.[1]

Dimerization of Co(II) tetrasulfo pc is also measured in ethanol and isopropanol solutions. For a given alcohol mole fraction, ethanol is more effective than methanol in promoting monomer formation.[2]

The monomer-dimer equilibrium measurements of Co(II) tetrasulfo pc in water at 5 to 75°C shows that the enthalpy of formation varies with temperature, with ΔC_p = 0.1 kcal/mol K.[3]

The aggregation of mononuclear Co(II) 2,9,16,23-tetrakis(neopentyloxy) pc and of related binuclear Co(II) pc derivatives in *o*-dichlorobenzene is studied by electron absorption spectroscopy. Dimerization constants are calculated.[4]

II. REACTION OF CO TETRASULFO PC WITH NITROGEN BASES

Co tetrasulfo pc adds NO, which oxidizes to Co pc NO_2, in aqueous solution. Co tetrasulfo pc forms an octahedral ammine complex with ammonia, which reacts with NO to form a mixed complex. Coordination occurs with pyridine, hexylamine, urea, hydrazine, and hydroxylamine.[5]

III. ACID-BASE REACTIONS

An electronic absorption spectroscopy study is made of protonation equilibria of pc in nonaqueous sulfuric acid mixtures.[6]

Stepwise protonation constants are determined for Nb(V) tetra-4-*tert*-butyl pc in anhydrous acetic acid-sulfuric acid mixtures. The dissociation kinetics of the dichloro-substituted complex are studied in aqueous sulfuric acid.[7]

Stepwise pK_2, pK_3, and pK_4 values are determined for Cu, Ni, Pd, and Zn tetra-4-*tert*-butyl pcs in anhydrous acetic acid-sulfuric acid mixtures.[8]

IV. PROTONATION

In a polemic "can water protonate porphyrazines?", the neglect of the Cu(Zn)-ligand π interaction in the discussion of substituted effects on the transition from substituted pcs to Cu or Zn pc is given.[9]

The first protonation constant of tetra-4-*tert*-butyl pc is measured spectrophotometrically at 25°C in weakly acidic solutions such as urea-sulfuric acid-acetic acid. The free ligand and its complexes of Al, Ca, Ga, Hf, In, Zr, or V are monoprotonated but the Zn complex is diprotonated.[10]

V. REACTION BETWEEN FE(II) PC AND CO IN DMSO IN THE PRESENCE OF PYRIDINE

Equilibrium and rate constants for axial substitution of Fe(II) pc·DMSO·pyridine or Fe(II) pc·DMSO·CO by CO and pyridine, respectively, are determined in DMSO at 20°C. The Fe DMSO bond has low stability and there seems to be a long-lived transient containing one CO and two pyridine molecules.[11]

VI. IONIZATION

The first-stage ionization of pc sulfonic acid is determined in DMSO at 298, 308, and 318 K, and the thermodynamic parameters are calculated.[12]

The first- and second-stage ionization constants of tetraazaporphin are also determined in DMSO at 298, 308, and 318 K. The electronic effects of the benzo groups in pc and tetrabenzoporphyrin are estimated.[13]

REFERENCES

1. **Ward, J. R. and Seiders, R. P.,** Aggregation of Cobalt(II) Tetrasulfonated Phthalocyanine in Methanol-water solutions, Report 1983, ARCSL-TR-83048, National Technical Information Service, Springfield, VA, 92 pp.
2. **Yang, Y. C., Ward, J. R., and Seiders, R. P.,** Dimerization of cobalt(II) tetrasulfonated phthalocyanine in water and aqueous alcoholic solutions, *Inorg. Chem.*, 24(12), 1765, 1985.
3. **Yang, Y. C., Ward, J. R., and Seiders, R. P.,** Dimerization of Cobalt(II) Tetrasulfonated Phthalocyanine in Water and Aqueous-Alcohol Solutions, Report 1985, CRDC-TR84098, National Technical Information Service, Springfield, VA, 68 pp.
4. **Nevin, W. A., Liu, W., and Lever, A. B. P.,** Dimerization of mononuclear and binuclear cobalt phthalocyanines, *Can. J. Chem.*, 65(4), 855, 1987.
5. **Zemtseva, L. I., Antipenko, V. R., and Ivasenko, V. L.,** Spectrophotometric study of nitrogen oxide reactions with aqueous solutions of cobalt tetrasulfophthalocyanines in the presence of nitrogen bases, *Izv. Sib. Otd. Akad. Nauk SSSR, Ser. Khim. Nauk,* (4), 62, 1986.
6. **Berezin, B. D., Khelevina, O. G., and Stuzhin, P. A.,** Spectroscopic study of the acid-base interaction of substituted tetraazaporphyrins in nonaqueous solvents, *Zh. Prikl. Spektrosk.*, 46(5), 809, 1987.
7. **Borovkov, N. Y. and Akopov, A. S.,** Acid-base characteristics and kinetic stability of niobium phthalocyanine in proton-donor media, *Izv. Vyssh. Uchenbn. Zaved., Khim. Khim. Tekhnol.*, 28(10), 34, 1985.
8. **Akopov, A. S., Borovkov, N. Y., and Shchigal, O. V.,** Acid-base properties of phthalocyanine complexes. Determination of stepwise equilibrium constants pK_{BH^+}, *Koord. Khim.*, 12(8), 1051, 1986.
9. **Akopov, A. S.,** Can water protonate porphyrazines? Comments, *Zh. Obshch. Khim.*, 55(10), 2402, 1985.
10. **Borovkov, N. Y. and Akopov, A. S.,** Protonation of tetra-4-tertiary-butylphthalocyanine and its complexes in buffer solutions, *Zh. Fiz. Khim.*, 60(3), 750, 1986.
11. **Pennesi, G., Ercolani, C., Rossi, G., Ascenzi, P., Brunori, M., and Monacelli, F.,** Equilibrium and kinetic study of the reaction between phthalocyaninatoiron(II), and carbon monoxide in dimethyl sulfoxide in the presence of pyridine. Evidence for the formation of a transient, *J. Chem. Soc., Dalton Trans.*, 6, 1113, 1985.
12. **Sheinin, V. B., Andrianov, V. G., Berezin, B. D., and Koroleva, T. A.,** Thermodynamics of acid ionization of porphine tetrabenzoporphyrin and a phthalocyanine in methyl sulfoxide, *Zh. Org. Khim.*, 21(7), 1564, 1985.
13. **Sheinin, V. B., Berezin, B. D., Khelevina, O. G., Stuzhin, P. A., and Telegin, F. Y.,** Acid ionization of tetraazaporphine in methyl sulfoxide, *Zh. Org. Khim.*, 21(7), 1571, 1985.

Chapter 8

RADIONUCLIDES

Am-242m and Am-242 are formed from Szilard-Chalmers irradiation of Am-241 dipc. The irradiated material is extracted in water.[1]

Cu phthalocyanine (pc) is bombarded by neutrons in the Szilard-Chalmers reaction to produce Cu-64, the specific activity of which is too low to be used in some *in vitro* tests for metabolic diagnosis.[2]

A target material for nuclear recoil enrichment of Cu-64 produced by simple chemical processing, instead of by tedious concentrated sulfuric acid processing, involves using water-soluble Cu pc octacarboxy pc, by solvent extraction or by precipitation. Extraction gives higher specific activity of Cu-64.[3]

Reactor-irradiated Co and Zn pcs are studied in the direct liquid-solid extraction procedure to obtain high specific activity radionuclides. Hot dilute acids such as 0.1 M $HClO_4$ give yields >50% at moderate enrichments from α-M pc targets. However, useful amounts of radionuclides from irradiated M pc targets are not obtained.[4]

The selectivity of isolation of short-lived isotopes isolated from a high-energy particle-irradiated target is increased by using degradation products of M pcs as the target.[5]

A Be-7 pc is made by Be-7 recoil. Net radiochemical yields of Be-7 pc are 5 to 10% for Cu and Zn pcs, and, for H pcs, yields >90% are obtained.[6]

Enriched isotopes of rare earth elements and americium-242 are made using M pcs.[7]

Modifications of the conventional precipitation method to purify α-M pcs gives a product essentially free of occluded impurities, significantly decreasing the total metallic content of the extracted radionuclide. The method is used to prepare an α-Co pc target.[8]

Pu, Th, and U pcs are fissionable semiconductor film in nuclear electric generators providing electric power without the necessity of intervening thermal energy. The fission fragment-produced electron pair charge carriers can be directly used to produce electric energy.[9]

REFERENCES

1. **Kalinichenko, B. S., Kaluzhko, V. G., and Shvetsov, I. K.,** The production of americium-242, -242m according to Szilard-Chalmers method using diphthalocyanine complex of americium-241, *Radiokhimiya,* 28(5), 644, 1986.
2. **Hetherington, E. L., Sorby, P. J., and Camakaris, J.,** The preparation of high specific activity copper-64 for medical diagnosis, *Appl. Radiat. Isot.,* 37(12), 1242, 1986.
3. **Sekine, T., Kimura, K., and Yoshihara, K.,** Application of nuclear recoil to radioisotope enrichment of copper-64. Feasibility of simplified chemical processing of water-soluble copper phthalocyanine, *J. Nucl. Sci. Technol.,* 23(12), 1064, 1986.
4. **Jardim, I. C. S. F., Collins, K. E., and Collins, C. H.,** Liquid-solid extraction of high specific activity radionuclides from reactor irradiated metal phthalocyanines, *J. Radioanal. Nucl. Chem.,* 106(6), 327, 1986.
5. **Golos, S. D., Lemeshko, G. G., Moskalev, P. N., and Tikhonov, V. I.,** U.S.S.R. Patent 948,006, 1985.
6. **Oki, Y., Shoji, H., Aihara, M., and Ikeda, N.,** Chemical behavior of beryllium-7 recoils formed through $^{12}C(\gamma,n\alpha)^7$ Be reaction in several phthalocyanine complexes, *J. Radioanal. Nucl. Chem.,* 106(4), 243, 1986.
7. **Shapkin, G. N., Mis'ko, Y. S., and Moskalev, P. N.,** Production of enriched isotopes of rare earth elements and americium-242 using metal phthalocyanines, *Poluch. i Vydelenie Radioaktiv. Izotopov. Tashkent,* 125, 1983.
8. **De Andrade, J. C., Collins, C. H., and Collins, E.,** An improved method for purifying α-metallophthalocyanines for use as radionuclide preparation targets, *J. Radioanal. Nucl. Chem.,* 96(6), 621, 1985.
9. **Garrett, D. L.,** U.S. Patent 4,415,526, 1983.

Chapter 9

CATALYSIS

I. CATALYST SUPPORTS

Co phthalocyanine (pc) and Cu pc are synthesized and characterized on γ-alumina and silica.[1]

A method is described to graft or to immobilize Cu pc on silica gel or zeolite Y.[2]

Catalysis by porphyrin and pc complexes occurs on oxide supports.[3]

The synthesis and characterization of Co, Ni, Cu, and Fe(II) catalytic pcs immobilized on silica and zeolite are described.[4]

A catalyst is made by impregnating a support with Co or Fe chloride solution and synthesizing the corresponding Co pc or Fe(II) pc directly on the support impregnated with phthalic anhydride, urea, and ammonium molybdate.[5]

The catalytic properties of Co and Fe pcs supported on silica gel are studied in the dehydration and dehydrogenation of isopropanol.[6]

The structure and catalytic properties of Co pc adsorbed on silica gel in the oxidation of oxalic acid and the dehydrogenation of isopropanol are studied by chromatography and spectrometry.[7]

Adsorption/desorption kinetics of Co pc from α-chloronaphthalene on γ-alumina are studied at 20°C. The catalytic activities for isopropanol conversion relate to the fact that Co pc modifies γ-alumina by lowering the dehydration rate to iso$(C_3H_7)_2O$ and promoting dehydrogenation to $(CH_3)_2CO$.[8]

A review with 40 references discusses catalysis by oxide-supported porphyrin and pc complexes.[9] The graphitized carbon support is significant in the dispersion and sintering behavior of Fe pc complex catalysts.[10] The dispersion of Fe pc on a carbon surface depends on the surface properties of the carbon support.[11]

II. CATALYSIS BY CO PC COMPLEXES

A. EMPHASIS ON SULFUR COMPOUNDS

The structural behavior of Co pc $(NaSO_3)_4$, cobalt pc tetrasodium sulfonate, a catalyst used in thiol oxidation, is studied in the presence of ionenes.[12]

Oxidation of sulfite ion by oxygen in the presence of Co tetrachloro pc (CoTSP) is catalyzed by traces of metal ions and not by CoTSP itself.[13]

A catalyst for oxidizing sulfur-containing compounds contains polyacrylonitrile, Co pc, and a polyethylene thermoplastic polymer.[14]

A catalyst is produced by mixing Co pc, or Co tetrasulfo pc, or Co tetrachloromethyl pc with a product from grafting acrylic acid to polyethylene, and by adding an aqueous solution of cobalt acetate.[15]

The Co and Fe complexes of tetrapyrazineporphyrazine are better catalysts for mercaptan oxidation than Co pc.[16]

Disodium sulfide is oxidized at 1 atm by oxygen in a 5% aqueous sodium hydroxide solution in the presence of Co disulfo pc.[17]

The simultaneous catalytic oxidation of C_4H_9SNa and sodium disulfide by oxygen is studied in sodium hydroxide solution containing Co disulfo pc.[18]

Monochloro- and dichloro-pc disulfonic acids are catalysts for the oxidation of sodium diethyldithiocarbamate, Na $(C_2H_5)_2NCS_2$.[19]

Co pc disulfonic acid is a catalyst for the oxidation of sodium pentamethylenedithiocarbamate.[20]

Co tetranitro- and tetrahydroxy-pc disulfonic acid and their sodium salts are catalysts for the oxidation of dithiocarbamic acids.[21]

Polycations promote the catalyzed oxidation of Co pc tetrasodium sulfonate.[22]

Autooxidation of thiols takes place with Co pc tetrasodium sulfonate attached to poly(vinyl amine). The polymeric catalyst exhibits an enzyme behavior and resembles the catalytic action of vitamin B12 in thiol oxidation.[23] This reaction also takes place with tetrasodium Co pc tetrasulfonate.[23-26]

The electrochemical oxidation of thiols and disulfides of cysteine, cystine, lipoic acid, and disulfiram is studied with a carbon paste electrode containing Co pc.[27]

The effect of the addition of basic macromolecules on autooxidation of hydrophilic and amphophilic thiols catalyzed by Co pc is studied.[28]

The effect of the polyamine and polyammonium salts ligand is studied in the oxidation of thiols catalyzed by Co pc tetrasodium sulfonate attached to the polyamines and polyammonium salts.[29]

The effect of the addition of basic compounds on the autooxidation of thio alcohols catalyzed by Co tetrasulfo pc tetrasodium salt is studied.[30]

The simultaneous catalytic oxidation of H_2S and mercaptans in an alkaline solution is studied with a Co disulfo pc catalyst.[31,32]

The oxidation of ethyl mercaptan in the presence of Co pc on a solid support is followed. The catalytic activity decreases in the order Amberlyst A21 > activated carbon > ZnO > SiO_2.[33]

The electrocatalytic response of Co pc containing carbon paste electrodes toward sulfhydryl compounds is studied.[34]

Autooxidation of thiols occurs with Co pc tetrasodium sulfonate attached to poly(vinylamine).[35]

B. EMPHASIS ON HYDROCARBONS

Monomeric and polymeric forms of metal pcs are studied as cocatalysts of the Co catalyzed oxidation of *p*-xylene and *p*-methyltoluate at 160°C and 0.7 mPa total pressure.[36]

Faujasite-encaged cobalt chelate complexes including Co pc catalyze the selective oxidation of propene but suffer from rapid deactivation.[37]

Cyclohexane is oxidized by cumene hydroperoxide catalyzed by Co tetra-4-*tert*-butyl pc.[38]

Co pcs are catalysts for the liquid phase oxidation of dicyclopentadiene.[39]

During ethylbenzene oxidation and cumyl hydroperoxide decomposition in the presence of Co pc, singlet oxygen is the cause of luminescence resulting from recombination of the corresponding ROO· in the presence of Co pc.[40]

Fluorene is oxidized by tetralin hydroperoxide, in the presence of Co pc, to fluorenone and fluorenol.[41-42] Styrene is epoxidized by peroxy acids in the presence of Co pc.[43]

Carbon paste electrodes modified with Co pc show an electrocatalytic response toward oxalic acid and α-keto acids.[44]

Co and Fe pcs are effective catalysts for the mild liquid phase oxidation of cyclohexane with cumene hydroperoxide.[45]

Co pc homopolymers are catalysts for the isomerization of quadricyclene-cyclodextrin as inclusion compounds.[46]

The oxidation of $Na_2C_2O_4$ is studied with a Co pc·2-pyridine catalyst.[47]

Exothermic isomerization of water-soluble quadricyclenes to norbornadienes is catalyzed by Co pc on activated carbon.[48]

The role of 2,6-di-*tert*-butyl-4-methylphenol in the decomposition of α-cumyl hydroperoxide on Co pc and on Fe pc is studied.[49]

A Cu-Zn-Al catalyst containing 4% Co pc is used to hydrate 1-butene to $C_2H_5COCH_3$ at 170 to 280°C in the absence of oxygen.[50]

Co pc is a catalyst for the nitrosation of arylalkenes.[51]

Many carbohydrates can be oxidized at low positive potentials at chemically modified carbon paste electrodes containing Co pc.[52]

The reaction products of Co pc 2,9,16,23-tetracarboxylic acid pc with chloromethylstyrene copolymers are a catalyst for isomerization of norbornadiene to quadricyclene.[53]

The catalytic properties of polymeric Co pc synthesized from Co 4,4′,4″,4‴-tetrachloromethyl pcs are studied.[54]

Co pc exerts an influence in the radical polymerization of methyl methacrylate in the presence of hydroquinone.[55]

Co *tert*-tetra-butyl pc is a catalyst for chain transfer in the radical polymerization of methyl methacrylate.[56]

Co 2,9,16,23-pc tetracarboxylic acid does not show catalytic activity in the photoreduction of methyl viologen but the Ni and Co pc complexes do.[57]

The catalytic activity of Cu, Co, and Fe(III) pcs on activated carbon are compared in oxidation of methylbenzene to benzene carboxylic acid. The activity is in the order Cu > Co > Fe.[58]

III. OXIDATION OF CO AND REDUCTION OF NO_x

The oxidation of CO with oxygen over Co, Cu, and Ni pcs is studied at 293 K. The order of reactivity is Ni pc > Co pc > Cu pc. A dual-site mechanism of CO oxidation on the supported metal pc may be involved.[59]

NO_x reduction to nitrogen in solution by ammonia takes place in the presence of water-soluble Co, Cr, Fe, and Mn complexes. The central metal ion has an effect on the catalytic activity.[60]

A catalyst based on metal pc complexes to reduce NO_x with ammonia is formed by impregnating porous alumina with an aqueous solution of Co, Cu, Fe, Mn, or Ni sulfo pc.[61]

IV. OXIDATION OF HYDRAZINE

In the oxidation of hydrazine, catalyzed by $Cu(OH)_2$ and copper oxide, replacement of the suspended copper catalyst by a homogeneous catalyst improves the method. Complexing agents, either ammonia or sulfonated pcs, prevent reduction of dissolved Cu(II) to solid Cu_2O.[62]

Cu tetrasulfo pc is the best of several catalysts tested vis-a-vis determination of gas-liquid mass transfer coefficients by hydrazine oxidation. The hydrazine reaction is suitable for study of large gas/liquid reactors.[63]

The oxidation kinetics are studied of aqueous hydrazine over Cu tetrasulfo pc at pH 11 to 13. The peak rate occurs at pH 12.4.[64]

Hydrazine oxidation is studied in a liquid phase in the presence of Co trisulfo pc and Cu tetrasulfo pc in a continuous stirred tank reactor. The reaction rate peaks at pH 12.4.[65]

V. CATALYSIS BY FE PC COMPLEXES

A. OXIDATION

The catalytic activity of Fe(II) pc derivatives in the oxidation of oxalic acid is in the order H pc FeCl > Fe pc· 2-pyridine > Co pc > μ-oxo complex H pc Fe-O-Fe pc H.[66]

Fe pc catalysts are studied in the oxidation of cyclic and noncyclic hydrocarbons.[67]

Aromatic olefins are oxidized to benzyl alcohols in the presence of Fe(II) pc.[68]

Triphenylphosphine is oxidized in the presence of Fe(II) pc.[69]

Alkanes are oxidized by iodosobenzene catalyzed by Fe pc encapsulated within large pore zeolites so that the product alcohols and ketones give improved yields and altered selectivities.[70]

B. OTHER REACTIONS

The catalytic activity of Fe(II) pc and of Fe(III) pc on benzene acylation is compared. Fe(III) pc is more effective. Also, α-Fe(II) pc is more effective than β-Fe(II) pc.[71]

Bis(trimethylsilyl)amides are converted into nitriles with high yields when treated with Fe pc.[72]

The polymer derivatives of iron-molybdenum pc complexes are catalysts for the acetoxylation of alkylcyclohexanes.[73]

Cumene is oxidized in the liquid phase on Fe(II) pc fixed in a sodium-Y zeolite.[74]

VI. HYDROPEROXIDE DECOMPOSITION

Tc Cl_2 pc and Tc N pc are catalysts in the epoxidation of cyclohexene with *tert*-butyl peroxide at 80°C as well as in the decomposition of *tert*-butyl peroxide.[75]

The decomposition of hydrogen peroxide is accelerated by Co, Fe, and Mn pcs.[76]

The catalase-like catalytic activity of ion-exchange resins modified with Co pc 2,9,16,23-tetrasulfonic acid in the decomposition of hydrogen peroxide is studied.[77] Heat-treated pc polymer complexes may exhibit catalytic activity in the decomposition of hydrogen peroxide in an acid medium.[78]

α-Cumyl hydroperoxide decomposes in the presence of Ni pc.[79] Co tetrasulfo pc on titania is a new and efficient electron relay for the photocatalytic formation and depletion of hydrogen peroxide in aqueous suspensions.[80]

Fe(III) 2,9,16,23-tetracarboxy pc-supported poly(2-vinylpyridine-CO-styrene) is a remarkably effective catalyst for hydrogen peroxide decomposition.[81]

VII. OXYGEN REDUCTION

Binuclear Co pcs connected by bridges containing one, two, four, or five atoms catalytically reduce oxygen more efficiently than mononuclear Co pcs.[82] Mg pc and Al pc Cl are the most efficient sensitizers in the photoreduction of methyl viologen from among several metal pcs tested.[83]

VIII. QUINOLINE HYDROGENATION

Sn(II) pc and Sn(IV) pc Cl_2 are active catalysts for hydrogenation of quinoline to tetrahydroquinoline at 200 to 250°C and 1000 psi.[84]

IX. ISOMERIZATION OF QUADRICYCLENES

An active carbon carrying Co pc polymer catalyzes the isomerization of quadricyclene in toluene.[85]

X. OXIDATION BY A VARIETY OF PCS

Metal poly pcs enhance the rate of cumene oxidation but decrease the selectivity to cumene hydroperoxide.[86]

Homopolymer metal pc complexes affect the oxidation of propane.[87]

The oxidation of fluorene to 9-fluorenol and 9-fluorenone is affected by Co, Cu, Fe, Ni, and Pd pcs.[88]

Co, Cu, Fe, and Mn pcs catalyze the oxygenolysis of 3-methylindole.[89]

Cumene peroxygenation takes place in the presence of a vanadyl (VO) pc,[90] as well as of $Sn(OH)_2$ pc,[91] and of Al pc Cl and Al pc OH.[92]

Tc pcs catalyze ring cleavage of epoxyoctane by *tert*-butyl hydroperoxide.[93]

Octene epoxidation is catalyzed by Tc pc nitride.[94]

XI. HYDROCARBON REACTIONS

Fe·pyridine$_2$, Co, Cu, Mn, Ni, and RhCl tetra-4-*tert*-butyl pc catalyze the cyclopropanation of styrene by diazoacetic acid ester.[95]

Cu Cl and Cu pc are carbonylation catalysts for 1-decene.[96]

Indole synthesis is affected by the dehydrocyclization of *O*-ethylaniline over poly Pt pc catalyst.[97]

Polyamic acids are cyclized in the presence of Cu pc. The properties of polyimides in the presence of Cu pc are studied.[98] The dehydrochlorination of trichloroethane takes place in the presence of H_2 pc catalysts in the preparation of vinylidene chloride.[99]

XII. PULPING PROCESS

The effectiveness of M pc tetrasulfonic acid as a catalyst for alkaline pulping of *Pinus radiata* is evaluated. NaFe complexes of pc tetrasulfonic acid cleave more guaiacol than anthraquinone from 1-(4-hydroxy-3-methoxyphenyl)-2-(2-methoxyphenoxy)ethanol as model for lignin.[100]

Chemicals are recovered from flue gases of sulfate pulping by two-stage gas spraying, trapping the dust particles, and removing the precipitate from spent liquor of the first stage, then trapping hydrogen disulfide in the second stage by treating the gases with a 0.001 to 0.05% alkaline solution of the sodium salt of Co sulfo pc. The spent solution is divided into a stream returned to the system to circulate the alkaline solution and into a stream that is fed for preparing pulp liquor.[101]

XIII. REVIEW

A review with 100 references relates to the catalytic activity of metal pcs with possible use in the catalytic oxidation of hydrocarbons, aldehydes, and sulfur compounds, and *NO* reduction, as well as in photochemical and electrocatalytic processes.[102]

A review with 64 references relates to the catalytic properties of transition metal complexes with sulfonated pcs in redox reactions of sulfur compounds.[103]

REFERENCES

1. **Woehrle, D., Huendorf, U., Schulz-Ekloff, G., and Ignatzek, E.,** Synthesis of cobalt(II) and copper(II) phthalocyanines on γ-alumina and silicon dioxide, *Z. Naturforsch., B,* 41(2), 179, 1986.
2. **Borisova, T. G., Izmailova, L. N., Kotov, E. I., and Romanovskii, B. V.,** Preparation and identification of copper(II) phthalocyanine grafted on an oxide support, *Zh. Fiz. Khim.,* 60(5), 1195, 1986.
3. **Romanovskii, B. V.,** Catalysis by porphyrin and phthalocyanine complexes fixed to oxide supports, *5 Mezhdunar. Simp. po Svyazimezhdu Gomogen.i Geterogen. Katal., Novobirsk, 15-19 Iyula, 1986 Dokl., Novosibirsk.* 2(CH 1), 40, 1986.
4. **Romanovskii, B. V.,** Immobilization of transition metal complexes on inorganic supports, *Proc. 8th Int. Congr. Catal., 1984,* 4, IV657, 1985.
5. **Zemtseva, L. I., Titov, V. I., Antipenko, V. R., and Ivasenko, V. L. (to Siberian Institute of Petroleum Chemistry and Tomsk Polytechnic Institute),** U.S.S.R. Patent 1,197,728, 1985.

6. **Borisenkova, S. A., Erokhin, A. S., and Novikov, V. A.,** Catalytic properties of cobalt and iron phthalocyanines on silica gel, *Zh. Fiz. Khim.*, 59(9), 2304, 1985.
7. **Daud, H., Borisenkova, S. A., and Pryakhin, A. N.,** Adsorptional Phthalocyanine Catalysts: Production Methods, Structure, and Catalytic Activity. Cobalt Phthalocyanine (PcCo)-Silicon Dioxide System, Deposited Doc. 1983; VINITI 5343, All Union Institute of Scientific and Technical Information, U.S.S.R, 21 pp.
8. **Daud, H. and Borisenkova, S. A.,** Adsorptional Phthalocyanine Catalysts: Production Methods, Structure, and Catalytic Properties. Cobalt Phthalocyanine (PcCo)-γ-Aluminum Oxide System, *Deposited Doc.* 1983, VINITI 5899, All Union Institute of Scientific and Technical Information, U.S.S.R., 11 pp.
9. **Romanovskii, B. V.,** Catalysis by oxide supported porphyrine and phthalocyanine complexes, *Homogeneous Heterog. Catal., Proc. Int. Symp. Relat. Homogeneous Heterog. Catal., 5th,* 343, 1986.
10. **Ehrburger, P. and Lahaye, J.,** Dispersion of metallic derivatives on carbon support, *Prepr.-Am. Chem. Soc., Div. Pet. Chem.*, 29(2), 400, 1984.
11. **Ehrburger, P. and Lahaye, J.,** Dispersion of metallic derivatives on carbon supports, *ACS Symp. Ser.*, 303, 310, 1986.
12. **Van Welzen, J., Van Herk, A. M., and German, A. L.,** Effects of ionenes on structure and catalytic activity of cobalt phthalocyanine. I. Visible light spectroscopic investigations, *Makromol. Chem.*, 188(8), 1923, 1987.
13. **Wagnerova, D. M.,** Catalytic activity of cobalt(II) tetrasulfophthalocyanine for the oxidation of sulfite ions and other substrates by dioxygen, *Collect. Czech. Chem. Commun.*, 52(5), 1223, 1987.
14. **Anan'eva, T. A., Al'yanov, M. I., Nikulina, T. A., Kalacheva, V. V., Akhmadullina, A. G., and Mazgarov, A. M., (Moscow Chemical Technological Institute; All Union Scientific Research Institute of Hydrocarbon Raw Materials),** U.S.S.R. 1,273,155, 1986.
15. **Kalacheva, V. V., Nikulina, T. A., Anan'eva, T. A., Al'yanov, M. I., Mazgarov, A. M., Akhmadullina, A. G., Benderskii, V. A., and Ponomarev, A. N. (Ivanovo Chemical Technological Institute; All Union Scientific Research Institute of Hydrocarbon Raw Materials; Institute of Chemical Physics, Chernogolovka),** U.S.S.R. Patent 1,181,706, 1985.
16. **Maizlish, V. E., Korzhenevskii, A. B., and Klyuev, V. N.,** Metal complexes of tetrapyrazinoporphyrazines as effective catalysts of mercaptan oxidation, *Khim. Geterotsikl. Soedin.*, 9, 1257, 1984.
17. **Ovsyannikova, L. V., Mazgarov, A. M., and Digurov, N. G.,** Kinetics of Liquid Phase Oxidation of Sodium Sulfide Catalyzed by Cobalt Disulfophthalocyanine, Deposited Doc. 1982, VINITI 6157, All Union Institute of Scientific and Technical Information, U.S.S.R., 13 pp.
18. **Mazgarov, A. M., Fomin, V. A., Ovsyannikova, L. V., Ostroumova, G. A., and Khrushcheva, I. K.,** Simultaneous Catalytic Oxidation of Hydrogen Sulfide and Mercaptans in an Alkaline Medium, Deposited Doc. 1982, VINITI 6158, All Union Institute of Scientific and Technical Information, U.S.S.R., 12 pp.
19. **Maizlish, V. E., Nikulina, T. A., Kalacheva, V. V., Smirnov, R. P., Al'yanov, M. I., and Titova, G. F. (Ivanovo Chemical Technological Institute),** U.S.S.R. Patent 1,046,245, 1983.
20. **Titova, G. F., Anan'eva, T. A., and Al'yanov, M. I.,** Catalytic oxidation of sodium pentamethylene dithiocarbamate by atmospheric oxygen in an aqueous medium, *Vopr. Kinet. i Kataliza. Ivanovo,* 21, 1982.
21. **Titova, G. F., Anan'eva, T. A., and Al'yanov, M. I.,** Catalytic oxidation of dithiocarbamic acids in the presence of macroheterocyclic compounds of cobalt, *Izv. Vyssh. Uchebn. Zaved., Khim. Khim. Tekhnol.*, 27(4), 397, 1984.
22. **Brouwer, W. M., Piet, P., and German, A. L.,** The promoting role of polycations in the cobalt(II) phthalocyanine tetrasodium sulfonate catalyzed oxidation of thiols, *J. Mol. Catal.*, 31(2), 169, 1985.
23. **Brouwer, W. M., Piet, P., and German, A. L.,** Autooxidation of thiols with cobalt(II) phthalocyanine tetrasodium sulfonate attached to poly(vinylamine). II. Kinetic measurements, *J. Mol. Catal.*, 22(3), 297, 1984.
24. **Brouwer, W. M., Piet, P., and German, A. L.,** Autooxidation of thiols with tetrasodium cobalt(II) phthalocyaninetetrasulfonate, bound to poly(vinylamine). III. Dependence on molecular weight, *Makromol. Chem.*, 185(2), 363, 1984.
25. **Brouwer, W. M., Piet, P., and German, A. L.,** IV. Influence of base density within the polymeric ligand, *J. Mol. Catal.*, 29(3), 335, 1985.
26. **Brouwer, W. M., Piet, P., and German, A. L.,** Autooxidation of thiols with cobalt(II) phthalocyanine tetra(sodium sulfonate) attached to poly(vinylamine). V. Effect of surfactant and thiol variation, *J. Mol. Catal.*, 29(3), 347, 1985.
27. **Linders, C. R., Patriarche, G. J., Kauffmann, J. M., and Guilbault, G. G.,** Electrochemical study of thiols and disulfides using modified electrodes, *Anal. Lett.*, 19(1—2), 193, 1986.
28. **Brouwer, W. M., Robierst, A. J. M. S., Piet, P., and German, A. L.,** Effect of the addition of basic macromolecules on autooxidation of hydrophilic and amphophilic thiols catalyzed by cobalt(II) phthalocyanine. *5 Mezhdunar. Simp. po Svyazi mezhdu Gomogen. i Geterogen. Katal., Novosibirsk, 15-19 Iyulya, 1986. Dokl., Novosibirsk,* 2(Ch. 2), 58, 1986.

29. **Brouwer, W. M.,** Macromolecular Effects on the Oxidation of Thiols: Catalyzed by Cobalt(II) Phthalocyanine Tetrasodium Sulfonate Attached to Polyamines and Polyammonium Salts, Report 1984, Order No. PB85-243764/GAR, National Technical Informational Service, Springfield, VA, 149 pp.
30. **Brouwer, W. M., Robierst, A. J. M. S., Piet, P., and German, A. L.,** Effect of the addition of basic macromolecules on autooxidation of hydrophilic and amphiphilic thio alcohols catalyzed by cobalt(II) phthalocyanine, *Mezhdunar. Simp. Svyazi Gomogennym Geterog. Katal., 5th,* 2(2), 58, 1986.
31. **Mazgarov, A. M., Fomin, V. A., Ovssyannikova, L. V., Ostroumova, G. A., and Khrushcheva, I. K.,** Simultaneous catalytic oxidation of hydrogen sulfide and mercaptans in an alkaline medium, *Khim. Prom-st. (Moscow),* 5, 281, 1985.
32. **Ovsyannikova, L. V., Digurov, N. G., Mazgarov, A. M., Golovanov, O. G., Akhmadullina, A. G., and Khrushcheva, I. K.,** Kinetics of simultaneous oxidation of sodium sulfide and mercaptide in the presence of cobalt disulfophthalocyanine, *Zh. Prikl. Khim. (Leningrad),* 59(4), 830, 1986.
33. **Prakhov, L. T., Andreev, A. A., and Shopov, D. M.,** Oxidation of ethyl mercaptan in the presence of cobalt phthalocyanine on a solid support, *Khim. Tekhnol. Topl. Masel,* 11, 33, 1986.
34. **Halbert, M. K. and Baldwin, R. P.,** Electrocatalytic and analytical response of cobalt phthalocyanine containing carbon paste electrodes toward sulfhydryl compounds, *Anal. Chem.,* 57(3), 591, 1985.
35. **Brouwer, W. M., Traa, P. A. M., De Weerd, T. J. W., Piet, P., and German, A. L.,** Autooxidation of thiols with cobalt(II) phthalocyanine tetrasodium sulfonate attached to poly(vinylamine). VI. Immobilized catalysts by crosslinking of poly(vinylamine), *Angew. Makromol. Chem.,* 128, 133, 1984.
36. **Hronec, M., Valachova, M., Masarovic, F., Cvengrosova, Z., and Ilavsky, J.,** Effect of metal catalysts on the oxidation of *p*-xylene and methyl *p*-toluate, *Chem. Zvesti,* 37(4), 519, 1983.
37. **Diegruber, H., Plath, P. J., and Schulz-Ekloff, G.,** Study on the structure, stability and propene oxidation capability of faujasite-encaged chelate complexes, *J. Mol. Catal.,* 24(1), 115, 1984.
38. **Barkanova, S. V., Zheltukhin, I. A., and Kaliya, O. L.,** Oxidation of saturated hydrocarbons by cumene hydroperoxide catalyzed by cobalt tetra-4-tert-butylphthalocyanine, *Zh. Org. Khim.,* 19(10), 2212, 1983.
39. **Schnurpfeil, D. and Lauterbach, G.,** Dicyclopentadiene oxidation. III. Variation of the catalyst in the liquid phase oxidation of dicyclopentadiene, *J. Prakt. Chem.,* 325(5), 848, 1983.
40. **Ivanov, S. and Kropf, H.,** Excited states during heterogeneous catalytic oxidation of ethylbenzene and decomposition of organic hydroperoxides, *Geterog. Katal., 5th,* Pt. 1, 93, 1983.
41. **Polishchuk, V. V. and Borisenkova, S. A.,** Oxidation of Fluorene by Tetralin Hydroperoxide in the Presence of Cobalt Phthalocyanine. Deposited Doc. 1983, VINITI 5410-83, All Union Institute of Scientific and Technical Information, U.S.S.R., 15 pp.
42. **Borisenkova, S. A. and Polishchuk, V. V.,** Combined Oxidation of Fluorene and Tetralin in the Presence of Cobalt Phthalocyanine. Deposited Doc. 1983, VINITI 5411-83, All Union Institute of Scientific and Technical Information, U.S.S.R., 14 pp
43. **Brook, M. A., Smith, J. R. L., Higgins, R., and Lester, D.,** Model systems for cytochrome P450 dependent mono-oxygenases., IV. The epoxidation of alkenes by peroxy acids in the presence of cobalt complexes, *J. Chem. Soc., Perkin Trans. 2,* 7, 1049, 1985.
44. **Santos, L. M. and Baldwin, R. P.,** Electrocatalytic response of cobalt phthalocyanine chemically modified electrodes toward oxalic acid and α-keto acids, *Anal. Chem.,* 58(4), 848, 1986.
45. **Backanova, S. V., Derkacheva, V. M., Zheltukhin, I. A., Kaliya, O. L., Kopranenkov, V. N., and Luk'yanets, E. A.,** Azaporphines. Effective catalysts for the mild liquid phase oxidation of cyclohexane with cumene hydroperoxide, *Zh. Org. Khim.,* 21(9), 2018, 1985.
46. **Taoda, H., Hayakawa, K., Yumoto, T., Kawase, K., and Yamakita, H. (Agency of Industrial Sciences and Technology),** Japan Patent 86,22,033, 1986.
47. **Borisenkova, S. A. and Gegele, F. F.,** Kinetics and mechanism of the catalysis by metal phthalocyanine complexes with saturated coordination spheres, *Vestn. Mosk. Univ., Ser. 2,* 27(3), 300, 1986.
48. **Maruyama, K., Tamiaki, H., and Kawabata, S.,** Exothermic isomerization of water soluble quadricyclanes to norbornadienes by soluble and insoluble catalysts, *J. Chem. Soc., Perkin Trans. 2,* 4, 543, 1986.
49. **Kropf, H., Ivanov, S., Debnar, K., Heinz-Becker, H., and Kotthaus, E.,** The role of 2,6-di-*tert*-butyl-4-methylphenol in the decomposition of α-cumyl hydroperoxide on phthalocyaninatocobalt and phthalocyanine iron, *J. Chem. Res., Synop.,* 7, 242, 1986.
50. **Devekki, A. V. and Devekki, A. B.,** Interaction of 1-butene with water activated with cobalt phthalocyanine, *Neftekhimiya,* 26(3), 382, 1986.
51. **Okamoto, T., Kobayashi, K., Oka, S., and Tanimoto, S.,** Cobalt catalyzed reaction of nitric oxide with aryl-substituted olefins in the presence of tetrahydroborate ion, *J. Org. Chem.,* 52(23), 5089, 1987.
52. **Santos, L. M. and Baldwin, R. P.,** Liquid chromatography/electrochemical detection of carbohydrates at a cobalt phthalocyanine containing chemically modified electrode, *Anal. Chem.,* 59(14), 1766, 1987.
53. **Taoda, H., Hayakawa, K., Kawase, K., Tazawa, M., and Yamakita, H.,** Synthesis and property of polymer-anchored cobalt catalysts for the isomerization of quadricyclene, *Kobunshi Ronbunshu,* 44(6), 437, 1987.

54. **Kalacheva, V. V., Bazanov, M. I., Smirnov, R. P., and Al'yanov, M. I.,** Synthesis and study of catalytic properties of polymeric cobalt phthalocyanine from cobalt 4,4′,4″,4‴-tetrachloromethylphthalocyanine, *Soversh. Protsessov Krasheniya i Metodov Sinteza Krasitelei, Ivanovo,* 74, 1983.
55. **Enikolopyan, N. S., Smirnov, B. R., and Askarov, K. A.,** Catalytic properties of metal complexes of porphyrins and related compounds in polymerization reactions, *Khim. Fiz.,* 3(1), 86, 1984.
56. **Smirnov, B. R., Mironychev, V. E., Golikov, I. V., Mogilevich, M. M., and Enikolopov, I. V.,** Catalytic Activity of a Cobalt Complex of Benzoporphyrin in Chain Transfer to a Monomer, Deposited Doc. 1982, SPSTL 598 Khp-D82, 9 pp. Avail. SPSTL.
57. **Yamaguchi, H., Fujiwara, R., and Kusuda, K.,** Water soluble substituted polystyrene with pendant tetrametal 2,9,16,23-phthalocyanine tetracarboxylate metal complexes, sensitizing the reduction of 1,1′-dimethyl-4,4′-bipyridinium dichloride, *Makromol. Chem., Rapid Commun.,* 7(4), 225, 1986.
58. **Milkowska, G. and Siedlewski, J.,** Catalytic activity of copper(II), cobalt(II), and iron(III) phthalocyanines deposited on activated carbon surfaces, *Chem. Stosow.,* 29(3—4), 271, 1985.
59. **Romanovskii, B. V. and Borisova, T. G.,** Low temperature oxidation of carbon monoxide by molecular oxygen on metal phthalocyanines, *Oxid. Commun.,* 6(1—4), 157, 1984.
60. **Ryazanova, V. S., Arkatova, N. A., and Ivasenko, V. L.,** Catalysts for homogeneous reduction of nitrogen oxides, *Kinet. Katal.,* 26(6), 1516, 1985.
61. **Velichko, A. V., Ivasenko, V. L., and Antipenko, V. R. (Tomsk Polytechnic Institute),** U.S.S.R. Patent 1,324,681, 1987.
62. **Sick, R., Weiland, P., and Onken, U.,** Determination of gas-liquid mass transfer by oxidation of hydrazine, *NATO ASI Ser., Ser. E,* 72, 351, 1983.
63. **Onken, U., Sick, R., and Weiland, P.,** Determination of gas-liquid mass transfer by oxidation of hydrazine, *Chem. Eng. Sci.,* 40(10), 1990, 1985.
64. **Gadooni, J. and Onken, U.,** Kinetics of homogenously catalyzed hydrazine oxidation, *Chem. Ing. Tech.,* 58(4), 320, 1986.
65. **Gadooni, J. and Onken, U.,** Catalysis of hydrazine oxidation by sulfonated phthalocyanines, *Ber. Bunsen-Ges. Phys. Chem.,* 90(2), 154, 1986.
66. **Borisenkova, S. A., Hegele, T. T., and Pozii, V. S.,** The catalytic properties of coordinated derivatives of ferrous phthalocyanine, *Vestn. Mosk. Univ., Ser. 2,* 27(1), 70, 1986.
67. **Nappa, M. J. and Tolman, C. A.,** Steric and electronic control of iron porphyrin catalyzed hydrocarbon oxidations, *Inorg. Chem.,* 24(26), 4711, 1985.
68. **Okamoto, T. and Oka, S.,** Oxygenation of olefins under reductive conditions. Cobalt catalyzed selective conversion of aromatic olefins to benzylic alcohols by molecular oxygen and tetrahydroborate, *J. Org. Chem.,* 49(9), 1589, 1984.
69. **Ercolani, C., Gardini, M., Pennesi, G., and Rossi, G.,** Dioxygen activation and catalytic oxidation of triphenylphosphine by iron phthalocyanine compounds, *J. Mol. Catal.,* 30(1—2), 135, 1985.
70. **Herron, N., Stucky, G. D., and Tolman, C. A.,** Shape selectivity in hydrocarbon oxidations using zeolite encapsulated iron phthalocyanine catalysts, *J. Chem. Soc., Chem. Commun.,* 20, 1521, 1986.
71. **Arai, Y. and Ogawa, S.,** Studies on phthalocyanines and related macrocyclic compounds possessing specific functions, *Kenkyu Hokoku-Asahi Garasu Kogyo Gijutsu Shoreikai,* 43, 83, 1983.
72. **Rigo, B., Lespagnol, C., and Pauly, M.,** Bis(trimethylsilyl)amide as a nitrile precursor, *Tetrahedron Lett.,* 27(3), 347, 1986.
73. **Ohkatsu, Y., Nakashima, S., Taguchi, K., and Inoue, S.,** Hydroxylation reactions of substituted cyclohexanes using iron(II) complex triethylamine oxide trifluoroacetic acid, *Sekiyu Gakkaishi,* 29(6), 463, 1986.
74. **Korol'kova, T. V. and Romanovskii, B. V.,** Liquid phase oxidation of cumene on iron(II) phthalocyanine fixed in a sodium-Y zeolite, *Neftekhimiya,* 26(4), 546, 1986.
75. **Rummel, S. and Schnurpfeil, D.,** Technetium compounds as catalysts in hydroperoxide decomposition and epoxidation processes, *Oxid. Commun.,* 6(1—4), 319, 1984.
76. **Pobedinskii, S. N., Trofimenko, A. A., and Zharnikova, M. A.,** Catalytic decomposition of hydrogen peroxide on metal complexes of anthraquinonecyanine and phthalocyanine in acid and basic electrolytes, *Izv. Vyssh. Uchebn. Zaved., Khim. Khim. Tekhnol.,* 28(12), 46, 1985.
77. **Saito, Y., Mifune, M., Kawaguchi, T., Odo, J., Tanaka, Y., Chikuma, M., and Tanaka, H.,** Catalase like catalytic activity of ion exchange resins modified with metalloporphyrins, *Chem. Pharm. Bull.,* 34(7), 2885, 1986.
78. **Pobedinskii, S. N., Trofimenko, A. A., Kolbashov, V. N., Koifman, O. I., and Shortmanova, L. P.,** Effect of heat treatment of macrocyclic complexes of cobalt on catalytic decomposition of hydrogen peroxide in an acid medium, *Izv. Vyssh. Uchebn. Zaved., Khim. Khim. Tekhnol.,* 29(7), 61, 1986.
79. **Kropf, H., Spangenberg, J., and Ivanov, S.,** Catalysis by phthalocyanines. XXVIII. Decomposition of α-cumyl hydroperoxide in the presence of phthalocyaninenickel, *J. Chem. Res., Synop.,* 10, 308, 1985.
80. **Hong, A. P., Bahnemann, D. W., and Hoffmann, M. R.,** Cobalt(II) tetrasulfophthalocyanine on titanium dioxide: a new efficient electron relay for the photocatalytic formation and depletion of hydrogen peroxide in aqueous suspensions, *J. Phys. Chem.,* 91(8), 2109, 1987.

81. **Shirai, H., Higaki, S., Hanabusa, K., Kondo, Y., and Hojo, N.,** Functional metal porphyrazine derivatives and their polymers. VIII. Catalase like activity of iron(III) and cobalt(II) 2,9,16,23-tetracarboxyphthalocyanine supported on poly(2-vinylpyridine-CO-styrene), *J. Polym. Sci., Polym. Chem. Ed.*, 22(6), 1309, 1984.
82. **Hempstead, M., Melnik, M., Nevin, A., Liu, W., and Lever, A. B. P.,** Binuclear Phthalocyanines as Model Electrocatalysts for the Reduction of Oxygen, Report 1985, TR-4, Order No. AD-A161758/8/GAR, National Technical Information Service, Springfield, VA, 11 pp.
83. **Ohtani, H., Kobayashi, T., Tanno, T., Yamada, A., Woehrle, D., and Ohno, T.,** Efficient photoreduction of methyl viologen by metallophthalocyanine sensitizers, *Photochem. Photobiol.*, 44(2), 125, 1986.
84. **Boucher, L. J. and Dadey, E. J.,** Supported tin phthalocyanines as catalysts in the hydrogenation of quinoline, *Prepr.-Am. Chem. Soc., Div. Pet. Chem.*, 27(3), 654, 1982.
85. **Dainippon Ink and Chemicals, Inc.; Kawamura Physical and Chemical Research Institute,** Japan Patent 85,48,145, 1985.
86. **Weerarathna, S., Hronec, M., Malik, L., and Vesely, V.,** Selectivity of metal polyphthalocyanine catalyzed oxidation of cumene, *React. Kinet. Catal. Lett.*, 22(1—2), 7, 1983.
87. **Enikolopov, N. S., Moshkina, R. I., Nikisha, L. V., Polyak, S. S., Promyslova, V. V., Sokolova, N. A., Sherle, A. I., and Epshtein, V. R.,** Gas phase oxidation of propane in the presence of azaporphyrin compounds, *Materialy 3 Vses. Konf. po Khimii i Biokhimii Porfirinov. 14-16 Sent., 1982, Samarkand,* 22, 1983.
88. **Polishchuk, V. V. and Borisenkova, S. A.,** Catalytic Oxidation of Condensed Aromatic Compounds in the Presence of Phthalocyanine Complexes, Deposited Doc. 1983, VINITI 3455-83, All Union Institute of Scientific and Technical Information, U.S.S.R., 17 pp.
89. **Ohkubo, K. and Takano, K.,** Catalytic action of iron(II) phthalocyanine as a tryptophan-2,3-dioxygenase model in oxygenolysis of 3-methyl indole with metallophthalocyanines, *J. Coord. Chem.*, 14(2), 169, 1985.
90. **Kropf, H. and Vogel, W.,** Peroxygenation of cumene in the presence of vanadyl phthalocyanine, *Geterog. Katal., 5th,* Pt. 1, 57, 1983.
91. **Kropf, H., Vogel, W., and Hopf, C.,** Catalysis in phthalocyanines. XXIX. Peroxygenation of cumene in the presence of phthalocyanine complexes of elements of main Group IV, *J. Chem. Res., Synop.*, 1, 1, 1986.
92. **Kropf, H. and Vogel, W.,** Catalysis by phthalocyanines. XXX. Peroxygenation of cumene in the presence of chloro- and hydroxy(phthalocyanine)aluminum, *J. Chem. Res., Synop.*, 1, 2, 1986.
93. **Rummel, S., Schnurpfeil, D., and Willecke, L.,** Technetium catalyzed radical epoxide ring opening reaction, *Z. Chem.*, 25(4), 150, 1985.
94. **Rummel, S. and Schnurpfeil, D.,** The formation of trans epoxides in the liquid phase oxidation of *cis/trans*-n-alkanes catalyzed by technetium, *J. Prakt. Chem.*, 329(1), 10, 1987.
95. **Pazynina, G. V., Luk'Yanets, E. A., and Bolesov, I. G.,** Transition metal phthalocyanine catalyzed cyclopropanation of styrene by diazoacetic acid ester, *Zh. Org. Khim.*, 20(4), 802, 1984.
96. **Hagen, J., Tschirner, E. G., Fink, G., and Lorenz, R.,** Carbonylation of 1-decene with modified copper(I) complex catalysts, *Chem.-Ztg.*, 109(1), 3, 1985.
97. **Takamiya, N., Furuhata, K., and Onikubo, T.,** Indole synthesis by dehydrocyclization of *o*-ethylaniline over poly(platinum phthalocyanine) catalyst, *Rikogaku Kenkyusho Hokoku Waseda Daigaku,* 114, 71, 1986.
98. **Ittiev, A. B. and Gazaev, M. A.,** Study of cyclization of polyamic acids and properties of polyimides in the presence of copper phthalocyanine, *Polikondensats. Protsessy i Polimery, Nal'chik,* 161, 1984.
99. **Toyo Soda Manufacturing Co., Ltd.,** Japan Patent 83,172,329, 1983.
100. **Wright, L. J. and Fullerton, T. J.,** Redox catalysis of alkaline pulping by organometallic complexes, *J. Wood. Technol.*, 4(1), 61, 1984.
101. **Dermanov, N. K., Naumov, A. V., Shvabskii, M. G., Levit, I. S., Prokhorov, B. V., and Burenin, A. M.,** U.S.S.R. Patent 1,146,346, 1985.
102. **Hronec, M. and Ilavsky, J.,** Phthalocyanines in catalysis, *Petrochemia,* 23(2—3), 89, 1983.
103. **Maizlish, V. E. and Borodkin, V. F.,** Catalytic properties of water soluble metal macroheterocyclic compounds, *Izv. Vssh. Uchebn. Zaved., Khim. Khim. Tekhnol.*, 27(9), 1003, 1984.

Chapter 10

PC AND M PC COLORANTS

I. RESINS, MOLDINGS, FOAMS, AND ADHESIVES

A. POLYESTERS

Titanium oxalate and C. I. pigment blue 15 (Cu phthalocyanine, pc) are mixed into a polyester parison which, blown molded into bottles, gives the bottles transparency and a green hue.[1]

Poly(ethylene terephthalate) contains ferric stearate and Cu pc for colored transparent bottle and plate moldings.[2]

A polyester for amber-colored containers and bottles is prepared with a titanium catalyst, Cu pc, and yellow and red colorants.[3]

Colored polyester parisons contain 75 ppm Cu pc; they are blown into bright amber 2-l bottles.[4]

Polyester fibers contain highly dispersed Cu pc added during kneading of polyester master batches.[5]

A polyester blend contains Cu pc colorants with uniform color dispersion.[6]

B. POLYOLEFINS

Cu pc pigments are mixed into granular polyolefin blends for master batches.[7]

Monofilaments are made from a blend of polyethylene, C. I. pigment yellow 73, pigment green 7, C. I. pigment white 6, and calcium carbonate.[8]

A colorant composition containing titania, Chromfina blue SR 5020 (Cu pc), Emanon 4110 polyethylene glycol monooleate, dioctyl phthalate, and liquid paraffin is mixed with Novatec JV070H polyethylene to give a uniformly colored material.[9]

Talc powder treated with Cu pc is dry blended with an ethylene-propylene block copolymer and injection molded into test pieces.[10]

Abrasion and impact-resistant ethylene-propylene block copolymer composite moldings are pigmented with Cu pc.[11]

Cu pc is dry mixed into a polyethylene-polypropylene mixture and injection molded to obtain a composition with good abrasion resistance and impact strength.[12]

Polypropylene is mixed with pc blue dye, calcium carbonate, and titania, kneaded, and compression molded.[13]

A glass fiber-reinforced propylene polymer molding contains Cu pc pigment.[14]

Polyethylene glycol ethers are dispersing agents for Cu pc in the manufacture of concentrates for coloring polyolefins.[15]

Coloring agents for moldings that retain good pigment dispersibility contain Cu pc.[16]

The drawing behavior of polypropylene fibers is affected by Cu pc in the polymer matrix. The resistance increases with Cu pc present.[17]

A master-batch composition containing Cu pc is used to uniformly color polypropylene.[18]

A composition to color low-pressure polyethylene includes paraffin oil and Cu pc.[19]

A water-based Cu pc paste containing 30% pigment, 40 parts polyethylene wax, and propylene glycol-adipic acid copolymer (10 parts) forms a pigment concentrate to color synthetic polymers.[20]

A master-batch concentrate for synthetic resins includes polyethylene, Cu pc, and water.[21]

A polyethylene-based concentrate of Cu pc has good dispersibility in polyolefins.[22]

A liquid colorant concentrate contains Cu pc. It is blended with polypropylene which is injection molded to give uniformly colored petals.[23]

A coloring agent for polyolefins includes Cu pc.[24]

A pigment dispersant is prepared by grafting Hiwax 400P polyethylene, styrene, and acrylonitrile with Cu pc and Toporex to give a master batch for aromatic polymers.[25]

A study evaluating thermal stability of FD & C lake pigments in low-density polyethylene compared with Cu pc pigment shows their applicability at $\leq$450 to 500°F.[26]

Di hydrogen pc may impart oxidative thermal stability to polyethylene.[27]

A sinterable low-density polyethylene powder contains Cu pc.[28]

Cu pc may have a photostabilizing action in polypropylene.[29]

Cu pc may have photostabilizing and antioxidant synergism with Cyasorb UV 2908 in polyethylene.[30]

Cu pc pigments are premixed with polymerization catalysts in the manufacture of polyethylene with improved pigment dispersion and compatibility.[31]

Carboxylated ethylene-propylene polymers containing Cu pc pigments are injected molded, giving nonwarping moldings.[32]

Heat- and abrasion-resistant propylene polymer compositions useful for automobile interior trim include Cu pc blue colorant.[33]

Glass fiber-reinforced polymers contain Cu pcs with low warping and good appearance.[34]

C. POLYCARBONATES

A colored polycarbonate resin is prepared with excellent pigment dispersion by using Escorez 5320, C. I. pigment blue 15:2 (Cu pc), and polycarbonate resin of 25,000 molecular weight.[35]

Colored polycarbonate (25,000 molecular weight) contains Cu pc and a suitable dispersant.[36]

D. POLYVINYLCHLORIDES

A pigment master batch that may contain pc pigments with good dispersibility is milled in acrylic acid-ethylene copolymer and n-pentanol, and dispersed in polyvinylchloride (PVC).[37,38]

A PVC that contains Cu pc, titania, and a tin mercaptide is extruded to give a molding with good IR transmittance and no heat distortion.[39]

PVC resin compositions with excellent heat and weather resistance contain Cu chloro pc in which the Cu pc macrocyclic ring contains one to eight chlorine atoms,[40] or in which the Cu pc macrocyclic ring may contain one to four chlorine atoms.[41]

Cu pc pigments are blended with titania to color PVC leather substitutes.[42]

Decorative material exhibiting different colors under a UV, fluorescent, or incandescent lamp, or by daylight are prepared by coloring a PVC with Cu pc, Epocolor EP3000G fluorescent green pigment, and dioctyl phthalate, and forming into a blue sheet that is yellowish green under a UV lamp.[43]

E. POLYSTYRENES

A coloring composition contains Cu pc for moldable polystyrene foam beads.[44]

A titania-Cu pc pigment mixture has no adverse effect on the mechanical and thermal properties of ABS and polypropylene.[45]

F. METHACRYLATE COPOLYMERS

Magnesium stearate is a dispersant for Cu pc pigments for methacrylic acid-styrene copolymers.[46]

A highly dispersible, heat-stable Cu pc pigment is coated with polyethylene glycol methacrylate copolymers for thermoplastic molding compositions.[47]

A pigment concentrate includes Cu pc and methyl methacrylate polymers for coloring plastics.[48]

Cu pc and Cu (bi pc) influence the aqueous polymerization of methyl methacrylate but the molecular weight with the latter is about 24% higher.[49]

G. POLYAMIDES

Molding compositions containing a polyamide, carbon black, and Cu pc exhibit improved toughness and resistance to shrinkage.[50]

H. EPOXY RESINS

Adding Cu pc in phthalic anhydride to an epoxy resin-anhydride mixture and curing at 120 to 130°C in a mold gives a transparent blue molding.[51]

Free-flowing epoxy resin microspheres contain Cu pc.[52,53]

I. ANILINE-PHENOL-FORMALDEHYDE RESINS

A molding composition contains an aniline-phenol-formaldehyde resin, with nitrile rubber and glass fiber, along with calcium stearate lubricant, hexachloro-*p*-xylene, talc, magnesium oxide, naphthalene, and Cu pc.[54]

J. PHENOL-FORMALDEHYDE OLIGOMERS

A molding composition with good tensile strength and wear resistance contains phenol-formaldehyde oligomers, along with chopped carbon fibers, graphite, polyalkylene polyamine, and Cu pc.[55]

K. POLYACRYLONITRILES

Co pc is included in a polyacrylonitrile mixture in the manufacture of rigid moldings.[56]

L. POLYURETHANES

Pc is a pigment in a pigment pack for polyurethane or poly(vinyl chloride) in leather substitutes or for unsaturated polyesters.[57]

Cu pc is a light-scattering filler for polyurethane-acrylic compositions hardened using UV light.[58]

Polyurethane foams contain thermally reversible colorants and Cu pc.[59]

Cu pc is tested dispersed in a polyurethane foam as a UV light stabilizer.[60]

Cu pc is a seeding additive in polyurethane contact adhesives.[61]

M. EQUIPMENT

A patented apparatus is designed to mix powdered pigments, such as Cu pc pigments, with powdered thermoplastic resins in a cylindrical container with rotating shaft with mixers and milling beads.[62]

II. SHEETS AND FILMS

A. INDUSTRIAL

Master batches contain bisphenol polyoxyalkylene ethers and pigments including Cu pc, kneaded with poly(butylene terephthalate) and injection molded to form a film.[63]

The particle size and optical efficiency of Cu pc pigment plasticized in PVC films are studied. Pigment dispersity increases when Cu pc is in a liquid formulation.[64]

Cu pc dispersions are used to color polyethylene film.[65]

A PVC sheet with plasticizer is printed with an ink composition containing Cu pc and cured to give an embossed surface.[66]

Thermoplastic compositions containing Cu pc and heat stabilizers are extruded to make a sheet.[67]

Cu pc monosulfonic acid, stearylamine, titania, Cu pc, and an unsaturated polyester resin are blended to make a color concentrate, and are mixed with unsaturated polyester resin to prepare a sheet molding composition.[68]

Unsaturated polyester resin compositions containing Cu pc are kneaded to make a sheet molding compound, the sheet from which exhibits good pigment dispersion.[69]

B. AGRICULTURAL

Plasticized, heat-stabilized PVC films contain Hostaperm blue AFN (Cu pc) and 2-(3′,5′-di-*tert*-butyl-2′-hydroxyphenyl)-5-chlorobenzotriazole used to improve corn growth.[70]

The growth of grass, chives, and cucumbers is favorably influenced in greenhouses covered with blue polyethylene film containing Cu pc.[71]

Blue plastic films contain C. I. pigment blue 15:1 (α-Cu pc, noncrystallizing) and 2,2′-dihydroxy-4-methoxybenzophenone. Plant growth yield is 120% in plants using this film.[72]

C. FILTERS

Optical filters that block out visible light and pass near-IR radiation are made from near-IR transparent polymers containing 0.01 to 1.0% Cu pc compounds.[73]

Optical filters may also be made to contain either di hydrogen pc or a M pc other than Cu pc.[74]

Colored films for microcolor filters contain Cu pc tetrasulfonic acid pyridium salts.[75]

III. ELASTOMERS

M(II) 4,4′,4″,4‴-tetraamino pcs are used to cure epoxy resins to form heat-resistant matrix polymers for composites. The effect of adding carboxy-terminated nitrile rubber is described.[76] An ethylene-propylene-diene-monomer (EPDM) vulcanized rubber composition contains Cu pc. The metal atom of the Cu pc pigment presumably bonds with the rubber.[77-79]

An ethylene-propylene-diene-terpolymer rubber contains, among other ingredients, a Cu pc pigment and sulfur compounds, which, upon heating, vulcanize to a colored rubber with the copper atom bonded to double bonds to other active sites in the rubber molecule.[80] Such a composition gives a 2-mm thick blue rubber sheet showing no discoloration or cracks even after 1000 h in a sunshine weatherometer.[81,82]

IV. SOFTENABLE POLYMERS

The particle growth of vanadyl pc is studied in softenable polymer substrates.[83,84]

REFERENCES

1. **Teijin Ltd.,** Japan Patent 83,125,743, 1983.
2. **Dainichiseika Color and Chemicals Manufacturing Co., Ltd.,** Japan Patent 83,180,552, 1983.
3. **Callander, D. D. and Massey, F. L. (to Goodyear Tire and Rubber Co.),** Europe Patent Application 128,109, 1984.
4. **Adams, L. J. (to Eastman Kodak Co.),** U.S. Patent 4,521,556, 1985.
5. **Japan Ester Co., Ltd.,** Japan Patent 84,230,029, 1984.
6. **Horie, S.,** Japan Patent 87,91,554, 1987.
7. **Toyo Printing Ink Manufacturing Co., Ltd.,** Japan Patent 83,167,624, 1983.
8. **Toyo Printing Ink Manufacturing Co., Ltd.,** Japan Patent 83,219,245, 1983.
9. **Dainichiseika Color and Chemicals Manufacturing Co., Ltd.,** Japan Patent 85,90,232, 1985.

10. **Mitsubishi Petrochemical Co., Ltd.**, Japan Patent 84,47,240, 1984.
11. **Mitsubishi Petrochemical Co., Ltd.**, Japan Patent 84,108,049, 1984.
12. **Mitsubishi Petrochemical Co., Ltd.**, Japan Patent 84,108,050, 1984.
13. **Showa Denko, K. K.**, Japan Patent 84,202,242, 1984.
14. **Mitsubishi Petrochemical Co., Ltd.**, Japan Patent 85,69,146, 1985.
15. **Mitroiu, M., Variu, C., Bader, B., Masala, F., Variu, A., and Grigorescu, A. (to Combinatul Petrochimic, Pitesti)**, Romania Patent 86,635, 1985.
16. **Kobayashi, H., Ando, T., Nakayasu, M., and Noguchi, K. (to Tokyo Printing Ink Manufacturing Co., Ltd.)**, Japan Patent 86,57,855, 1986.
17. **Li, P. R.**, The effect of various pigments on the rheological behavior and morphology of polypropylene, *Angew. Makromol. Chem.*, 147, 113, 1987.
18. **Toyo Printing Ink Manufacturing Co., Ltd.**, Japan Patent 84,06,225, 1984.
19. **Talpus, A. and Onofrei, L. (to Combinatul Petrochimic, Pitesti)**, Romania Patent 81,231, 1983.
20. **Matysiak, R., Kosmowski, H., Sekula, W., Dalek, M., and Halaburda, M. (to Osrodek Badawczo-Rozwojowy Przemyslu Barnikow "Organika")**, Poland Patent 124,088, 1984.
21. **Kobayashi, H. and Ando, T. (to Toyo Printing Ink Manufacturing Co., Ltd.)**, Japan Patent 86,126,139, 1986.
22. **Dinu, G., Mitroi, M., Variu, C., Turcu, A., Gagiu, E., Bader, B., Grigorescu, A., and Musetescu, I. (to Combinatul Petrochimic, Pitesti)**, Romania Patent 87,013, 1985.
23. **Jarzombek, R. E. and Moeller, R. J. (to Bee Chemical Co.)**, U.S. Patent 4,571,416, 1986.
24. **Shiraiwa, N., Sato, S., and Kumabe, J. (to Dainippon Ink and Chemicals, Inc.)**, Japan Patent 86,228,044, 1986.
25. **Ueda, T. and Aoki, M. (to Mitsui Petrochemical Industries, Ltd.)**, Japan Patent 86,53,364, 1986.
26. **Lykens, D. N.**, Thermal stability of FD & C lake pigments, *Plast. Compd.*, 9(7), 35, 1986.
27. **Gilroy, H. M. and Chan, M. G.**, Effect of pigments on the aging characteristics of polyolefins, *Org. Coat. Appl. Polym. Sci. Proc.*, 46, 293, 1981.
28. **Asahi Chemical Co., Ltd.**, Japan Patent 84,197,439, 1984.
29. **Allen, N. S., Chirinos-Padron, A., and Appleyard, J. H.**, Photostabilizing action of metal chelate stabilizers in polypropylene. VIII. Pigment interactions, *Polym. Degrad. Stab.*, 9(1), 15, 1984.
30. **Allen, N. S., Kotecha, J. L., Parkinson, A., Loffelman, F. F., Rauhut, M. M., and Susi, P. V.**, Photo-stabilizing action of a *p*-hydroxybenzoate light stabilizer in polyolefins. III. Antioxidant behavior and additive/pigment interactions in high density polyethylene, *Polym. Degrad. Stab.*, 10(1), 1, 1985.
31. **Hayashi, H., Matono, K., Asahi, S., and Uoi, M. (to Idemitsu Kosan Co., Ltd.)**, Europe Patent Application 142,143, 1985.
32. **Sobashima, Y. and Ichikawa, Y. (to Mitsui Petrochemical Industries, Ltd.)**, Japan Patent 85,212,440, 1985.
33. **Toyota Motor Co., Ltd.**, Japan Patent 84,172,540, 1984.
34. **Sobashima, Y., Goto, S., and Yoshikawa, M. (to Mitsubishi Petrochemical Co., Ltd.)**, Japan Patent 86,123,642, 1986.
35. **Toyo Printing Ink Manufacturing Co., Ltd.**, Japan Patent 83,174,438, 1983.
36. **Shiraiwa, N., Sumibe, J., and Komaki, M. (to Dainippon Ink and Chemicals, Inc.)**, Japan Patent 86,106,660, 1986.
37. **Jaffe, E. E. (to Du Pont Co.)**, Europe Patent Application 122,122, 1984.
38. **Jaffe, E. E. (to Ciba-Geigy Corp.)**, U.S. Patent 4,548,968, 1985.
39. **Taniguchi, M., Hiromi, T., Awano, N., Sakamoto, K., and Takeda, T. (to Sekisui Chemical Co., Ltd.)**, Japan Patent 85,30,347, 1985.
40. **Nisshin Steel Co., Ltd.**, Japan Patent 84,147,038, 1984.
41. **Nisshin Steel Co., Ltd.**, Japan Patent 84,147,039, 1984.
42. **Zharikova, V. G., Fil'chikov, A. S., Fedotov, V. V., Porvatova, L. M., and Andrianova, G. P.**, Study of optical and chromatic characteristics of pigmented PVC films and coatings for synthetic leathers using mathematical models, *Kozh.-Obuvn. Prom-st.*, 7, 39, 1987.
43. **Dainichiseika Color and Chemicals Manufacturing Co., Ltd.**, Japan Patent 85,67,545, 1985.
44. **Maxwell, P. J. (to Monsanto Australia Ltd.)**, Australia Patent 533,848, 1983.
45. **Dragaun, H., Radax, M., and Veith, M.**, Effect of pigment preparations on polymer structure and properties, *Oesterr. Kunstst.-Z.*, 16(1—2), 9, 1985.
46. **Fukuoka, M. and Takenaka, H. (to Dainippon Ink and Chemicals, Inc.)**, Japan Patent 86,69,850, 1986.
47. **Sakurai, M., Umemura, T., Mizutani, Y., Kawada, T., Kobashi, R., and Takase, S. (to Mitsubishi Gas Chemical Co., Inc. and Nippon Oils and Fats Co., Ltd.)**, Japan Patent 86,145,254, 1986.
48. **Krieg, M., Lichtenstein, H., Hosch, L., and Ittmann, G. (to Roehm GmbH)**, German Patent 3,323,951, 1985.

49. **Moustafa, A. B., Abd El-Ghaffar, M. A., and Badran, A. S.,** Effect of copper metal powder and some of its salts on the aqueous polymerization of methyl methacrylate and on the tenacity of obtained polymers, *Acta Polym.,* 35(1), 68, 1984.
50. **Pollard, R. E. (to Monsanto Co.),** U.S. Patent 4,518,728, 1985.
51. **Fedtke, M. and Keck, R. (to Technische Hochschule "Carl Schorlemmer" Leuna-Merseburg),** East German Patent 218,110, 1985.
52. **Oka, K. (to Toray Industries, Inc.),** Japan Patent 87,62,826, 1987.
53. **Oka, K. (to Toray Industries, Inc.),** Japan Patent 87,62,828, 1987.
54. **Aleksandrovich, I. R., Sokolov, A. D., Galkina, N. M., Zherebtsov, I. P., Kozakov, L. M., Krakhmalets, I. A., and Balyatina, E. V.,** U.S.S.R. Patent 1,208,052, 1986.
55. **Mokienko, R. L., Zemlyanoi, A. A., Prokopenko, B. N., Shatrovskii, V. V., Pedan, V. P., Zakharov, A. V., Azarova, M. T., Rykov, V. A., Lavrenko, S. I., et al. (to Dnepropetrovsk Chemical-Technological Institute),** U.S.S.R. Patent 1,211,266, 1986.
56. **Nakamura, K. and Miwatsukino, T. (to Agency of Industrial Sciences and Technology),** Japan Patent 86,89,017, 1986.
57. **Mos, V., Popa, D., Negreanu, D., Antonie, M., Goldenberg, S., Calciu, L., and Dimitriu, E. (to Intreprinderea Chimica "Victoria"),** Romania Patent 89,434, 1986.
58. **Szycher, M., Dempsey, D. J., and Rolfe, J. L. (to Thermo Electron Corp.),** German Patent 3,323,581, 1984.
59. **Shimizu, G. and Hayashi, Y. (to Matsui Shikiso Chemical Co., Ltd.),** Japan Patent 87,15,236, 1987.
60. **Sutton, R. G. and Steever, J. A.,** Upgrading the long term stability of polyurethanes by the use of pigments, *Polyurethane: New Paths, Progress, Marketing, Technology, Proc. 6th SPI Int. Technology/ Marketing Conf.,* 6th, 145, 1983.
61. **Kujawa-Penczek, B., Penczek, P., and Zajezierski, J.,** Effect of seeding additives on the initial bonding strength of polyurethane contact adhesives, *Adhaesion,* 29(6), 26, 1985.
62. **Igarashi, N. and Shimizu, H. (to Toyo Ink Manufacturing Co., Ltd.),** Japan Patent 85,58,248, 1985.
63. **Sanyo Chemical Industries, Ltd.,** Japan Patent 84,12,934, 1984.
64. **Jevtic, L.,** Effect of pigment dispersibility on optical efficiency in PVC films, *Plast. Guma,* 5(4), 152, 1985.
65. **Necas, M. and Klimesova, J.,** Czech Patent 229,111, 1986.
66. **Kuroda, H. (to Bando Chemical Industries, Ltd. and Dainippon Printing Co., Ltd.),** Japan Patent 86,04,578, 1986.
67. **Gaku, M. and Kimbara, H. (to Mitsubishi Gas Chemical Co., Inc.),** German Patent 3,608,253, 1986.
68. **Dainichiseika Color and Chemicals Manufacturing Co., Ltd.,** Japan Patent 84,187,051, 1984.
69. **Dainichiseika Color and Chemicals Manufacturing Co., Ltd.,** Japan Patent 83,111,820, 1983.
70. **Nippon Carbide Industries Co., Inc.,** Japan Patent 83,101,625, 1983.
71. **Coufal, J., Stach, P., and Kopcil, M.,** Colored sheeting for greenhouse covering, *Plasty Kauc.,* 21(5), 142, 1984.
72. **Harasawa, I. and Hariki Y. (to Nippon Carbide Industries Co., Inc.),** Japan Patent 85,56,462, 1985.
73. **Mitsubishi Rayon Co., Ltd.,** Japan Patent 84,10,909, 1984.
74. **Ida, K. (to Mitsubishi Rayon Co., Ltd.),** Japan Patent 85,139,757, 1985.
75. **Fuji Photo Film Co., Ltd.,** Japan Patent 85,67,544, 1985.
76. **Achar, B. N., Fohlen, G. M., and Parker, J. A.,** Curing of epoxy resins with metal(II) 4,4′,4″,4‴-tetraaminophthalocyanines — modification with elastomers, *J. Polym. Sci., Polym. Chem. Ed.,* 23(2), 389, 1985.
77. **Toa Gomu Kogyo K. K.,** Japan Patent 85,35,041, 1985.
78. **Toa Gomu Kogyo K. K.,** Japan Patent 85,31,543, 1985.
79. **Toa Gomu Kogyo K. K.,** Japan Patent 85,47,042, 1985.
80. **Toa Gomu Kogyo K. K.,** Japan Patent 85,31,544, 1985.
81. **Tanimoto, Y., Ikeda, K., Hisada, H., and Kawakami, K. (to Sumitomo Chemical Co., Ltd. and Toa Gomu Kogyo K. K.),** Japan Patent 86,36,345, 1986.
82. **Tanimoto, Y. and Ikeda, K. (to Sumitomo Chemical Co., Ltd.),** Japan Patent 86,26,642, 1986.
83. **Kovacs, G. J. and Vincett, P. S.,** Subsurface particle monolayer and film formation in softenable substrates: techniques and thermodynamic criteria, *Thin Solid Films,* 111(1), 65, 1984.
84. **Kovacs, G. J. and Vincett, P. S.,** Particles in polymers: surface chemistry of their nucleation, growth, configuration, and interactions with the matrix, *Can. J. Chem.,* 63(1), 196, 1985.

Chapter 11

COLORING CEMENT, MORTAR, SAND, AND GYPSUM

Cu phthalocyanine (pc) is added to decorative or ordinary portland cement to give colored cements with acceptable properties. Technical Cu pc is almost as satisfactory as Cu pc pigment for this purpose.[1]

White cements with good structural properties are made from local limestone, clays, and sand, and helicopter factory wastes in the northern Caucasus, U.S.S.R. Adding colorfast Cu pc and Pc Green pigments does not affect the strength or other cement properties.[2]

A mortar contains cement, clay, sand, water, and 0.04 to 0.1 wt % atmosphere-resistant Cu pc pigment to color its surfaces.[3]

A sand composition is mixed with Cu pc for decorative wall finishes.[4]

Cu pc pigments are incorporated in electron beam-cured resin coatings on gypsum molded panels.[5]

REFERENCES

1. **Nikiforov, Y. V., Astakhova, M. A., Makhotin, M. A., and Pestina, R. A.,** Colored cements with technical copper phthalocyanine, *Tsement,* (3), 12, 1987.
2. **Arutyunov, I. M., Gaidzhurov, P. P., Golovanova, S. P., and Didkovskii, V. P.,** Development of the technology for producing decorative cements from local raw materials, *Izv. Sev.-Kavk. Nauchn. Tsentra Vyssh. Shk., Tekh. Nauki,* (3), 71, 1983.
3. **Sushko, V. A., Kruglitskii, N. N., Arpaksyd, M. A., Belov, P. A., and Mironenko, A. A.,** U.S.S.R. Patent 1,244,125, 1986.
4. **Miyatake, M. and Oishi, K.,** Japan Patent 87,59,673, 1987.
5. **Sugimito, K., Maruyama, T., and Ogawa, M.,** Australia Patent 549,233, 1986.

Chapter 12

PCS IN DETERGENT AND DISPERSANT COMPOSITIONS

An Al phthalocyanine (pc) Cl photoactivator is prepared as follows. Al pc Cl is chlorosulfonated, suspended in water containing cyanamide, aqueous sodium hydroxide added, dried, treated with aqueous hydrochloric acid, and dried to give a blue powder

$$\mathrm{Al} \left| \begin{array}{l} -\mathrm{Cl} \\ \\ -(\mathrm{SO_2NCNNa})_3 \end{array} \right.$$

which is used as an antimicrobial agent and to remove stains from cotton textile when the textile is IR irradiated. Maximum absorption is at 675 nm.[1]

A photobleach system based on Al chloro pc tetrasulfonic acid sodium salt is used with a detergent composition for washing or treating fabrics.[2]

A solution or suspension of the photobleach compound sulfonated Al pc Cl is sprayed onto a spray-dried or granulated laundry detergent.[3]

The sodium salt of Al pc Cl is mixed with sodium bicarbonate and citric acid, dried, and sieved, giving speckles to be added to a powdered laundry detergent.[4]

A slurry of sulfonated aluminum pc, formaldehyde-naphthalenesulfonic acid condensate, sodium dodecylbenzenesulfonate, and water is spray dried and granulated to add to a powdered laundry detergent to give uniform bleaching of fabrics during laundering.[5]

Bleach speckles contain the fast dissolving sodium salt of sulfonated aluminum pcs.[6]

Gelation of concentrated aqueous flow activator solutions of sulfonated zinc or aluminum pcs, used as photobleaches in washing agents, is prevented by concentrating the starting solution in a semipermeable porous membrane, the basis of which is a modified cellulose acetate.[7]

Laundry bars may contain a photobleach compound such as sulfonated aluminum pc. The bars have good bleach stability in storage and use.[8]

Pc pigments are easily dispersed in aqueous solutions of anionic oligosoaps.[9]

Ge(IV) pc $(SO_3Na)_{2.8}$ is incorporated at 0.005% in an anionic detergent bath and used to wash cotton stained with tea under an IR lamp to give cotton with a high whiteness.[10]

Cu chloro pc incorporated in a liquid detergent for kitchen use shows good detergency for oily soil.[11]

Cu pc $Cl_{8\text{-}16}$ pigments are incorporated in colored detergent tablets.[12]

Surfactants containing Cu pc are prepared from poly(12-hydroxystearic acid)-aminoethanesulfonic acid reaction products.[13]

Maleated liquid polybutadiene and its half-ester derivatives are neutralized with alkaline solution to give an aqueous solution in which Cu pc pigment is easily dispersed.[14]

A series of polycarboxylic biosurfactants are good dispersants for α-Cu pc.[15]

Formaldehyde-lignosulfonic acid salt-naphthalene sulfonic acid reaction products are dispersing agents for Cu pc.[16]

The relation between the polarity of organic pigments including Cu pc vs. their wettability with nonionogenic surfactants is determined.[17]

A solid cake containing 99.5% trichloroisocyanuric acid and 0.5% sodium stearate and a solid cake containing nonionic surfactants 47.78, sodium sulfate 49.22, and C.I. direct blue 86 (Cu pc disulfonate disodium salt) 3.0% are used in a dispenser in a flush toilet.[18]

REFERENCES

1. **Hoelzle, G., Miotto, M., Reinert, G., and Polony, R. (to Ciba-Geigy AG),** Europe Patent Application 81,462, 1983.
2. **Finch, T. D. and Beavan, S. W. (to Unilever PLC),** British Patent Application 2,115,027, 1983.
3. **Finch, T. D. and Wilde, R. J. (to Unilever PLC),** Europe Patent Application 119,746, 1984.
4. **Eckhardt, C. and Will, H. (to Ciba-Geigy AG),** German Patent 3,430,773, 1985.
5. **Balliello, P., Bruttel, B., and Will, H. (to Ciba-Geigy AG),** Europe Patent Application 236,270, 1987.
6. **Eckhardt, C. and Will, H. (to Ciba-Geigy AG),** Swiss Patent 659,082, 1986.
7. **Preiswerk, W. and Bruttel, B. (to Ciba-Geigy AG),** German Patent 3,410,620, 1984.
8. **Finch, T. D., Postlethwaite, D., and Powers, P. J. (to Unilever PLC),** British Patent Application 2,172,301, 1986.
9. **Suzuki, S., Ishigami, Y., and Takemoto, K.,** Dispersibility of pigments in aqueous solutions of anionic oligosoaps. Synthesis of anionic oligosoaps having dye structures and their properties, *Shikizai Kyokaishi,* 57(10), 541, 1984.
10. **Graf, G., Hoelzle, G., and Reinhert, G. (to Ciba-Geigy AG),** Europe Patent Application 153,278, 1985.
11. **Lion Corp.,** Japan Patent 85,51,795, 1985.
12. **Shiozaki, R. and Kato, H. (to Kao Corp.),** Japan Patent 86,264,098, 1986.
13. **Imperial Chemical Industries PLC,** Japan Patent 85,215,651, 1985.
14. **Araki, Y., Ohtsuki, Y., Aoyama, K., and Kaiya, A.,** Reactive oligomer type surfactants. I. Surface properties of maleated liquid polybutadiene (LPB) and its derivatives, *Yukagaku,* 35(11), 945, 1986.
15. **Ishigami, Y. and Yamazaki, S.,** Surface active properties and application of polycarboxylic biosurfactants, *Kagaku Gijutsu Kenkyusho Hokoku,* 80(6), 231, 1985.
16. **Sugiwaki, T., Yamaki, M., and Tamao, M. (to Sanyo-Kokusaku Pulp Co., Ltd.),** Japan Patent 87,168,531, 1987.
17. **Shvets, V. I., Popov, E. V., Karpov, V. V., and Slepchenko, V. I.,** Relation between the polarity of organic pigments and their wettability with noniogenic surfactants, *Zh. Prikl. Khim. (Leningrad),* 58(7), 1556, 1985.
18. **Brummer, B. J., Gandhi, A. C., Tutschek, P. C., and Weinstein, B. (to American Home Products Corp.),** Canada Patent 1,222,923, 1987.

Chapter 13

ADSORPTION

The adsorption of Co phthalocyanines (pcs) is studied vs. the nature of the surface, solvent, and Co pc coordination compound. Adsorption on silica and γ-alumina is reversible; there is strong adsorption on graphitized active carbon.[1]

The high-order structure and dye adsorption of a porous alunite is studied with Cu pc adsorbed horizontally on the surface.[2]

The state of Cu pcs adsorbed on silica and alumina gel and zeolite NaY is studied by ESR and diffuse reflectance spectra.[3]

The adsorption of Al pcs with different axial ligands and of Si pc $(OH)_2$ on α-alumina is studied at 293 K.[4,5]

Carbochrom B is modified with Cu pc by adsorption from 1-bromonaphthalene solution and with sodium salts of the sulfonic acids of Cu pc, Cu pc $(SO_3Na)_4$, by adsorption from aqueous solutions. Efficient adsorbents with high selectivity and suitable for gas-solid chromatography are studied.[6]

An adsorbent for extracting iodine from steam and gas media contains silica gel and Lu di pc (Lu pc_2). The Lu pc_2 assures repeated sorption-desorption cycles without deterioration of the adsorbent. The component amounts are 92.8 to 99.8 wt % silica gel and 0.2 to 7.2 wt % Lu di pc.[7]

A molecular sieve for gas separation by pressure swing adsorption is made by sealing an active carbon-Fe pc mixture at 0.01 torr, heating at 450 to 550°C for 1 h, and cooling. This molecular sieve exhibits a good adsorption selectivity for oxygen in air separation.[8]

Co chloromethyl pc (4.5 to 5.5 wt %) is added to improve the selectivity of a sorbent to separate polar aromatic compounds. The sorbent contains silica.[9]

The adsorption of methyl methacrylate-2-vinylpyridine copolymer and acrylic acid-methyl methacrylate copolymer in titania and β-Cu pc from toluene solutions is determined.[10]

The affinity of aromatic compounds having three fused rings to Cu pc trisulfonate is studied in trisodium phosphate buffer at pH 7.4. Cu pc trisulfonate is bound to cellulose in the form of blue cotton. The adsorptions are reversible as the original compounds can be quantitatively recovered by eluting the treated blue cotton with a mixture of methanol and concentrated ammonia.[11]

REFERENCES

1. **Borisenkova, S. A., Daud, G., Denisova, E. P., and Plate, A. F.,** Study of Adsorption of Cobalt Phthalocyanine and its Coordination Derivatives on the Surface of Different Types of Supports, Deposited Doc., VINITI 1958—83, All Union Institute of Scientific and Technical Information, U.S.S.R., 14 pp, 1983.
2. **Inouye, K., Koyama, T., Kaneko, K., and Ozeki, S.,** The high-order structure and dye adsorption of a porous alunite, *Nippon Kagaku Kaishi,* (2), 156, 1985.
3. **Borisova, T. G., Izmailova, L. N., Kotov, E. I., Romanovskii, B. V., and Semin, V. V.,** Dispersion of copper(II) phthalocyanine deposited on an oxide support surface, *Zh. Fiz. Khim.,* 60(3), 745, 1986.
4. **Borisenkova, S. A., Klimova, T. E., and Mokshin, V. M.,** Adsorption of aluminum and silicon phthalocyanines on graphitized carbon black, *Zh. Fiz. Khim.,* 60(10), 2607, 1986.
5. **Borisenkova, S. A., Klimova, T. E., and Mokshin, V. M.,** Adsorption of aluminum and silicon phthalocyanines on oxide supports, *Zh. Fiz. Khim.,* 60(10), 2631, 1986.
6. **Hille, J., Prochazka, M., Feltl, L., Smolkova-Keulemansova, E., Kiselev, A. V., Kovaleva, N. V., and Zagorevskaya, E. V.,** Silica gel and Carbochrom B modified with phthalocyanines as stationary phases in gas-solid chromatography, *J. Chromatogr.,* 283, 77, 1984.

7. **Moskalev, P. N.,** U.S.S.R. Patent 571,958, 1985.
8. **Tsumura, M., Ora, H., and Ichida, T.,** Japan Patent 87,132,543, 1987.
9. **Aratskova, A. A., Gvozdovich, T. N., Yashin, Y. I., Snegireva, F. P., Lukina, V. B.,** U.S.S.R. Patent 1,230,670, 1986.
10. **Dulog, L.,** Adsorption of polymers on pigment surfaces, *Makromol. Chem., Suppl.,* 12, 265, 1985.
11. **Hayatsu, H., Kobayashi, H., Michiue, A., and Arimoto, S.,** Affinity of aromatic compounds having three fused rings to copper phthalocyanine trisulfonate, *Chem. Pharm. Bull.,* 34(2), 944, 1986.

Chapter 14

GAS AND WATER PURIFICATION

I. GAS PURIFICATION

A. H_2S AND MERCAPTAN REMOVAL

Metal phthalocyanines (pcs) are used as catalysts in the elimination of thiosulfate and oxidation of H_2S in a procedure to remove H_2S and CO from natural gas and fuel gas streams.[1]

H_2S is removed from gases by absorption with dimethyl sulfoxide or dimethylacetamide with subsequent regeneration of a spent absorbent by passing an oxygen containing gas through the absorbent. Regeneration time is decreased and degree of purification is increased by regenerating spent absorbent in the presence of Co sulfo pc and Mannich base.[2]

H_2S is removed from gases by absorption by an organic solvent and an alcohol; spent absorbent is regenerated by stripping by an oxygen containing gas in the presence of Co sulfo pc in *N*-dimethylaminopropionitrile or allylamine.[3]

A waste gas catalyst is prepared by loading an alkali salt obtained by carboxylating or sulfonating pc in an aqueous weak alkali solution onto a ceramic support, and then soaking the support in an acid to fix the alkali salt in the pc. The catalyst has good heat resistance and gives high removals of acetaldehydes, alcohols, and mercaptans.[4]

Air containing ≤ 1 vol % of H_2S from degassing liquid sulfur at the Orenburg gas processing plant in the U.S.S.R. is purified by sorbing H_2S in an alkaline medium and catalytic H_2S oxidation to sulfur with Co tetrasulfo pc catalyst. Oxidation is about 90%.[5]

A gas stream containing 97.5 CO_2 and 2.5 H_2S vol % is mixed with air at an O/S mole ratio of 5.9:1, and mixed with water at 115 to 125 psig (pounds per square inch - gauge) at 132°C with Co pc disulfonate catalyst on a carbon support. The effluent gas contains a negligible concentration of H_2S.[6]

Ventilation air from synthetic fiber plants is purified by absorption and oxidation of H_2S in alkaline solutions with Co pc catalyst.[7]

The catalytic oxidation of H_2S by O_2 in air in the presence of Co disulfo pc and poly(Co pc) is studied.[8]

B. CO, NO_x, SO_x, AND HYDROCARBON REMOVAL

A flue or waste gas is scrubbed with Co pc sulfonic acid or with a solution of Co pc sodium sulfonate containing sulfamic acid to remove NO.[9]

The electrochemical conversion of CO and NO is activated by a chemically modified electrode; effective electrode coatings are Everitt's salt for CO and Co pc or H_2 pc for NO.[10]

Cu pc pigment on a cordierite monolithic support is used to reduce NO to N_2 and to simultaneously oxidize CO to CO_2 in model combustion gases. The catalyst can be activated with a 1:3 NO-CO mixture at about 600°C.[11]

NO_x, CO, and hydrocarbons are removed from waste gases with Co, Cu, or Ni pc supported on a porous ceramic at 0 to 700°C and $\leq 10^6$/h space velocities. The catalyst converts 50% of a mixture containing NO 1000 and 3000 ppm at 440°C and 90% at 520°C.[12]

Power plant flue gas containing about 1000 ppm NO_x is treated with $\geq 90\%$ efficiency with a ten phase contact plate column and a counter stream of contact liquid of 200 l carrier liquid and 10 kg Fe pc catalyst.[13]

An amino group containing polyurethane foam is impregnated with an aqueous solution containing Fe pc carboxylic acid derivative to obtain a deodorant removing 300 ppm H_2S from a gas in 10 h.[14]

C. DEODORANTS

A nonwoven fiber is impregnated with 20% of a benzene solution containing the reaction product of 1:3 weight ratio of Fe pc tetracarboxylic acid and polystyrene to form a deodorant that effectively removes fish odor from air in a closed container.[15]

Wool yarns are soaked in an aqueous solution containing Co pc trisulfonic acid at 70°C for 10 h, washed, dried, and punched to give a nonwoven fabric filter, which oxidizes mercaptans generated in a pressure test of a propane bomb, to odorless materials.[16]

A stabilized ClO_2 aqueous solution is blended with Cu pc and polyoxyethylene sorbitan polyoleate surfactant to give a deodorant solution effective to treat odorous air from water closets or garbage tanks.[17]

Alkali treated wood pulp is treated with a solution containing 1.0% Co pc octacarboxylic acid octasodium salt for 5 h, dewatered, and used to prepare paper which deodorizes ammonia, amines, H_2S, and mercaptans.[18]

Air cleaner filters effective to remove cigarette smoke from closed rooms contains a treated cotton containing Cu pc disulfonic acid which is used to remove tar 36% and mutagen 43% from cigarette smoke in a control test.[19]

Arginine CM-cellulose fibers are soaked in a Co octaamine pc solution: KOH is added to adjust the pH to 5. The fibers are filtered and washed with water, and dried. They give a fibrous deodorant for air fresheners.[20]

Rayon treated with a Fe(III) pc derivative is >20 times effective in the deodorization of H_2S, mercaptans, ammonia, and amines than coconut shell activated charcoal. The treated fiber is used in futon.[21]

Acrylonitrile-methylacrylate-sodium methallylsulfonate copolymer gelled fibers are treated with an emulsion containing Fe pc tetracarboxylic acid, dried, treated with [(laurylamido)ethyl]dimethylamine sulfate, and heat treated to give odor resistant fibers.[22]

A rayon staple containing 2.5% Fe pc polycarboxylic acid and staple containing 2.5% copper acetate mixture and polyvinyl alcohol binder are mixed to give a crepe paper showing a good deodorizing effect.[23]

Viscose rayon is soaked in a solution containing 0.3% Fe pc tetracarboxylic acid, and dried to give fibers containing 2% Fe pc tetracarboxylic acid. The treated fibers catalyze oxidation of ammonia, amines, H_2S, and mercaptans with high efficiency.[24]

Nylon 6 fibers are soaked in a solution containing 4.0% Fe pc tetracarboxylic acid, soaked in an acidic solution, and dried to deposit 1% Fe pc tetracarboxylic acid. The fibers decompose ammonia, amine, H_2S, and mercaptans.[25] Acrylic fibers are soaked in an HCl solution of Co pc octaamine, then neutralized by sodium hydroxide, washed by water, and dried to give deodorant fibers containing fiber containing 7% Co pc octaamine, which decomposes ammonia, amines, H_2S, and mercaptans.[26]

Alkali-treated degreased cotton is soaked in 1.0% solution of Co pc octacarboxylic acid octasodium salt and bonded together with a polyvinyl acetate emulsion to yield a nonwoven fabric which decomposes ammonia, amines, H_2S, and mercaptans.[27]

Polyvinyl alcohol fibers are immersed in a 2% disodium sulfide solution of Fe pc tetrathiocyanate and dried to give deodorant fibers containing about 10% Fe pc tetrathiocyanate, which decomposes ammonia, amines, H_2S, and mercaptans.[28]

Polyester fibers are soaked in a solution containing Co pc octaamine to give deodorant fibers containing about 5% Co pc octaamine, which decomposes ammonia, amines, H_2S, and mercaptans.[29]

The deodorant properties and application of futons to fibers made from styrene-2-vinylpyridine copolymer and M pcs are reviewed with 11 references.[30]

II. WATER PURIFICATION

Mutagens are removed from solution by absorbent cotton dyed with Cu pc disodium disulfonate. The dye treated absorbent cotton adsorbs 98% *Trp-P-1* by shaking brine containing *Trp-P-1*.[31]

Pc blue B (Cu pc) and diethylene glycol are kneaded, ground, and added to brine containing *Trp-P-1*. The adsorptivity is 95%.[32]

Several methods including blue cotton absorption are compared for concentrating mutagens from humus containing drinking water. A water-soluble pc derivative is used to adsorb mutagens from humus containing drinking water in water analysis.[33]

Co pc disulfonic acid dissolved in a water sample through a soil bed for $\leq$32 d did not biodegrade.[34]

A Co pc catalyst to oxidize disodium sulfide-containing alkaline wastewater and to remove mercaptans is prepared by adding -Cl, $-CH_2Cl$, or $-C(CH_3)_3$ substitutents into one of the Co pc benzene rings. A catalyst containing 20 wt % Co pc was stable for 300 h in oxidizing disodium sulfide and hydrogen sodium sulfide in a 5 to 15% sodium hydroxide solution.[35]

Polyethylene shavings that contain 10% Co pc are effective catalysts to oxidize H_2S in stratal waters.[36]

Pc is an activation agent to remove oxygen from cooling water circulating in closed circuits, such as in electric generators.[37]

Catalysis of the autooxidation of sulfur(IV) by Co tetrasulfo pc is characterized. Oxidation by oxygen or hydrogen peroxide is a primary mode to remove pollutants from industrial and domestic wastewaters.[38]

The sensitized photooxidation of bromacil to remove bioaccumulated pesticides and herbicides from wastewater is optimum at alkaline pH. Cu pc disulfonic acid is evaluated as a photosensitizer.[39]

Carbon powders are treated with Fe pc sulfonate solution and heated under N_2 at 700°C to immobilize nitrifying bacteria for wastewater treatment.[40]

III. GAS SENSORS

A. OF GASES OTHER THAN OXYGEN

A gas sensor designed primarily for low NO_2 measurement contains a Pb pc film.[41] The Pb pc film sensor can detect from 1 ppb to 10 ppm NO_2 in air, and is based on electrical conductivity changes in the film brought about by NO_2 adsorption.[42]

A solid-state sensor for sensitive gas detection is based on laying a thin layer of a transition M pc complex on substrates.[43]

A Langmuir-Blodgett film of the asymmetrically substituted Cu pc has an electrical conductivity extremely sensitive to NO_2 ambients.[44]

Co pc is deposited on the gate electrode of a field effect transistor as the basis for a sensor for CO.[45]

H_2 pc is a surface acoustic wave delay-line oscillator chemosensor for NO_2 concentration determination.[46]

A planar type gas sensor for a reducing gas (SiH_4, B_2H_6, AsH_3, PH_3) used in the manufacture of a semiconductor contains H_2pc.[47]

A sensor for determining arsine comprises a substrate with a semiconductor layer and electrodes. To increase the sensitivity of the sensor, the semiconductor layer is made from chlorinated Cu pc.[48]

The effect of humidity on the response of Pb pc films to NO_2 is studied.[49]

When SO_2, benzene, air, ethanol, water, di-methyl methylphosphonate, and ammonia vapors are introduced into helium flowing over thin layers of Co, Cu, Fe, Ni, and Pb pc complexes sublimed onto the surface of interdigital microelectrodes, the conductivities depend on the M pc and on the vapor.[50]

A sensor having improved sensitivity toward low vapor concentrations to monitor gases is made by depositing films of aryloxy-, arylthio-, alkyloxy-, or alkylthio-substituted pcs on a nonconducting substrate between interdigital electrodes.[51]

NO_2 sensitive field effect transistors (FETs) are prepared by subliming a Pb pc film on a gate oxide region of the FET. The change in threshold voltage with NO_2 concentration is linear at 20 to 1000 ppb.[52]

Co pc films are deposited on metal-insulator-metal tunnel junctions in chemical sensors.[53]

A composite substrate contains H_2 pc derivatives in the analysis of gases or liquids.[54]

Time-of-flight secondary ion mass spectrometer (SIMS) is used to monitor the chemical state of a pc gas detector.[55]

A Cu pc complex multilayer film deposited by using the Langmuir-Blodgett technique onto planar microelectrode arrays can easily detect ammonia at sub-parts-per-million concentrations.[56]

The electrical conductive properties of thin films of conjugated macrocycles, such as pc, are studied to find a reversible sensor for N_2O_4 in industrial gases that can operate at 10 to 10,000 ppm at room temperature.[57]

A method to measure ammonia concentration in air involves adsorption of ammonia on a Cu pc film.[58]

A procedure produces Langmuir-Blodgett films that include H_2, Bi, Co, Cu, Mg, Ni, Pb, Pd, Pt, and Zn tetracumylphenoxy pcs to produce rapid, reversible, and reproducible electrical responses to parts-per-million concentrations of ammonia in the atmosphere.[59]

Sensitivity to a few parts per million of NO_2 in nitrogen is demonstrated by an acoustic wave gas sensor based on changes in film conductivity of Pb pc.[60]

Sensors based on electrical conductivity changes in Pb pc are used to provide a measure of NO_2 and NO_x in air.[61]

A proton conductor gas sensor for detecting H_2 or CO in air at room temperature has a reference electrode that includes a metal pc.[62]

A chemoresistive gas sensor for detecting and measuring gaseous contaminants in industrial purification processes, anesthetic gases in operating rooms, chemical warfare agents on battlefields, and air pollutants has pc or its derivatives as the preferred sensor layer between two electrodes.[63] Hydrogen gas sensors that sense by optical absorption include Lu pc electrochromic coating materials absorbing electrolytes in an electrolyte probe. Hydrogen concentration at 10 to 2000 ppm can be detected in gases or in the atmosphere with 15% error.[64,65]

B. OF OXYGEN

A battery-operated oxygen sensor has a positive electrode that contains Co, Cu, Fe, or Ni pc.[66] The cathode of an oxygen sensor contains a Co, Cu, or Fe pc complex.[67]

The anode of an oxygen sensor is composed of a porous conductor containing Co, Cu, Mn, Mo, Ni, or Sn pc polymer.[68]

The anode of an oxygen sensor contains polymeric Fe pc.[69]

An oxygen sensor anode contains films of H_2 pc and aluminum.[70]

An Fe pc complex is a working electrode in an electrochemical oxygen sensor.[71]

A gravure ink contains cobalt Schiff base complexes and Cu pc as colorimetric indicators for oxygen, changing colors in the presence of 0.1% O_2 after 1 h.[72]

A potentiometric oxygen sensor contains a Fe, Pd, or Pt pc-sensing electrode.[73]

An oxygen sensor with solid electrolyte cells sensing from 20 to 250°C contains an M pc electrode. The effect of Fe pc preparation methods is studied.[74]

IV. PURIFICATION OF CU PC PIGMENTS

Pigments including Cu pc are recovered by treating a waste ink in a dispersion column with an upward flow of a solvent such as a toluene-ethyl acetate mixture having good resin solubility and poor pigment dispersibility.[75]

Polychlorinated biphenyls (PCBs) are extracted from Cu pc pigment in concentrated sulfuric acid followed by hexane. PCBs are detected in pigments manufactured using trichlorobenzene but not in those made with nonchlorinated solvents.[76]

Cu pc pigments are not hazardous and have no known adverse effects on human health. Cu pc pigments are insoluble in water and the copper atom is strongly bound to the pc ligand so that the copper atom is not available to aquatic or mammalian life, such that seasonal environmental regulations relating to Cu pc should differ from those applicable to soluble or bioavailable copper substances.[77]

REFERENCES

1. **Urban, P. and Johnson, R. W. (UOP Inc.),** U.S. Patent 4,496,371, 1985.
2. **Mazgarov, A. M., Vil'danov, A. F., Fakhriev, A. M., Sattarov, U. G., Neyaglov, A. V., and Latypova, M. M. (All-Union Scientific Research Institute of Hydrocarbon Raw Materials),** U.S.S.R. Patent, 1,150,007, 1985.
3. **Mazgarov, A. M., Vil'danov, A. F., Fakhriev, A. M., Sattarov, U. G., Neyaglov, A. V., and Latypova, M. M. (All-Union Scientific Research Institute of Hydrocarbon Raw Materials),** U.S.S.R. Patent 1,159,603, 1985.
4. **Kitajima, H., Ito, M., Shiraki, H., Kato, Y., and Shirai, H. (Toyoda Automatic Loom Works, Ltd.),** Japan Patent 87,53,741, 1987.
5. **Alekseenko, B. A., Brolinskii, G. L., Kolesnikova, S. A., and Faddenkova, G. A.,** Removing hydrogen sulfide from the air during degasification of liquid sulfur of the Orenburg gas processing plant using cobalt phthalocyanine, *Soversh. Tekhnol. Protsessov Poluch. Sery.*, 111, 1982.
6. **Massie, S. N. and Broyles, A. R. (UOP Inc.),** U.S. Patent 4,487,753, 1984.
7. **Rybakov, L. A., Afanas'ev, Y. M., Shimko, I. G., and Chinennaya, S. K.,** Development of waste-free technology for the removal of hydrogen sulfide from ventilation air of synthetic fiber plants, *Khim. Volokna,* 6, 12, 1984.
8. **Agaev, G. A., Mikhtarov, M. M., Aliev, N. M., Ruvinskaya, T. I., Veliev, A. A., Meerovich, T. P., and Amirgulyan, N. S.,** Study of catalytic oxidation of hydrogen sulfide by oxygen of air in the presence of cobalt phthalocyanines, *Izv. Vyssh. Uchebn. Zaved. Neft Gaz.*, 28(6), 41, 1985.
9. **Kawasaki Steel Corp.,** Japan Patent 83,128,125, 1983.
10. **Ogura, K.,** Carbon monoxide and nitric oxide conversion reactions using homogeneous metal complex-electrode system catalysts, *Nippon Kagaku Kaishi,* 11, 1774, 1984.
11. **Steinbach, F. and Thrams, N.,** Waste gas catalysts based on metal phthalocyanines. I. Reduction of nitric oxide with carbon monoxide in a functional model on modified CuPc, *Chem.-Ing.-Tech.*, 57(6), 522, 1985.
12. **Steinbach, F., Thrams, N., and Jesse, D.,** German Patent 3,522,637, 1986.
13. **Schmid, P. J., Kranebitter, F., and Panning, P.,** (Envirotech AG), Europe Patent Application 218,574, 1987.
14. **Shirai, H., Tsunoda, K., Sugiura, K., Inamura, S., Shirasaka, H., and Ishii, A. (Earth Clean K. K.),** Japan Patent 86,232,856, 1986.
15. **Shirai, H., Kakegawa, E., Ito, Y., Kamimura, M., Yokozeki, T., Shimizu, K., Sakaguchi, I., and Hojo, N. (Nippon Carbide Industries Co., Inc., Nissetsu K. K., and Earth Clean K.K.),** Japan Patent 87,97,555, 1987.
16. **Fukamachi, K. and Shirai, H. (Earth Clean K.K.),** Japan Patent 87,74,424, 1987.
17. **Kawachi, T. (Teisan Seiyaku K.K.),** Japan Patent 87,170,252, 1987.
18. **Fukamachi, K. and Shirai, H. (Earth Clean K.K.),** Japan Patent 87,33,900, 1987.
19. **Sumitomo Chemical Co., Ltd.,** Japan Patent 85,58,219, 1985.
20. **Fukamachi, K. and Shirai, H. (Earth Clean K.K.),** Japan Patent 86,211,371, 1986.
21. **Shirai, H.,** Deodorization with artificial enzymes, *Senshoku Kogyo,* 34(10), 482, 1986.
22. **Ono, M. and Yamamoto, T. (Kanebo, Ltd.),** Japan Patent 87,141,128, 1987.
23. **Shirai, H., Hasebe, K., Minamide, N., Higaki, S., and Yokozeki, T. (Dawa Spinning Co., Ltd. and Earth Clean K.K.),** Japan Patent 87,07,000, 1987.
24. **Fukamachi, K. and Shirai, H. (Earth Clean K.K.),** Japan Patent 86,258,805, 1986.

25. **Fukamachi, K. and Shirai, H. (Earth Clean K.K.)**, Japan Patent 86,258,078, 1986.
26. **Fukamachi, K. and Shirai, H. (Earth Clean K.K.)**, Japan Patent 87,06,985, 1987.
27. **Fukamachi, K. and Shirai, H. (Earth Clean K.K.)**, Japan Patent 87,33,854, 1987.
28. **Fukamachi, K. and Shirai, H. (Earth Clean K.K.)**, Japan Patent 87,06,984, 1987.
29. **Fukamachi, K. and Shirai, H. (Earth Clean K.K.)**, Japan Patent 87,06,986, 1987.
30. **Shirai, H. and Yokozeki, T.**, Deodorizing activity and effect of deodorizing fibers, *Fragrance J.*, 13(3), 83, 1985.
31. **Sumitomo Chemical Co., Ltd.**, Japan Patent 84,98,789, 1984.
32. **Sumitomo Chemical Co., Ltd.**, Japan Patent 84,98,788, 1984.
33. **Vartiainen, T., Liimatainen, A., Jaaskelainen, S., and Kauranen, P.**, Comparison of solvent extractions and resin adsorptions for isolation of mutagenic compounds from chlorinated drinking water with high humus content, *Water Res.*, 21(7), 773, 1987.
34. **Suzuki, Y., Iimoto, T., and Katayama, M.**, Biodegradation of water soluble dyes, *Kenkyu-Tokyo Kasei Daigaku*, 24(2), 189, 1984.
35. **Akhmadullina, A. G., Khrushcheva, I. K., Mazgarov, A. M., and Abramova, N. M.**, Heterogeneous phthalocyanine catalyst for liquid-phase oxidation of sodium sulfide and mercaptide, *Khim. Tekhnol. Topl. Masel*, 8, 42, 1985.
36. **Kochetkova, R. P., Shpilevskaya, L. I., Akhmadullina, A. G., Latyshev, V. P., Eppel, S. A., and Mazgarov, A. M.**, Removing hydrogen sulfide from stratal water by oxidation with atmospheric oxygen in the presence of a heterogeneous catalyst, *Zh. Prikl. Khim. (Leningrad)*, 58(4), 916, 1985.
37. **Kuhn, A. and Mueller, R. (Brown, Boveri und Cie.)**, Europe Patent Application 190,637, 1986.
38. **Hoffmann, M. R. and Hong, A. P. K.**, Catalytic oxidation of reduced sulfur compounds by homogeneous and heterogeneous cobalt(II) phthalocyanine complexes, *Sci. Total Environ.*, 64(1—2), 99, 1987.
39. **Eisenberg, T. N., Middlebrooks, E. J., and Adams, V. D.**, Sensitized photooxidation of bromacil: a dye survey of potential photosensitizing agents, *Proc. Ind. Waste Conf. 1986, 325, 1987.*
40. **Endo, K. and Watanabe, A. (Ebara-Infilco Co., Ltd.)**, Japan Patent 86,70,987, 1986.
41. **Jones, T. A., Bott, B., Hurst, N. W., and Mann, B.**, Solid state sensors: zinc oxide single crystals and metal phthalocyanine films, *Anal. Chem. Symp. Ser.*, 17, 90, 1983.
42. **Bott, B. and Jones, T. A.**, A highly sensitive nitrogen dioxide sensor based on electrical conductivity changes in phthalocyanine films, *Sens. Actuators*, 5(1), 43, 1984.
43. **Scheffler, W. (Siemens AG)**, German Patent 3,217,883, 1983.
44. **Baker, S., Roberts, G. G., and Petty, M. C.**, Phthalocyanine Langmuir Blodgett film gas detector, *IEE Proc., Part I*, 130(5), 260, 1983.
45. **Matsushita Electric Industrial Co., Ltd.**, Japan Patent 83,23,585, 1983.
46. **Barendsz, A. W., Vis, J. C., Nieuwenhuizen, M. S., Nieuwkoop, E., Vellekoop, M. J., Ghijsen, W. J., and Venema, A.**, A SAW-chemosensor for nitrogen dioxide gas-concentration measurement, *Ultrason. Symp. Proc.*, 1, 586, 1985.
47. **Shimizu, T., Mihira, H., and Yamamoto, N. (Estech K. K.)**, Japan Patent 86,116,650, 1986.
48. **Soborover, E. I., Pakhomov, L. G., and Fel, Y. A. (Scientific Research Institute of Chemistry, Gorki)**, U.S.S.R. Patent 1,249,425, 1986.
49. **Chadwick, A. V., Dunning, P. B. M., and Wright, J. D.**, Applications of organic solids to chemical sensing, *Mol. Cryst. Liq. Cryst.*, 134(1—4), 137, 1986.
50. **Barger, W. R., Wohltjen, H., Snow, A. W., Lint, J., and Jarvis, N. L.**, Microsensor vapor detectors based on coating films of phthalocyanine and several of its metal complexes, *ACS Symp. Ser.*, 309, 155, 1986.
51. **U.S. Department of the Navy**, U.S. Patent Application 768,004, 1986.
52. **Burr, P. M., Jeffery, P. D., Benjamin, J. D., and Uren, M. J.**, A gas sensitive field effect transistor utilizing a thin film of lead phthalocyanine as the gate material, *Thin Solid Films*, 151(1), L111, 1987.
53. **Khanna, S. K., Lambe, J., LeDuc, H. G., and Thakoor, A. P.**, Thin Film Chemical Sensors Based on Electron Tunneling, Report 1985, DOE/ER/13007-1, JPL-PUB-85-85, Order No. 87004646, N.T.I.S., Springfield, VA, 37 pp.
54. **Barendsz, A. W. and Nieuwenhuizen, M. S. (Nederlandse Centrale Organisatie voor Toegepast-Natuurwetenschappelijk Onderzoek)**, PCT International Application 87,02,135, 1987.
55. **Vickerman, J. C.**, Static SIMS — a technique for surface chemical characterization in basic and applied surface science, *Surf. Sci.*, 189, 7, 1987.
56. **Wohltjen, H., Barger, W. R., Snow, A. W., and Jarvis, N.**, A vapor sensitive chemiresistor fabricated with planar microelectrodes and a Langmuir-Blodgett organic semiconductor film, *IEEE Trans. Electron Devices*, ED-32(7), 1170, 1985.
57. **Honeybourne, C. L., Ewen, R. J., and Hill, C. A. S.**, Use of thin films of conjugated organic macrocycles as the active elements in toxic gas sensors operating at room temperature, *J. Chem. Soc., Faraday Trans. I*, 80(4), 851, 1984.

58. **Szczurek, A. and Lorenz, K.,** Copper phthalocyanine film as gas detector, *Mater. Sci.*, 10(1—2), 271, 1984.
59. **Snow, A., Barger, W., Jarvis, N., and Wohltjen, H.,** Semiconducting Langmuir-Blodgett films of oligomeric phthalocyanine complexes for use in chemical microsensors, *Natl. SAMPE Tech. Conf.*, 16, 388, 1984.
60. **Ricco, A. J., Martin, S. J., and Zipperian, T. E.,** Surface acoustic wave gas sensor based on film conductivity changes, *Sens. Actuators,* 8(4), 319, 1985.
61. **Bott, B. and Jones, T. A.,** The use of multisensor systems in monitoring hazardous atmospheres, *Sens. Actuators,* 9(1), 19, 1986.
62. **Yamazoe, N. and Miura, N. (Figaro Engineering Inc.),** Europe Patent Application 231,766, 1987.
63. **Hirschfeld, T. B. (U.S. Department of Energy),** U.S. Patent Application 781,543, 1986.
64. **Sudo, E. and Nishizawa, K. (Agency of Industrial Sciences and Technology),** Japan Patent 86,201,140, 1986.
65. **Sudo, E. and Nishizawa, K. (Agency of Industrial Sciences and Technology),** Japan Patent 86,204,545, 1986.
66. **Nippon Telegraph and Telephone Public Corp.,** Japan Patent 83,200,157, 1983.
67. **Nippon Telegraph and Telephone Public Corp.,** Japan Patent 84,26,052, 1984.
68. **Nippon Telegraph and Telephone Public Corp.,** Japan Patent 84,138,943, 1984.
69. **Nippon Telegraph and Telephone Public Corp.,** Japan Patent 84,138,942, 1984.
70. **Kuwata, S., Kudou, A., Yano, K., Sadaoka, Y., and Sakai, Y.,** The development of solid state oxygen sensor operative at room temperature, *Niihama Kogyo Koto Senmon Gakko Kiyo Rikogaku-hen,* 21(2), 62, 1985.
71. **Fouletier, M., Siebert, E., and Le Moigne, J. (Societe Nationale Elf Aquitaine SA),** French Patent 2,550,344, 1985.
72. **Toyo Ink Manufacturing Co., Ltd.,** Japan Patent 84, 87, 364, 1984.
73. **Kuwata, S., Miura, N., Yamazoe, N., and Seiyama, T.,** Potentiometric oxygen sensor with fluoride ion conductors operating at lower temperatures, *Nippon Kagaku Kaishi,* 8, 1232, 1984.
74. **Siebert, E., Fouletier, J., and Kleitz, M.,** Oxygen sensing with solid electrolyte cells from room temperature up to 250°C, *J. Electrochem. Soc.*, 134(6), 1573, 1987.
75. **Mitsubishi Heavy Industries, Ltd.,** Japan Patent 83,124,505, 1983.
76. **Buchta, R. C., Wyles, H. F., Hensler, C. J., Van Lenten, F. J., Westerberg, R. B., and Williams, L. A.,** Determination of polychlorinated biphenyls in copper phthalocyanine pigments, *J. Chromatogr.*, 325(2), 456, 1985.
77. **Webb, P. G.,** Environmental regulations of copper phthalocyanine, *Am. Ink Maker,* 65(2), 11, 14, 1987.

Chapter 15

PC DYES AND PIGMENTS AS WATER POLLUTANTS

The chemical characteristics, health risks, and environmental effects, including toxicity and biodegradability, for 41 types of additives including Cu phthalocyanine (pc), used in paper manufacture are enumerated.[1]

Studies of changes in COD, BOD, nitrate-N, ammonia-N, dissolved oxygen, and other indicators show that dye concentrations of 0.001 g/l of Cu pc disulfonic acid (Direct Turquoise Lightfast) have no effect on oxygen, COD, and BOD, and little effect on pH.[2]

A method is described that allows identification of about 370 compounds including pc that may represent a hazard in the agricultural use of sewage sludges.[3]

A preliminary study of the manufacture of pc dyes and pigments is conducted to determine if process waste streams might contain hazardous material. Six companies that produce representative dyes and pigments are recommended for a sampling program to confirm the composition and amounts of these waste streams.[4]

Waste waters containing 1800 and 5000 mg/l of copper from the manufacture of Cu pc and cuprous chloride, respectively, are treated with 20 and 50 mg/l of $Na_2S_2O_4$, respectively, at 20 to 25°C and pH >7 to reduce copper concentrations to <1 mg/l.[5]

Copper is recovered from wastewater containing 550 mg/l copper, sourced from producing Cu pc pigment, with a 0.1 *M* solution of alkylamine in ligroin, and then reextracted into an aqueous 10% sodium hydroxide solution, followed by flotation with ethylxanthogenate to give an effluent containing 0.25 mg/l copper.[6]

Ammonium ion is removed from wastewater from pc manufacture by adding sulfuric acid to form ammonium sulfate followed by addition of ethanol to precipitate the ammonium sulfate.[7]

Surfactants and dyes including Cu pc trisulfonic acid are removed from spent activated carbon from adsorption treatment of industrial wastewater by extraction with low boiling solvents such as methanol followed by steam stripping.[8]

Mechanisms of oxidation are studied of 1 m*M* solution of organic dyes including Cu tetrasulfo pc in relation to detoxification and biodegradation of the dyes and purification of wastewater from the dye industry.[9]

REFERENCES

1. **Bjoerndal, H., Duering, L., and Solyom, P.,** Environmental effects of supplementary chemicals used widely in paper manufacturing, *Inst. Vatten-Luftvaardsforsk.*, [Publ.] B, IVL B 739, 202 pp, 1984.
2. **Timofeeva, S. S., Kashina, N. F., and Shivernovskaya, O. A.,** Effect of Arylmethane and Benzidine Dyes on Self-Purification of Water, Deposited Doc. 1982, VINITI 1975—83, All Union Institute of Scientific and Technical Information, U.S.S.R., 93 pp.
3. **Arendt, G., Eggersdorfer, R., Faltin, M., Frische, R., Haag, F., Lichtwer, L., Rippen, G., and Steinsiek, E.,** Determination of sources of selected substances and their retention in sewage sludge, *Forschungsber.-Bundesminist. Forsch. Technol., Technol. Forsch. Entwickl,* BMFT-FB-T 83-281, 1983, 136 pp.
4. **Bomberger, D. C.,** Wastes from Manufacture of Dyes and Pigments, Report 1984, EPA-600/2-84-114, Order No. PB84-200948, National Technical Information Service, Springfield, VA, 36 pp.
5. **Kwiatkowska, E., Zawadzka, A., and Stach, H.,** Removing cupric ions from wastes formed during phthalocyanine production, *Biul. Inf.: Barwniki Srodki Pomocnicze,* 29(1), 17, 1985.
6. **Formanek, J., Mrnka, M., Prochazka, J., Sladkovska, J., Heyberger, A., Bizek, V., Javorek, P., and Sychra, V.,** Czech Patent 218,295, 1985.

7. **Gradecka, L.,** Removal of ammonium from wastewater formed in phthalocyanine preparation by sintering, *Biul. Inf.: Barwniki Srodki Pomocnicze,* 30(3), 61, 1986.
8. **Mamontova, A. A., Kondratova, T. B., and Klimenko, N. A.,** Removal of surfactants and dyes from activated carbons by extraction with solvents, *Khim. Tekhnol. Vody,* 9(2), 113, 1987.
9. **Matrka, M. and Pipalova, J.,** Chemical models of biotransformation. I. Study of the degradation of organic dyes, *Chem. Prum.,* 35(2), 86, 1985.

Chapter 16

CORROSION PREVENTION

Di hydrogen phthalocyanine is included in a corrosion-inhibiting thixotropic gel. The gel is used to coat multistranded electric conductors and wire ropes or cables.[1]

REFERENCE

1. **Parkinson, A. R.,** Europe Patent Application 108,536, 1984.

Chapter 17

PHYSICAL CHEMISTRY

I. PHYSICAL PROPERTIES

The thermal decomposition of Fe(II) phthalocyanine (pc), 1:1 Fe(II) pc-O_2 and 1:1 Fe(II) pc-O are studied by differential thermal analysis, thermogravimetry, and exit gas analysis.[1]

Al pc O heptyl, Al pc OH, and Al pc CH_2CH_2OH are studied by DTA and thermogravimetry. At ~800°C, the pc is completely decomposed and alumina is formed.[2]

Thermogravimetric studies of Sn pc_2·n(α-chloronaphthalene) indicate that the α-chloronaphthalene is lost at 403 K.[3]

In the preparation of Cu complexed polymer from pyromellitonitrile and cuprous chloride, 400°C cannot be exceeded, but the polymer retains its integrity at ≤500°C.[4]

The orders of thermal stabilities of M pc under nitrogen and oxygen are Mg > VO > Ni > Cu > Co > Zn > Mn > Cr > Fe and Mg > Zn > Co > Ni > Cu > VO > Mn > Cr > Fe, respectively. They are more stable in nitrogen.[5]

Temperatures at which conversions of 2, 5, 50, and 100% occur for 12 M pcs are tabulated for a 20/min heating rate in air at 300 to 1000 K.[6]

Ultrafine particles of di hydrogen pc metal derivatives are prepared by heating and evaporating the M pcs in a vacuum vessel into which an inert gas is introduced at a desired pressure, then condensing and depositing the vaporized compound on a cooled recovering surface.[7]

Thin films of Cu pc and Pb pc are prepared at 0.0001 Pa. The vapor pressure of Pb pc is determined at 542 to 663 K.[8]

The absorption and velocity of a zero-order shear wave in β-Cu pc treated under high pressure (0.5 and 2 GPa), and subsequent shear deformation (20 and 200° deflection angle) are determined vs. temperature (<200°C).[9]

Structural and physical properties are made of a series of partially oxidized transition M pcs.[10]

II. ELECTRONIC STRUCTURE

The electronic structure of pc and of Lu pc is studied using an INDO method.[11]

A quantum chemical study is made of the electronic structure of Si dihydroxy pc and of its oligomers with siloxane bonds.[12] The electronic structure of pc is studied by the restricted Hartree-Fock method.[13]

The geometric configurations and electronic structure of lanthanide pc Ln(III) pc Cl complexes are calculated by the INDO method.[14]

A prescription for the stabilization of ferromagnetic exchange in molecular solids via admixing of the ground state with a virtual charge transfer excited state devolves upon the Hubbard-McConnell model and relates to pc metal complexes.[15]

Electron microscope high-resolution images of chlorinated Cu pc are obtained from which the degree of ionization can be determined.[16]

It is proposed that the chloride ion in Fe pc · HCl is coordinated to Fe and the HCl proton is located on one of the nitrogen atoms coordinated to Fe.[17]

III. MOLECULAR ORBITAL (MO) CALCULATIONS

The MO theory of metal ligand interaction is examined in terms of charge and spin transfer between metal and ligand atoms for Co pc.[18]

MO calculations and reaction field calculations of solvation by di hydrogen pc is made, using the improved two-parameter omega technique (ITPOT).[19]

MO calculations are made in relation to tautomerism and the electron configuration of di hydrogen pc.[20]

Closed shell crystal MO calculations are made on di hydrogen pc.[21]

The electronic and redox properties of stacked ring Si pcs are studied from MO theory. Cyclic voltammograms for Si pc monomer-oligomer sets can be predicted from initial stage MO energy levels if field shifts are used to allow for the charges of the anions and the cations.[22]

IV. EXCIPLEXES

Formation of exciplexes in the deactivation of the triplet state of Mg pc by nitroxyl radicals is studied. Triplet quenching of Mg tetra-4-*tert*-butyl pc by nitroxyls is examined.[23]

The lifetime of the triplet exciplexes of porphyrins with electron acceptors in toluene solution is measured. The quenching kinetics of complexes of tetra-4-*tert*-butyl pc with pyridyl Mg, AlOH, Al Cl, GaOH, InOH, and TiO are studied.[24]

V. LANGMUIR-BLODGETT (LB) FILMS

True LB monolayers and multilayers of a lightly substituted pc are prepared. The films are tough, adherent, and ultrathin of about 17 Å thickness. Such layers may permit the first really practical electronic devices that contain LB films.[25]

LB films of tetra-4-*tert*-butyl pc silicon dichloride are prepared. Their uniformity is reflected in their ability to sustain high electric fields.[26]

Cu and Zn tetra-*tert*-butyl pcs form LB monolayers on a water surface. In-plane conduction is ohmic with a resistivity of about 10,000-Ω-m.[27]

Pc derivatives are prepared for deposition as thin films by the LB technique. They include tetraphenoxy, dicumylphenoxy, tetracumylphenoxy, tetraoctadeoxy, and tetraneopentoxy pcs.[28]

The preparation and characterization of LB monolayers of Cu octa(dodecoxymethyl) pc are discussed. The monolayers of this pc on water are stable to film pressures of >30 dynes/cm.[29]

Ultrathin layers from polymeric pcs are prepared by the LB technique. Films of tetramethoxytetraoctyloxy pc siloxane are prepared by ferric chloride catalyzed monomer condensation in a chloroform layer spread over a water surface.[30]

The preparation, association in solution, and LB monomolecular layer formation are described for di hydrogen, Bi, Co, Cu, Mg, Ni, Pb, Pd, Pt, and Zn pc complexes (see Figure 1). These compounds form stable LB molecular layers to a film pressure of >20 mN/m.[31]

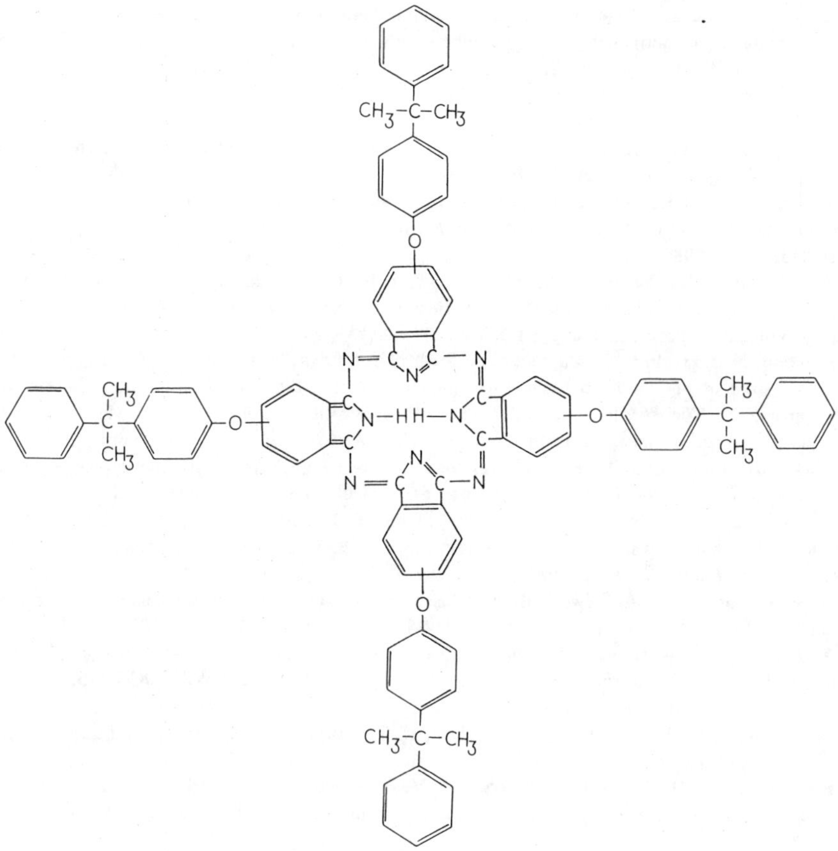

FIGURE 1. M pcs which form stable LB monomolecular layers.

REFERENCES

1. **Schneider, O., Moll, J., Hanack, M., and Krug, D.,** Thermoanalytical investigations on release of oxygen from a heme model, *Therm. Anal. Proc. 7th Int. Conf.*, 2, 836, 1982.
2. **Volodina, G. B., Borodkin, V. F., and Mityushina, V. I.,** Thermal Stability of Some Axial Complexes of Aluminum Phthalocyanines, Deposited Doc. 1982, SPSTL 1326 Khp-D82, 6 pp. Avail. SPSTL.
3. **Starke, M.,** IR spectroscopic and thermogravimetric studies of tin(IV) phthalocyanine, *Z. Chem.*, 24(12), 449, 1984.
4. **Woehrle, D. and Schulte, B.,** Polymeric phthalocyanines and their precursors. X. Thermal stability of polymeric phthalocyanines and their low molecular analogs, *Makromol. Chem.*, 186(11), 2229, 1985.
5. **Zhan, M., Zheng, B., and Gu, X.,** Thermal stabilities of metallophthalocyanines, *Xiamen Daxue Xuebao, Ziran Kexueban,* 25(2), 192, 1986.
6. **Starke, M.,** Thermogravimetric studies of selected metal phthalocyanines, *Wiss. Z.-Tech. Hochsch. Karl-Marx-Stadt,* 28(2), 238, 1986.
7. **Toyotama, H. (to Research Development Corp. of Japan),** Europe Patent Application 209,403, 1987.
8. **Mrwa, A., Starke, M., and Hamann, C.,** Determination of the vapor pressure of lead phthalocyanine, *Z. Phys. Chem. (Leipzig),* 265(3), 577, 1984.
9. **Volkova, A. V.,** Acoustic properties of copper phthalocyanine, *Zh. Fiz. Khim.*, 59(6), 1553, 1985.
10. **Stanton, J. L.,** Structural and Physical Studies of a Series of Partially Oxidized Transition Metal Phthalocyanines, Ph.D. thesis, Northwestern University, Evanston, IL, 1985.
11. **Li, Z., Ni, J., Li, L., and Xu, G.,** Electronic structure of phthalocyanine and its complex with lutetium, *Fenzi Kexue Yu Huaxue Yanjiu,* 4(3), 281, 1984.

12. **Semenov, S. G. and Sigolaev, Y. F.,** Quantum chemical study of the electronic structure of dihydroxo(phthalocyaninato)silicon and its oligomers with siloxane bonds, *Koord. Khim.*, 11(12), 1635, 1985.
13. **Kuz'mitskii, V. A., Klimko, G. T., Mestechkin, M. M., and Poltavets, V. N.,** Calculation of the electronic structure of metalloporphyrin dianions by the restricted Hartree-Fock method, *Teor. Eksp. Khim.*, 22(2), 153, 1986.
14. **Li, Z., Ni, J., Xu, G., and Li, L.,** Molecular configuration and electronic structure of Ln(III)PcCl complexes, *Wuji Huaxue*, 2(2), 8, 1986.
15. **Miller, J. S. and Epstein, A. J.,** Prescription for stabilization of ferromagnetic exchange in molecular solids via admixing of the ground state with a virtual charge-transfer excited state, *J. Am. Chem. Soc.*, 109(13), 3850, 1987.
16. **Fujiyoshi, Y., Ishizuka, K., Tsuji, M., Kobayashi, T., and Uyeda, N.,** Charge density distribution from high-resolution molecular images, *Lawrence Berkeley Lab. [Rep.] LBL* LBL-16031, Proc. 7th Int. Conf. High Voltage Electron Microscopy, 1983, 21.
17. **Bundina, N. I., Gulina, T. Y., Iodko, S. S., Mekhryakova, N. G., Kaliya, O. L., and Luk'yanets, E. A.,** Structure of "chloro-containing iron phthalocyanine", *Zh. Obshch. Khim.*, 54(5), 1199, 1984.
18. **Coppens, P., Koritsanzky, T., and Becker, P.,** Transition metal complexes: what can we learn by combining experimental spin and charge densities?, *Chem. Scripta*, 26(3), 463, 1986.
19. **Villar, H. O., Cachau, R. E., and Castro, E. A.,** An improved two-parameter omega technique for molecular orbital calculations. VII. Pi-reaction field calculations, *Acta Sud Am. Quim.*, 3(2), 67, 1983.
20. **Abeysekera, A. M., Grigg, R., Malone, J. F., King, T. J., and Morley, J. O.,** Chemical and theoretical studies of tautomerism in meso-aza- and meso-2,3,7,8,12,18-hexamethylporphyrinatocobalt(II), *J. Chem. Soc., Perkin Trans. 2*, 3, 395, 1985.
21. **Honeybourne, C. L. and Ewen, R. J.,** Closed-shell crystal orbital calculations on columnar stacks of conjugated macrocyclic ligands, *J. Phys. Chem. Solids*, 45(4), 433, 1984.
22. **Anderson, A. B., Gordon, T. L., and Kenney, M. E.,** Electronic and Redox Properties of Stacked-Ring Silicon Phthalocyanines from Molecular Orbital Theory, Report 1984, CWRU/DC/TR15, Order No. AD-A147161/4/GAR, National Technical Information Service, Springfield, VA, 28 pp.
23. **Degtyarev, L. S., Kapinus, E., and Skuridin, E. Y.,** Formation of exciplexes in the deactivation of the triplet state of magnesium phthalocyanine by nitroxyl radicals, *Khim. Vys. Energ.*, 18(1), 56, 1984.
24. **Kapinus, E. I., Aleksankina, M. M., Staryi, V. P., Boghillo, V. I., and Dilung, I. I.,** Lifetime of the triplet exciplexes of porphyrins with electron acceptors in toluene solution, *J. Chem. Soc., Faraday Trans. 2*, 81(5), 631, 1985.
25. **Kovacs, G. J., Vincett, P. S., and Sharp, J. H.,** Stable, tough, and adherent Langmuir-Blodgett films: preparation and structure of ordered, true monolayers of a phthalocyanine, *Can. J. Phys.*, 63(3), 346, 1985.
26. **Hua, Y. L., Roberts, G. G., Ahmad, M. M., Petty, M. C., Hanack, M., and Rein, M.,** Monolayer films of a substituted silicon phthalocyanine, *Philos. Mag. B*, 53(2), 105, 1986.
27. **Hann, R. A., Gupta, S. K., Fryer, J. R., and Eyres, B. L.,** Electrical and structural studies on copper tetra-*tert*-butyl phthalocyanine Langmuir-Blodgett films, *Thin Solid Films*, 134(1—3), 35, 1985.
28. **Barger, W. R., Snow, A. W., Wohltjen, H., and Jarvis, N. L.,** Derivatives of phthalocyanine prepared for deposition as thin films by the Langmuir-Blodgett technique, *Thin Solid Films*, 133, 197, 1985.
29. **Kalina, D. W. and Crane, S. W.,** Langmuir-Blodgett films of soluble copper octa(dodecoxymethyl)phthalocyanine, *Thin Solid Films*, 134(1—3), 109, 1985.
30. **Orthmann, E. and Wegner, G.,** Preparation of ultrathin layers with molecularly controlled buildup from polymeric phthalocyanines with the Langmuir-Blodgett technique, *Angew. Chem.*, 98(12), 1114, 1986.
31. **Snow, A. W. and Jarvis, N. L.,** Molecular association and monolayer formation of soluble phthalocyanine compounds, *J. Am. Chem. Soc.*, 106(17), 4706, 1984.

Chapter 18

COATINGS

I. DISPERSIONS

Pigment dispersions contain Cu phthalocyanine (pc), epoxy resins, xylene, a xylene solution of Epikote 1001, and dispersing agents mixed with a varnish to give a clear, glossy coating.[1]

Similar pigment dispersions contain Cu pc, a dispersing assistant derived from aromatic compounds, Monopet SB, sucrose benzoate, and xylene which are mixed with a varnish to prepare a clear, glossy coating.[2] Cu pc is mixed with Tohesta L010 polyester oligomer, a dispersing assistant based on aromatic compounds, xylene, and compounded with an alkyd resin varnish to give a clear, glossy coating.[3]

Polyether polyol derivatives are excellent dispersants for finely powdered Cu pc pigments in water.[4]

Pigment coating compositions containing partially oxidized nitrocellulose and pc derivatives exhibit an improved dispersibility.[5]

A mixture of cobalt oxide and alumina is mixed with Cu pc and with vinyltriethoxyethane and water to give a composition which, when mixed with Beckosol J524 IM60 alkyd resin and Super-Beckamine J820 melamine resin, gives a paint with excellent dispersion.[6]

β-Cu pc is added to polyester-melamine lacquer, combined with white polyester-melamine lacquer containing titania, and diluted with solvents, to give a violet-brown coating with a metallic luster, the color of which changes from violet to blue-gray at different observation angles.[7]

Cu pc Cl_8 $[SO_2NHCH_2CH_2N(C_2H_5)_2]_2$, highly chlorinated Cu pc, and an alkyl melamine varnish when ball milled give a nonagglomerating pigment composition for coatings.[8]

Pigment compositions for coatings with good dispersibility and resistance to flocculation contain Cu pc, chloro Cu pc, Cu pc monosulfonic acid, and calcium resinate.[9]

Epoxidized polyester derivatives are effective dispersing agents for Cu pc in thermosetting paints.[10]

An aqueous dispersion contains Cu pc colorants, carbon black, and humic acid mixed with a resin solution.[11]

Cu pc pigments and dispersions contain lactone-modified vinyl polymer which provide a mill base with good fluidity and dispersion stability.[12]

Acrylic acid-lauryl methacrylate-vinyl pyrrolidone copolymer salts are dispersing agents for Pc Green 6YK pigment in water thinned coatings.[13]

Epoxidized poly(ricinoleic acid) derivatives are dispersing agents for Co pc and Cu pc in organic liquids.[14]

A nonagglomerating, noncrystallizing pigment dispersant is prepared by reaction of Cu pc tetraamine with cyanuric chloride and amino alcohols or diamines.[15]

Pigment flushing of Cu pc from aqueous suspensions is actuated by trichlorotrifluoroethane.[16]

Anionic oligosoaps, having carboxylic-hydroxyethylamide groups, provide excellent dispersibility for α-Cu pc.[17]

A stable aqueous suspension of Cu pc is made by mixing Cu pc and N-methylpyrrolidone with Jar 80 (oxyethylated fatty acids and an anion-active surfactant), and diluted with water.[18]

Cationic copolymers are dispersing agents for Cu pc pigments in aqueous dispersions.[19]

Acrylic polymers are dispersants for Cu pc pigments in the manufacture of aqueous coatings with high gloss and good adhesion.[20]

Acrylated unsaturated polyester high gloss coatings contain highly dispersed Cu pc.[21]

Block copolymer polyoxyethylene oxypropylene is a dispersing agent for Cu pc pigments.[22]

Cu pc is easily dispersible in water-based vehicles with the Cu pc pigment adsorptively coated with 2 to 15% bis[poly(oxyethylene)]1,1,4,4-tetraalkyl-2-butyn-1,4-diol ether.[23]

Chlorosulfonated Cu pc reaction products with aminotriazine derivatives are dispersants for pigments for inks and coatings.[24]

Cu pc pigment preparations compatible with dispersion paints contain nonionic surfactants such as polyoxyethylated condensate of cyclohexylamine.[25]

Pc blue (Cu pc) wet cake is homogenized with sodium hydroxide containing N-lauroyl lysine, neutralized with hydrochloric acid, filtered, washed, and dried. The resulting pigment shows improved dispersibility in linseed oil varnish.[26]

Cu pc increases its dispersibility in water with a low temperature plasma.[27-29]

Nonaqueous Cu pc pigment dispersions, useful in paints and inks, are manufactured by mixing aqueous slurries containing 0.1 to 20% pigment and a polymeric or nonpolymeric dispersant, with nonaqueous media under vigorous agitation.[30]

Aminotriazine derivatives are dispersing agents for Cu pc pigments. The dispersant and dispersed pigment form the basis for preparing inks and paints.[31]

Cu pc is a pigment in fluorocarbon coatings.[32]

Silicone polymer-coated Cu pc pigments exhibit improved hydrophobicity and dispersibility for coatings and cosmetics.[33]

A paint-dispensing apparatus is designed to disperse Cu pc pigments in polymeric media.[34]

A review with 43 references emphasizes the dispersion of Cu pc and titania pigments in apolar media.[35]

II. FORMULATIONS

Dispersed Cu pc pigments are sealed in small packs and added to aqueous white coating materials to give desired colors.[36]

The resistance of enamels to aging as a function of pigmentation decreases in the order irgazin yellow > heliogen blue (Cu pc) > monastral green.[37]

The influence of pigments on tinter performance is illustrated using Cu pc pigment. The relationship is stressed between grinding intensity and tinter strength and hue.[38]

Compositions of polyolefins, Cu pc, and metal deactivators exhibit excellent thermal degradation resistance, and are especially useful as wire coatings.[39]

A lacquer for covering metal parts of a vehicle chassis contains β-Cu pc.[40]

Decontaminable epoxy resin flooring containing Cu pc is used in a nuclear installation.[41]

Nondusting pigment-solid resin concentrates contain melamine resin and Cu pc.[42]

BSD coated blue pigment consists of mica and talc as the core and Cu pc and inorganic salts as the coating.[43]

Modifying the surface of titania pigment with dimethyldichlorosilane increases the colorfastness of alkyd resin coatings containing Cu pc.[44]

Cu pc pigments useful for cosmetics are prepared by precipitating polymeric binders containing the Cu pc on flaky mica supports in suspension.[45]

Cyclic organosiloxane polymers impart hydrophobicity to powdered Cu pc.[46]

Resins colored with Cu pc are vacuum plated with 20 to 250 Å thick metals and topcoated with an acrylic urethane.[47]

Pc blue and Pc green are pigments in cationic acrylic electrophoretic coatings on stainless steel sheets.[48]

Discoloration and fade-resistant UV cured acrylic coatings contain Cu pc sulfonic acid.[49]

The antiflocculating capacity of modified Cu pc additives in pc-pigmented alkyd or melamine resin coatings is higher when the additives are nonassociated and can interact with the vehicle.[50]

Colored flaky pigments are made by forming a dispersion of mica and Cu pc.[51]

A preformed melamine resin container is printed with a composition comprising urea-guanamine resin, lactic acid, Cu pc, and ethyl hydroxyethyl cellulose, and compression molded to give a final product.[52]

Pc blue, montan wax acid ethylene glycol ester, and powdered carbonate are mixed to prepare a free flowing granulated concentrate.[53]

An aqueous paste useful to coat natural and synthetic leather and useful in textile printing contains Cu pc.[54]

High-solids oil-free polyester industrial protective coatings for metals coated by hot tank dipping contain Co pc.[55]

Colored unsaturated polyester or acid-curable amino alkyd coating compositions in which the curing agents and catalysts do not cause the color to change or fade contain M pc dyes such as Cu pc -$[SO_2NHCH_2CH(C_2H_5)(C_4H_9)]$.[56]

A nickel-plated mild iron is electrodeposited with an aqueous anionic acrylic coating containing Cu pc to form a transparent blue coating.[57]

A nonpolluting and safe acrylic aerosil coating may contain a Cu pc colorant.[58]

Granular mineral or ceramic aggregates are colored by Cu pc in a synthetic resin emulsion.[59]

Cu pc pigments have their surface modified with aluminum chelate and diketones for improved dispersibility in organic coatings.[60]

Heat-resistant Cu pc pigment concentrates contain silane-treated fillers for polyethylene moldings.[61]

Metallic coating compositions contain Cu pc and photocurable resin and photocuring catalyst. The composition gives a metallic pale blue coating.[62]

Pc green pigment particles are covered with crosslinked polymers, useful as electrical insulating colorants for PVC wire coatings.[63]

Cu pc pigment particles coated with cured polymers are electrically insulating and are useful to color PVC wire coatings.[64]

Carbon fibers are immersed in an aqueous solution of Cu pc trisulfonic acid as a basis to make electrically coated fibers.[65]

Colored pigments and fillers with diverse shades are made by coating titania with 1 to 5% Cu pc and 2 to 20% metal or silica.[66]

UV-curable electrophoretic coatings contain Cu pc pigments and acrylic polymers.[67]

A coating suitable for coating wood, plastic, paper, or cardboard includes Cu pc and a mixture of two or more liquids from among glossy lacquer, turpentine oil, aromatic petroleum solvents, and/or flat wall paint.[68]

A steel panel is coated with an opalescent coating including pc green and pc blue, titania, and lampblack.[69]

A sharp blue color filter is made from polyimide precursor solution from 4,4′-diaminodiphenyl ether, 3,3′,4,4′-biphenyltetracarboxylic dianhydride in *N*-methyl-2-pyrrolidone mixed with Cu pc and coated on a glass plate.[70]

An optical filter is also made from polyimide precursor solution from 4,4′-diaminodiphenyl ether and benzophenonetetracarboxylic dianhydride mixed with Cu pc.[71]

Heat-resistant colored pastes for optical color filters are made from 4,4′-diaminodiphenyl ether, Cu pc, *N*-methyl-2-pyrrolidone, and benzophenonetetracarboxylic dianhydride to form a varnish coated on a glass plate.[72]

A heat-resistant varnish for color filters contains benzophenonetetracarboxylic dianhydride-3,3′-diaminophenyl sulfone copolymer and Cu pc.[73]

III. CU PC PIGMENTS AND DYES

Pigment grade Cu pc is prepared by milling crude Cu pc in the presence of 0.5 to 30% Cu trichlorophenoxy pc.[74]

Cu pc agglomerates with primary particles <0.1 m are stirred with n-pentane at 90°C and 5 bar. The pentane is distilled off to give finely divided β-Cu pc.[75]

Crude Cu pc is conditioned by milling in the presence of 5 to 15% phthalimidomethyl pc or sulfonated phthalimidomethyl pc but without salt or solvent. The product can be used directly in printing ink or paint formulations.[76]

Crude β-Cu pc, phthalimidomethylated Cu pc, sodium chloride, diethylene glycol, and Perclene are milled, taken up in 1% sulfuric acid, filtered, washed, and dried to give a pigment with coloring power compared with pigment prepared with phthalimidomethylated Cu pc.[77]

Chlorinated Cu pc is ground in the presence of corn syrup and in the absence of salts.[78]

β-Cu pc from solvent-free production is refined by heating in 5 to 70% sulfuric acid at ≥50°C to improve its dispersibility and brilliance, and to remove a greenish shade.[79]

Crude Cu pc pigment conditioned under nitrogen gives purified β-Cu pc with better coloring power and clearness than a control conditioned in air.[80]

Crude chlorinated Cu pc pigment in a low-crystalline state is milled in an aqueous medium at pH 7 to 14 containing a separate organic phase to obtain pigment grade product.[81]

A greenish sky blue pigment is prepared by mixing Cu pc with highly chlorinated Cu pc. It is much less costly to prepare than H_2 pc but its tinting strength is 15 to 20% lower.[82]

Mill base letdown and tinting problems, changes in storage, and color separation in paints and tinters containing Cu pc are discussed.[83]

Crude Cu pc is dry pulverized with >60% reduction in bulk to give a crude Cu pc that can be conditioned readily.[84]

Crude Cu pc is ground in an aqueous medium containing a pc derivative such as Cu pc -$[CH_2N(C_3H_7)_2]_3$ and a polyester chain containing compound as the dispersing agent to give the pigment.[85]

An aqueous paste of a pc pigment is dispersed with rosin derivatives to produce a commercial dry pigment.[86]

An aqueous solution of 70:30 sodium chloride-polyethylene glycol is spray dried to give a powder with 4 to 6μ average particle size, suitable for grinding Cu pc.[87]

Cu pc coatings are applied to the walls of polymerization reactors for scale prevention of unsaturated compounds containing chelating agents and metal compounds.[88]

The walls of a polymerization reactor are coated with Cu pc disulfonic acid containing phosphorus compounds to prevent scale formation.[89]

Scale-preventing coatings on reactor walls contain H_2 pc; the reactor is involved in vinyl chloride polymerization.[90,91]

IV. TESTS AND MEASUREMENTS

The dispersibility of pigments including Cu pc and powdered PVC in vehicles is controlled by wettability. The color stability of paints in storage is related to the stability of pigment dispersions in vehicles.[92]

Sorenson's acid-base concept is useful to explain Cu pc pigment dispersibility in vehicles containing vinyl acetate-vinyl chloride copolymer.[93]

PVC coatings pigmented with Cu pc or pc green baked on phosphated steel are subjected to outdoor weathering. Degradation is observed on the surface of Cu pc without crystal structure changes, while pc green undergoes similar surface degradation with crystal structure changes by addition reaction with chlorine.[94]

The weatherability of alkyd-melamine metallic coatings containing aluminum and organic pigments shows that the changes in coatings containing a tetrachloroisoindolinone pigment are minimal compared with rapid deterioration in coatings containing Cu pc.[95]

Stable aqueous dispersions of pc, Cl pc, and M pc pigments suitable for spectrophotometer evaluation of the pigment color are prepared by adding a solution of pigment in 90 to 100% sulfuric acid into an aqueous solution of a nonionic dispersing agent.[96]

The UV-visible absorption spectra of Cu pc and polychlorinated Cu pc (C. I. pigment green 17, with 15 to 16 chlorine atoms per pc molecule) are determined in potassium bromide tablets and in 1-hydroxy-2-ethoxy ethane solution.[97]

An apparatus is designed to determine the flocculation gradient of titania in alkyd resin coatings containing Cu pc.[98]

IR spectroscopy is used to determine the adsorption of alkyd resin on Cu pc and pc green.[99]

Pc pigments in paints are identified easily by solution spectrophotometry.[100]

The light resistance of epoxy resin coatings containing Cu pc pigments is measured by an accelerated test by a gas laser at 0.5 to 0.6 μ.[101]

The effect of surface plasmon resonance in a silver island on the absorption and Raman scattering of a Cu pc dye coating is discussed.[102]

A computer program is developed to study the light scattering-induced color fading in latex paint films by using titania colored Cu pc blends as references.[103]

The particle size distribution of Cu pc pigments is determined by Fraunhofer diffraction.[104]

V. GLASS

Five parts glass fibers are slurried in water at 90°C and 0.15 parts of a 25% solution of Cu pc -$[CH_2N^+(C_2H_5)_3Cl^-]_3$ are added. After 20 min the mixture is diluted with cold water to 0.01% fibers, forming a washfast turquoise blue fibers mat.[105]

A useful method to form an electrically conductive fiber on silica particles that is useful, in turn, as fillers with controlled electrical properties or as flowing electrodes involves mixing hollow, fused silica spheres with Zn pc, heating to 800°C, and maintaining at 800°C in a nitrogen atmosphere, and cooling. This yields carbon coated spheres containing a film with carbon 77.20, hydrogen 1.77, nitrogen 15.94, and zinc 5.3%, with a bulk resistance of 385 Ω.[106]

VI. PAPER PRODUCTS

Addition of

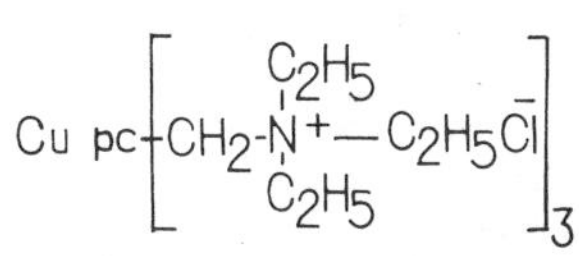

cationic dyes to inorganic pigment dispersions containing binders gives products to coat paper, such as 80 g/m^2 basis weight, with good lightfastness.[107]

Mixtures of turquoise pc and substantive violet azo dyes containing cationic or basic groups give clear, neutral blue dyeing with brilliance on paper. Such a dye is Cu -$[SO_2NHCH_2CH_2N(CH_3)_2]_{3.1}$.[108]

A colorimetric analysis of wallpaper dyes with suspensions of VS0-brand organic pigments includes Cu pc.[109]

Colored paper is made by coloring resin modified paper with Cu pc disulfonic acid.[110]

Dyes for coloring paper include

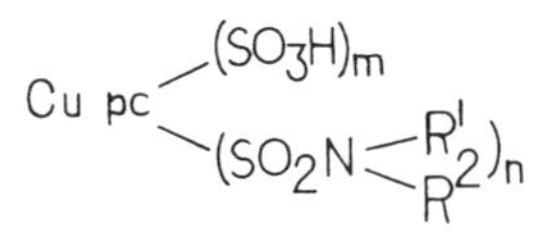

where R′,R may = H, and m,n = 1.[111]

On-line control of color including Cu pc disulfonic acid with direct dyeing of paper in closed systems is discussed.[112]

Bagasse pressboard coated with a pc green pigment containing resin has good quality surface finish properties.[113]

Cu pc -(SO_2NRZN^+-$R'R^2$ R^3 X^-)$_{2.2}$ where $N^+R'R^2R^3$ = *N*-methylmorpholinium, R = H, Z = alkylene bridge, and X = CH_3 SO_4, dyes paper in clear turquoise shades with good bleachability.[114]

REFERENCES

1. **Too Ink Manufacturing Co., Ltd.,** Japan Patent 83,225,159, 1983.
2. **Toyo Ink Manufacturing Co., Ltd.,** Japan Patent 83,225,160, 1983.
3. **Toyo Ink Manufacturing Co., Ltd.,** Japan Patent 83,145,762, 1983.
4. **Kao Corp.,** Japan Patent 84,19,530, 1984.
5. **Ishino, T., Ueda, I., Kai, R., and Nagatomo, S. (to Asahi Chemical Industry Co., Ltd.),** Europe Patent Application 107,867, 1984.
6. **Ukima Color and Chemicals Manufacturing Co., Ltd.,** Japan Patent 84,223,757, 1984.
7. **Fabian, W., Bock, G., and Radtke, V. (to BASF AG),** German Patent 3,306,400, 1984.
8. **Toyo Ink Manufacturing Co., Ltd.,** Japan Patent 84,168,070, 1984.
9. **Smrchek, V., Sinitsyna, I. N., Sergeeva, Z. I., Makarovskaya, G. M., Kurakin, E. M., Kiselev, A. V., and Orel, N. I.,** East German Patent 210,547, 1984.
10. **Schofield, J. D. (to Imperial Chemical Industries PLC),** Europe Patent Application 127,325, 1984.
11. **Ricoh Co., Ltd.,** Japan Patent 84,100,168, 1984.
12. **Nippon Oils and Fats Co., Ltd.,** Japan Patent 84,174,620, 1984.
13. **Kansai Paint Co., Ltd.,** Japan Patent 85,92,360, 1985.
14. **Schofield, J. D. (to Imperial Chemical Industries PLC),** Brazil Patent 84,02,502, 1985.
15. **Toyo Ink Manufacturing Co., Ltd.,** Japan Patent 85,88,185, 1985.
16. **Murakami, T. and Ishikawa, N. (To Daikin Kogyo Co., Ltd., Daido Kasei Kogyo Ltd., Daido Kasei Kogyo Co., Ltd.),** German Patent 3,439,488, 1985.
17. **Suzuki, S. and Ishigami, Y.,** Dispersability of pigments in aqueous solutions of anionic oligosoaps. Synthesis of anionic oligosoaps having the carboxylic-hydroxyethylamide groups and their properties, *Shikizai Kyokaishi,* 58(11), 633, 1985.
18. **Gemzova, I. and Novak, A.,** Czech Patent 218,370, 1985.
19. **Aihara, T. and Nakayama, Y. (To Kansai Paint Co., Ltd.),** Japan Patent 85,217,271, 1985.
20. **Aihara, T., Nakayama, Y., and Umeyama, K. (to Kansai Paint Co., Ltd.),** Japan Patent 85,188,472, 1985.
21. **Kumada, H. and Maruyama, K. (to Dainippon Ink and Chemicals, Inc.),** Japan Patent 86,23,652, 1986.
22. **Kawashima, Y. (to Tokyo Ink Manufacturing Co., Ltd.),** Japan Patent 85,260,665, 1985.
23. **Abe, K., Miki, T., and Maekawa, K. (to Sanyo Color Works, Ltd.),** Japan Patent 85,262,861, 1985.
24. **Ehashi, S., Sakamoto, M., and Hikosaka, M. (to Tokyo Ink Manufacturing Co., Ltd.),** Japan Patent 86,246,261, 1986.

25. **Aign, V., Pusch, N., Wienkenhoever, M., Paulat, V., and Schubert, K. (to Bayer AG),** German Patent 3,617,010, 1987.
26. **Sagawa, K., Yokota, H., and Takehara, M. (to Ajinomoto Co., Inc.),** Japan Patent 86,136,555, 1986.
27. **Ihara, T., Ito, S., Tanaka, M., and Kiboku, M.,** Oxidation treatment of organic pigments by low temperature plasma, *Shikizai Kyokaishi,* 59(12), 732, 1986.
28. **Kinki University, Nippon Paint Co., Ltd.,** Japan Patent 84,199,764, 1984.
29. **Ihara, T., Ito, S., and Kiboku, M.,** Low temperature plasma oxidation treatment of several organic pigments, *Chem. Lett.,* 5, 675, 1986.
30. **Vinther, A., Olsen, S. C., Larsen, B. E. H., Madsen, E. S., and Madsen, B. (to Kemisk Vaerk Koege A/S),** Europe Patent Application 219,323, 1987.
31. **Katsura, H., Ehashi, S., Kashioka, M., and Sakamoto, M. (to Tokyo Ink Manufacturing Co., Ltd.),** British Patent Application 2,173,812, 1986.
32. **Kuwamura, S., Oka, M., and Ozawa, H.,** Japan Patent 86,275,364, 1986.
33. **Fukui, H., Namba, R., Saito, T., Ohtsu, T., Kimura, A., Nakano, M., Nakata, O., Tomita, K., Tokubo, K., et al. (to Shiseido Co., Ltd.),** Europe Patent Application 212,870, 1987.
34. **Nippon Paint Co. Ltd.,** Japan Patent 85,122,033, 1985.
35. **McKay, R. B.,** Pigment dispersion in apolar media, *Surfactant Sci. Ser.,* 21, 361, 1987.
36. **Kanpe Katei Toryo, K. K.,** Japan Patent 84,04,660, 1984.
37. **Loev, A. M., Katsadze, S. V., and Pipaishvili, M. N.,** Atmospheric aging of metal effect enamels, *Atmosferostoikie Lakokras, Pokrytiya Prognozirovanie Srokov Ikh Sluzhby, Mater. Semin.,* 87, 1982.
38. **Cohen, E. and Pineiro, R.,** Factors governing tinter performance, *J. Coat. Technol.,* 55(704), 69, 1983.
39. **Dainichi Nippon Cables, Ltd.,** Japan Patent 84,147,036, 1984.
40. **Graetz, H. J. and Stranghoener, D. (to Herberts GmbH),** German Patent 3,314,777, 1984.
41. **Kunze, S., Loesch, G., and Schmidt, H. R.,** Decontaminability of an epoxy coating containing various pigments before and after abrasion and after radiation exposure, *Ind. Lackierbetr.,* 52(10), 379, 1984.
42. **Kelch, W., Reiner, K., Haegele, G. (to BASF AG),** German Patent 3,402,440, 1984.
43. **Shan, R.,** BSD-coated blue pigment, *Tuliao Gongye,* 1, 17, 1987.
44. **Shtanov, N. P., Shelyapin, O. P., Moroz, V. A., Kovalenko, S. A., Savishchenko, O. V., and Bondarenko, E. A.,** Fading of polymer coatings pigmented with organic pigments and dimethyldichlorosilane-modified titania, *Lakokras. Mater. Ikh Primen.,* 4, 63, 1987.
45. **Noguchi, T. and Watanabe, T. (to Merck GmbH),** Europe Patent Application 220,617, 1987.
46. **Fukui, H., Nanba, R., and Saito, T. (To Shiseido Co., Ltd.),** Japan Patent 86,268,763, 1986.
47. **Uchiyama, H. (to Kanto Seiki Co., Ltd.),** Japan Patent 87,13,565, 1987.
48. **Sakakibara, T., Inagaki, A., and Takeda, S.,** Japan Patent 87,104,880, 1987.
49. **Yokoshima, M. and Nakayama, K. (to Nippon Kayaku Co., Ltd.),** Japan Patent 86,31,449, 1986.
50. **Sabitov, M. V., Vil'ner, S. G., Shabanova, S. A., Smrchek, V. A., Tolstaya, S. N., Bulgakova, R. A., and Sokolova, N. I.,** Selection of a polycyclic dye-based modifier for stabilization of paint and varnish systems, *Lakokras. Mater. Ikh Primen.,* 3, 15, 1986.
51. **Nedorost, M., Palffy, A., Halamova, K., Jirakova, D., and Sladecek, L.,** Czech Patent 224,792, 1984.
52. **Yoshida, H. and Fukuhara, I.,** Japan Patent 86,141,520, 1986.
53. **Hoechst Holland NV,** Netherlands Application 84,02,535, 1986.
54. **Matysiak, R., Sekula, W., Zemlo, T., Gajda, W., Rybak, E., Organa, G., and Lecki, H.,** Poland Patent 127,761, 1985.
55. **Mostafa, Y. A.,** U.S. Patent 4,597,998, 1986.
56. **Hodogaya Chemical Co., Ltd.,** Japan Patent 85,47,073, 1985.
57. **Shimizu, S.,** Japan Patent 85,70,200, 1985.
58. **Atom Chemical Paint Co., Ltd.,** Japan Patent 84,232,151, 1984.
59. **Kikusui Kagaku Kogyo Co., Ltd.,** Japan Patent 85,106,865, 1985.
60. **Kawaken Fine Chemicals Co., Ltd.,** Japan Patent 85,79,068, 1985.
61. **Dainichiseika Color and Chemicals Manufacturing Co., Ltd.,** Japan Patent 84,49,268, 1984.
62. **Dainichiseika Color and Chemicals Manufacturing Co., Ltd.,** Japan Patent 84,15,459, 1984.
63. **Dainichiseika Color and Chemicals Manufacturing Co., Ltd.,** Japan Patent 84,75,936, 1984.
64. **Dainichiseika Color and Chemicals Manufacturing Co., Ltd.,** Japan Patent 84,75,937, 1984.
65. **Naarmann, H., Haaf, F., and Fahrbach, G. (to BASF AG),** German Patent 3,531,019, 1987.
66. **Nedorost, M., Palffy, A., Jirakova, D., Svoboda, L., and Halamova, K.,** Czech Patent 232,424, 1986.
67. **Shimizu, Y. and Fukuda, M. (to Shimizu Construction Co., Ltd.),** Japan Patent 85,208,376, 1985.
68. **Tscheligi, Z.,** Europe Patent Application 159,667, 1985.
69. **Panush, S. (to Inmont Corp.),** U.S. Patent 4,539,258, 1985.
70. **Miura, Y., Hiramoto, Y., and Maeda, K. (to Toray Industries, Inc.),** Japan Patent 85,184,203, 1985.
71. **Miura, Y., Hiramoto, Y., and Eguchi, M. (to Toray Industries, Inc.),** Japan Patent 85,184,204, 1985.

72. **Miura, Y., Hiramoto, Y., and Hirono, T. (to Toray Industries, Inc.**, Japan Patent 85,184,205, 1985.
73. **Miura, Y., Hiramoto, Y., and Saito, S. (to Toray Industries, Inc.)**, Japan Patent 85,184,202, 1985.
74. **Ciba-Geigy AG**, Japan Patent 86,296,071, 1986.
75. **Graser, F. and Wickenhaeuser, G. (to BASF AG)**, U.S. Patent 4,451,654, 1984.
76. **Johnson, S. L., McLaren, G., and Robertson, G. H. (to Sun Chemical Corp.)**, German Patent 3,331,998, 1984.
77. **Sumitomo Chemical Co., Ltd.**, Japan Patent 84,102,958, 1984.
78. **Flatt, T. R. (to Mobay Chemical Corp.)**, U.S. Patent 4,493,734, 1985.
79. **Stepanek, F., Velesovsky, F., Sebesta, J., and Stepanek, I.**, Czech Patent 235,666, 1986.
80. **Sakaguchi, I. and Okazaki, H. (to Sumitomo Chemical Co., Ltd.)**, Japan Patent 86,190,566, 1986.
81. **E. I. du Pont de Nemours & Co., Inc.**, Japan Patent 84,06,261, 1984.
82. **Kissin, B. I. and Volkova, G. N.**, *Lakokras. Mater. Ikh Primen.*, 5, 16, 1984.
83. **Daniel, F. K.**, The obstacle course from mill base to finished coating, *J. Coat. Technol.*, 56(708), 38, 1984.
84. **Funatsu, T., Hayashi, M., and Fujigamori, T. (to Tokyo Ink Manufacturing Co., Ltd.)**, Japan Patent 87,72,758, 1987.
85. **Dainippon Ink and Chemicals, Inc.**, Japan Patent 83,168,661, 1983.
86. **Aristov, B. G., Smrchek, V. A., Feizulova, R. K., Gorenko, V. N., Poznyakevich, A. I., Berezin, V. A., and Sosnin, G. A.**, U.S.S.R. Patent 1,079,655, 1984.
87. **Rakusan, J., Smrcek, V. A., Velesovsky, F., Ladyzenskij, J. B., Stepanek, I., Korobov, S. A., Murcek, V., Titov, V. V., Ruzicka, K., and Golubev, V. M.**, Czech Patent 238,764, 1987.
88. **Shimizu, T., Kaneko, I., and Shimakura, Y. (to Shin-Etsu Chemical Industry Co., Ltd.)**, Europe Patent Application 126,991, 1984.
89. **Shin-Etsu Chemical Industry Co., Ltd.**, Japan Patent 84,170,102, 1984.
90. **Koyanagi, S., Kitamura, H., Shimizu, T., and Kaneko, I. (to Shin-Etsu Chemical Co., Ltd.)**, Europe Patent Application 172,427, 1986.
91. **Koyanagi, S., Kitamura, H., Shimizu, T., and Kaneko, I. (to Shin-Etsu Chemical Industry Co., Ltd.)**, Japan Patent 86,51,001, 1986.
92. **Satoh, T.**, Correlation between color stability of paint systems and pigment-solvent wettability, *Org. Coat.*, 5, 147, 1983.
93. **Fujitani, T.**, Application of acid-base concepts in pigment dispersion, *Toso Kogaku*, 19(5), 178, 1984.
94. **Takeshima, E., Kawano, T., Takamura, H., and Awa, K.**, Study on the method of predicting service life of coated steel sheets. X. Degradation of phthalocyanine pigments in poly(vinyl chloride) resin coating, *Shikizai Kyokaishi*, 57(5), 239, 1984.
95. **Aslanova, N. V., Andryushchenko, E. A., and Baenkevich, V. V.**, Weatherability of alkyd-melamine metallic coatings, *Atmosferostoikie Lakokras. Pokrytiya Prognozirovanie Srokov Ikh Sluzhby, Mater. Semin.*, 15, 1982.
96. **Latinak, J., Kozeny, M., and Cepicky, M.**, Czech Patent 208,644, 1983.
97. **Fedeli, E. and Brillo, A.**, UV-VIS solid state spectrophotometric analysis: identification of organic pigments in coatings, *Riv. Ital. Sostanze Grasse*, 63(4), 221, 1986.
98. **Rutherford, D. J. and Simpson, L. A.**, Control of pigment dispersion in wet or dry paint films by a flocculation gradient measurement device, *Double Liaison-Chim. Peint.*, 31(348), 407, 1984.
99. **Feizulova, R. K., Aristov, B. G., and Frolov, Yu. G.**, Use of an infrared spectroscopy method to determine adsorption of alkyd resin on organic pigments, *Kolloidn. Zh.*, 48(2), 311, 1986.
100. **Kumar, R., Billmeyer, F. W., Jr., and Saltzman, M.**, Identification of organic pigments in paints, *J. Coat. Technol.*, 57(720), 49, 1985.
101. **Odinokov, B. N. and Khananina, T. S.**, Laser for studying light resistance of colored polymer coatings, *Izv. VUZOV. Stroit. Arkhit.*, 3, 118, 1984.
102. **Hayashi, S.**, Optical study of electromagnetic surface modes in microcrystals, *Jpn. J. Appl. Phys., Part 1*, 23(6), 665, 1984.
103. **Ghisolfi, G.**, Coloring latex paints of medium and high pigment volume concentration, *Pitture Vernici*, 60(12), 21, 1984.
104. **Odinokov, B. N.**, Diffraction method for the dispersion analysis of pigments using a laser, *Lakokras. Mater. Ikh Primen.*, 6, 33, 1983.
105. **Troesch, J. (to Sandoz GmbH)**, German Patent 3,412,762, 1984.
106. **Rickborn, S. F. and Rogers, D. Z.**, U.S. Patent 4,686,116, 1987.
107. **Troesch, J. and Portmann, W. (to Sandoz GmbH)**, German Patent 3,311,091, 1983.
108. **Beck, U., Stoehr, F. M., and Nickel, H. (to Bayer AG)**, German Patent 3,300,896, 1984.
109. **Smirnova, T. V., Smirnyagina, N. A., Karimova, A. Z., Belyaev, A. N., and Zaporozhets, N. Ya.**, Study of some colorimetric characteristics of wallpapers, *Issled. Fiz.-Khim. Svoistv Bum.*, 58, 1983.
110. **Shcherbakova, L. I. and Gorbushin, V. A.**, Manufacture of water-resistant and colored paper by surface treatment, *Issled. Fiz.-Khim. Svostv Bum.*, 51, 1983.

111. **Degen, H. J. and Kraus, H. (to BASF AG),** German Patent 3,622,363, 1987.
112. **Markert, J., Haener, A., and Traser, G.,** The use of direct dyes for coloring in closed systems, *Wochenbl. Papierfabr.,* 115(10), 424, 1987.
113. **Puig, J. L. and Leal, J. A.,** Pigment addition to resins for decorative boards, *Sobre Deriv. Cana Azucar,* 20(1), 26, 1986.
114. **Groll, M. and Ockelmann, D. (to Bayer AG),** German Patent 3,248,466, 1984.

Chapter 19

INKS

I. INK JET INKS

An ink for jet recording contains Cu phthalocyanine (pc) disulfonic acid and metal dithiocarbamate light stabilizers.[1,2]

A jet printing ink contains Cu *N,N,N′,N′*-tetrakis(29-hydroxy-3,6,9,12,15,18,21,24,27-nonaoxanonacos-1-yl) pc disulfonamide dye.[3]

A jet printing ink contains hydroxycarboxylic acid metal chelates and Cu pc disulfonic acid.[4]

A cyan jet printing ink contains tetra-Na Cu pc tetrasulfonate.[5]

A jet printing ink that has excellent solubility stability and eliminates nozzle plugging contains C. I. direct blue 86 Cu pc disulfonic acid 3, a water soluble phthalimide derivative 2, polyethylene glycol 200 10, amadine sodium stabilizer 0.2, and water 84.8 wt %.[6]

A recording ink for ink jet contains mainly water-soluble dyes, wetting agents, water, and xanthines. One such dye is Cu pc trisulfonic acid. The inks increase injection stability, equipment reliability, ink dropping effectiveness, and granulation stability.[7]

A jet printing ink contains Cu pc dyes substituted with alkali metal sulfonate groups.[8]

Another clogging-resistant jet printing ink that gives prints with excellent color tone contains Cu pc sodium tetrasulfonate and Ni pc sodium tetrasulfonate.[9]

A jet printing ink contains Cu pc -($SO_2NHCH_2CH_2OH)_3$, -(SO_3Na) water-miscible organic solvents, and water.[10]

A jet printing ink contains Cu pc disulfonic acid for multicolor images.[11]

Jet printing inks containing 0.1, 0.3, 0.7, 1.0, 2.0, and 4.0% Cu pc disulfonic acid having a minimum viscosity-maximum viscosity ratio of 0.87:1 give dots having a 184.0 μm minimum and 195.0 μm maximum value.[12]

A jet printing ink contains Cu pc disulfonic acid with <5 ppm calcium ions for good storage stability.[13]

A jet printing ink consists of an aqueous solution of reaction products of sulfonated Cu pc containing halotriazinyl groups with polyethylenimines.[14]

A jet printing ink contains Cu pc disulfonic acid, *N*-methyl-2-pyrrolidone, ethylene glycol, $CH_3OCH_2CH_2OH$, glycerol, and water.[15]

Anticlogging jet printing inks with high reflection density contain Cu pc disulfonic acid.[16]

A nonclogging ink for an ink jet printer contains water-soluble Cu pc disulfonic acid with an alkanolammonium compound.[17]

Fast drying, nonclogging, water-thinned jet printing inks contain oil-soluble dyes and Cu pc trisulfonic acid.[18]

A receptor-coated embossed paper is jet printed with an ink composition comprising ethylene glycol, water, and pc disulfonic acid to give a sharp pattern at <1 min drying time.[19]

A composition containing C. I. direct blue 87 (Cu pc trisodium trisulfonate) can be jet printed for 12 h without clogging and forms prints which dry in <3 s.[20]

A blue jet printing ink resistant to printer clogging during intermittent use contains

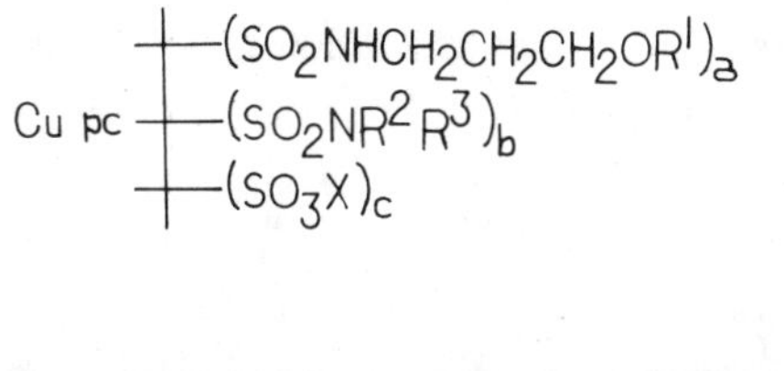

where R, R^1, R^2 may be H, Z^1 may be $CH_2(CH_2)_q$, R^3 may be $C_{6\text{-}18}$ (alkoxy)alkylamino, R^4 may be $C_{1\text{-}18}$ alkanolamino, m,n = 0 to 3, p = 1 to 4, m + n + p = 2 to 4, q = 1 to 4.[21]

A jet printing ink containing Cu pc disulfonic acid is treated with sound waves to remove dissolved air.[22]

A jet printing ink that forms a mark that dries in 3 s contains Cu pc disulfonic acid.[23]

A water-based jet printing ink contains Cu tetrasulfo pc. The ink forms sharp, fast-drying prints.[24]

A clogging-resistant aqueous ink for ink jet printing contains Cu pc disulfonic acid. It is storage stable for 2 years at 0 to 60°C.[25]

A jet printing ink containing Cu pc disulfonic acid (C. I. direct blue 86) containing a trace amount of inorganic salt impurities is printed, stopped 0.5 year, and reused without plugging a nozzle. At a higher percentage of inorganic salts, a nozzle is plugged after the printing is stopped 4 months.[26]

A jet printing ink contains Cu pc disulfonic acid.[27]

A jet printing ink containing digallic acid alkanolamine salts and Co pc disulfonic acid gives a stable printed image.[28]

Blotting-resistant aqueous inks contain Cu pc disulfonic acid for dot matrix printing.[29]

Cyan jet printing inks contain Cu pc disulfonic acid.[30-36]

Jet recording inks that contain water-soluble Cu pc disulfonic acid and fornyl (hydroxyethyl) hydrazine are stable and yield a dry image.[37]

Inks for wire-dot printers contain Cu pc -$(SO_3Na)_n$ where n = 2 to 4, and $\geqslant$1 triphenylmethane acid dye. Water blue 117-L, (mixture of Cu pc Na disulfonate and Cu pc trisulfonate) and Daiwa Green 70 (C. I. acid green 16) are among the components of a cyan ink.[38]

Jet printing inks contain pc among other dyes, with diethylene glycol, polyethylene glycol, *N*-methyl-2-pyrrolidone, and water.[39]

Full color images are formed from yellow and magenta dyes, and C. I. direct blue 86 (Cu pc disulfonic acid), diethylene glycol, glycerol, water, sodium dihydroacetate, and $Na_2S_2O_3$.[40]

Jet printing inks containing pc sulfonates show zero light fading.[41]

Quick drying jet printing inks contain C. I. direct blue 86.[42]

Inks for jet printing contain alkali metal salts of sulfonated Cu pc. The inks give characters that can be easily read after 50 h of xenon testing.[43]

Water-based storage stable jet recording inks contain pc dyes prepared from Cu pc, chlorosulfonic acid, thionyl chloride, and ethanolamine.[44]

Cu sulfo pc 1.5, Cu disulfo pc 0.5, Cu polysulfo pc 1.5, ethylene glycol 27, and water 73 are mixed to form a jet printing ink that retains >65% of their color intensity when immersed in water for 5 min 2 d after printing.[45]

Storage stable cyan jet printing inks contain Cu pc disulfonic acid.[46-48]

Cu pc tetrasulfonic acid dyes are components of aqueous jet inks.[49]

Jet printing inks containing a Cu pc dye and nonionic surfactants for jet printing of cellulosic fibers are storage stable and anticlogging.[50]

Phosphorus free storage stable and clogging-resistant jet printing inks contain Cu pc disulfonic acid.[51]

A jet printing ink giving good blue images and causing no nozzle clogging after long intermittent use contains a water-soluble Cu pc dye, polyethylene glycol, glycerol, and N-methylpyrrolidone.[52]

A cyan jet printing ink giving high-resolution inks contains Cu pc disulfonic acid 30 and water 70 parts. The surface tension is 54 dynes/cm[53] and the viscosity is $\leqslant$3.5 cp.[54]

Jet printing inks with excellent color tone contain a tetrasulfone pc and trisulfone pc dye.[55]

Cyan jet printing inks containing Cu pc disulfonic acid, diethylene glycol, N-methylpyrrolidinone, and water give a storage stable ink, the images of which are resistant to light, weathering, and water.[56]

Cu pc tetrasulfonate is extracted with methanol and used as a component in a jet printing ink.[57]

Jet printing inks contain C. I. acid yellow 23, C. I. acid red 92, and C. I. direct blue 86, to prepare yellow, magenta, and cyan inks.[58]

C. I. reactive blue 15 is treated with morpholine, hydrochloric acid, and sodium hydroxide to prepare a jet printing ink.[59]

A jet printing ink prepared from Cu pc disulfonic acid dye purified by a coagulating agent exhibits long-term storage stability and good image sharpness.[60]

Cu pc disulfonic acid quaternary ammonium salts are a component of clogging-resistant water thinned inks.[61]

Inorganic impurities are removed from Cu pc disulfonic acid and jet printing inks by contacting aqueous solutions of the dye with aminoalkylphosphonic acid chelating resins.[62]

Water-resistant dot images are formed on ink jet printing paper by ink jet printing with Cu pc trisulfonic acid. The paper incorporates controlled sizing, surface pH, and salt content.[63]

An ink contains Cu pc disulfonic acid for ink jet recording on heat fusible porous receptors for transparency and photomask preparation.[64]

An ink containing Cu pc disulfonic acid is used for ink jet recording on a receptor sheet.[65]

Ink jet recording materials with an ink absorbing layer contain Cu pc disulfonic acid for projection slide production.[66]

An ink contains Cu pc disulfonic acid for transparency production using an ink jet recording receptor sheet.[67-69]

An ink contains diethylene glycol and Cu pc disulfonic acid for ink jet recording on polyamide-coated polyester sheets for transparency preparation.[70]

Ink jet recording inks containing Cu pc disulfonic acid are prepared for continuous tone images.[71]

An ink jet recording sheet has a high ink absorbing capacity for inks including Cu pc disodium disulfonate.[72]

An impact transfer ink sheet contains liquid ink in its pores containing Cu pc complex.[73]

Cu pc disulfonic acid is purified for use in jet printing ink.[74]

A vinyl acetate-vinyl pyrrolidone copolymer is an ink-accepting layer in receptor material for ink containing Cu pc disulfonic acid for ink jet recording.[75]

An ink contains Cu pc disulfonic acid for ink jet recording.[76-78]

Cu pc disodium disulfonate is used in 2% concentration with 20% diethylene glycol and 10% polyethylene glycol for ink jet recording.[79]

A powdered ink contains copolymer and Cu pc for powdered ink jet printing.[80]

An ink contains Cu pc disulfonic acid for ink jet printing on amine containing receptor sheets.[81]

An ink contains Cu pc disulfonic acid for ink jet printing with a hydrophilic water-insoluble polymer ink receiving layer in the printing material.[82]

An ink jet printing ink contains Cu pc.[83]

An ink jet recording medium containing cationic polymers and amino alcohols is used for ink jet recording using an ink containing Cu pc disulfonic acid.[84]

An ink jet printing method includes Cu pc tetrasulfonic acid in the ink.[85]

Cu pc disulfonic acid is used in multicolor ink jet printing.[86]

Light transmitting substrates for ink jet recording contains cationically modified polyvinyl alcohol. The inks contain Cu pc disulfonic acid.[87]

An ink jet printing ink containing Cu pc disulfonic acid is printed on paper coated with both fine silica and an amino acid,[88] or coated with both fine silica and an alkali metal containing salt.[89]

II. MARKING INKS

A. PERMANENT INKS

A marking ink composition with excellent storability contains titania 100, kaolin 10, Cu pc 8, Haikoto FL-200 resin 60, ethanol 20, ethylene glycol 5, and Emulgen A-90 surfactant 5 parts dispersed in water. The mixture in a felt-tipped marker gives excellent hiding power, water resistance, and light stability.[90]

Vivid drying-resistant marking inks contain alcohol and/or glycol solvents, Pc Green NY, and sucrose esters which dry in <1 min.[91]

A triple color marking ink contains blocking-resistant Cu pc disulfonic acid.[92]

Cu pc pigment for a marking pen is coated with a phosphoric acid ester to improve the redispersibility. A pen filled with the ink composition exhibits smooth ink flow without blocking after 1-month storage.[93]

B. ERASABLE MARKING INKS

A rubber erasable ink contains Cu pc 5, poly(ethylene oxide) 15, polyoxyethylene alkylphenyl ether 0.5, glycerin 15, and water 64.5 g.[94]

An erasable marking ink contains Cu pc, alcohols, alcohol-soluble and silicone resins, and monobasic esters.[95]

An erasable marking ink contains Cu pc, alcohols, alcohol-soluble resins, and silicones soluble in the alcohols.[96]

An erasable marking ink contains Cu pc with betaine type methacrylate polymer binder.[97]

Erasable water-thinned inks contain Cu pc, fatty acid potassium salts, and surfactants.[98]

An erasable marking ink contains Cu pc, resins, solvents, and additives containing aliphatic dibasic acid esters and ≥ 1 compounds selected from sulfuric or phosphoric acid esters of polyoxyethylene alkyl ethers or polyoxyethylene alkylphenyl ethers.[99]

A marking ink showing good erasure from ceramic surfaces contains Cu pc, lower alcohols, alcohol-soluble resin, isopropyl myristate, a saturated hydrocarbon, and polyoxyethylene alkylether phosphate.[100]

III. WRITING INKS

Water-thinned inks contain Cu pc with improved writing properties for ball point pens with good flowability when exposed to air.[101]

Concentrated Cu pc sulfonic acid dyes which do not form precipitates over time are especially useful for ball point pen inks in tropical areas.[102]

Cu pc 15, ethylene glycol 15, propylene glycol 5, bleached shellac 1, diethanolamine 10, antiseptic 0.1, and water 53.9 parts are mixed and applied to a cotton containing ball point pen which, after standing at 65% humidity for 10 d without a cap, shows good writing properties.[103]

A water-based writing ink contains Cu pc. When the ink is used in a pen having a plastic point, it has excellent storage stability, coverage, and writing properties.[104]

A writing ink contains a Cu pc trisulfonic acid and an amine-vinyl reaction product as a waterproofing additive. Flowing from a felt tipped pen, it exhibits excellent water and heat resistance.[105]

A fine point pen with a stainless steel pen point tube containing Cu pc, oil-soluble polymers, and oxidized paraffin wax, shows no blurring and good ink delivery on polypropylene.[106]

A water-thinned writing ink contains Cu pc and polyelectrolyte pigment dispersants with miscible organic solvents. Felt pens filled with the ink do not dry out when left uncapped 5 h at 20°C and 65% humidity.[107]

Cu pc is a pigment for water-based ball point pen smooth writing inks containing benzotriazole.[108-109]

Aqueous inks contain dispersants and Cu pc for felt-tipped pens. The dispersants include 50:40:10 styrene/methacrylic acid/diethylene glycol methacrylate copolymer methylamine salt.[110]

Water-thinned ball point pen writing inks enabling longer writing and reduced ball sinking comprise pigments including Cu pc, dispersing agents, EDTA, alkali metal salts, and water.[111]

An aqueous ink that is stable on standing 9 months in a ball point pen contains pc sulfonic acid dyes of the formula where R^1 may = alkyl, $R^{2,3}$ may = H, X may = H, a = 1 or 2, b = 0 to 2, c = 1 to 3, and $2 \leq a + b + c \leq 4$. The ink also does not clog the nozzle of a jet printer.[112]

Aqueous black inks for ball point pens containing carbon black 3 to 15, pc compounds 0.15 to 6.0, and protective colloids 0.3 to 15%. The pc compounds are adsorbed on the carbon black and are dispersed in an aqueous vehicle by the protective colloids.[113]

Multipurpose stable, rapid drying aqueous inks contain both pigments such as Cu pc, and dyes, solvents, lubricants, wetting agents, dissolving aids, pH regulators, corrosion inhibitors, permeating agents, and/or antiseptics. The inks are useful for pens, porous stamps and rollers, and stamp pad and ink jet printers.[114]

A solvent ink-based vinyl polymer or cellulose based ink which gives a continuously strippable writing contains polyoxyethylene-polyoxypropylene block copolymer and Cu pc.[115] Another such ink contains Cu pc 5, water 70, dispersant 0.5, Polysol PS 20, Eftop EF-122A water-soluble poly(oxyethylene)fluoro surfactant, and butyl carbitol acetate 2 parts.[116]

A writing ink comprises a predispersed Cu pc pigment in an alkali soluble resin and an alkaline material which solubilizes the resin.[117]

Aqueous ink pastes containing Cu pc are prepared by dispersing the pigment in an aqueous medium without a surfactant. Letters or drawings written on a paper using the ink are easily erased by commercial erasers.[118]

Colored inks based on disperse dyes and shellac may comprise, for example, a pc dye 15, ethylene glycol 5, shellac 8, $Na_2B_4O_7$ 6, phenol 1, urea 5, and water 60 parts. These inks are suitable for lettering pens and markers with hollow fiber tips.[119]

Water-thinned opaque pigment inks that may contain Cu pc with excellent storability and drying resistance of the pen tip contain finely powdered silica and fluorine-containing or silicone surfactant as pigment dispersion stabilizers.[120]

IV. PRINTING INKS

Pc blue presscake is fluidized for processing into water flexo inks by mixing with rosin dissolved in aqueous alkali metal or ammonium hydroxide, yielding a pourable, high-solids dispersion exhibiting good dispersibility in water flexo ink and producing ink with good gloss and color strength.[121]

Surface modification of pc and azo pigments with rosin derivatives decreases the grinding time, increases the resistance to flocculation, and increases the thinning capacity of the pigments in aqueous printing inks for flexographic printing on packaging paper and cardboard.[122]

A continuous method produces aqueous Cu pc pigment dispersions suitable for aqueous flexographic ink formulations.[123]

A pc anionic salt pc $(NH_2^+ RR^1)_{1-4}$ where R and R^1 are branched alkyl- and/or alkyl-substituted cycloalkyl group, is a useful colorant in flexography.[124]

The effect of pigments, including Cu pc, on the moisturizer uptake in offset printing inks is determined by measuring the amount of moisturizer in the printing machines.[125]

The optimal processing of Cu pc pigment granules for offset printing inks is discussed.[126]

A colorimetric evaluation is made of gloss enhancement by Cu pc of carbon black pigmented offset printing inks.[127]

Cu pc pigment granules for offset inks are included in a discussion of the use of pigment granulates in offset inks.[128]

Cu pc filter cake is flushed with a phosphate ester dispersant in petroleum solvent to transfer the pigment to the oily dispersant phase. An ink containing this flushed color provides a bright cyan color in offset printing on uncoated paper.[129]

Cu pc tetrakis$[CH_2N(C_{17}H_{35})_2]$ is a component in a printing ink for intaglio or offset printing.[130]

A pigment composition for gravure printing ink contains Cu pc $[Z(CH_2)R]_m$ where Z may = a direct bond, R may = cyclic amino, and m = 1 to 4.[131]

A gravure ink for magazines is prepared from crude Cu pc 15, sulfonated pc 0.6, zinc rosinate 21.9, Triton X-15 0.3, solvent 112.2, and 0.75-mm glass powder 200 parts.[132]

A polyester amide-dispersing agent may be modified with monsulfonated Cu pc or disulfonated disazo dyes for use as dispersants for pc or disazo pigments, suitable for use in lithographic and rotogravure printing inks.[133]

Inks for high-quality gravure printing contain a suspension of 400 g Cu pc in 2.4 l water, mixed with a solution of *N,N*-bis(2-hydroxyethyl)stearamide 30, *N,N*-bis(2-hydroxyethyl)oleamide 20, and acetic acid 250 g.[134]

β-Cu pc is coated with monoethanolamide of stearic or palmitic acid and the diethanolamide or dipropanolamide of oleic acid and is suitable for gravure printing.[135]

β-Cu pc is milled with an aqueous rosin acid potassium salt solution and acidified to pH 5 to 6 to give a pigment concentrate easily dispersible in organic media for gravure printing inks.[136]

Organic pigments including Cu pc are examined colorimetrically for optimum selection for flexographic and gravure printing on packaging paper and cellulose acetate.[137]

The effects of pH and isopropanol on the properties of blue (including Cu pc) and red pigments for water-based inks are discussed.[138]

A continuous method produces pc pigment dispersions in an organic solvent. Crude Cu pc pigment, a sulfonated pc, zincated rosin, Triton X, and lactol spirits are ground to prepare a dispersion suitable for gravure ink.[139]

Cu pc is reacted with rosin derivatives to prepare a dispersed pigment for nonaqueous offset or gravure inks.[140]

A solution of Cu pc in sulfuric acid is stirred with methylol melamine and poured into iced water to give a dispersant. A PVC gravure varnish containing C. I. pigment yellow 34 is stirred with this dispersant to give a gravure ink.[141]

Chlorosulfonated Cu pc reaction products with (aminophenylamino)bis[(diethylamino)propylamino]triazine are dispersants for Cu pc printing ink pigments useful in offset or gravure inks.[142]

β-Cu pc treated with nonionic surfactants and emulsifiers are useful for water-based gravure ink.[143]

Intaglio printing inks for illustrations contain sulfonate and sulfonamide derivatives of Cu pc. Strong, turquoise blue printings with little breakthrough except on very fine paper are obtained.[144]

Intaglio inks for color filters for liquid crystal display devices contain epoxy resins, solvents, and Cu pc.[145]

Mixtures of Cu pc and Fe(III) hexacyanoferrate pigments are milled, making them completely dispersible in binders for intaglio printing.[146]

Cu pc is dissolved in concentrated sulfuric acid, adjusted to 71 to 90% concentration, heat treated, and hydrolyzed to α-Cu pc of 20 to 40 m^2/g of surface area for intaglio printing inks with excellent wiping properties on printing plates.[147,148]

Compositions for ink outline marking with several adjacent colors contain resins, solvents, and colorants including Cu pc.[149]

Near IR absorbing inks that give prints that can be identified by IR contain Al pc Cl, halogenated Al pc Cl, and brominated Cu pc.[150]

Pc Green is a component in a solder resist ink.[151]

Rheological measurements are made of the cure rate of UV inks containing Cu pc.[152]

An etching-resistant ink containing Cu pc is printed on a copper-clad laminate and UV cured to give a film showing good hardness and chemical resistance.[153]

The heat stability of yellow pigments containing Cu pc, in a bead mill, varies depending on the heat, milling time, and machine type.[154]

Aqueous compositions containing Cu pc basic dyes are used in the preparation of bleedfast printing inks.[155]

Using static and dynamic test methods, it is shown that the recrystallization stability of organic pigments including Cu pc in printing inks depends first on particle size distribution then on chemical constitution.[156]

Halogenated Cu pc pigments for copper plate printing inks with excellent heating strength and fluidity are prepared by adding β-naphthol to halogenated Cu pcs, then heating and milling in the presence of water.[157]

A cosolvent in a water-based printing ink containing, for example, Cu pc, such as isopropanol gives more efficient pigment wetting, higher attainable color strength, and better ink transfer and drying than water alone.[158]

Abietyl resin in a viscous printing ink containing β-Cu pc weakens face-to-face and edge-to-edge contact in the crystals.[159]

The effect is shown of acrylic resins in the dispersibility of organic (including Cu pc) and carbon black pigments, and on the rheology of aqueous printing inks.[160]

The dispersibility of organic pigments, including Cu pc, in the manufacture of printing inks, is improved by using a dispersant which comprises a mixture of alkyl derivatives of a pigment with poly(12-hydroxystearic acid) octyl ester.[161]

The use of physical instrumental methods to determine pigment particle size and size distribution are discussed and are demonstrated on Cu pc C. I. pigment blue 15:3 among other pigments.[162]

An aqueous pigment ink includes titania and Cu pc. After storage for 20 d it can be easily redispersed by light shaking due to the presence of borax in the dispersion.[163]

Powder materials including Cu pc pigments with a hydrophilic surface are made dispersible in organic media by treating with an aluminum chelate composition and then with a β-ketone. The treated pigments disperse well in oil-based inks.[164]

Storage-stable oil-based inks contain resin particles with an adsorbed Cu pc derivative.[165]

Pigment compositions with good dispersibility in solvent born inks are prepared by high shear mixing of premilled crude Cu pc, tris(diethylaminomethyl) Cu pc or its organic salt, and a phase director, and digesting the mixture in water or in an aqueous mineral acid solution.[166]

Aqueous pigment inks contain Cu pc, benzotriazole, and acrylic acid-styrene copolymer. They have good dispersibility and corrosion resistance.[167]

Pigments including Cu pc for inks are coated with 0.01 to 1% higher fatty acid modified silicone oil and are used to prepare inks with good abrasion resistance when dry.[168]

Cu pc particles for use in electrophotographic toners and printing inks are prepared by coating with waxy substances and polymers insoluble in liquid aliphatic hydrocarbons.[169]

Printing inks containing α-Cu pc flakes have different rheology than printing inks containing cubic α-Cu pc.[170]

Inks containing Cu pc tris-N-substituted sulfonamides dissolved in a chlorinated $C_{1\text{-}2}$-hydrocarbon solvent are used for color coding vinyl electric insulators.[171]

The effect of temperature on the rheology of printing inks containing α-Cu pc with various sizes and shapes dispersed in alkyd GFL-2 or rosin-modified phenolic resin-linseed oil varnish is discussed.[172]

A mixture of Cu pc, sodium chloride, monophthalimidomethyl Cu pc, diethylene glycol, and xylene are kneaded to prepare an ink having greater coloring power than an ink containing only Cu pc.[173]

Storage-stable Cu pc disulfonic acid reaction products with 3-amino-4-hydroxybenzene-sulfonamide and 4-vinylpyridine are coloring compositions for inks containing water or water-soluble organic solvents.[174]

The tinctorial strength of pigments, e.g., pcs for printing inks, is developed by particle size reduction.[175]

Inks for ink ribbons contains Cu pc and thickening agents such as metallic soaps, hydrogenated castor oil, aluminum chelates, organic bentonite, and so on.[176]

Inks for heat-sensitive nonimpact printers contain Cu pc which is mixed with solvents and coated on a polyethylene film to prepare a ribbon giving satisfactory printing at 40 letters per second.[177]

Inks with metallic luster contain gold or silver pigments and pc blue FG.[178]

The energetic parameters of a variety of organic ink pigments including Cu pc and solvents are calculated.[179]

β-Cu pc and Cu pc $[CH_2NHCOCH_2N^+(C_4H_9)_2\ CH_2\ CH_2OH\ OH^-]_{1.4}$ are mixed with nitrocellulose lacquer to form a flocculation-resistant printing ink.[180]

Cu pc cake is stirred in *N*-2-ethylhexylaspargic acid dodecyl ester in methanol to give a product that disperses well in oil to give bright printing inks.[181]

A commercial paper is printed with a wood grain pattern by using an aqueous printing ink containing Cu pc, printing again with a darker ink of the same composition, but containing a varnish repelling agent; after 2 months the printed surface is varnished to give a decorative print with a three-dimensional effect.[182]

Highly transparent and dust-free pigments with high tinctorial strength useful in printing inks are prepared by adding 10 to 30% alkaline resin solution to an aqueous suspension of Cu pc, acidifying, and flushing.[183]

Cu pc is a component in a screen printing ink.[184]

V. THERMAL TRANSFER INKS

An ink containing Cu pc that produces high-resolution nonimpact prints with low dot expansion comprises an electrothermal transfer ink layer, a high resistance layer, and an electric anisotropic support layer having selective conductivity in the thickness direction.[185]

A transfer ink medium for nonimpact printers contains Cu pc in an electrically conductive ink layer on an electrical anisotropic support formed by filling a porous sheet substrate with an electrically conductive material.[186]

A thermal transfer printing method includes coating a liquid ink containing Cu pc pigment, in a thermoplastic resin, and a solvent on a support, drying the ink, and subsequently printing the ink on a substrate.[187]

A mixture of 85 parts diethyl maleate-styrene copolymer and 15 parts Cu pc is coated on a 6μ polyester-based base sheet to prepare a thermal transfer sheet.[188]

A mixture of 85 parts $C_{30\text{-}55}$ α-olefin dimethyl maleate copolymer and 15 parts Cu pc is coated on a 6μ polyester-based film to prepare a thermal transfer sheet.[189]

Hot melt ink layers contain Cu pc reactive dyes for thermal transfer.[190]

A wax-impregnated paper is coated with 1% aqueous methyl cellulose, electrically flocked with colorless nylon short fibers, screen printed with an ink containing Cu pc and an ethylene-vinyl acetate copolymer to form a flower pattern, coated with poly(vinyl isopropyl ether) adhesive, coated scatteringly with powdered talc to form a nonsticking layer, and transferred to a urethane leather bag.[191]

Cu pc pigments are included in thermal transfer inks containing dyes and fatty amides.[192]

A gradational thermal transfer printing sheet contains an ink layer consisting of Cu pc 24, paraffin wax 11.1, candelilla wax 9.3, Arkon P70 9.3, Arkon P125 9.3, and alumina 30 parts in xylene which is coated on a polyethylene terephthalate film to give a thermal transfer printing sheet.[193]

Thermal transfer printing sheets capable of gradational printing comprise a heat-resistant base film coated with an ink composition containing Cu pc, Neopolymer NP aromatic petroleum resin, carnauba wax, and xylene.[194]

Paper is printed with thermal transfer inks containing Cu pc coloring agent.[195]

A thermal transfer material contains wax powder and Cu pc.[196]

Heat-sensitive ink ribbons contain Cu pc for high speed thermal printers.[197]

A multicolor ink jet or thermal transfer recording includes an aqueous cyan ink containing diethylene glycol and Cu pc disulfonic acid.[198,199]

A thermal transfer recording material for gravure printing contains Cu pc.[200]

A color thermal transfer recording ink sheet contains Cu pc. The ink sheet has high opacity to prevent undesirable color mixing.[201]

REFERENCES

1. **Fuji Photo Film Co., Ltd.,** Japan Patent 83,152,072, 1983.
2. **Ricoh Co., Ltd.,** Japan Patent 83,215,467, 1983.
3. **Canon, K. K.,** Japan Patent 83,191,772, 1983.
4. **Dainippon Ink and Chemicals, Inc.,** Japan Patent 83,134,164, 1983.
5. **Fuji Photo Film Co., Ltd.,** Japan Patent 83,12,786, 1983.
6. **Dainippon Ink and Chemicals, Inc.,** Japan Patent 83,164,665, 1983.
7. **Ricoh Co., Ltd.,** Japan Patent 84,215,369, 1984.
8. **Canon, K. K.,** Japan Patent 84,191,775, 1984.
9. **Ricoh Co., Ltd.,** Japan Patent 84,161,471, 1984.
10. **Koike, S., Eida, T., Kobayashi, M., Yokoyama, Y., and Komori, T.,** German Patent 3,411,476, 1984.
11. **Kanekiyo, K., Kobayashi, M., Yokoyama, Y., Eida, T., Komori, T., and Koike, S.,** U.S. Patent 4,505,749, 1985.
12. **Canon, K. K.,** Japan Patent 84,215,889, 1984.
13. **Canon, K. K.,** Japan Patent 84,147,064, 1984.
14. **Allen, D. A. (Xerox Corp.),** U.S. Patent 4,664,708, 1987.
15. **Miura, K., Takimoto, H., and Taki, T. (Mitsubishi Chemical Industries Co., Ltd.),** Japan Patent 86,285,277, 1986.
16. **Saot, H. (Alps Electric Co., Ltd.),** Japan Patent 87,00,567, 1987.
17. **Meuter, A. R. and Palmer, D. J., (Hewlett-Packard Co.),** Europe Patent Application 224,909, 1987.
18. **Aoki, S. and Tanaka, M. (Konishiroku Photo Industry Co., Ltd.),** Japan Patent 86,53,373, 1986.
19. **Arai, R., Sakaki, M., and Matsui, S. (Canon, K. K.),** Japan Patent 86,141,585, 1986.
20. **Aoki, S. and Tanaka, M. (Konishiroku Photo Industry Co., Ltd.),** Japan Patent 86,53,372, 1986.
21. **Kawashita, H., Ikeda, T., and Sakaguchi, S. (Taoka Chemical Co., Ltd. and Sumitomo Chemical Co., Ltd.),** Japan Patent 86,02,772, 1986.
22. **Yamagishi, Y., Narisawa, T., Saruwatari, N., and Ko, K. (Fujitsu Ltd.),** Japan Patent 86,81,471, 1986.
23. **Aoki, S., Oya, H., and Itano, M. (Konishiroku Photo Industry Co., Ltd.),** Japan Patent 86,36,380, 1986.

24. **Ricoh, Co., Ltd.**, Japan Patent 85,92,369, 1985.
25. **Chiba, M., Iwata, K., Nakousai, T., Ogawa, H., and Ohba, T. (Canon, K. K.)**, German Patent 3,438,421, 1985.
26. **Nippon Telegraph and Telephone Public Corp.**, Japan Patent 85,58,477, 1985.
27. **Canon, K. K.**, Japan Patent 83,196,274, 1983.
28. **Ricoh Co., Ltd.**, Japan Patent 83,183,769, 1983.
29. **Pentel Co., Ltd.**, Japan Patent 83,196,275, 1983.
30. **Canon, K. K.**, Japan Patent 84,25,853, 1984.
31. **Canon, K. K.**, Japan Patent 84,25,847, 1984.
32. **Canon, K. K.**, Japan Patent 84,25,849, 1984.
33. **Canon, K. K.**, Japan Patent 84,25,851, 1984.
34. **Canon, K. K.**, Japan Patent 84,25,852, 1984.
35. **Canon, K. K.**, Japan Patent 83,224,785, 1983.
36. **Canon, K. K.**, Japan Patent 84,25,854, 1984.
37. **Ricoh Co., Ltd.**, Japan Patent 83,196,273, 1983.
38. **Pentel Co., Ltd.**, Japan Patent 84,01,580, 1984.
39. **Ricoh Co., Ltd.**, Japan Patent 83,217,569, 1983.
40. **Ricoh Co., Ltd.**, Japan Patent 83,222,166, 1983.
41. **Aruga, T., Fujii, T., Murakami, K., and Akutsu, E. (Ricoh Co., Ltd.)**, German Patent 3,344,648, 1984.
42. **Ricoh Co., Ltd.**, Japan Patent 83,201,863, 1983.
43. **Yokoyama, Y., Komori, T., Eida, T., Koike, S., and Kobayashi, M. (Canon K. K.)**, German Patent 3,329,081, 1984.
44. **Taoka Chemical Co., Ltd. and Sumitomo Chemical Co., Ltd.**, Japan Patent 84,30,874, 1984.
45. **Fujitsu Ltd.**, Japan Patent 84,51,959, 1984.
46. **Canon K. K.**, Japan Patent 84,25,846, 1984.
47. **Canon K. K.**, Japan Patent 84,25,845, 1984.
48. **Canon K. K.**, Japan Patent 84,25,844, 1984.
49. **Ariga, T., Kawanishi, T., Shimada, M., Murakami, K., and Kamimura, H. (Ricoh Co., Ltd.)**, Japan Patent 86,183,370, 1986.
50. **Nagashima, S. (Mitsubishi Chemical Industries Co., Ltd.)**, Japan Patent 86,241,370, 1986.
51. **Shiotani, M., Sakaeda, T., and Kobayashi, M. (Canon K. K.)**, Japan Patent 86,113,673, 1986.
52. **Kawashita, H. and Ota, M. (Takaoka Chemical Manufacturing Co., Ltd.)**, Japan Patent 86,185,576, 1986.
53. **Haruta, M., Togano, S., and Arai, R. (Canon K. K.)**, Japan Patent 85,197,778, 1985.
54. **Haruta, M., Togano, S., and Arai, R. (Canon K. K.)**, Japan Patent 85,197,776, 1985.
55. **Akutsu, H., Murakami, K., and Ariga, T. (Ricoh Co., Ltd.)**, Japan Patent 85,199,078, 1985.
56. **Canon K. K.**, Japan Patent 84,25,848, 1984.
57. **Ricoh Co., Ltd.**, Japan Patent 85,51,760, 1985.
58. **Haruta, M., Togano, S., and Arai, R. (Canon K. K.)**, Japan Patent 85,197,777, 1985.
59. **Matsura, T., Yamanaka, N., and Wakai, S. (Nippon Kayaku Co., Ltd.)**, Japan Patent 85,208,365, 1985.
60. **Sakaeda, T., Shiotani, M., and Kobayashi, M. (Canon K. K.)**, Japan Patent 87,167,375, 1987.
61. **Murakami, K., Shimada, M., Ariga, T., Kamimura, H., and Nagai, K. (Ricoh Co., Ltd.)**, Japan Patent 87,149,770, 1987.
62. **Canon K. K.**, Japan Patent 85,15,462, 1985.
63. **Ricoh Co., Ltd.**, Japan Patent 84,78,885, 1984.
64. **Canon K. K.**, Japan Patent 84,178,288, 1984.
65. **Canon K. K.**, Japan Patent 84,178,290, 1984.
66. **Toganoh, S. and Arai, R. (Canon K. K.)**, German Patent 3,410,828, 1984.
67. **Canon K. K.**, Japan Patent 84,209,189, 1984.
68. **Canon K. K.**, Japan Patent 84,215,890, 1984.
69. **Canon K. K.**, Japan Patent 84,214,684, 1984.
70. **Canon K. K.**, Japan Patent 84,227,483, 1984.
71. **Suzuki, E. (Canon K. K.)**, German Patent 3,432,823, 1985.
72. **Mori, H., Hida, M., and Eto, N., (Canon K. K.)**, Japan Patent 85,248,387, 1985.
73. **IBM Corp.**, Japan Patent 86,03,778, 1986.
74. **Eida, T., Shioya, M., and Kobayashi, M. (Canon K. K.)**, Europe Patent Application 181,226, 1986.
75. **Sakaki, M., Arai, R., and Togano, S. (Canon K. K.)**, Japan Patent 86,63,474, 1986.
76. **Akiya, T., Toganoh, S., and Arai, R. (Canon K. K.)**, French Patent 2,570,988, 1986.
77. **Akiya, T., Toganoh, S., and Arai, R. (Canon K. K.)**, Europe Patent Application 174,859, 1986.
78. **Arai, R., Sakaki, M., and Matsui, S. (Canon K. K.)**, Japan Patent 86,118,285, 1986.

79. **Sakaki, M., Arai, R., and Togano, S. (Canon K. K.),** Japan Patent 86,63,475, 1986.
80. **Azuma, K. (Tomoegawa Paper Manufacturing Co., Ltd.),** Japan Patent 87,74,686, 1987.
81. **Akitani, T., Morohoshi, N., and Togano, S. (Canon K. K.),** Japan Patent 87,85,983, 1987.
82. **Sakamoto, M. and Arai, R. (Canon K. K.),** Japan Patent 86,217,280, 1986.
83. **Iida, Y. and Takahashi, Y. (Toyo Ink Manufacturing Co., Ltd.),** Japan Patent 86,164,836, 1986.
84. **Togano, S., Akitani, T., Morohoshi, N., Kobayashi, M., Senoo, S., and Yamazaki, T. (Canon K. K.),** Japan Patent 87,37,183, 1987.
85. **Murakami, K., Kawanishi, T., Shimada, M., Ariga, T., Kamimura, H., and Nagai, K. (Ricoh Co., Ltd.),** Japan Patent 87,37,184, 1987.
86. **Aoki, A. (Canon K. K.),** Japan Patent 87,55,186, 1987.
87. **Sakaki, M., Arai, R., and Togano, S. (Canon K. K.),** Japan Patent 86,228,984, 1986.
88. **Akitani, T., Morohoshi, N., and Togano, S. (Canon K. K.),** Japan Patent 87,85,982, 1987.
89. **Akitani, T., Morohoshi, N., and Togano, S. (Canon K. K.),** Japan Patent 87,85,981, 1987.
90. **Mikuni Shikiso, K. K. and Mitsubishi Pencil Co., Ltd.,** Japan Patent 84,217,776, 1984.
91. **Orient Chemical Industries, Ltd.,** Japan Patent 85,84,369, 1985.
92. **Hamamoto, H. and Sano, H. (Pentel Co., Ltd.),** Japan Patent 85,181,179, 1985.
93. **Nagasawa, T. and Kitamura, T. (Orient Chemical Industries, Ltd.),** Japan Patent 85,190,480, 1985.
94. **Adger Industry Co., Ltd.,** Japan Patent 85,67,577, 1985.
95. **Pentel Co., Ltd.,** Japan Patent 85,86,183, 1985.
96. **Pentel Co., Ltd.,** Japan Patent 85,86,182, 1985.
97. **Konuki, I. (Pentel Co., Ltd.),** Japan Patent 85,161,471, 1985.
98. **Iijima, Z. (Adger Industry Co., Ltd.),** Japan Patent 86,106,681, 1986.
99. **Kawaguchi, K. (Pilot Pen Co., Ltd.),** Japan Patent 85,55,553, 1985.
100. **Kobayashi, Y. and Saito, S. (Pentel Co., Ltd.),** Japan Patent 86,197,662, 1986.
101. **Inoue, S. and Yamamoto, H. (Sakura Color Products Corp.),** German Patent 3,337,866, 1984.
102. **Hodogaya Chemical Co., Ltd.,** Japan Patent 85,53,566, 1985.
103. **Pentel Co., Ltd.,** Japan Patent 85,76,576, 1985.
104. **Daini Seikosha Co., Ltd.,** Japan Patent 85,26,070, 1985.
105. **Akiyama, K. and Ono, T.,** Europe Patent Application 147,667, 1985.
106. **Sano, H. and Hamamoto, H. (Pentel Co., Ltd.),** Japan Patent 85,203,682, 1985.
107. **Kanbara, T. (Fuji Shikiso Co., Ltd.),** Japan Patent 85,161,469, 1985.
108. **Otaguro, K., Toyama, T., Takahashi, H., and Sasage, D. (Pentel Co., Ltd.),** Japan Patent 86,55,175, 1986.
109. **Takahashi, H., Hamamoto, H., and Ishiguro, A. (Pentel Co., Ltd.),** Japan Patent 86,223,075, 1986.
110. **Toyama, T., Sano, H., and Sasage, D. (Pentel Co., Ltd.),** Japan Patent 86,223,074, 1986.
111. **Hamamoto, H., Onuki, T., and Ishiguro, A. (Pentel Co., Ltd.),** Japan Patent 86,130,384, 1986.
112. **Kawashita, H. and Ota, M. (Taoka Chemical Co., Ltd. and Sumitomo Chemical Co., Ltd.),** Europe Patent Application 196,901, 1986.
113. **Ozaki, K. (Pilot Ink Co., Ltd.),** Japan Patent 86,200,183, 1986.
114. **Shachihata Industrial Co., Ltd.,** Japan Patent 84,129,272, 1984.
115. **Okabe, E. (Pentel Co., Ltd.),** Japan Patent 85,206,881, 1985.
116. **Okabe, E. (Pentel Co., Ltd.),** Japan Patent 85,219,273, 1985.
117. **Ferguson, E. J. (Gillette Co.),** PCT International Application 86,00,327, 1986.
118. **Adger Industry Co., Ltd.,** Japan Patent 83,167,659, 1983.
119. **Bartusek, T. and Curda, L.,** Czech Patent 230,740, 1986.
120. **Murai, A. and Goto, K. (Shachihata Industrial Co., Ltd.),** Japan Patent 87,91,574, 1987.
121. **Donegan, T. E. (BASF Wyandotte Corp.),** U.S. Patent 4,589,922, 1986.
122. **Aristov, B. G., Satushev, S. A., Dzyuba, V. M., Mischenko, V. N., and Gorenko, V. N.,** Modified organic pigments for waterborne ink systems, *Lakokras. Mater. Ikh Primen.*, 1, 14, 1986.
123. **Chisvette, D., Lattimore, J., and Patterson, D. M. (Inmont Corp.),** U.S. Patent 4,427,810, 1984.
124. **Loeffler, H. and Dyllick-Brenzinger, R. (BASF AG),** Europe Patent Application 163,113, 1985.
125. **Stark, E. G. and Frank, E.,** Effect of pigments on moisturizer uptake of offset printing inks, *Farbe Lack,* 92(1), 8, 1986.
126. **Braun, F.,** Pigment granules for offset printing inks, *Farbe Lack,* 92(2), 108, 1986.
127. **Schumacher, W.,** Principles and use of carbon black. XIII. Colorimetric evaluation of high-gloss black printing inks, *Farbe Lack,* 90(6), 454, 1984.
128. **Braun, F.,** Pigment granules. Pigment in a novel form for offset inks, *Am. Ink Maker,* 64(10), 28, 30, 32, 35, 38, 42, 50, 1986.
129. **Nakamura, M., Takeuchi, H., Takahashi, T., Takizawa, M., and Horiguchi, S. (Dainichiseika Color and Chemicals Manufacturing Co., Ltd.),** Europe Patent Application 202,394, 1986.

130. **Bugrov, G. A., Vil'ner, S. G., Gorin, V. P., Oknin, Y. A., Poznyakevich, A. L., Smrchek, V. A., and Tikhonov, V. P. (Moscow Institute of Engineers of Geodesy, Aerial-Photographic Surveying and Cartography),** U.S.S.R. Patent 1,171,495, 1985.
131. **Tokyo Ink Manufacturing Co., Ltd.,** Japan Patent 83,167,654, 1983.
132. **Inmont Corp.,** Japan Patent 86,163,978, 1986.
133. **Canestri, G. (Bergvik Kemi AB),** Europe Patent Application 189,385, 1986.
134. **Necas, M., Plechacek, V., Pridal, J., Stastny, L., Kral, J., and Hejtmanek, M.,** Czech Patent 234,629, 1985.
135. **Necas, M. and Plechacek, V.,** Czech Patent 235,187, 1987.
136. **Pridal, J., Lustig, J., Stastny, V., Hejtmanek, M., Skvrnova, A., Znamenacek, M., and Salek, J.,** Czech Patent 231,433, 1986.
137. **Da Dalt, V., Gremmo, E., and Saccomano, L.,** Colorimetric evaluation of organic pigments in solvent inks, *Pitture Vernici,* 58(10), 13, 1982.
138. **Olschner, H.,** Water and organic dyes: possible dilution in rotogravure printing?, *Pitture Vernici,* 62(5), 31, 1986.
139. **Chisvette, D., Lattimore, J., and Sova, S. G. (Inmont Corp.),** U.S. Patent 4,522,654, 1985.
140. **Hayashi, M., Mochizuki, A., and Kawashima, Y. (Toyo Ink Manufacturing Co., Ltd.),** Japan Patent 87,195,053, 1987.
141. **Toyo Ink Manufacturing Co., Ltd.,** Japan Patent 84,96,175, 1984.
142. **Katsura, H., Ehashi, S., Kashioka, M., and Sakamoto, M. (Toyo Ink Manufacturing Co., Ltd.),** German Patent 3,514,077, 1986.
143. **Fujigamori, T. (Toyo Ink Manufacturing Co., Ltd.),** Japan Patent 86,111,365, 1986.
144. **Lorenz, M., Groll, M., and Haus, A. (Bayer AG),** German Patent 3,600,348, 1987.
145. **Sawamura, M., Oshima, S., Masaki, T., and Tachibana, S. (Toyo Ink Manufacturing Co., Ltd. and Toppan Printing Co., Ltd.),** Japan Patent 87,54,774, 1987.
146. **Fechner, W. D., Polster, R., Kranz, J., and Hartmann, E. (BASF AG),** German Patent 3,211,165, 1983.
147. **Nippon Business K. K. and Sumitomo Chemical Co., Ltd.,** Japan Patent 83,160,364, 1983.
148. **Nippon Business K. K. and Sumitomo Chemical Co., Ltd.,** Japan Patent 83,160,363, 1983.
149. **Sakura Color Products Corp.,** Japan Patent 84,06,271, 1984.
150. **Toyo Ink Manufacturing Co., Ltd.,** Japan Patent 84,138,284, 1984.
151. **Hitachi, Ltd.,** Japan Patent 83,176,254, 1983.
152. **Otsubo, Y., Amari, T., and Watanabe, K.,** Rheological measurements of cure rate of UV inks, *J. Appl. Polym. Sci.,* 31(7), 2099, 1986.
153. **Nippon Shokubai Kagaku Kogyo Co., Ltd.,** Japan Patent 84,170,165, 1984.
154. **Leveque, J. C.,** Thermal stability of yellow pigments in heat-set inks, *Double Liaison-Chim. Peint.,* 32(360), XXIV-XXVII, 317, 1985.
155. **Degen, H. J. (BASF AG),** German Patent 3,329,340, 1985.
156. **Kaiser, W.,** Resistance of organic pigments in printing inks to recrystallization, *Farbe Lack,* 91, 600, 1985.
157. **Mori, A., Tsuchiya, K., Horiguchi, S., (Dainichiseika Color and Chemicals Manufacturing Co., Ltd.),** Japan Patent 86,66,757, 1986.
158. **Olschner, H.,** Water and organic colorants, *Am. Ink Maker,* 64(6), 25, 28, 33, 39, 42, 44, 46, 48, 50, 112, 1986.
159. **McKay, R. B. and Mather, R. R.,** The influence of aggregate structure and wetting on the dispersibility of β-copper phthalocyanine pigments in viscous printing inks, *Colloids Surf.,* 27(1), 175, 1987.
160. **Watson, W. M.,** Designing ink dispersing resins, *Am. Ink Maker,* 61(10), 19, 22, 24, 26, 28, 30, 60, 1983.
161. **Smrcek, V. A., Vil'ner, S. G., Gorin, V. P., and Bugrov, G. A.,** Improvement of dispersibility of organic pigments in ink systems, *Lakokras. Mater. Ikh Primen.,* 2, 11, 1984.
162. **Vernardakis, T. G.,** Pigment particle size and its effect on printing ink performance, *Am. Ink Maker,* 62(2), 24, 28, 30, 32, 34, 42, 109, 1984.
163. **Pentel, Co., Ltd.,** Japan Patent 84,59,756, 1984.
164. **Saito, S. and Kuramitsu, T. (Kawaken Fine Chemicals Co., Ltd.),** Japan Patent 86,09,460, 1986.
165. **Takagishi, I., Watabe, H., and Kawabata, K. (Pentel Co., Ltd.),** Japan Patent 87,54,775, 1987.
166. **Donegan, T. E. and Leary, T. G. (BASF Corp.),** Europe Patent Application 221,466, 1987.
167. **Sano, H. and Toyama, T. (Pentel Co., Ltd.),** Japan Patent 86,200,182, 1986.
168. **Toyo Ink Manufacturing Co., ltd.,** Japan Patent 84,15,455, 1984.
169. **Ricoh Co., Ltd.,** Japan Patent 83,108,256, 1983.
170. **Sinitsyna, I. N., Aristov, B. G., and Gilyazetdinov, L. P.,** Effect of the shape of particles on the coloring and rheological properties of α-phthalocyanine pigments, *Lakokras Mater. Ikh Primen.,* 6, 13, 1983.

171. **Ulry, D. C., Cosner, C. G., and Ryan, T. D.,** U.S. Patent 4,451,264, 1984.
172. **Aristov, B. G., Gilyazetdinov, L. P., and Sinitsina, I. N.,** Effect of temperature on the rheology of printing inks from phthalocyanine pigments, *Kolloidn. Zh.*, 47(1), 126, 1985.
173. **Takahashi, J., Asano, H., and Kamiya, K., (Dainippon Ink and Chemicals, Inc.),** Japan Patent 85,133,065, 1985.
174. **Akiyama, K. and Ono, T. (Orient Chemical Industries, Ltd.),** Europe Patent Application 142,867, 1985.
175. **Ford, J. A. (Blu-Max Pigments, Inc.),** Europe Patent Application 154,498, 1985.
176. **Shino, M., Ando, K., and Omori, Y. (Oki Electric Industry Co., Ltd.),** Japan Patent 86,297,184, 1986.
177. **Olivetti SpA,** Japan Patent 84,196,295, 1984.
178. **Sotojima, M. (Seiko Seisakusho Co., Ltd.),** Japan Patent 84,191,778, 1984.
179. **Schroeder, J.,** Contribution to energetic characterization of pigment dispersions: pigments and solvents, *Org. Coat.*, 6, 455, 1984.
180. **Patsch, M., Ruske, M., and Hahn, E.,** (BASF AG), German Patent 3,207,330, 1983.
181. **Kobayashi, T., Sagawa, K., Takehara, M., and Kawashima, H. (Ajinomoto Co., Inc.),** Japan Patent 86,183,362, 1986.
182. **Diesel, H. D., Schmidt, H. J., and Sauer, B. (Letron GmbH),** German Patent 3,247,677, 1984.
183. **Deubel, R., Schilling, C., and Marx, W. (Hoechst AG),** Indian Patent 156,867, 1985.
184. **Pilot Ink Co., Ltd.,** Japan Patent 84,108,067, 1984.
185. **Ricoh Co., Ltd.,** Japan Patent 83,162,675, 1983.
186. **Ricoh Co., Ltd.,** Japan Patent 83,157,872, 1983.
187. **Hasegawa, T. (Canon K. K.),** Japan Patent 86,280,993, 1986.
188. **Hayashi, S., Yamaguchi, Y., Takei, K., Taniguchi, M., and Usui, T. (Seiko Epson K. K.),** Japan Patent 86,169,287, 1986.
189. **Hayashi, S., Yamaguchi, Y., Takei, K., Taniguchi, M., and Usui, T. (Seiko Epson K. K.),** Japan Patent 86,169,286, 1986.
190. **Suzuki, A. and Mochizuki, A. (Ricoh Co., Ltd.),** Japan Patent 86,143,196, 1986.
191. **Matsui Shikiso Kgaku Kogyosho Co., Ltd.,** Japan Patent 83,33,831, 1983.
192. **Fujitsu, Ltd.,** Japan Patent 84,165,693, 1984.
193. **Onishi, H., Ezaki, H., and Kobashi, T. (Matsushita Electric Industrial Co., Ltd.),** Japan Patent 86,123,587, 1986.
194. **Kobashi, T., Ezaki, H., and Onishi, H. (Matsushita Electric Industrial Co., Ltd.),** Japan Patent 86,123,589, 1986.
195. **Dainippon Printing Co., Ltd., and Morohoshi Printing Ink Co., Ltd.,** Japan Patent 83,43,277, 1983.
196. **Pilot Pen Co., Ltd.,** Japan Patent 84,24,693, 1984.
197. **Olivetti SpA,** Japan Patent 84,207,287, 1984.
198. **Canon K. K.,** Japan Patent 85,25,784, 1985.
199. **Canon K. K.,** Japan Patent 85,30,386, 1985.
200. **Toyo Ink Manufacturing Co., Ltd.,** Japan Patent 85,49,993, 1985.
201. **Yokoi, N. (Fuji Xerox Co., Ltd.),** Japan Patent 86,108,587, 1986.

Chapter 20

SPECTROSCOPY

I. RAMAN SPECTROSCOPY

Thin solid films of Co, Mg, and Zn phthalocyanine (pc) complexes are studied using Raman spectroscopy.[1]

Raman scattering is measured from Cu pc thin films deposited on silicon and glass surfaces and evaporated silver films.[2]

Raman spectra of aqueous solutions of Co and Fe tetrasulfonated pcs at various pH are correlated with published data.[3]

Preresonance and resonance Raman spectra of vanadyl pc films about 100 nm thick deposited on glass and quartz substrates at -180 to $+100°C$ are studied.[4]

The resonance Raman spectrum of Fe(II) oxy pc, Fe pc O_2, is measured in O matrices at about 15 K.[5]

Metal island films of silver and indium are used to study the surface-enhanced Raman scattering (SERS) of Cu and Zn pc complexes. Metal island films are emerging as a powerful spectroscopic technique to identify very thin films and monolayers.[6]

The angular dependence of scattering intensity in total reflection Raman spectra of Cu pc films is measured.[7]

The suitability of SERS as a diagnostic tool in preparing semiconductor films is proven using pcs as an example.[8]

Raman spectra are obtained for a solid Fe pc complex and a one millionth molar solution of its tetrasulfonated derivative with Kr laser excitation at 6471 Å.[9]

SERS activity of silver particles is studied. The enhancement factor is determined by measuring the Raman intensity of Cu pc molecules deposited on the particle layer.[10]

Raman spectra are determined for Cu pc in a compacted metal disk.[11]

Mounting a sodium nitrate crystal on a disk surface made by compacting silver powder and Ni and Zn pcs enables their Raman spectra to be measured such that relative peak intensities in any one disk can be quantified using sodium nitrate as a standard.[12]

Changes in the electrical and optical characteristics of molecules adsorbed on a conductor or semiconductor caused by the electrical or optical excitation of electrons within the molecules can be used as signals to carry information which can be varied by varying optical inputs to the molecules. The optical response can be determined by Raman spectroscopy. The information can be manipulated by ion (4-) H_2 pc metal complexes 2,10,16,24-tetrasulfonic acid.[13]

SERS of a true Langmuir-Blodgett monolayer is observed for the first time. Metal island films of indium and silver are used to obtain SERS of a monolayer of tetra-*tert*-butyl pc.[14]

Spontaneous and resonance Raman spectra of Pb and Sn pcs are measured.[15]

Di hydrogen, Co, and Fe tetrasulfonated pcs adsorbed on a silver electrode are studied by resonant and surface enhanced Raman spectroscopy, and by cyclic voltammetry.[16]

Raman scattering excitation profile spectra are obtained for symmetrical and nonsymmetrical vibration modes of pc molecules adsorbed on silver.[17]

Raman spectra of amorphous Cu pc thin films are studied.[18]

Attempts to prepare Fe pc O do not succeed since the resonance Raman spectra of Fe pc O_2 are observed.[19]

Resonance Raman spectra of α-Cu pc are obtained at room temperature and at 10 K between 457 to 714 nm.[20]

The molecular orientation of submicron films of pc is studied by polarization Raman techniques.[21]

The Raman spectra of chloroaluminum, chlorogallium, and chloroindium pc thin 200-nm thick solid films are studied at excitation frequencies in resonance and near resonance.[22]

Molecular electronic devices are based on the electrooptical behavior of a pc monolayer on a silver electrode, exhibiting unusually strong Raman-emitted light.[23]

A study is made of the coverage frequency dependence of SERS of Co pc on calcium fluoride-roughened silver films.[24]

Raman spectroscopic studies of the molecular orientation and crystal structure are made of thin films of Al pc Cl.[25]

SERS from Cu pc thin films deposited onto silver films roughened by underlayers of gas evaporated silicon particles is studied.[26]

SERS of thin Cu pc films on evaporated silver films is studied.[27]

SERS spectra are obtained from Fe pc tetrasulfonic acid over a silver-silver bromide underlayer.[28]

The Raman enhancement factor is determined for Cu pc molecules deposited onto random rough silver surfaces.[29]

Langmuir-Blodgett films of arachidic acid are used as spacer layers to control the separation distance between evaporated indium and silver metal island films and Langmuir-Blodgett monolayers of a lightly substituted pc. The distance dependence of the SERS enhancement factor for the pc is long range.[30]

There is a correlation between SERS and photographic processes. Analysis of adsorbed photoactive silver bromide and Fe-tetrasulfonated pc indicates a possible charge-transfer pathway that may be involved in the high gain mechanism of the SERS effect.[31]

A compacted electrode technique is used to obtain a layer of α-Cu pc on a silver surface. A selective enhancement of some Raman bands of the α-Cu pc at the surface indicates a SERS effect.[32]

Resonance Raman scattering (RRS) spectra support a Langmuir-Blodgett film structure in which pc molecules are stacked with only the bottom members of the stacks contacting the surface.[33]

II. MASS SPECTROMETRY

The mass spectra of di hydrogen, Co, Cu, Fe, Ni, Pt, Zn, $GeCl_2$ and GeH_2 phthalocyanines (pcs) are measured by electron impact ionization-mass spectrometry (EI-MS).[34] EI-MS of Al pc Cl is also studied.[35]

Fast atom bombardment mass spectra of two series of M pcs are determined. The first series includes di hydrogen, Be, Co, Cu, Fe, Mg, Ni, Pb, and Zn pcs. The second series includes dihydrogen, Be, Co, Cu, Fe, Mg, Ni, Pb, Pd, Pt, and Zn tetrakis (cumylphenoxy) pcs. The technique is a facile method to obtain molecular weight information.[36]

Mass spectrometric fragmentation reactions are studied of Mg tetrakis (1,1-dimethylethyl) pc.[37]

Mass spectroscopic determinations are made at various temperatures for DEN 438 epoxylated novolac resin alone and after curing with Co, Cu, or Ni pc tetraamines. Possible modes of fragmentation are suggested.[38]

Charge compensation is studied in SIMS analysis of polymer foils containing Cu pc. Low-energy electron beams are used with 8-keV Xe^+ ion beams.[39]

III. MOESSBAUER AND PENNING IONIZATION ELECTRON SPECTROSCOPY

Partially oxidized Fe pc (pyrazine) $I_{0.26}$ and Fe pc (bipyridine) $I_{1.06}$ are prepared and their Fe-57 Moessbauer spectra and magnetic moments are obtained.[40]

Moessbauer spectra are obtained for Fe(III) μ-oxobis pc.[41]

Moessbauer spectroscopy is used to study Fe pc homopolymer linked to activated carbon by imidazole.[42]

Moessbauer spectra of tetracoordinate Fe(II) pc complexes with electron withdrawing substituents are obtained.[43]

Moessbauer and ESR spectroscopic studies are made of Fe(II) 2,3,9,10,16,17,23,24-octacarboxy-, 2,9,16,23-tetracarboxy-, 2,9 (10)-, and 2,16 (17)-bis(3′,4′-dicarboxybenzoyl) pc.[44]

Local electron distributions of individual molecular orbitals of Fe pc are studied by Penning ionization electron spectroscopy (PIES).[45]

Fe pc adsorbed on a graphite substrate is studied by PIES. For monolayers and island structures, the molecules are oriented flat to the substrate.[46]

Moessbauer and magnetic susceptibility measurements are made of Fe(II) pc at 77 to 300 K.[47]

The calculated Moessbauer shift agrees with the spectroscopically observed value in Fe pc.[48]

Moessbauer spectroscopy and cyclic voltammetry show that heat treatment of Fe pc at 300°C leads to decomposition of the Fe pc to yield FeOOH as the only Fe-containing species detected by Moessbauer spectroscopy at room temperature.[49]

A range of (Fe(II) pc-pyrazine)$_n$ I$_x$ and (Fe(III) pc-pyrazine)$_n$ I$_x$ complexes among other bidentate bridging base polymer complexes are studied.[50]

Magnetic susceptibility, Moessbauer, and ESR measurements over a wide temperature range are used to determine the ground states and electronic structure of a range of Fe(III) pc complexes.[51]

(Fe pc·*s*-tetrazine)$_n$, Fe nc·(*s*-tetrazine)$_2$, and [Fe nc (*s*-tetrazine)0.5CHCl$_3$]$_n$ are characterized by Moessbauer spectroscopy (nc = naphthalocyanine).[52]

Thermal condensation of [(*tert*-butyl)$_4$pc]Sn(OH)$_2$ gives the cofacially stacked polymer {[(*tert*-butyl)$_4$pc]SnO}$_n$ which is observed in a variable temperature Moessbauer spectroscopic experiment to be rigid in structure.[53]

IV. IR SPECTROSCOPY

Matrix isolation IR spectra of Fe pc O_2 are made.[54] Assignment of the far IR spectrum of μ-oxo-(Fe pc)$_2$ suggests a linear arrangement of the Fe-O-Fe with iron atoms in the plane of the pc ring.[55]

A process is developed for fabricating a vapor deposition film having an absorption maximum in the near-IR range of Co, Cu, Er, Lu, Ni, Pb, Pt, Sn, Ti, and Zn pcs, optionally chlorinated, in an inert atmosphere.[56]

Optical sensors have a photoelectric conversion layer containing In nc. The product is very sensitive to near-IR radiation at 600 to 1000 nm.[57]

V. MAGNETIC CIRCULAR DICHROISM

The temperature dependence of the dichroism of absorption of dyes including Ga pc Cl is studied in a liquid crystal matrix.[58]

The bleaching and dichroism of Ga pc Cl is studied in nematic liquid crystals.[59]

An analysis is made of the absorption and magnetic circular dichroism (MCD) spectra of Zn pc and the π-cation-radical species [Zn pc (1-)]$^{\cdot +}$.[60]

Vacuum UV MCD and absorption spectra are measured for Zn pc isolated in an argon matrix using synchrotron radiation at about 37,000 to 74,000 1/cm.[61]

VI. PICOSECOND PROCESSES

Excited sodium Cu pc tetrasulfonate adsorbed on titania injects an electron in about 20 to 50 ps. The conduction band electron is trapped in titania localized sites in about 500 ps.[62]

Subpicosecond excitation of polycrystalline thin film samples of β-pc results in excitonic saturation of the crystal lattice.[63,64]

Subpicosecond time-resolved absorption spectroscopy is used to measure the characteristic annihilation rate for singlet excitons on adjacent lattice sites. Singlet exciton fusion is determined in solid pc.[65]

Direct picosecond time-resolved absorption measurements are made which determine excited singlet state lifetimes for Co pc and Ni pc in 1-chloronaphthalene solution to be 2.1 and 2.8 ps, respectively.[66]

Picosecond measurements are made of the fluorescence kinetics of Lu pc.[67]

VII. IODINE-DOPED PCS

UV and IR absorption and ESR spectra of iodine complexes with di hydrogen, Co, Cu, Fe, Mn, Ni, VO, and Zn pcs are drawn. The spectra are discussed in terms of the electronic spectra of the complexes.[68]

The electron paramagnetic (EPR) spectra of crystalline Cu pc I_x, $0.2 \leqslant x \leqslant 1.7$, are determined. The place of localization of I^- is established.[69]

Photoconductive effect spectra of an evaporated film of α-Cu pc I are measured.[70]

The stable doping of Al pc F polymer thin film by ion implantation of I^- is studied by UV, visible, and IR spectra.[71]

VIII. STACKED STRUCTURES

Electronic and redox properties of stacked ring silicon pcs are determined from molecular orbital theory. For the dimers, a 45° staggered conformation is predicted to be slightly more stable than the eclipsed conformation.[72]

Stacked or bridge stacked polymeric structures of metalloporphyrins and pcs are studied by difference EXAFS spectroscopy. Recent use of the method includes structural characterization of β-Ni pc and of Ni pc I.[73]

IX. DIMERS AND BINUCLEAR PCS

The role of oxygen is studied by visible spectroscopy in the redox chemistry of Lu pc_2.[74]

Absorption spectra show that some substituted pcs form dimers in less polar solvents, and, since they are not photoactive, they cannot be used as sensitizers in photoinduced redox reactions.[75]

The electronic absorption and emission spectra of mononuclear and binuclear di hydrogen pcs are studied. The binuclear species are linked through a benzene ring by a bridge of zero, one, two, four, or five atoms.[76]

The spectroelectrochemistry of iron porphyrins and pc complexes is investigated including the interaction with chalcocarbon ligands and dimerization with single- and multiatom bridging ligands.[77]

Near IR absorption spectra of Dy, Lu, Tm, and Yb pc_2 show two relatively intense bands.[78]

Magnetooptical studies are made of substituted di pcs of rare earth elements.[79]

X. DI HYDROGEN PCS

The McLachlan approximation is applied to π-electron open shell crystal orbital calculations on columnar stacks of macrocyclic radical ions, including that of H_2 pc.[80]

The temperature dependence of the hole-burning linewidth of H_2 pc is studied in polystyrene and polyethylene matrices.[81]

Photochemical hole burning at 1.5 K is used to study the influence of uniaxial pressure on the homogeneous line profile of the $S_1 \leftarrow S_0$ transition of H_2 pc on polyethylene and polymethylmethacrylate matrices.[82]

A RATIO method is developed to obtain physical structure information from thin H_2 pc films on air-oxidized aluminum using reflection absorption spectroscopy (RAIR).[83]

Frequency modulation polarization spectroscopy is used to detect persistent anisotropic spectral holes in the 683-nm absorption in pc-doped polyethylene.[84]

Photoemission spectra of pc are obtained over the wavelength range $5.4 \leq h\nu \leq 14.4$ eV.[85]

The optical properties of pc films are determined by spectroellipsometry.[86]

The electronic absorption and emission spectra of a series of mono- and binuclear pc species are obtained; the binuclear species are linked through a benzene ring by a bridge of zero, one, two, four, or five atoms.[87]

A correlation is obtained between pressure-induced optical shifts and compressibilities of polymer doped with pc molecules.[88]

Laser hole burning, a spectroscopic method based on site selection by laser irradiation experiments are made in polymer doped with pc.[89]

Hole-burning experiments are made with $(CH_2)_x$, polymethylmethacrylate, and polystyrene doped with pc.[90]

A laser spectral luminescent apparatus allows a study of spectral characteristics of complex molecules in a supersonic molecular jet. The apparatus is checked with control measurements of the luminescence excitation of pc molecules.[91]

Photoemission partial cross-section spectra of pc are measured and interpreted in terms of Auger processes.[92]

A global least-squares fit of the extinction coefficient of triplet-triplet absorption spectra of pc is developed.[93]

The electronic absorption spectrum of H_2 tetra-*tert*-butyl pc is studied in various solvents. All bands are slightly red shifted on increasing polarity of the solvents, indicating π-π^* transitions.[94]

The proton dynamics in crystalline pc are studied by two-dimensional carbon-13 NMR spectroscopy. Rates of two-site proton exchange are determined.[95]

ESR detection of charge exchange is made in redox reactions in dispersed pc particles. It is indicated that interactions between the molecules are limited and only local interactions or clusters may exist.[96]

XPS is used to study the role of oxygen adsorption on pc thin film ammonia sensors. It is indicated that for pc films previously exposed to air oxygen is primarily adsorbed on the surface as water vapor.[97]

The reaction between polycrystalline β-pc and sodium vapor, as a result of which $Na_x(H_2$ pc) with $0 < x \leq 8.5$ are formed, is studied. ESR data show that the electronic structure of the H_2 pc molecule is a function of the level of sodium doping.[98]

Electron beam radiation damage is studied in crystalline films of H_2 pc. A mechanism for radiation damage in the electron microscope is proposed.[99]

XI. CU PCS

Pc and glow discharge polymers are studied using inelastic tunnel spectroscopy. Special emphasis is placed on Cu and Pb pc.[100] Electroreflectance spectra are measured on β-Cu pc single crystals at 77 K.[101]

The visible reflectance spectrum of a Cu pc single crystal is measured at 30 and 300 K in the region of the Q band.[102]

The Fourier Transform IR spectrum of Cu pc is measured using an electrothermal atomizer for vaporization.[103]

High-resolution soft X-ray excitation spectra of 3d metal pcs, Cu and Mn pc, are determined. The results show that high-resolution soft X-ray spectroscopy has possibilities as a site specific local probe of empty electronic states in complex organic molecules.[104]

An XPS study is made on α-, β-, γ-, and x-Cu pc.[105]

A near-IR absorber that does not shield visible light and thus is useful in silicon photodiodes is made from Cu tetraamino pc as a starting point.[106]

Transmission electron diffraction data for Cu tris(CH_2NH iso-C_3H_7) pc show that realistic monomolecular areas can be achieved for a number of compounds in terms of making electronic devices incorporating stable Langmuir-Blodgett films.[107]

Thermoelectroluminescence studies are made on Zn and Cu pc films deposited on transparent and conducting plates.[108]

Electroabsorption spectra are measured on sublimed films of α- and β-Co, Cu, Ni, and Zn pc.[109]

Solid solutions of Cu octadodecyl pc in H_2 octadodecyl pc are formed and their luminescence properties are studied.[110]

Luminous layers containing Cu pc are prepared for an apparatus to evoke color.[111]

Dispersions of Cu pc in N-methylpyrrolidone are used for the electron microscopic determination of particle size distribution.[112]

Electron beam radiation damage is studied in crystalline films of Cu pc.[113]

The temperature shift of the long-range hyperfine structure of the ESR line of Cu pc (O- and N-containing organic bases) is studied at 77 to 363 K.[114]

Monolayers of Cu *tert*-tetrabutyl pc are spread at 20° from xylene solution and deposited on an electron microscope grid. The molecules seem to lie in columns in the plane of the film with intermolecular separation of 0.33 nm and intercolumnar separation of 1.9 nm.[115]

Hyperfine coupling constants of coordinating nitrogen atoms in Cu pc with a pseudo-planar coordination array are collected by the nitrogen-14 ENDOR spectroscopy.[116]

The molecular orientation of Cu *tert*-tetrabutyl pc (Figure 1) in a Langmuir-Blodgett film is determined using ESR spectroscopy. The molecules are aligned with the plane of the ring at 80 ± 10° to the substrate surface.[117]

The ESR spectral characteristics of delocalized metal-ligand bonds in Cu pc complexes are calculated.[118]

An X-ray photoelectron spectroscopic study is made of Cu pc, showing π-electron delocalization over metal and ligand.[119]

Using scanning tunneling microscopic (STM) images of individual Cu pc molecules adsorbed on polycrystalline silver under ultra-high-vacuum conditions are made.[120]

XII. CO, FE, MN, NI, AND ZN PCS

The spectral behavior of Co pc-loaded zeolite single crystals is compared with that of Co pc films produced at ≤65 kbar pressure in a diamond anvil cell.[121]

The fluorescence spectra quantum yields for Mg, Zn, and Co pcs are 0.7, 0.3, and 0.025, respectively.[122]

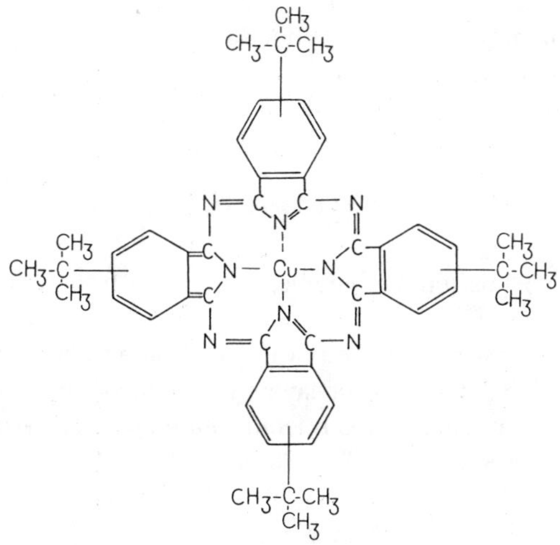

FIGURE 1. Cu *tert*-tetrabutyl pc.

The reflectance spectra of Ni pc films on glass are determined vs. Ni pc thickness.[123]

ESR tests on conductive columnar stacks of Ni pc partially oxidized with iodine indicate that carrier electrons are primarily localized onto the ligands of Ni pc.[124]

An ESR study is made of the formation of a molecular oxygen adduct of the Mn(II) tetrasodium 3,10,17,24-tetrasulfo pc. An Mn(II)-O_2 complex is observed in slightly alkaline aqueous solution.[125]

Superoxo- and μ-peroxo di cobalt complexes are identified by ESR and reflection spectroscopy on silica containing different concentrations of Co pc pyridine complexes chemically bonded to the surface.[126]

Reflection spectroscopy shows that Co pc complexes on a modified aerosil surface react with oxygen to form superoxides and superoxide-bridged derivatives.[127] The complexes are also studied by ESR.[128]

Gas phase oxidative addition reactions of Co pc with alkyl radicals are studied by negative chemical ionization mass spectrometry.[129]

Polycrystalline samples of Co pc derivatives including Co 3,9,16,23-tetrabenzenethiotetra-*tert*-butyl pc are studied by ESR and magnetic susceptibility measurements.[130]

Absorption spectra of reduced pc prepared by cocondensation under high vacuum of Co pc or Fe pc with potassium show four reduction states corresponding to the mono-, di-, tri-, and tetraanion of Co pc.[131]

In part 5 of spectroscopic and electrochemical studies of the transition metal tetrasulfonated pcs, voltammetric studies are made of adsorbed tetrasulfonated pcs in aqueous solutions. Cyclic voltammetric measurements of adsorbed Fe- and Co-tetrasulfonated pc and of Co pc are carried out on ordinary pyrolytic graphite and silver electrodes at pH 1 to 13. Many voltammetry peaks are pH dependent.[132]

The data are explained in part 6 by a model of the adsorbed macrocycle which has the plane of the ligand perpendicular to the electrode surface.[133]

ESR studies are made on Fe poly pcs. The spin concentration of paramagnetic centers in the polymer, prepared by polymerization of pyromellitic dianhydride with urea and ferrous chloride, is affected by the ferrous chloride-anhydride ratio in the polymerization.[134]

A spectrophotometric and X-ray absorption edge study of complexation of cobalt with Fe(III) tetrasulfo pc in alkaline solution indicates that the combined effect of CO and hydroxide on the Fe center enables reduction of Fe(III) to Fe(II) by the macrocyclic ring.[135]

X-ray absorption spectroscopy indicates *in situ* formation of pc-bis(pyridine) Fe(II). It is indicated that the complex is formed without a net transfer of electron charge between the central atom and the pyridine ligands.[136]

The interaction between thionyl chloride, $SOCl_2$, and Fe or Mo pc, or Fe or Mo nc is studied by XPS. Thionyl chloride is reduced and chlorine is rapidly fixed on the Fe or Mo.[137]

XIII. PB PC

Angle-resolved photoemission measurements are made from oriented films of Pb pc on a copper (100) surface.[138]

Thickness dependent absorption spectra (350 to 1100 nm) are made of Pb pc films.[139]

The UV-visible absorption spectra of layers of Pb pc sublimed onto surfaces at 77 K are amorphous but those sublimed at room temperature are polycrystalline.[140]

ESR signals are observed in crystalline Pb pc.[141]

XIV. MG PC

CNDO/S calculations are made of the electronic spectra of Mg pc and of its ionic forms.[142]

The absorption and fluorescence of Mg pc and of Al pc Cl, and of Mg and Al Cl *tert*-butyl pcs are studied in solvents of varying polarity.[143]

The phosphorescence quenching of Mg pc is studied by silver porphyrins.[144]

XV. PT FAMILY PCS

Co, Cu, Ni, and Ru pcs are fixed in Y zeolites by surface assembling them or by exchanging Na^+ for $[RuNO(NH_3)_4OH]^{2+}$ and $[Rh(NH_3)_5Cl]^{2+}$ cations; they are studied by XPS.[145]

XPS spectra are obtained of Ru(II) pcs.[146]

Pt pc is a nonlinear material that may have application in novel laser-controlled laser-electron beam generation.[147]

A Zeeman study is made of Pt pc in Shpol'skii matrices at 25 to 77 K.[148]

The spectral and transient behavior of the first excited singlet and the lowest triplet state of randomly oriented Pd pc molecules in a Shpol'skii matrix is studied at liquid helium temperatures.[149]

The Zeeman splitting in Shpol'skii matrices is studied at 25 K.[150] The effects of three different mixed solvent Shpol'skii matrices at 4.2 K are studied on site formation and on vibronic progressions of Pd pc.[151]

The dyeing and diffusion of Pt chlorosulfo pc in wool fibers is studied by transmission electron microscopy. The dye enters the fiber at the junctions between the cuticle cells.[152]

XVI. RARE EARTH PCS

The absorption spectra and deactivation kinetics of triplet state molecules of tetra-4-*tert*-butyl pc and its metal complexes in toluene are studied by flash photolysis. The metals include erbium, gallium, hafnium, indium, lanthanum, lutetium, neodymium, samarium, scandium, thulium, tin, magnesium, ytterbium, yttrium, and zinc.[153]

XVII. VANADYL (VO) PCS

An X-ray valence photoelectron spectrum of VO pc is obtained.[154]

High-resolution vanadium K-edge absorption spectra are recorded for a number of vanadium compounds by synchrotron radiation available at the Stanford Synchrotron Radiation Laboratory. The compounds include VO pc.[155]

High-resolution vanadium K-edge absorption spectra are recorded for a number of vanadium compounds including vanadyl pc.[156]

Electroabsorption spectra of vanadyl pc films are obtained under electric fields close to one million volts per centimeter in the visible and IR regions with significant electric dipole moment and polarizability increases.[157]

The structure of binuclear associations of VO^{2+} and Cu complexes of substituted pcs in acidic and basic solutions is studied by ESR. Substituted pcs with VO^{2+} in solution form dimerize the associates in which the metal-metal internuclear distance is 4.1 to 12 Å.[158]

The binding and decomposition of VO(IV) pc, tetraphenylporphyrin, and etioporphyrin on hydrotreating catalysts is studied by X-ray photoelectron and UV-visible spectrometry.[159]

VO tetra (3-nitro) pc and VO hexadecachloro pc are prepared. The protonation of the latter by sulfuric acid causes a displacement of the Q band into the near-IR region.[160]

The Q band electronic absorption of VO pc is determined.[161]

XVIII. AL AND CR PCS

Enhanced S-T transitions of pc coordinating to paramagnetic chromium are studied. The absorption bands are assigned to ^{3}LE (LE = ligand excitation). The lowest excited state of Cr pc OH is ^{3}LE of the pc with 35-ns lifetime.[162]

Cr(III) pc complexes exhibit a fine structure absorption band with a moderate intensity below the strong band of ^{1}LE.[163]

Transient absorption spectra are obtained using diffuse reflectance laser flash photolysis for Al-sulfonated pc on water-saturated cotton fibers and for titania and cadmium sulfide powders.[164]

Third-harmonic generation is performed in reflection from thin polycrystalline films of Al pc Cl, Al pc F, and gallium halo pcs.[165]

The IR spectrum of Al pc Cl is studied in the presence of aluminum bromide. A drastic increase in aluminum bromide concentration increases the shift in the absorption band.[166]

An electronic absorption spectral study is made of the reaction of Al pc Cl with acetyl chloride.[167]

XIX. OXIDATION-REDUCTION

The oxidation and reduction of M pcs give rise to at least six oxidation states the electronic spectra of which are rationalized by SCF energies and Coulomb and exchange integrals. The variation in the Q band energies provides a means to identify whether it is the metal or the pc ligand that is reduced upon addition of an electron to the metal pc.[168,169]

The redox potentials of a series of Zn pcs are measured by cyclic voltammetry. The HOMO, LVMO levels, and redox potential of the excited state are calculated and confirmed with fluorescence quenching of the Zn pcs by electron acceptors having different potentials.[170]

XX. LABORATORY EXPERIMENT

A laboratory experiment uses Cu-tetrasulfonated pc to introduce students to spectrophotometry and Beer's Law.[171]

XXI. REVIEW

A review with 37 references discusses "remarkable organic optical and electronic materials — phthalocyanines."[172]

REFERENCES

1. **Jennings, C., Aroca, R., Hor, A. M., and Loutfy, R. O.,** Raman spectra of solid films. III. Magnesium, copper and zinc phthalocyanine complexes, *J. Raman Spectrosc.*, 15(1), 34, 1984.
2. **Hayashi, S. and Samejima, M.,** Surface-enhanced Raman scattering from copper phthalocyanine thin films, *Surf. Sci.*, 137(2—3), 442, 1984.
3. **Simic-Glavaski, B., Zecevic, S., and Yeager, E. B.,** Spectroscopic and electrochemical studies of transition metal tetrasulfonated phthalocyanines. II. Resonant Raman spectra of aqueous solutions of cobalt and iron tetrasulfonated phthalocyanines, *J. Raman Spectrosc.*, 14(5), 338, 1983.
4. **Aroca, R. and Loutfy, R. O.,** Raman spectra of solid films. II. Vanadyl phthalocyanine, *Spectrochim. Acta, Part A*, 39A(10), 847, 1983.
5. **Bajdor, K., Oshio, H., and Nakamoto, K.,** Observation of two oxygen-isotope sensitive bands in the low-frequency resonance Raman spectrum of oxy(phthalocyaninato)iron(II), *J. Am. Chem. Soc.*, 106(23), 7273, 1984.
6. **Jennings, C., Aroca, R., Hor, A. M., and Loutfy, R. O.,** Surface-enhanced Raman scattering from copper and zinc phthalocyanine complexes by silver and indium island films, *Anal. Chem.*, 56(12), 2033, 1984.
7. **Ohsawa, M., Hashima, K., and Suetaka, W.,** Angular dependence of scattering intensity from thin films in total reflection Raman spectroscopy, *Appl. Surf. Sci.*, 20(1—2), 109, 1984.
8. **Saad, E., Lippitsch, M. E., Leitner, A., and Aussenegg, F. R.,** Surface-enhanced Raman scattering as a diagnostic method in preparing organic semiconductor films, *Springer Ser. Chem. Phys.*, 39, 486, 1984.
9. **Melendres, C. A. and Maroni, V. A.,** Raman spectra and normal coordinate analysis of the planar vibrations of iron phthalocyanine, *J. Raman Spectrosc.*, 15(5), 319, 1984.
10. **Hayashi, S. and Samejima, M.,** SERS activity of gas-evaporated silver particles, *Solid State Commun.*, 55(12), 1085, 1985.
11. **Nimmo, J. A., Bovill, A. J., McConnell, A. A., and Smith, W. E.,** The use of compacted metal disks to measure Raman spectra, *J. Raman Spectrosc.*, 16(4), 245, 1985.
12. **Bartholomew, C. R., McConnell, A. A., and Smith, W. E.,** Crystal standard/metal disk method for the measurement of low temperature resonance Raman spectra in the solid state, *J. Raman Spectrosc.*, 18(4), 277, 1987.
13. **Simic-Glavaski, B. (H.S.G. Venture),** PCT International Application 85,01,617, 1985.
14. **Aroca, R., Jennings, C., Kovac, G. J., Loutfy, R. O., and Vincett, P. S.,** Surface enhanced Raman scattering of Langmuir-Blodgett monolayers of phthalocyanine by indium and silver island films, *J. Phys. Chem.*, 89(19), 4051, 1985.
15. **Jennings, C., Aroca, R., Hor, A. M., and Loutfy, R. O.,** Raman spectra of solid films. IV. Lead and tin phthalocyanine complexes, *Spectrochim. Acta, Part A*, 41A(9), 1095, 1985.
16. **Simic-Glavaski, B., Zecevic, S., and Yeager, E.,** Spectroscopic and electrochemical studies of transition-metal tetrasulfonated phthalocyanines. III. Raman scattering from electrochemically adsorbed tetrasulfonated phthalocyanines on silver electrodes, *J. Am. Chem. Soc.*, 107(20), 5625, 1985.
17. **Brotman, A. and Burstein, E.,** Raman scattering-excitation profiles of molecules adsorbed on metal surfaces: free-base phthalocyanine on silver, *Phys. Scr.*, 32(4), 385, 1985.
18. **Aroca, R.,** Raman spectra of phthalocyanines, *J. Mol. Struct.*, 143, 131, 1986.
19. **Proniewicz, L. M., Bajdor, K., and Nakamoto, K.,** *J. Phys. Chem.*, 90(9), 1760, 1986.
20. **Bovill, A. J., McConnell, A. A., Nimmo, J. A., and Smith, W. E.,** Resonance Raman spectra of α-copper phthalocyanine, *J. Phys. Chem.*, 90(4), 569, 1986.
21. **Aroca, R., Jennings, C., Loutfy, R. O., and Hor, A. M.,** The structure of organic thin films: Raman polarization studies, *J. Phys. Chem.*, 90(21), 5255, 1986.
22. **Jennings, C., Aroca, R., Hor, A. M., and Loutfy, R. O.,** Raman spectra of solid films. V. Chloroaluminum, chlorogallium and chloroindium phthalocyanine complexes, *Spectrochim. Acta, Part A*, 42A(9), 991, 1986.
23. **Simic-Glavaski, B.,** Molecular electronic devices based on electrooptical behavior of heme-like molecules, *Proc. SPIE-Int. Soc. Opt. Eng.*, 634, 195, 1987.
24. **Zeman, E. J., Carron, K. T., Schatz, G. C., and Van Duyne, R. P.,** A surface enhanced resonance Raman study of cobalt phthalocyanine on rough silver films, *J. Chem. Phys.*, 87(7), 4189, 1987.
25. **Aroca, R., Jennings, C., Loutfy, R. O., and Hor, A. M.,** Raman spectra of thin solid films. VI. Davydov splitting, *Spectrochim. Acta, Part A*, 43A(6), 725, 1987.
26. **Koh, R., Hayashi, S., and Yamamoto, K.,** Optimum surface roughness for surface enhanced Raman scattering, *Solid State Commun.*, 64(3), 375, 1987.
27. **Osawa, M. and Suetaka, W.,** Surface enhanced Raman scattering on evaporated silver films: the importance of short-ranged electromagnetic fields in the film, *Surf. Sci.*, 186(3), 583, 1987.
28. **Simic-Glavaski, B., Simic, M. G., Junger, P. D., Doyle, D. M., and Jaros, S. T.,** Europe Patent Application 216,551, 1987.

29. **Hayashi, S.,** SERS on random rough silver surfaces: evidence of surface plasmon excitation and the enhancement factor for copper phthalocyanine, *Surf. Sci.*, 158(1—3), 229, 1985.
30. **Kovacs, G. J., Loutfy, R. O., Vincett, P. S., Jennings, C., and Aroca, R.,** Distance dependence of SERS enhancement factor from Langmuir-Blodgett monolayers on metal island films: evidence for the electromagnetic mechanism, *Langmuir*, 2(6), 689, 1986.
31. **Simic-Glavaski, B.,** Spectroscopic and electrochemical studies of transition-metal tetrasulfonated phthalocyanines. VII. A correlation between surface-enhanced Raman scattering and photography, *J. Phys. Chem.*, 90(17), 3863, 1986.
32. **Bovill, A. J., McConnell, A. A., Nimmo, J. A., and Smith, W. E.,** Raman scattering of α-copper phthalocyanine and other complexes formed at a silver surface using a compacted electrode, *Surf. Sci.*, 158(1—3), 333, 1985.
33. **Dilella, D. P., Barger, W. R., Snow, A. W., and Smardzewski, R. R.,** Resonance Raman spectra of Langmuir-Blodgett monolayers, *Thin Solid Films*, 133, 207, 1985.
34. **Isa, K., Sasaki, K., and Mizuta, K.,** EI-MS [electron impact ionization-mass spectrometry] of mixture of various phthalocyanine complexes and dipotassium hexachloroplatinate, *Shitsuryo Bunseki*, 32(3), 305, 1984.
35. **Isa, K., Sasaki, K., Murano, I., Fukui, K., and Mizuta, K.,** Mass spectra of phthalocyanine complexes. II. Electron ionization mass spectra of phthalocyanine complexes, *Nippon Kagaku Kaishi*, 3, 615, 1985.
36. **Freas, R. B. and Campana, J. E.,** Fast atom bombardment mass spectrometry of phthalocyanines, *Inorg. Chem.*, 23(26), 4654, 1984.
37. **Budzikiewicz, H., Taraz, K., Muenster, H., Wagestian, F., and Heucher, R.,** Mass spectrometric fragmentation reactions. XXX. Peculiarities of the mass spectroscopic fragmentation behavior of alkyl substituted porphyrin and phthalocyanine metal(II) complexes, *Monatsh. Chem.*, 116(4), 525, 1985.
38. **Achar, B. N., Fohlen, G. M., Hsu, M. S., and Parker, J. A.,** Mass spectroscopy of epoxylated novolac resin cured with phthalocyanine tetraamines, *J. Polym. Sci., Polym. Chem. Ed.*, 22(6), 1471, 1984.
39. **Wittmaack, K.,** Charge compensation in SIMS analysis of polymer foils using negative secondary ions, *SIA, Surf. Interface Anal.*, 10(6), 311, 1987.
40. **Molins, E., Labarta, A., Tejada, J., Caubet, A., and Alvarez, S.,** Moessbauer studies on one-dimensional iron phthalocyanines, *Transition Met. Chem. (Weinheim, Ger.)*, 8(6), 377, 1983.
41. **Frampton, C. S. and Silver, J.,** A Moessbauer investigation of the interaction of (phthalocyaninato)iron(II) with molecular oxygen. Evidence for only one form of (μ-oxo)bis(phthalocyaninato)iron(III) is found, *Inorg. Chim. Acta*, 96(2), 187, 1985.
42. **Blomquist, J., Helgeson, U., Moberg, L. C., Johansson, L. Y., Larsson, R., Tov, B. Y., and Joensson, L.,** On the nature of the imidazole link between activated carbon and polymeric iron phthalocyanine studied by Moessbauer spectroscopy, *Spectrosc. Lett.*, 18(8), 575, 1985.
43. **Ohya, T., Kobayashi, N., and Sato, M.,** Bonding character in tetracoordinate (phthalocyaninato)iron(II) complexes with electron-withdrawing substituents as studied by Moessbauer spectroscopy, *Inorg. Chem.*, 26(15), 2506, 1987.
44. **Kobayashi, N., Shirai, H., and Hojo, N.,** Iron(III) phthalocyanines: oxidation and spin states of iron in iron phthalocyanines with carboxyl groups, *J. Chem. Soc., Dalton Trans.*, 10, 2107, 1984.
45. **Ozaki, H. and Harada, Y.,** Local electron distribution of individual molecular orbitals studied by Penning ionization electron spectroscopy: iron phthalocyanine, *J. Am. Chem. Soc.*, 109(3), 949, 1987.
46. **Harada, Y., Ozaki, H., Ohno, K., and Kajiwara, T.,** Observation of structural change in organic monolayer film by Penning ionization electron spectroscopy: Fe-phthalocyanine on graphite, *Surf. Sci.*, 147(2—3), 356, 1984.
47. **Labarta, A., Molins, E., Vinas, X., Tejada, J., Caubet, A., and Alvarez, S.,** Electronic structure determination of iron(II) phthalocyanine via magnetic susceptibility and Moessbauer measurements, *J. Chem. Phys.*, 80(1), 444, 1984.
48. **Coppens, P. and Li, L.,** Electron density studies of porphyrins and phthalocyanines. III. The electronic ground state of iron(II) phthalocyanine, *J. Chem. Phys.*, 81(4), 1983, 1984.
49. **Scherson, D. A., Fierro, C. A., Tryk, D., Gupta, S. L., Yeager, E. B., Eldridge, J., and Hoffman, R. W.,** *In situ* Moessbauer spectroscopy and electrochemical studies of the thermal stability of iron phthalocyanine dispersed in high surface area carbon, *J. Electroanal. Chem. Interfacial Electrochem.*, 184(2), 419, 1985.
50. **Wei, H. H. and Shyu, H. L.,** Moessbauer spectra and electric conductivities of bidentate base bridged iron(II) phthalocyanine, *Polyhedron*, 4(6), 979, 1985.
51. **Kennedy, B. J., Murray, K. S., Zwack, P. R., Homborg, H., and Kalz, W.,** Spin states in iron(III) phthalocyanines studied by Moessbauer magnetic susceptibility, and ESR measurements, *Inorg. Chem.*, 25(15), 2539, 1986.
52. **Keppeler, U., Deger, S., Lange, A., and Hanack, M.,** 1,2,4,5-Tetrazine as ligand for monomer and bridged macrocyclic complexes, *Angew. Chem.*, 99(4), 349, 1987.

53. **Molloy, K. C. and Quill, K.**, Variable-temperature tin-119 Moessbauer study of tin phthalocyanines, *Polyhedron,* 5(11), 1771, 1986.
54. **Watanabe, T., Ama, T., and Nakamoto, K.**, Matrix-isolation infrared spectra of dioxygen adducts of iron(II) porphyrins and related compounds, *J. Phys. Chem.*, 88(3), 440, 1984.
55. **Hutchinson, B., Spencer, B., Thompson, R., and Neill, P.**, Metal-nitrogen stetching assignments in some metallophthalocyanines and μ-oxo-(iron phthalocyanine)$_2$, *Spectrochim. Acta, Part A,* 43(5), 631, 1987.
56. **Kawaguchi, T. and Takeya, Y.**, Japan Patent 87,77,455, 1987.
57. **Umibe, K., Nishioka, Y., Kato, M., and Kawamura, K.**, Japan Patent 87,47,169, 1987.
58. **Kovalev, A. A., Nekrasov, G. L., and Serak, S. V.**, *Vesti Akad. Navuk BSSR, Ser. Fiz.-Mat. Navuk,* 5, 68, 1984.
59. **Kovalev, A. A., Nekrasov, G. L., and Serak, S. V.**, Bleaching of some dyes in nematic liquid crystals, *Zh. Prikl. Spektrosk.*, 45(3), 400, 1986.
60. **Nyokong, T., Gasyna, Z., and Stillman, M. J.**, Analysis of the absorption and magnetic circular dichroism spectra of zinc phthalocyanine and the π-cation-radical species [ZnPc(1-)]·+, *Inorg. Chem.*, 26(7), 1087, 1987.
61. **Williamson, B. E., Boyle, M. E., Schatz, P. N., Marks, J. P., and Snyder, P. A.**, Vacuum ultraviolet MCD and absorption spectra of zinc phthalocyanine isolated in an argon matrix using synchrotron radiation, *Chem. Phys. Lett.*, 140(5), 483, 1987.
62. **Kirk, A. D., Langford, C. H., Saint Joly, C., Lesage, R., and Sharma, D. K.**, Sub-nanosecond processes in the quenching of a copper phthalocyanine dye adsorbed on titanium oxide, *J. Chem. Soc., Chem. Commun.*, 15, 961, 1984.
63. **Greene, B. I. and Millard, R. R.**, Singlet-exciton fusion in molecular solids: a direct subpicosecond determination of time-dependent annihilation rates, *Phys. Rev. Lett.*, 55(12), 1331, 1985.
64. **Greene, B. I. and Milard, R. R.**, Subpicosecond spectroscopic studies of singlet exciton fusion in molecular solids, *J. Phys. Colloq.*, C7, 371, 1985.
65. **Millard, R. R. and Greene, B. I.**, Singlet exciton fusion in molecular solids, *Springer Ser. Chem. Phys.*, 46, 178, 1986.
66. **Millard, R. R. and Greene, B. I.**, Direct determination of nonradiative relaxation rates in nonfluorescent metallophthalocyanines, *J. Phys. Chem.*, 84(14), 2976, 1985.
67. **Aaviksoo, J., Savikhin, S. F., Stel'makh, G. F., Freiberg, A. M., and Tsvirko, M. P.**, Picosecond measurements of the fluorescence kinetics of metalloporphyrins, *Zh. Prikl. Spektrosk.*, 44(2), 238, 1986.
68. **Vinogradskii, A. G. and Sidorov, A. N.**, Electronic structure and spectra of complexes of phthalocyanines with iodine, *Khim. Fiz.*, 3(3), 380, 1984.
69. **Grigoryan, L. S. and Sharoyan, E. G.**, EPR of iodide-containing copper phthalocyanine compounds, *Fiz. Tverd. Tela (Leningrad),* 27(1), 39, 1985.
70. **Liu, T.**, Photoconductive effect spectra of the evaporated film of αI-phthalocyanine copper, *Ganguang Kexue Yu Kuang Huaxue,* 1, 42, 1986.
71. **Dann, A. J., Fahy, M. R., Willis, M. R., and Jeynes, C.**, Ion implantation of polymeric phthalocyanines, *Synth. Met.*, 18(1—3), 581, 1987.
72. **Anderson, A. B., Gordon, T. L., and Kenney, M. E.**, Electronic and redox properties of stacked-ring silicon phthalocyanines from molecular orbital theory, *J. Am. Chem. Soc.*, 107(1), 192, 1985.
73. **Loos, M., Friant, P., Ascone, I., Goulon, J., Barbe, J. M., Coutsolelos, A., and Guilard, R.**, Stacked or bridge stacked polymeric structures of metalloporphyrins and phthalocyanines investigated by difference EXAFS spectroscopy, *J. Phys. Colloq.*, 8(2), C8/641,C 1986.
74. **Nicholson, M. M. and Weismuller, T. P.**, The role of oxygen in the redox chemistry of lutetium diphthalocyanine, *J. Electrochem. Soc.*, 131(10), 2311, 1984.
75. **Xu, H., Zhou, Q., and Shen, S.**, Photoinduced redox reactions sensitized by metal phthalocyanines. I. Absorption spectra, *Ganguang Kexue Yu Kuang Huaxue,* 1, 32, 1984.
76. **Dodsworth, E. S., Lever, A. B. P., Seymour, P., and Leznoff, C. C.**, Intramolecular coupling in metal-free binuclear phthalocyanines, *J. Phys. Chem.*, 89(26), 5698, 1985.
77. **Gorce, J. N.**, Spectroelectrochemistry of Iron Porphyrin and Phthalocyanine Complexes: Interaction with Chalcocarbonyl Ligands and Dimerization with Single and Multi-Atom Bridging Ligands, Ph.D. thesis, Georgia Institute of Technology, Atlanta, 1986.
78. **Markovitsi, D., Thu Hoa Tran Thi, Even, R., and Simon, J.**, Near infrared absorption spectra of lanthanide bis-phthalocyanines, *Chem. Phys. Lett.*, 137(2), 107, 1987.
79. **Burkov, V. I., Talanina, I. B., Mokshin, V. M., Tomilova, L. G., and Luk'yanets, E. A.**, Magnetooptical studies of substituted di phthalocyanines of rare earth elements, *Opt. Spektrosk.*, 61(4), 786, 1986.
80. **Honeybourne, C. L.**, The application of the McLachlan approximation to π-electron open shell crystal orbital calculations on columnar stacks of macrocyclic radical ions, *Mol. Phys.*, 50(5), 1045, 1983.
81. **Gutierrez, A., Castro, G., Schulte, G., and Haarer, D.**, Dynamical linewidth effects of hold burning of free base phthalocyanine in polymers: spectral diffusion and exchange narrowing, *Springer Ser. Solid-State Sci.*, 49, 206, 1983.

82. **Richter, W., Schulte, G., and Haarer, D.,** Pressure-induced spectral diffusion as investigated by linewidth studies of polymers doped with dye molecules, *Opt. Commun.*, 51(6), 412, 1984.
83. **Debe, M. K.,** Extracting physical structure information from thin organic films with reflection absorption infrared spectroscopy, *J. Appl. Phys.*, 55(9), 3354, 1984.
84. **Romagnoli, M., Levenson, M. D., and Bjorklund, G. C.,** Frequency-modulation polarization-spectroscopy detection of persistent spectral holes, *J. Opt. Soc. Am. B*, 1(4), 571, 1984.
85. **Mikryukov, S. E. and Zagrubskii, A. A.,** Autoionization Processes and Shake-Up Satellites in Photoemission Spectra of Phthalocyanine, Deposited Doc. 1983, VINITI 6764-83, All Union Institute of Scientific and Technical Information, U.S.S.R., 95.
86. **Arwin, H. and Aspnes, D. E.,** Determination of optical properties of thin organic films by spectroellipsometry, *Thin Solid Films*, 138(2), 195, 1986.
87. **Dodsworth, E. S., Lever, A. B. P., Seymour, P., and Leznoff, C. C.,** Intramolecular Coupling in Metal Free Binuclear Phthalocyanines, Report 1985, TR-3, Order No. AD-A160784/5/GAR, National Technical Information Service, Springfield, VA, 44 pp.
88. **Sesselmann, T., Richter, W., and Haarer, D.,** Correlation between pressure-induced optical shifts and compressibilities of polymers doped with dye molecules, *Europhys. Lett.*, 2(12), 947, 1986.
89. **Kador, L., Personov, R., Richter, W., Sesselmann, T., and Haarer, D.,** Laser photochemistry and hole burning spectroscopy in polymers and glasses: external field effects, *Polym. J. (Tokyo)*, 19(1), 61, 1987.
90. **Sesselmann, T., Richter, W., and Haarer, D.,** Hole-burning experiments in doped polymer under uniaxial and hydrostatic pressure, *J. Lumin.*, 36(4—5), 263, 1987.
91. **Asimov, M. M., Portnov, E. V., Rubinov, A. N., and Stepanov, A. A.,** Pulsed-continuous supersonic jet forming for laser spectroscopy of complex molecules, *Zh. Prikl. Spektrosk.*, 47(2), 286, 1987.
92. **Schaefer, K., Kaul, M., and Kampfrath, G.,** Studies of organic substances by inelastic tunneling spectroscopy, *Wiss. Z.-Tech. Hochsch. Karl-Marx-Stadt*, 28(2), 243, 1986.
93. **Mikryukov, S. E., Petrov, V. V., Lovcius, V., and Zagrubskii, A. A.,** Autoionization structure of phthalocyanine, *Elektron Org. Mater.*, 165, 1985.
94. **Carmichael, I., Helman, W. P., and Hug, G. L.,** Extinction coefficients of triplet-triplet absorption spectra of organic molecules in condensed phases: a least-squares analysis, *J. Phys. Chem. Ref. Data*, 16(2), 239, 1987.
95. **Jerwin, K. and Wasgestian, F.,** Solvent dependence of the spin-allowed transitions in free base tetra-*tert*-butylphthalocyanine, *Spectrochim. Acta, Part A*, 40A(2), 159, 1984.
96. **Meier, B. H., Storm, C. B., and Earl, W. L.,** Two dimensional chemical exchange NMR in the solid: proton dynamics in phthalocyanine, *J. Am. Chem. Soc.*, 108(19), 6072, 1986.
97. **Harbour, J. R. and Walzak, M. J.,** Spectroscopic detection of charge exchange in phthalocyanine dispersions, *Langmuir*, 2(6), 788, 1986.
98. **Wilson, A. and Collins, R. A.,** Gas-surface reactions on phthalocyanine thin films, *Phys. Status Solidi A*, 98(2), 633, 1986.
99. **Arutyunyan, A. R., Grigoryan, L. S., and Sharoyan, E. G.,** EPR study of the electronic structure of sodium doped metal-free phthalocyanine (H_2Pc), *Izv. Akad. Nauk Arm. SSR, Fiz.*, 22(2), 109, 1987.
100. **Fryer, J. R.,** Radiation damage in organic crystalline films, *Ultramicroscopy*, 14(3), 227, 1984.
101. **Tokura, Y., Koda, T., Iyechika, Y., and Kuroda, H.,** Electroreflectance spectra of charge-transfer excitations in copper phthalocyanine single crystals, *Chem. Phys. Lett.*, 102(2—3), 174, 1983.
102. **Iyechika, Y., Yakushi, K., and Kuroda, H.,** Temperature dependence of the visible absorption band of copper phthalocyanine single crystal, *Chem. Phys.*, 87(1), 101, 1984.
103. **Tittarelli, P., Zerlia, T., and Ferrari, G.,** Fourier transform infrared spectrometer coupled with electrothermal atomizer for high-temperature vapor-phase investigations, *Appl. Spectrosc.*, 38(5), 715, 1984.
104. **Koch, E. E., Jugnet, Y., and Himpsel, F. J.,** High-resolution soft x-ray excitation spectra of 3d-metal phthalocyanines, *Chem. Phys. Lett.*, 116(1), 7, 1985.
105. **Wang, D. and Shi, Z.,** XPS study on different crystalline forms of copper phthalocyanine, *Kexue Tongbao (Foreign Language Edition)*, 30(5), 612, 1985.
106. **Hasegawa, M., Tanaka, N., and Kono, T. (Dainichiseika Color and Chemicals Manufacturing Co.),** Japan Patent 85,260,674, 1985.
107. **Roberts, G. G., Petty, M. C., Baker, S., Fowler, M. T., and Thomas, N. J.,** Electronic devices incorporating stable phthalocyanine Langmuir-Blodgett films, *Thin Solid Films*, 132, 113, 1985.
108. **Tripathi, L. N. and Misra, Y.,** Thermoelectroluminescence in thin films of zinc and copper phthalocyanines, *Bull. Mater. Sci.*, 8(3), 407, 1986.
109. **Yoshida, H., Tokura, Y., and Koda, T.,** Charge-transfer excitation bands in electroabsorption spectra of metal (cobalt, nickel, copper, zinc)-phthalocyanine films, *Chem. Phys.*, 109(2—3), 375, 1986.
110. **Blanzat, B., Barthou, C., Tercier, N., Andre, J. J., and Simon, J.,** Energy transfer in solid phases of octasubstituted phthalocyanine derivatives, *J. Am. Chem. Soc.*, 109(20), 6193, 1987.

111. **Dijkslag, H. H., Groen, D. W., Rienks, A. M., and Schoenmaker, A. T.**, PCT International Application, 87,03,123, 1987.
112. **Nepras, M. and Novak, A.**, Czech Patent 211,770, 1983.
113. **Fryer, J. R.**, Radiation damage in organic crystalline films, *Ultramicroscopy,* 14(3), 227, 1984.
114. **Vrublevskii, A. I., Glazkov, Y. V., and Kachura, T. F.**, Certain features of the extracoordination reaction on copper porphyrins, *Zh. Prikl. Spektrosk.*, 41(4), 620, 1984.
115. **Fryer, J. R., Hann, R. A., and Eyres, B. L.**, Single crystal organic monolayer imaging by electron microscopy, *Nature,* 313(6001), 382, 1985.
116. **Iwaizumi, M., Kudo, T., and Kita, S.**, Correlation between the hyperfine coupling constants of donor nitrogens and the structures of the first coordination sphere in copper complexes as studied by nitrogen-14 ENDOR spectroscopy, *Inorg. Chem.*, 25(10), 1546, 1986.
117. **Cook, M. J., Daniel, M. F., Dunn, A. J., Gold, A. A., and Thomson, A. J.**, Determination of molecular orientation of a copper phthalocyanine in a Langmuir-Blodgett film using ESR spectroscopy, *J. Chem. Soc., Chem. Commun.*, 11, 863, 1986.
118. **Murav'ev, V. I.**, EPR spectral characteristics of delocalized metal-ligand bonds in tetragonal copper(2+) complexes, *Zh. Strukt. Khim.*, 27(3), 56, 1986.
119. **Roe, S. P., Hill, J. O., and Liesegang, J.**, An x-ray photoelectron spectroscopic study of some copper(II) tetraaza complexes, *Polyhedron,* 5(3), 723, 1986.
120. **Gimzewski, J. K., Stoll, E., and Schittler, R. R.**, Scanning tunneling microscopy of individual molecules of copper phthalocyanine adsorbed on polycrystalline silver surfaces, *Surf. Sci.*, 181(1—2), 267, 1987.
121. **Diegruber, H. and Plath, P. J.**, Pressure on cobalt phthalocyanines in zeolite cavities, *Z. Phys. Chem. (Leipzig),* 266(4), 641, 1985.
122. **Xu, H., Zhou, Q., and Shen, S.**, Photoinduced redox reactions sensitized by metal phthalocyanines. II. Fluorescence spectra characteristics and photosensitized reduction of viologen, *Ganguang Kexue Yu Kuang Huaxue,* 1, 40, 1984.
123. **Godlewski, J., Kalinowski, J., Stizza, S., Davoli, I., and Bernardini, R.**, Asymmetries in the optical properties of vacuum-deposited organic films illuminated at the substrate and non-substrate surfaces, *Thin Solid Films,* 146(2), 115, 1987.
124. **Kutzler, F. W. and Ellis, D. E.**, A comparison of the one-dimensional band structures of nickel tetrabenzoporphyrin and phthalocyanine conducting polymers, *J. Chem. Phys.*, 84(2), 1033, 1986.
125. **Oliver, S. W., Smith, T. D., and Pilbrow, J. R.**, An electron spin resonance study of the formation of a molecular oxygen adduct of the manganese(II) chelate of tetrasodium 3,10,17,24-tetrasulfonatophthalocyanine, *Inorg. Chim. Acta,* 117(2), 123, 1986.
126. **Tashkova, K. and Andreev, A.**, Cobalt(II) phthalocyanine complex with pyridine bonded to silica and its interaction with oxygen, *Inorg. Chim. Acta,* 84(1), 31, 1984.
127. **Tashkova, K., Kitova, R., and Martinova, N.**, Reflection spectroscopic study of the interaction with oxygen of cobalt(II) phthalocyanine synthesized complexes with pyridine attached to an Aerosil surface, *God. Vissh. Khim.-Tekhnol. Inst. ''Prof. d-r As. Zlatarov'', gr. Burgas,* 18(1), 41, 1984.
128. **Andreev, A., Tashkova, K., and Lyubchev, L.**, Cobalt(II) phthalocyanine complex with pyridine bound to the surface of modified inorganic fibers and its interaction with oxygen, *Dokl. Bolg. Akad. Nauk,* 37(11), 1483, 1984.
129. **Dillow, G. W., Gregor, I. K., and Guilhaus, M.**, Gas phase oxidative-addition reactions of alkyl radicals at complexed cobalt(II) centers, *J. Organomet. Chem.*, 294(1), 131, 1985.
130. **Minin, V. V., Larin, G. M., Rakitin, Y. V., Rotov, A. V., Derkacheva, V. M., Kaliya, O. L., and Luk'yanets, E. A.**, ESR, adduct formation and oxygen addition to cobalt(II) complexes with substituted phthalocyanines, *Zh. Neorg. Khim.*, 32(5), 1135, 1987.
131. **Le Moigne, J. and Even, R.**, Spectroscopic properties and conductivity of thin films of partially reduced metallo phthalocyanines, *J. Chem. Phys.*, 83(12), 6472, 1985.
132. **Zecevic, S., Simic-Glavaski, B., Yeager, E., Lever, A. B. P., and Minor, P. C.**, Spectroscopic and Electrochemical Studies of the Transition Metal Tetrasulfonated Phthalocyanines. V. Voltammetric Studies of Adsorbed Tetrasulfonated Phthalocyanines (MTsPc) in Aqueous Solutions, Report 1985, TR-56, Order No. AD-A161357/9/GAR, National Technical Information Service, Springfield, VA, 23 pp.
133. **Adzic, R., Simic-Glavaski, B., and Yeager, E. B.**, Spectroscopic and Electrochemical Studies of Transition Metal Tetrasulfonated Phthalocyanines. VI. The Adsorption of Iron Tetrasulfonated Phthalocyanine on Single Crystal Silver Electrodes, Report 1985, TR-58, Order No. AD-A161640/8/GAR, National Technical Information Service, Springfield, VA, 20 pp.
134. **Kreja, L. and Rozploch, F.**, ESR studies of some iron poly(phthalocyanines), *Angew. Makromol. Chem.*, 121, 151, 1984.
135. **Linkous, C. A., O'Grady, W. E., Sayers, D., and Yang, C. Y.**, Spectrophotometric and x-ray absorption edge study of complexation of carbon monoxide with ferric tetrasulfophthalocyanine in alkaline solutions, *Inorg. Chem.*, 25(21), 3761, 1986.

136. **Frank, K. H., Koch, E. E., and Biester, H. W.,** Observation of *in situ* formation of phthalocyanine-bis(pyridine) iron(II) by x-ray absorption spectroscopy, *J. Phys. Colloq.*, (C8, Vol. 2), C8/653-C8/657, 1986.
137. **Riga, J., Savy, M., and Verbist, J. J.,** Metal-solvent interactions of iron and naphthalocyanines with thionyl chloride: an XPS approach, *Bull. Soc. Chim. Belg.*, 94(4), 267, 1985.
138. **Permien, T., Engelhardt, R., and Feldmann, C. A.,** Angle-Resolved Photoemission from Oriented Films of Lead Phthalocyanine on a Copper(100) Surface, Report 1983, DESY-SR-83-06, Internation Nuclear Information System, 12 pp.
139. **Hoenig, J.,** Thickness dependent absorption in lead phthalocyanine thin films, *Phys. Status Solidi A*, 92(1), K7—K10, 1985.
140. **Vartanyan, A. T. and Sidorov, A. N.,** Spectral properties of lead phthalocyanine layers, *Zh. Fiz. Khim.*, 59(9), 2221, 1985.
141. **Simonyan, M. V., Kafadaryan, E. A., Nuriyanyan, M. C., Petrosyan, A. K., and Sharoyan, E. G.,** ESR, optical absorption and reflection spectra of monoclinic and triclinic modifications of lead phthalocyanine, *Phys. Status Solidi A*, 101(1), 143, 1987.
142. **Maslov, V. G.,** CNDO/S calculations of electronic spectra of magnesium porphyrin and magnesium phthalocyanine and their ionic forms, *Teor. Eksp. Khim.*, 20(3), 288, 1984.
143. **Freyer, W., Daehne, S., Le Quoc, M., and Teuchner, K.,** On the fluorescence of *tert*-butyl-substituted phthalocyanines, *Z. Chem.*, 26(9), 334, 1986.
144. **Sapunov, V. V. and Solov'ev, K. N.,** Mechanism of quenching of chlorophyll-like molecule triplet states by silver porphyrins, *Opt. Spektrosk.*, 61(6), 1235, 1986.
145. **Shpiro, E. S., Antoshin, G. V., Tkachenko, O. P., Gudkov, S. V., Romanovskii, B. V., and Minachev, K. M.,** XPS study of transition metal complexes in zeolites, *Stud. Surf. Sci. Catal.*, 18, 31, 1984.
146. **Rivera-Rodriguez, P., Ruiz-Ramirez, L., Zaitsev, B. E., and Ivanova, T. M.,** X-ray photoelectron spectra of ruthenium(II) phthalocyanines, *Transition Met. Chem. (Weinhein, Germany)*, 11(12), 447, 1986.
147. **Massey, G. A. and Bowersox, S. H.,** Organic nonlinear materials for laser-electron beam generation, *Proc. SPIE-Int. Soc. Opt. Eng.*, 682, 113, 1987.
148. **Chen, W. H., Rieckhoff, K. E., and Voigt, E. M.,** The first excited states of platinum phthalocyanine in magnetic fields, *Mol. Phys.*, 59(2), 355, 1986.
149. **Chen, W. H., Rieckhoff, K. E., and Voigt, E. M.,** The dynamics of palladium phthalocyanine excited states in magnetic fields at low temperatures, *Chem. Phys.*, 101(1—2), 193, 1986.
150. **Chen, W. H., Rieckhoff, K. E., and Voigt, E. M.,** Zeeman study of the phosphorescent state of palladium phthalocyanine in Shpol'skii matrixes, *Chem. Phys.*, 95(1), 123, 1985.
151. **Chen, W. H., Rieckhoff, K. E., and Voigt, E. M.,** Thermostable sites of palladium phthalocyanine in Shpol'skii matrixes of mixed solvents, *Can. J. Chem.*, 62(11), 2264, 1984.
152. **Leeder, J. D., Rippon, J. A., Rothery, F. E., and Stapleton, I. W.,** Use of the transmission electron microscope to study dyeing and diffusion processes, *Proc. 7th Int. Wool Text. Res. Conf.*, 5, 99, 1985.
153. **Kapinus, E. I., Staryi, V. P., and Dilung, I. I.,** Effect of the nature of the central metal atom on deactivation processes of triplet states of tetrapyrrole pigments, *Teor. Eksp. Khim.*, 21(1), 61, 1985.
154. **Baran, E. J., Etcheverry, S. B., Diemann, E., and Jostes, R.,** Photoelectron spectrum of vanadyl phthalocyanine, *An. Asoc. Quim. Argent.*, 72(1), 27, 1984.
155. **Wong, J., Lytle, F. W., Messmer, R. P. and Maylotte, D. H.,** K-edge absorption spectra of selected vanadium compounds, *Phys. Rev. B*, 30(10), 5596, 1984.
156. **Wong, J., Lytle, F. W., Messmer, R. P., and Maylotte, D. H.,** Study of the K-Edge Absorption Spectra of Selected Vanadium Compounds, Report 1984, TR-1, Order No. AD-A149334/5/GAR, National Technical Information Service, Springfield, VA, 55 pp.
157. **Huang, T. H.,** Electroabsorption spectra of the vanadyl phthalocyanine films, *J. Phys. Soc. Jpn.*, 56(3), 1213, 1987.
158. **Shklyaev, A. A., Selyutin, G. E., Shakot'ko, N. I., and Mikhalenko, S. A.,** EPR study of complexes of copper and vanadium with substituted phthalocyanines, *Izv. Akad. Nauk SSSR, Ser. Khim.*, 6, 1280, 1984.
159. **Mitchell, P. C. H., Scott, C. E., Bonnelle, J. P., and Grimblot, J. G.,** Binding and decomposition of oxovanadium(IV) phthalocyanine, tetraphenylporphyrin, and etioporphyrin on hydrotreating catalysts studied by x-ray photoelectron and ultraviolet-visible spectroscopies. Relevance to catalytic demetalization, *J. Chem. Soc., Faraday Trans. 1*, 81(4), 1047, 1985.
160. **Freyer, W.,** Electronic spectra of tetra-[2,3-(9,10-anthraquino)]tetraazaporphin and of related compounds, *Z. Chem.*, 26(6), 216, 1986.
161. **Freyer, W.,** Absorption behavior of ring-annellated metallotetraazaporphines, *Z. Chem.*, 26(6), 217, 1986.
162. **Ohno, T., Kato, S., Kaizaki, S., and Hanazaki, I.,** Enhanced S-T transitions of aromatic compounds coordinating to paramagnetic chromium(III), *Chem. Phys. Lett.*, 102(5), 471, 1983.
163. **Ohno, T., Kato, S., Kaizaki, S., and Hanazaki, I.,** Singlet-triplet transitions of aromatic compounds coordinating to a paramagnetic chromium(III) ion, *Inorg. Chem.*, 25(22), 3853, 1986.

164. **Wilkinson, F. and Willsher, C.,** Detection of transient absorption in a dyed cotton fabric and in semiconductor powders by diffuse reflectance laser flash photolysis, *J. Chem. Soc., Chem. Commun.*, 3, 142, 1985.
165. **Ho, Z. Z., Ju, C. Y., and Hetherington, W. M., III,** Third harmonic generation in phthalocyanines, *J. Appl. Phys.*, 62(2), 716, 1987.
166. **Freyer, W. and L. Q. Minh,** Influence of Lewis acids on the absorption behavior of substituted metallotetraazaporphines, *J. Prakt. Chem.*, 329(2), 365, 1987.
167. **Freyer, W.,** Interaction of metallotetraazaporphyrins with a mixture of aluminum bromide and acetyl chloride, *Z. Chem.*, 26(7), 263, 1986.
168. **Minor, P. C., Gouterman, M., and Lever, A. B. P.,** Electronic spectra of phthalocyanine radical anions and cations, *Inorg. Chem.*, 24(12), 1894, 1985.
169. **Minor, P. C., Gouterman, M., and Lever, A. B. P.,** Electronic Spectra of Phthalocyanine Radical Anions and Cations, Report 1985, TR-34, Order No. AD-A152328, National Technical Information Service, Springfield, VA, 42 pp.
170. **Shen, T., Yuan, Z., and Xu, H.,** Redox potentials of zinc phthalocyanines, *Huaxue Xuebao*, 43(9), 865, 1985.
171. **Joesten, M. D. and Higgins, M. M.,** The use of copper(II) tetrasulfonated phthalocyanine to introduce spectrophotometry and Beer's law, *J. Chem. Educ.*, 63(12), 1097, 1986.
172. **Miyata, S.,** Remarkable organic optical and electronic materials — phthalocyanines, *Oputoronikusu*, 51, 103, 1986.

Chapter 21

PC USE IN ENERGY SYSTEMS

I. HYDROCARBO FUELS, GREASES, AND LUBRICANTS

A. METALS IN CRUDE OIL AND TAR SANDS

Metals in crude oil and tar sands are analyzed by ESR and FTIR spectroscopy, providing information on their occurrence in porphyrins and other organic complexes, including a vanadyl phthalocyanine (pc) associated with clay in tar sands.[1]

B. COAL LIQUID UPGRADING

M pcs supported on silica, alumina, or silica/alumina are effective catalysts in the hydrogenation of nitrogen-containing heterocycles in coal liquids.[2,3] These heterocycles include quinoline, isoquinoline, quinaldine, 1,2,3,4-tetrahydroquinoline, 2,3-cyclohexanopyridine, *o*-ethyl aniline, pyridine, acridine, pyrrole, indole, carbazole, naphthalene, and phenanthrene. M pcs include Mg, V(IV), Cr(III), Mn(II), Fe(II), Co, Ni, Cu, Zn, Al(III), Si(IV), Mo(II), Ru(II), Ag(II), Sn(II), Sn(IV), Pt(II), and Pb(II).

For example, quinoline is hydrogenated to 1,2,3,4-tetrahydroquinoline in decane at 200 to 350°C and 500 to 3000 psi hydrogen pressure.[3]

C. EFFECTS ON LIQUID FUEL STORAGE STABILITY

Iron pc[4] and Cu pc[5] are studied in relation to their effect on oxygen absorption, light scattering, and storage stability in liquid fuels such as SRC II, H-Coal middle distillates, their hydrogenated liquid products, and petroleum-derived JP-5 jet fuels against sediment and gum formation. The degradation products obtained in the presence of Fe pc and Cu pc indicate that Cu pc interacts with JP-5 jet fuel with added 2,5-dimethyl pyrrole more strongly than Fe pc.[6]

D. ANTIFOULING AGENTS TO PREVENT COKE FORMATION

Alcoholic aqueous M pc dispersions prevent coke deposition or fouling on the metal surfaces of petroleum processing equipment. In hydrovisbreaking of a topped crude oil, 10,000 ppm Co pc is blended with the crude oil and reacted at 400°C and 1500 psig hydrogen for 5 h without coke deposition vs. 14.0 wt % for the same oil without Co pc.[7]

E. DESULFURIZATION OF GASEOUS AND LIQUID STREAMS

A fuel gas from oil shale gasification is purified of H_2S by absorption with an arsenic-soda solution in a horizontal hollow absorber; the spent solution is regenerated with air with 3 mg/l of Co sulfo pc. The H_2S fuel gas content is reduced from 0.7 vol % to 600 ppm.[8]

A raw C_3 unsaturated gas containing 3 to 3.5% H_2S, mercaptans, and COS is purified sequentially by passage through a deethanizer, stabilizing column, and 10 to 15% sodium hydroxide solution containing 1 to 2% organic solvent and 0.05% Co disulfo pc. The product is fractionated to obtain propylene with $\leq$0.003% sulfur.[9]

In a study of polysulfides in absorptive solution for coke oven gas cleaning, polysulfides containing two-four sulfur atoms form when aqueous $(NH_4)_2S$ is oxidized with air in the presence of Co disulfo pc.[10]

In H_2S and HCN removal from coke oven gas by absorption followed by CNS^- oxidation with Co pc catalyst, the oxidation ceases as CN^- concentration increases.[11]

Mercaptans are removed from hydrocarbon streams in which oxygen-containing gas reacts with the mercaptans to form disulfides in the presence of a M pc mercaptan oxidation catalyst.[12]

The mercaptan sulfur content of a light naphtha is reduced from 206 to 8 ppm in air with 5 to 15 wt % aqueous caustic solution containing 50 to 150 ppm Cu, Co, or Ni pc sulfonamides.[13]

Mercaptans in a naphtha stream are oxidized to disulfides. The naphtha contains 0.1 to 10 wt % of a M pc supported on charcoal.[14]

A cracked gasoline is sweetened by oxidation in a charcoal bed containing Co pc monosulfonate and alkylbenzyldimethylammonium chloride. The mercaptan content decreases from 550 to <10 ppm; the catalyst remains active for 200 h.[15]

A 5:1 molar ratio of iso-sodium propionate-mercaptan is added to kerosine and oxidized over Co pc on activated charcoal: the mercaptan concentration drops from 68 to <10 ppm.[16]

Mercaptans are converted to disulfides in liquid fuels by mixing the fuel with an aqueous alkaline solution and air, oxidizing the mercaptans over Co pc sulfonates at <550°F and >120 psia(pounds per square inch absolute), and depressurizing to remove dissolved N_2 and O_2 at <5 psig(pounds per square inch gauge), thereby reducing safety hazards and emission problems in storage.[17]

Mercaptans in gasoline and kerosine are oxidized in an alkaline solution by 50 to 150 ppm polymetallic pc sulfonamide catalyst promoted by $HgCl_2$ at ambient temperature and 1 atm.[18]

An isooctane solution containing phenyl mercaptans is sweetened with air at 20°C, through a Co or Mn pc catalyst supported on charcoal and a zeolite 5A desiccant. There is a 74% oxidation of phenyl mercaptan in <40 min.[19]

A cracked gasoline is sweetened by oxidation in a charcoal bed loaded with Co pc sulfonate and dimethylbenzylalkylammonium chloride, with sufficient activity to reduce mercaptan concentrations from 550 to <10 ppm for 25 h. About 100 ppm anhydrous ammonia in the gasoline increases the time to 60 h.[20]

A sweetening catalyst to remove mercaptans from petroleum products is prepared by dispersing di-sodium Co pc disulfonate on activated carbon in aqueous methanol. Such a catalyst 13.3 parts is mixed with 5 parts sodium hydroxide solution and gasoline containing 810 ppm mercaptans. After mixing for 30 min, the mercaptan content decreases to 4.1 ppm.[21]

A mercaptan containing petroleum fraction such as naphtha is sweetened by mixing with air in an alkaline solution at ≥8.4 kg/cm^2 and passing the mixture through a fixed bed of 0.1 to 10 wt % Co pc monosulfonate on charcoal to oxidize the mercaptans to disulfide compounds which are separated at ≤0.35 kg/cm^2 and washed with water.[22]

Toxic sulfur containing residues from cracked gasoline processing are neutralized by air oxidation at 40 to 50°C and 0.3 to 0.5 MPa in the presence of a sodium hydrosulfide catalyst containing H_2 pc. The COD is reduced 40 to 50% and sodium hydrosulfide is converted to sodium thiosulfate and sodium sulfate, and all sodium mercaptides are completely reacted.[23]

A cap covering the opening of propane containers contains a deodorant of a needle-punched nonwoven fabric impregnated with 5 wt % Co pc sulfonic acid adjusted to pH 2 to 4, eliminating the mercaptan odor in the fuel gas.[24]

Light oils are deodorized by introducing a prepared catalyst-activator-alkaline solution at the top of the reactor and contacting countercurrently with the light oil-air mixture. The 3 to 10% sodium hydroxide solution contains 100 ppm sulfonated Co pc catalyst and a quarternary ammonium salt activator of 0.005 *M* concentration. Reductions of 60 to 70% in alkali consumption and 80% in catalyst consumption are achieved in gasoline deodorized compared with a conventional liquid-liquid deodorization method.[25]

Light petroleum fractions containing ≤750 ppm sulfur are sweetened by passing through a fixed bed of 0.1 to 10 wt % alkali metal salt containing preferably sulfonated Co pc catalyst on an activated carbon support. Kerosine containing 150 ppm sulfur is passed through a fixed bed of a 4 wt % sodium hydroxide solution containing catalyst at 20°C, 1 atm, and 1/h space velocity. The resulting kerosine contains ≤10 ppm sulfur.[26]

Mercaptans are removed from olefinic hydrocarbons by extraction with an aqueous alkaline solution, with or without methanol or ethanol, to increase mercaptan solubility in the solution. The aqueous solution can be regenerated by oxidizing the extracted mercaptans to disulfides with a Co pc catalyst.[27]

Sweetening of a sour distillate to remove mercaptans in two oxidation zones occurs in one vessel. The distillate is contacted with air in the presence of a fiber bundle and then reacted over a bed of a Co pc oxidation catalyst on activated carbon with 5 to 60 min residence time.[28]

The purity of C_5 fractions prior to isomerization in gasoline manufacture is improved by their desulfurization over Co sulfo pc to a 0.003 to 0.006% sulfur content.[29]

Mercaptans in lightweight hydrocarbons are removed by oxidation in alkaline solutions catalyzed by disodium Co disulfo pc promoted by glycols or, more effectively, by Co poly pc, which, however, is more costly. Mercaptans removed in tests are C_2H_5SNa, C_3H_8SNa, and C_4H_9SNa.[30]

Increasing order of activity of Co(II) complexes anchored to the surface of amorphous silica by a coordination bond with pyridine in the oxidation of ethyl hydrosulfide to diethyl disulfide at 298 K is Co(II) bis(salicyl-o-phenylenediamine) $>>$ Co(II) pc $>$ Co pc sulfonate sodium salt $\sim$ Co(II) acetylacetonate.[31]

F. GREASES AND LUBRICANTS

A grease consists of 76 wt % petroleum aviation oil from sulfuric acid refining thickened with 16 wt % lithium soaps of a 12:4.5:0.7 mixture of stearic acid, castor oil, and rosin, and 4 wt % potassium soaps of a 12:4.5:0.7 mixture of stearic acid, castor oil, and rosin, and contains 2 wt % Cu pc.[32]

Adding 3.8 to 4.2 wt % molybdenum disulfide and 1.5 to 2.2 wt % Cu pc to a grease containing a castor oil sodium soap, castor oil calcium soap, diphenyl amine, zinc dialkyl dithiophosphate, oleic acid, polyethylene, polyisobutylene, and axle grease increases the wear resistance of splined joints.[33]

Increase in the strength of lubricating greases obtained by thickening a perfluoropolymer or a siloxane with ammeline is less by ten times vs. thickening with H_2 pc, upon heating at 200 to 300°C.[34]

Bulk poly(phenylene sulfide) has a high coefficient of friction but its composites with graphite, molybdenum disulfide, antimony oxide, lead oxide, Cu pc, polytetrafluoroethylene, or their mixtures have a lower coefficient of friction than bulk poly(phenylene sulfide).[35]

II. PROPELLANTS

Binder thermal degradation in HTPB(AP) (AP = ammonium perchlorate) propellant combustion is rate limiting at low pressures (2 to 7 MPa) when catalyzed by copper chromite, Cr(Salen-N-decyl)$_3$, and Cu pc. At higher pressures (8 to 14 MPa), gas diffusion appears to control the burning rate.[36]

M pc burning rate enhancement in an HTPB(AP) propellant is Fe $>$ Cr(F) $>$ Cu $>$ V(O) $>$ Mn $>$ Ni $>$ Zn, Na, Mg. M pcs appear to catalyze binder thermal decomposition, so condensed phase reactions can be rate limiting at lower pressures during combustion. The reversible oxidation potentials of the M pc complexes can provide a rationale for the observed order of catalytic burning rate efficiencies.[37]

The pot life of diisocyanate-cured binders in propellants is increased by incorporating maleic or oxalic acids and Cu pc burning rate catalyst-ballistic modifiers.[38]

Fe pcs are combustion rate catalysts in propellant igniter compositions with controlled viscosity and volatile component migration. The igniters are for rocket motors comprising hydroxyl terminated polybutadiene binder prepolymer among other ingredients.[39]

III. SOLAR OR PHOTOELECTRIC CELLS AND PHOTOELECTRIC CAPACITORS

A. SOLAR CELLS

A review with 43 references of organic solar cells includes a notation to pc in photoelectric cells.[40]

A review with 32 references covers pc solar cells, including recent improvements in the performance of pc solar cells and obstacles relating to their further development.[41]

A review with 29 references discusses the structure and electronic properties of pc, as well as its applications in gas sensors and photosensitive materials for electrophotography and solar cells.[42]

The effects of binder polymer on the performance of pc photovoltaic cells are studied. The photoelectric cells are made by dispersing pc particles in a binder polymer and sandwiching them between aluminum and indium oxide electrodes.[43]

Cells containing α-pc sandwiched between two gold electrodes are prepared and their photoelectric properties at 0.0005 Pa studied. For weakly absorbed light the quantum generation efficiency is about 1.3%.[44]

Two photoelectric cells containing Ni pc are fabricated in the fashion Nesa (an indium-tin coated transparent glass)/Ni pc/aluminum (MS) and Nesa/Ni pc/polyethylene/aluminum (MIS). The MS cell has good rectification characteristics and photovoltaic properties such as 0.68-V open circuit voltage and 4.2×10^{-8} A/cm^2 short circuit current under illumination (630 nm, 80 μW/cm^2). The open circuit voltage increases to 0.83 V for the MIS cell and the short circuit photocurrent decreases, increasing photovoltaic energy conversion efficiency.[45]

Pc films sublimed onto glass substrates yield the α-phase which when heated convert to the β-phase. They are diamagnetic in the dark; on exposure to light with photon energy above the band gap, paramagnetism is observed.[46]

Surfactant Al pc is electrodeposited on Nesa glass, antimony-doped stannic oxide and on gold electrodes, with 270- to 4000-Å films obtained. The maximum quantum efficiency is 4% at 630 nm and 0.26 mW/cm^2 with about 0.01% power efficiency under white light at 0.5 to 100 mW/cm^2.[47]

Electrochemical and photoelectrochemical studies are made of thin film pc electrodes.[48]

Energy conversion is measured in indium-tin oxide/Pb pc/Ni solar cells; best efficiencies are about 0.01% for air-mass-1 illumination.[49]

Gallium phosphide/Cu pc/silver solar cells have a maximum power conversion efficiency of 0.1%.[50]

Photoelectrochemical measurements are made using 5,10,15,20-tetra(4-pyridyl)porphyrin and Zn pc sublimed on indium-tin oxide (ITO) coated glass with 0.4 *M* KCl electrolyte containing about 0.2 mmol oxygen.[51]

An indium-tin oxide/Cd S/surfactant Al pc/gold solar cell is made by sequential electrodeposition of CdS and surfactant Al pc and analyzed.[52]

A solar cell is made by sequential electrodeposition of CdS and surfactant Al pc on indium-tin oxide (ITO) (Nesatron) conductive glass. Optimum thickness for the Al pc layer is 2500 Å. The photovoltaic characteristics of the optimized cell under 50 mW/cm^2 illumination are short-circuit current 0.33 mA/cm^2, 0.46 V, open circuit voltage, and 0.09% efficiency.[53]

A thin film of a n-type semiconductor (ZnO, CdS, CdSSe) is deposited on a transparent conducting indium-tin oxide followed by sequential vacuum deposition of M pc and gold layers. Among the seven divalent pcs, Zn and Mg pc exhibit the highest photovoltaic activity. The trivalent pcs Al pc Cl, Cl Al pc Cl, and In pc Cl exhibit even higher photoactivity. Under air-mass-2 illumination of 75 mW/cm^2, the power conversion efficiencies of CdS/Al pc Cl and CdS/Mg pc n-p-heterojunction solar cells are 0.22 and 0.13%, respectively.[54]

Metal/Al pc F/gold sandwich cells rectify; forward bias makes the metal electrode negative with respect to gold. With light of >480 nm shining on the tin electrode of the cell, the short-circuit photocurrent is 32μ A and the open circuit photovoltage is 320 mV. Fill factor is 0.3 and the energy conversion efficiency is 0.4%.[55]

A p-n junction photovoltaic cell is formed from Zn pc and 5,10,15,20-tetra(3-pyridyl) porphyrin. The energy conversion efficiency of the cell is about 2% for monochromatic light at 430 nm.[56]

Photoanodes of n-titania sensitized with Cu pc are tested. n-Si solar cells coated with titania films show a photoresponse in the visible range, depending on the thickness of the titania layer.[57]

The photoelectrochemistry of thin film Ga pc Cl electrodes for solar energy conversion is the subject of a dissertation.[58]

The performance of solar cells of a thin metal (Cu, Zn, Mg, Mn) layer pc between gold and aluminum contacts yields a maximum energy conversion efficiency at 2000-Å thickness, and is 0.01 to 0.1% for Mg pc. Doping with iodine improves the Mg pc cell performance.[59]

Slurries of a dispersed Cu pc powder in a polymer binder are sandwiched between an aluminum layer on a glass substrate, forming a Schottky barrier, and by an indium-tin oxide coated glass plate, forming an ohmic contact. The Cu pcs tested are α-, β-, and ε-Cu pc, Cu chloro pc, as well as H_2 pc. Pc-binder layer thicknesses are 10 to 100 μ. Moderate cell performance is obtained with α-Cu pc and H_2pc.[60]

A photoelectric device has a photoactive layer of a poly(vinylidene fluoride) dispersed with x-H_2 pc between a conductive glass plate and a lead electrode to form a solar cell. The energy conversion efficiency of the cell is 0.027% under a 75.8 mW/cm^2 irradiation of white light.[61]

A slurry of x-H_2 pc dispersed in a poly(vinylidene fluoride) binder and tetramethyl urea is added to a conductive glass plate and an aluminum film is deposited on the formed film to form a solar cell, which gives energy conversion efficiencies of 16.2, 10.0, 8.0, and 6.5% under 617-nm light of 0.001, 0.007, 0.01, and 0.017 mW/cm^2, respectively vs. 8.7, 5.2, 4.6, and 3.7% for a cell using a poly(vinyl acetate) binder and cyclohexanone as solvent.[62]

A photovoltaic cell is made from Cu pc and a perylene tetracarboxylic derivative. An energy conversion efficiency of about 1% is achieved under simulated air-mass-2 illumination. The fill factor is 0.65, high because the charge generation efficiency is relatively independent of the bias voltage. The interface between the Cu pc and the perylene tetracarboxylic derivative is crucial in developing the cell photovoltaic properties.[63]

A photovoltaic cell is prepared by mixing 60% Ni pc dicarboxylic acid didecyl ester and polyurethane in a solvent, applying a 200-Å layer on a 1.5-μ indium oxide layer, drying, and sputter depositing an aluminum electrode on the Ni pc derivative-containing layer. Under 100 mW/cm^2 illumination, the efficiency of the cell is 0.095%.[64]

A lightweight flexible fibrous solar cell is prepared by depositing a photoelectric active layer of a semiconducting pc, an aluminum or indium layer, and an indium oxide layer successively on conductive nylon fibers containing brass fibers. Polyacetylene and polythiazine may be used as conductive fibers.[65]

GaP/Cu pc/Ag cells exhibit photovoltaic effects due to both the heterojunctions at the GaP/Cu pc interface and Schottky junctions at the Cu pc/Ag interface. The maximum energy conversion efficiency is 0.1%.[66]

A solar cell is prepared from a layer containing 60% tetrakismethacryloyloxyethyl Co pc tetracarboxylate and polyurethane on a 1.5μ indium oxide layer and by sputter depositing an aluminum electrode. On exposing to 100 mW/cm^2 illumination, the cell has an energy conversion efficiency of 0.056% vs. 0.0015% for a converter containing Ni pc and PTFE.[67]

Photoelectrochemical cells are made from a thin 600-Å film of Al pc OH on stannic oxide, by spin coating, electrodeposition, and sublimation. Under white light irradiation of 10 mW/cm^2, the stannic oxide/Al pc OH/I/I^-/Pt cells made by the three methods have photocurrents of 18, 11, and 13 $\mu A/cm^2$; photovoltages of 66, 54, and 72 mV fill factors of 0.35, 0.36, and 0.41, and power conversion efficiencies of 0.004, 0.002, and 0.004%, respectively.[68]

p-n heterojunction solar cells with improved response are made by vacuum evaporation of thin (200 to 400 Å) dye layers on glass covered with tin oxide. Al pc Cl is one such dye tested. Overall solar cell efficiencies are about 0.1%, about twice as high as previously reported values for some p-n junction solar cells with organic semiconductors that do not take dye hierarchy and energy loss through internal molecular transitions into account.[69]

A combination of 5,10,15,20-tetraphenyltetrabenzoporphyrin (TPTBP) and H_2 pc yields the best spectral match to the solar spectrum, but the photocurrents from an In-stannic oxide/ H_2 pc, TPTBP/Al cell are about 0.1 of those from an In-stannic oxide/H pc/Al cell.[70]

A preferred solar cell configuration is stannic oxide/bis-methylimide of perpylenetetracarboxylic acid (DMP)/Al pc Cl/Ag. This p-n heterojunction solar cell, thus made with a pair of photoconductive organic dyes, provides short-circuit current yields of >19%.[71]

Powdered α-Cu pcs coated with a layer of a plasma-polymerized hydrocarbon are mixed with a charge-transporting material and a binder to form a photoelectron converting layer for an organic solar cell. The coating has a preferred thickness of 20 to 80 Å.[72]

Thin sandwich cells of metal/Al pc F/Au, the barrier electrode of which, M, is Al, In, Sn, or Zn, rectifies. Illumination with >780-nm light on the tin electrode side yields a short current photocurrent of 32 μA and open circuit photovoltage of 320 mV. The fill factor is 0.3 and energy conversion efficiency is 0.4%.[73]

A 2.5-cm^2 solar cell prepared from a semitransparent gold film covered with a 3000-Å film of K-doped Zn pc gives about 2μ A at 2.0 V when irradiated by a halogen lamp at 1.6 mW/cm^2.[74]

When the surface of Ga pc Cl is modified by the addition of submonolayer amounts of platinum, photoassisted H_2 evolution occurs. The power conversion efficiency is low, <0.1% compared with other semiconductor materials.[75]

Numerous organic fluorescent dyes including Mg pc in polymethylmethacrylate are screened for potential use in solar collectors.[76]

Two types of Ga pc Cl thin films are prepared by vacuum deposition on gold-coated plastic substrates. Type-A Ga pc Cl films consist of small crystallites and type-B Ga pc Cl films consist of larger crystallites isolated on the exposed gold substrate. Type A films cannot support an appreciable photopotential but type B film exhibits a photopotential of −220 mV vs. a platinum cathode in the hydroquinone solutions.[77]

The Schottky barrier with a small concentration of the surface states is formed at the contact of solar-cell Pb pc films with bottom Ni electrode. Ni/Pb pc/Au, Ni/Pb pc/Ni, and Au/Pb pc/Ni structures are also studied.[78]

Al pc complexes and Cr pc complexes are studied as thin film photoelectrodes by electron microscopy, UV-visible spectrophotometry, and electrochemical and photoelectrochemical techniques. They exhibit both positive and negative photopotentials, depending on the potential of the redox couple.[79]

B. PHOTOELECTRIC CAPACITORS

Photoelectric capacitors with an In-stannic oxide/M pc/Al structure have an open circuit voltage of 250 to 600 mV and 0.1- to 0.5-μF capacity.[80]

Photoelectric capacitors of a transparent conductive Al/H_2 pc/Au structure have a photoelectric voltage of 300 to 650 mV and 0.07- to 0.15μF capacity.[81]

C. SOLAR ENERGY CONVERSION IN HOMOGENEOUS SOLUTION

An Fe tetraphenyl/pc complex, (Fe tetraphenyl pc)$_2$O, photodisproportionates in solution to yield high oxidation state species. Photodisproportionation is a new approach to solar energy conversion.[82]

IV. BATTERIES AND FUEL CELLS

A. CATHODES WITH EMPHASIS ON AIR CATHODES

Carbon is impregnated with Li pc and is then treated with a metal salt solution such as nickel chloride. Thus, a mixture of active carbon 10, graphite 10, and PVC 5 parts is extruded to a rod and heated to 1000°C, then immersed in 10% Li$_2$ pc/n-amyl alcohol, and then treated in silver nitrate-ethanol solution to form Ag$_2$ pc and to prepare an air cathode.[83]

An air cathode with good polarization properties is prepared by impregnating carbon with

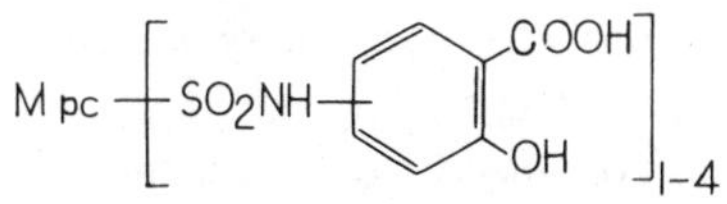

(I)

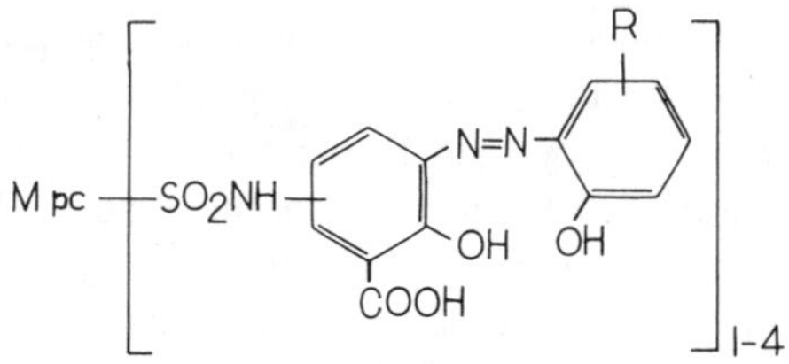

(II)

where M = Fe, Ni, Co, Cr, or Cu; R = H or C$_1$ to C$_4$ alkyl, and chelating with a polyvalent metal. For example, an air cathode is prepared by extruding a mixture of active carbon 10, graphite 10, and PVC 5 parts to prepare a 1.0-cm rod, heated to 1000°C, immersed in a 10% aqueous alkaline solution of (II), dried, and warmed in 2% aqueous ferric chloride. The electrode has a current density of about 50 mA/cm^2 at an electric field change of -1.0 V/s vs. about 30 mA/cm^2 for a cathode obtained from immersing the heat-treated rod in 10% aqueous Cu pc tetrasulfonate and drying.[84]

An air cathode is prepared by extruding a mixture of active carbon 10, graphite 10, and a thermoplastic resin 5 parts to form a 1.0-cm diameter rod which is heated to 1000°C to decompose the resin, immersed for 1 h, in 5% aqueous disodium sulfide solution of 4,4′-diamino-3″,3‴-dimercapto Fe pc, dried, and oxidized with air. The cathode has a current density of ~50 mA/cm^2 at an electric field change of -0.8 V/s vs. about 5 mA/cm^2 for an air cathode made by immersing the heat-treated rod in a saturated Fe pc aqueous solution and drying.[85]

After heating to 1000°C, the rod may be immersed 1 h in a 10% suspension of Cu pc-3,3′,3″,3‴-tetrakis[(N-hydroxymethyl)sulfamide]. The current density of the electrode is ~50 mA/cm^2 at an electric field change of -1.0 V/s vs. ~5 mA/cm^2 for a cathode formed by immersing the rod in a 1% quinoxaline solution of Cu pc.[86]

An air cathode is prepared by extruding a tube (1.0 cm ID and 0.4 cm wall thickness) from activated charcoal 10, graphite 10, and a thermoplastic resin 10 parts, heating at 200°C and decomposing 5 parts PVC on the tube, coating the tube with activated charcoal 40, graphite 10, and a thermoplastic resin 5 parts, sintering, immersing the tube in an alkaline hydrosulfite solution containing sulfonated Co pc, oxidizing the tube surface, and impregnating the tube with a polystyrene-benzene solution.[87]

An air cathode for fuel cells or batteries uses activated charcoal and graphite containing a metal polyhydrazino pc M pc $(NHNH_2)_{2\text{-}4}$ where M = Co, Cr, Cu, Fe, or Ni. The air cathode has high polarization.[88]

Air cathodes are prepared from carbonaceous substrates impregnated with metal pc sulfinic acid M pc $(SO_2H)_{2\text{-}4}$ where M = Co, Cr, Cu, Fe, or Ni. These cathodes have excellent polarization characteristics.[89]

Using a condensed transition M pc as catalyst improves durability and activity of the cathode. For example, an air cathode containing Fe pc condensation product is prepared by coating a carbonaceous grid bar with 2% Fe 4,4′,4″,4‴-tetraamino pc and treating it in 2% adipic acid dichloride in benzene.[90]

Air electrodes with carbonaceous substrates containing metal (Co, Cr, Cu, Fe, Ni) pc hydrazides show excellent polarization characteristics. For example, coconut shell active carbon, graphite, thermoplastic resin, and Fe 4,4′,4″,4‴-tetracarbohydrazide are extrusion molded and heated to give an air electrode.[91]

Catalytic activity of Co or Cr pc in air cathodes is enhanced by complexation with amines, and especially, with polyamines. An air cathode from coconut husk active carbon, graphite, thermoplastic resin, and Cr pc pyrazine complex formed into a 1.0-cm diameter rod shows good polarization performance.[92]

An air cathode for batteries contains dimetal pc homopolymer. The polymeric structure increases the activity and stability of the cathode. Such an air cathode is prepared from active carbon from coconut husk, graphite, thermoplastic resin, and Fe-Mn pc homopolymer, and extruding into a 1.0-cm diameter rod. Tests show good polarization performance.[93]

Air cathodes contain M pcs with polymeric linkages and provide improved polarization characteristics, higher current density, and longer service life. Fe pc 4,4′,4″,4‴-tetracarboxylic acid is reacted with chlorosulfonic acid, and then with polystyrene in the presence of aluminum trichloride to obtain an Fe pc derivative-polystyrene product. Coconut shell activated charcoal, acetylene black, and the modified Fe pc mixture are treated with PTFE emulsion, dried, and pressed to obtain a 0.95-mm thick air cathode which shows improved performance in button-type alkaline zinc batteries.[94]

K_4 Mn pc 3,3′,3″,3‴-tetrasulfonate is added to 40% potassium hydroxide to 1% concentration and kneaded with 0.5% CMC. The electrolyte is kneaded with zinc-mercury. An air cathode is formed from coconut active carbon and thermoplastic resin binder. The anode is zinc. The caustic electrolyte containing the alkali salt of Mn pc tetrasulfonic acid is the oxygen carrier. The assembled battery shows excellent discharge properties at high current density.[95]

The manufacture of a carbonaceous material for air cathodes includes a metal (Co, Cr, Cu, Fe, or Ni) pc $(SO_2NHCH_2OH)_{1\text{-}4}$ or M pc $[SO_2NHCOC(R){:}CH_2]_{1\text{-}4}$ which is polymerized on the carbon by condensation, forming the basis for cathodes for button-type zinc-air batteries.[96]

A small partial polarization air cathode for fuel cells or batteries contains powdered carbon, acetylene black, and Fe pc-Co_2 V_2O_7 catalyst.[97]

A cathode for a fuel cell or air battery is prepared from a polymeric metal (Co, Cu, Mn, Mo, Ni, Sn) pc on the carbonaceous collector material. An air battery is constructed by using the cathode, zinc anode, and potassium hydroxide electrolyte, and it shows reduced polarization.[98]

A cathode for fuel cells or air batteries contains an Fe pc polymer. A zinc-air battery is prepared using this cathode and potassium hydroxide electrolyte, and shows low polarization.[99]

A cathode for fuel cells and air batteries is formed by loading the electrode grid with polymeric metal (Fe and ≥ 1 selected from among Co, Cu, Mn, Mo, Ni, and Sn) simultaneous to the polymerization. A zinc-air battery using this cathode and potassium hydroxide electrolyte shows a low-voltage drop at high current density.[100]

Cathodes prepared for fuel cells and air batteries are made by mixing conductive material with a mixture of phthalonitrile, $FeCl_2{\cdot}nH_2O$, urea, carbon powder, acetylene black, and Ketjen black EC. The cathode shows smaller polarization than for an air cathode made from carbon and presynthesized Fe pc.[101]

Oxygen reduction occurs at cathodes containing carbon black and pyrolyzed Co pc for alkaline air fuel cells.[102]

Pd pc 2,9,16,23-tetrasulfonic acid is a chemically regenerable fuel cell redox catalyst.[103]

Moessbauer spectroscopy of an oxygen-fed cathode with preadsorbed Fe pc in alkaline solution indicates that the isomer shift and quadrupole splitting are the same as those obtained at open circuit without oxygen and correspond to adsorbed Fe pc presumably at monolayer coverage.[104]

A methanol fuel cell contains a cathode using Co, Cu, Fe, or Ni pc as catalyst.[105]

A secondary fuel cell is based on oxygen reduction on a platinum electrode modified by polymers and Co, Cu, Fe(III), or Ni pc.[106]

In a secondary fuel cell oxygen evolved by water electrolysis in the charging process is stored in the polymer of Co, Cu, Fe(III), or Ni 2,9,16,23-tetracarboxy pc on a platinum electrode.[107]

The electrocatalytic properties of a Co, Cu, Fe, Mn, or Ni pc fuel cell catalyst are studied for electrooxidation of hydrazine and hydroxylamine, cystein, and cystine.[108]

The electroreduction of oxygen is examined in aqueous solution containing Co or Fe 4,4′,4″,4‴-tetracarboxy pc, a possible fuel cell catalyst.[109]

A cathode for reducing oxygen in a fuel cell has a low ash coconut shell charcoal substrate loaded with platinum pc or Pb-Ru oxide catalyst.[110]

An air battery using a cathode based on 100 g active carbon mixed with 0.02 g Co pc and covered with PTFE film has a discharge capacity of 420 mA-h vs. 350 mA-h for a battery using catalyst loaded active carbon only.[111]

Air cathodes are made from active carbon impregnated with Co pc homopolymer for aluminum-air batteries.[112]

Carbon gas-diffusion air electrodes in saline electrolyte contain two layers: a hydrophobic layer (gas side) and hydrophilic layer (solution side) with the current collector (grid) pressed into the hydrophobic layer. The catalyst is Norit NK active carbon into which Co Ni oxide, Co pc, and Co tetramethoxyphenylporphyrin are introduced.[113]

Acetylene black is chemically modified by bonding with polymerized Fe tetrachloro pc catalyst, followed by heat treatment. An electrode made with this catalyst shows significant improvement in activity and stability vs. a similar unmodified catalytic electrode.[114]

An electrolyte contains di Li pc for a sealed metallic oxide-hydrogen battery using a hydrogen storage alloy with anodes containing hydrogen absorbing Co, Cu, or Fe pc.[115]

B. SECONDARY BATTERIES

A secondary battery consists of an electrode of a Cu or Fe pc-coated base and a second electrode. The first electrode is charged as anode and discharged as cathode. The electrolyte is an aqueous alkaline solution. Co pc is deposited on a platinum plate in 5000-Å thickness to form a first electrode and charged in 30% potassium hydroxide with a platinum counterelectrode.[116]

A secondary battery comprises an alkaline electrolyte, a platinum anode, and a cathode grid coated with a hydrophobic polymer linked to a Co or Fe pc.[117]

A secondary battery cathode is made by laminating a platinum plate with Co pc layers and a styrene resin.[118]

A styrene-vinylpyridine copolymer with Fe pc groups bonded to the styrene benzene ring by a CO group is coated on a platinum electrode which with a zinc electrode and 0.05 M zinc iodide saturated with iodine yields a secondary battery with good charge-discharge.[119]

Iodine secondary batteries are prepared using electrodes modified by Co, Cu, Fe, or Ni pc bound to 2-vinylpyridine-styrene copolymer in metal iodide solution saturated with iodine.[120]

Galvanostatic curves are drawn of the cathodic reduction of SO_2 from nonaqueous electrolyte on electrodes containing Fe pc catalyst. The results lead to prospects for using high catalytic activity for increasing the discharge voltage in batteries of the Li-SO_2 system.[121]

C. POLY PCS

A principal element in a novel photovoltaic cell is a thin film of a metal polypc dispersed in a polymer binder coated on a conductor substrate. A significant increase in the electric conductivity of pc $Ge(OH)_2$ occurs on polymerization to (Ge pc $O)_n$.[122]

A catalyst for fuel cells contains Fe polypc deposited on carbon or soot.[123]

Crude M pc polymers such as Co pc polymer are isomerization catalysts for photochemical energy storage with improved activity and regenerability.[124]

Fe polypcs made from ferrous chloride and pyromellitic dianhydride in a 6:4 ratio are suitable for use as a fuel cell oxygen catalyst.[125]

V. LITHIUM CELLS

The electrochemical reduction of $SOCl_2$ at the baseline cathode in a Li-$SOCl_2$ battery under the influence of Fe pc and Co pc polymer (Co $pc)_n$ proceeds slowly and cathode doping with Fe pc or (Co $pc)_n$ catalyst improves both the rate and the specific catalyst cathode capacity.[126]

The feasibility of a spin-activated high-rate Li-SO Cl_2 reserve battery containing a stack of bipolar plates is demonstrated. Stability and discharge lives are improved with a Fe pc polymer catalyst.[127]

The effects of cell design and M pc catalysts on the safety of Li-$SOCl_2$ reserve cells are evaluated. When the 1.5 *M* $LiAlCl_4/SOCl_2$ is replaced with 1.5 *M* $LiAlCl_4/SO_2Cl_2$ electrolyte in the cathode limited design cells, both the cell discharge and cell safety improve.[128]

High-rate and high-capacity Li-SO_2 Cl_2 cells are obtained with cathodes fabricated with suitable carbon substrates. Surface groups such as Co pc complexes catalyze the electrochemical reduction of SO_2Cl_2.[129]

Gas evolution from a M pc electrode in lithium batteries during discharge is proportional to the discharge capacity. The activity decreases in the order Fe $>$ Co $>$ Mn $>$ Ni $>$ CrF $>$ Ag pc. H_2, Li_2, Na_2, GaCl, and Cd pc do not have much activity.[130]

The intercalation of lithium in H_2 pc is studied in a propylene carbonate electrolyte containing $LiClO_4$. The theoretical specific capacity and energy of the H_2 pc/Li system are very high (870 A-h/kg and 1190 W-h/kg, respectively).[131]

The electrocatalytic activity of pcs in secondary lithium cells is measured. Specific discharge capacities of β-pc vs. the ratio of acetylene black to β-pc in the cathode mixture and specific discharge capacities of cathode material containing β-Cu pc, Cu octacyano pc, and Cu polypc are given.[132]

An electrolyte contains Fe pc in a Li-$SOCl_2$ battery.[133]

A battery containing an acetylene black cathode with nickel collector, a lithium anode, and $SOCl_2$ containing 4.5 $AlCl_3$, LiCl, and 2 mg Fe pc/ml show in a discharge test higher average voltage, a smaller discharge curve, and a higher efficiency than a control battery not containing Fe pc.[134]

$SOCl_2$ containing 2.0 *M* $GaCl_3$ and 2.5 mg Fe pc/ml is used in a battery containing a Ni-acetylene black cathode collector, a lithium anode, and a glass fiber separator. The battery has a greater capacity and higher average voltage than a control battery without Fe pc.[135]

$Li\text{-}SOCl_2$ batteries with Co pc polymer-catalyzed cathodes generally show increased life depending on the discharge temperature and current load.[136]

A nonaqueous lithium battery contains a halocarbon cathode depolarizer and a cathode catalyst of Co, Cu, Fe, Ni, Pd, Rh pc among other macrocycles. The higher middischarge voltages of the batteries containing the catalyzed cathode demonstrate the effectiveness of the catalyst.[137]

Severe voltage delays occur in active $Li\text{-}SOCl_2$ batteries containing additives after storing at 71°C for ⩾10 d, except for those containing Co pc-catalyzed cathodes.[138]

Tetracarboxylated phenylporphyrin and Co tetracarboxylated pc are added to the electrolyte of a $Li\text{-}SOCl_2$ battery to prevent anode passivation and premature battery failure.[139]

VI. SOLAR ENERGY STORAGE

Co pc adsorbed on activated carbon is an insoluble catalyst for heat gain in a water-soluble solar energy storage system. A quadricyclane derivative is isomerized to a norbornadiene derivative with heat release in an alkaline solution.[140]

Solar energy is stored as chemical energy by photoisomerizing a compound such as quadricyclane from a lower energy state to a higher energy state and released as heat by the reverse reaction catalyzed by a Co pc catalyst-containing polymer.[141]

Solar energy storage occurs through catalyzed reversible photovalence isomerization of 9-*tert*-butylanthracene with Zn pc catalyst.[142]

REFERENCES

1. **Graham, W. R. M.,** Analysis of metal species in petroleum and tar sands using the electron paramagnetic resonance and Fourier transform infrared techniques, *ACS Symp. Ser.*, 344, 358, 1987.
2. **Boucher, L. J., Dadey, E. J., and Spears, D. R.,** Metallophthalocyanines: new catalysts for coal liquid upgrading, *Proc.-Int. Conf. Coal Sci.*, 763, 1983.
3. **Boucher, L. J.,** New Catalysts for Coal Liquid Upgrading. Final Technical Report, Report 1984, DOE/PC/40812-T11; Order No. DE84015921, 112 pp.
4. **Yaggi, N. F., Lee, S. H., Ge, J., and Li, N. C.,** Oxygen absorption and light scattering studies of the stability of liquid fuels, *Prepr. Pap.-Am. Chem. Soc., Div. Fuel Chem.*, 29(1), 178, 1984.
5. **Lee, S. H., Ge, J. H. S., and Li, N. C.,** Light scattering, oxygen uptake and deposit formation in studies of stability of coal-derived and petroleum fuels, *Prepr.-Am. Chem. Soc., Div. Pet. Chem.*, 29(4), 996, 1984.
6. **Li, J. and Li, N. C.,** Storage stability of jet fuels, *Fuel*, 64(8), 1041, 1985.
7. **Urban, P. and Stolfa, F. (UOP Inc.),** U.S. Patent 4,647,367, 1987.
8. **Afanas'nev, Y. M., Golyand, S. M., Belyuga, Y. V., and Rybakov, L. A.,** Removal of hydrogen sulfide from shale processing generator gas with the production of final products, *Khim. Neft. Mashinostr.*, 2, 16, 1984.
9. **Akhmadullina, A. G., Gorodilova, S. I., Kopylov, Y. P., and Mazgarov, A. M.,** Purifying unsaturated gases of petroleum processing, *Khim. Tekhnol. Topl. Masel*, 6, 12, 1984.
10. **Panferova, G. D., Lebedeva, G. N., Lipatova, L. F., and Butakova, V. I.,** Study of polysulfides in absorptive solutions for coke oven gas cleaning, *Koks Khim.*, 5, 23, 1984.
11. **Andrzheevskii, M. Y. and Lyapkin, A. A.,** Industrial verification of oxidative removal of hydrogen sulfide and hydrogen cyamide from coke-oven gas, *Koks Khim.*, 9, 33, 1987.
12. **Asdigian, G. (UOP Inc.),** U.S. Patent 4,412,912, 1983.
13. **Agarwal, B. B., Bahl, J. S., Fish, S. F., Puri, S., and Gulati, I. B. (Council of Scientific and Industrial Research (India)),** Indian Patent 152,541, 1984.
14. **Verachert, T. A. (UOP Inc.),** U.S. Patent 4,481,106, 1984.
15. **Frame, R. R. (UOP Inc.),** U.S. Patent, 4,498,977, 1985.
16. **Urban, P. (UOP Inc.),** U.S. Patent 4,481,107, 1984.
17. **Verachtert, T. A. (UOP Inc.),** U.S. Patent 4,490,246, 1984.

18. **Aggarwal, B. B., Bahl, J. S., Fish, S. F., Puri, S. N., and Gulati, I. B. (Council of Scientific and Industrial Research (India)),** Indian Patent 153,190, 1984.
19. **Frame, R. R. (UOP Inc.),** U.S. Patent 4,498,978, 1985.
20. **Frame, R. R., Johnson, R. W., and Staehle, B. E. (UOP Inc.),** U.S. Patent 4,502,949, 1985.
21. **Angelescu, E., Zirna, I., Auslander, G., Dobrescu, D., Toma, M., and Barbul, M. A. (Intreprinderea de Coloranti "Colorom"),** Romania Patent 85,477, 1984.
22. **UOP Inc.,** Japan Patent 86,157,582, 1986.
23. **Akhmadullina, A. G., Mazgarov, A. M., Khrushcheva, I. K., and Nurgaheva, G. M.,** Detoxification of sulfur containing alkaline wastes on a heterogeneous phthalocyanine catalyst, *Khim. Tekhnol. Topl. Masel,* 5, 36, 1985.
24. **Fukamachi, K. and Shirai, H. (Earth Clean K. K.),** Japan Patent 87,75,198, 1987.
25. **Su, Y.,** China Patent 85,103,713, 1986.
26. **Institut Francais du Petrole,** Japan Patent 87,39,693, 1987.
27. **Verachtert, T. A. (UOP Inc.),** U.S. Patent 4,626,341, 1986.
28. **Maple, R. E. and Suarez, F. J. (Merichem Co.),** Europe Patent Application 203,574, 1986.
29. **Fedotov, Y. I., Mikheeva, L. Y., and Abramov, G. I.,** *Neftepererab. Neftekhim. (Moscow),* 2, 23, 1987.
30. **Fomin, V. A., Komleva, T. I., Izmailova, N. R., and Zainullin, R. A.,** Catalytic agent for the sweetening of lightweight hydrocarbon raw material, *Khim. Tekhnol. Topl. Masel,* 9, 14, 1984.
31. **Andreev, A. A. and Tashkova, K.,** Oxidation of ethyl mercaptan on cobalt(II) complexes immobilized on silicon dioxide surface, *Dokl. Bolg. Akad. Nauk,* 39(4), 59, 1986.
32. **Kut'kov, A. A., Goncharov, S. I., Loginov, V. T., Mamaev, N. M., Bashkirov, O. I., Derlugyan, P. D., Mshvenieradze, T. G., Vakar, A. A., Voronov, N. G., and Semerinov, V. M.,** U.S.S.R. Patent 1,161,521, 1985.
33. **Yalyshev, R. G., Konovalov, V. V., and Yalysheva, G. R.,** U.S.S.R. Patent 1,074,896, 1984.
34. **Chepurova, M. B. and Kobzova, R. I.,** Thermal stability of ammeline greases, *Khim. Tekhnol. Topl. Masel,* 5, 41, 1985.
35. **Tewari, U. S., Sharma, S. K., and Vasudevan, P.,** Effect of lubricant additives on friction and wear studies of poly(phenylene sulfide), *Addit. Schmierst. Arbeitsfluessigkeiten. 5th Int. Kolloq.,* 2, 6/7/1-6/7/7, 1986.
36. **Fong, C. W. and Hamshere, B. L.,** The mechanism of burning rate catalysis in composite HTPB-AP [ammonium perchlorate] propellant combustion, *Combust. Flame,* 65(1), 61, 1986.
37. **Fong, C. W. and Hamshere, B. L.,** The mechanism of burning rate catalysis in composite propellants by transition metal complexes, *Combust. Flame,* 65(1), 71, 1986.
38. **Ducote, M. E. (U.S. Department of the Army),** U.S. Patent 4,597,811, 1986.
39. **Shaw, G. C. (Morton Thiokol, Inc.),** U.S. Patent 4,658,578, 1987.
40. **Zhao, B. and Xu, H.,** Organic solar cells, *Huaxue Tongbao,* 8, 22, 50, 1983.
41. **Minami, N.,** Phthalocyanine solar cells, *Denshi Shashin Gakkaishi,* 22(2), 121, 1984.
42. **Takahashi, Y.,** Dark and photo electric properties of phthalocyanine and its applications, *Kagaku Kogyo,* 35(10), 895, 1984.
43. **Minami, N., Sasaki, K., and Tsuda, K.,** Improvement of the performance of particulate phthalocyanine photovoltaic cells by the use of polar polymer binders, *J. Appl. Phys.,* 54(11), 6764, 1983.
44. **Nespurek, S.,** Photovoltaic properties of metal-free phthalocyanine, *Czech. J. Phys.,* 34(3), 222, 1984.
45. **Misoh, K., Tasaka, S., Miyata, S., and Yamada, A.,** Photovoltaic cells fabricated by nickel phthalocyanine and evaporated polyethylene, *Kobunshi Ronbunshu,* 41(4), 183, 1984.
46. **Sasseville, R. L., McIntosh, A. R., Bolton, J. R., and Harbour, J. R.,** Flash photolysis electron paramagnetic resonance studies of charge-carrier production in sublimed films of phthalocyanine, *J. Phys. Chem.,* 88(14), 3139, 1984.
47. **Belanger, D., Dodelet, J. P., Dao, L. H., and Lombos, B. A.,** Photoelectrochemical characteristics and behavior of a surfactant aluminum phthalocyanine cell, *J. Phys. Chem.,* 88(19), 4288, 1984.
48. **Tachikawa, H.,** Energy Collection and Charge Transfer Processes in Thin Film Photocells and Photoelectrochemical Cells, Final technical report, July 1, 1980—June 30, 1981, Report 1983, DOE/CS/83115-T1, Order No. DE84005763, 4 pp.
49. **Hamann, C., Hoehne, H. J., and Naegler, M.,** Parameters of lead phthalocyanine solar cells and their improvement, *Mater. Sci.,* 10(1—2), 105, 1984.
50. **Nakata, J. and Imura, T.,** Photovoltaic effect of gallium phosphide/dye/metal cells, *Oyo Butsuri,* 54(1), 65, 1985.
51. **Harima, Y. and Yamashita, K.,** Spectral cosensitization effect observed at a photoelectrode based on an organic p-n junction, *J. Electroanal. Chem. Interfacial Electrochem.,* 186(1—2), 313, 1985.
52. **Lawrence, M. F., Dodelet, J. P., and Dao, L. H.,** Electrodeposited cadmium sulfide/SAlPc heterojunction cell: origin of the p-type semiconducting character of the surfactant aluminum phthalocyanine, *J. Phys. Chem.,* 89(8), 1395, 1985.

53. **Lawrence, M. F., Dodelet, J. P., and Dao, L. H.,** Photoelectric properties of ITO/cadmium sulfide/ surfactant aluminum phthalocyanine/gold solar cells, *J. Electrochem. Soc.*, 131(12), 2977, 1984.
54. **Loutfy, R. O. and Hor, A. M.,** Sensitization of n-type semiconductors by phthalocyanines, *Photochem. Photobiol., Proc. Int. Conf.*, 2, 759, 1983.
55. **Shimura, M. and Toyoda, A.,** Photovoltaic properties of sandwich cells prepared with fluoroaluminum phthalocyanine, *Jpn. J. Appl. Phys., Part 1,* 23(11), 1462, 1984.
56. **Harima, Y., Yamashita, K., and Suzuki, H.,** Spectral sensitization in an organic p-n junction photovoltaic cell, *Appl. Phys. Lett.*, 45(10), 1144, 1984.
57. **Hartig, K. J. and Getoff, N.,** An overview of different polycrystalline n-titanium dioxide anodes for photoelectrochemical water splitting, *Adv. Hydrogen Energy,* 4, 1085, 1984.
58. **Rieke, P. C.,** Photoelectrochemistry of Thin Film Chlorogallium Phthalocyanine Electrodes for Solar Energy Conversion, Ph.D. thesis, University of Arizona, Tucson, 1984.
59. **Khelifi, M., Mejatty, M., Berrehar, J., and Bouchriha, H.,** Photovoltaic effect in the thin layers of phthalocyanines, *Rev. Phys. Appl.*, 20(7), 511, 1985.
60. **Takeuchi, M., Ohtsu, S., Takeoka, H., Naganuma, S., and Nagasaka, H.,** Photovoltaic properties of copper phthalocyanine binder layer solar cells, *Tech. Dig. 1st -Int. Photovoltaic Sci. and Eng. Conf.*, 263, 1984.
61. **Minami, S., Sasaki, K., Tsuda, K., Yonehara, Y., and Kinoshita, S. (Agency of Industrial Sciences and Technology, Kawamura Physical and Chemical Research Institute),** Japan Patent 86,03,473, 1986.
62. **Minami, S., Sasaki, K., Tsuda, K., Yonehara, Y., and Kinoshita, S. (Agency of Industrial Sciences and Technology, Kawamura Physical and Chemical Research Institute),** Japan Patent 85,201,672, 1985.
63. **Tang, C. W.,** Two layer organic photovoltaic cell, *Appl. Phys. Lett.*, 48(2), 183, 1986.
64. **Shimada, S. (TDK Corp.),** Japan Patent 85,250,679, 1985.
65. **Kawamura, K., Kato, M., and Umibe, K. (Oki Electric Industry Co., Ltd.),** Japan Patent 86,29,179, 1986.
66. **Nakata, J. and Imura, T.,** Photovoltaic effect of gallium phosphide/dye/metal cells, *Oyo Butsuri,* 54(1), 65, 1985.
67. **Shimada, S. (TDK Corp.),** Japan Patent 85,250,678, 1985.
68. **Perrier, G. and Dao, L. H.,** Aluminum hydroxide phthalocyanine photoelectrochemical cells: deposition techniques and chemical treatment, *INTERSOL 85, Proc. 9th Bienn. Congr. Int. Sol. Energy Soc.*, 3, 1887, 1985, (Pub. 1986).
69. **Panayotatos, P., Parikh, D., Sauers, R., Bird, G., Piechowski, A., and Husain, S.,** Improved p-n cheterojunction solar cells employing thin film organic semiconductors, *Sol. Cells,* 18(1), 71, 1986.
70. **Yamashita, K., Harima, Y., Kubota, H., and Suzuki, H.,** Improvement of porphyrin photovoltaic cells in panchromaticities and quantum yields by use of tetrabenzoporphyrin skeleton, *Bull. Chem. Soc. Jpn.*, 60(2), 803, 1987.
71. **Panayotatos, P., Bird, G., Sauers, R., Piechowski, A., and Husain, S.,** An approach to the optimal design of p-n heterojunction solar cells using thin film organic semiconductors, *Sol. Cells,* 21, 301, 1987.
72. **Yasutomi, H. (Minolta Camera Co., Ltd.),** Japan Patent 87,102,569, 1987.
73. **Shimura, M., Toyoda, A., and Bara, H.,** Preparation and photovoltaic behavior of fluoroaluminum phthalocyanine sandwich type solar cells, *Mem. Fac. Technol., Tokyo Metrop. Univ.*, 33, 3357, 1983.
74. **Le Moigne, J. and Even, R. (Centre National de la Recherche Scientifique),** French Demande 2,518,585, 1983.
75. **Rieke, P. C. and Armstrong, N. R.,** Light assisted, aqueous redox reactions at chlorogallium phthalocyanine thin film photoconductors: dependence of the photopotential on the formal potential of the redox couple and evidence for photoassisted hydrogen evolution, *J. Am. Chem. Soc.*, 106(1), 47, 1984.
76. **Iden, R., Seybold, G., Stange, A., and Eilingsfeld, H.,** Fluorescent dyes for solar collectors, *Forschungsber.-Bundesminist. Forsch. Technol., Technol. Forsch. Entwickl.*, BMFT-FB-T 84, 115 pp, 1984.
77. **Rieke, P. C., Linkous, C. L., and Armstrong, N. R.,** Chlorogallium phthalocyanine thin film photoelectrochemistry. Effect of surface coverage and physical distribution of microcrystallites on the photoelectrochemical response, *J. Phys. Chem.*, 88(7), 1351, 1984.
78. **Vertsimakha, Y. I., Koval'chuk, A. V., Mueller, M., and Hamann, C.,** Role of surface states in the appearance of photoemf. at the contact of metals with lead phthalocyanine films, *Ukr. Fiz. Zh. (Russ Ed.),* 29(9), 1370, 1984.
79. **Klofta, T. J., Rieke, P. C., Linkous, C. A., Buttner, W. J., Nanthakumar, A., Mewborn, T. D., and Armstrong, N. R.,** Tri- and tetravalent phthalocyanine thin film photoelectrodes: comparison with other metal and demetalated phthalocyanine systems, *J. Electrochem. Soc.*, 132(9), 2134, 1985.
80. **Yanagiuchi, K., Iwata, T., and Kosho, N. (Fuji Electric Co., Ltd.),** Japan Patent 87,169,374, 1987.
81. **Yanagiuchi, K., Iwata, T., and Kosho, N. (Fuji Electric Co., Ltd.),** Japan Patent 87,169,375, 1987.
82. **Richman, R. M.,** Solar Energy Conversion in Homogeneous Solution, Report 1984, DOE/ER/10457-4-Rev., Order No. DE85006883, 5 pp.

83. **Pentel Co., Ltd.,** Japan Patent 83,209,867, 1983.
84. **Pentel Co., Ltd.,** Japan Patent 83,131,658, 1983.
85. **Pentel Co., Ltd.,** Japan Patent 83,131,659, 1983.
86. **Pentel Co., Ltd.,** Japan Patent 83,131,657, 1983.
87. **Pentel Co., Ltd.,** Japan Patent 84,201,370, 1984.
88. **Pentel Co., Ltd.,** Japan Patent 84,79,977, 1984.
89. **Pentel Co., Ltd.,** Japan Patent 84,98,468, 1984.
90. **Pentel Co., Ltd.,** Japan Patent 84,194,362, 1984.
91. **Pentel Co., Ltd.,** Japan Patent 84,98,467, 1984.
92. **Pentel Co., Ltd.,** Japan Patent 85,74,356, 1985.
93. **Pentel Co., Ltd.,** Japan Patent 85,74,357, 1985.
94. **Deguchi, K., Machama, Y., and Sasage, D. (Pentel Co., Ltd.),** Japan Patent 87,76,256, 1987.
95. **Pentel Co., Ltd.,** Japan Patent 85,117,558, 1985.
96. **Deguchi, K. (Pentel Co., Ltd.),** Japan Patent 87,176,062, 1987.
97. **Nippon Telegraph and Telephone Public Corp.,** Japan Patent 83,163,174, 1983.
98. **Nippon Telegraph and Telephone Public Corp.,** Japan Patent 84,132,570, 1984.
99. **Nippon Telegraph and Telephone Public Corp.,** Japan Patent 84,90,365, 1984.
100. **Nippon Telegraph and Telephone Public Corp.,** Japan Patent 84,138,066, 1984.
101. **Hirai, T., Yamamoto, J., and Okada, T. (Nippon Telegraph and Telephone Public Corp.),** Japan Patent 85,154,468, 1985.
102. **Huang, J. C.,** Structural aspects of pyrolyzed cobalt macrocyclic complexes for alkaline air cathodes, *Proc.-Electrochem. Soc.*, 87-12, 224, 1987.
103. **Kummer, J. T. and Oei, D. G. (Ford Motor Co.),** U.S. Patent 4,407,902, 1983.
104. **Scherson, D. A., Fierro, C., Yeager, E. B., Kordesch, M. E., Eldridge, J., Hoffman, R. W., and Barnes, A.,** *In situ* Moessbauer spectroscopy on an operating fuel cell, *J. Electroanal. Chem. Interfacial Electrochem.*, 169(1—2), 287, 1984.
105. **Hitachi Ltd.,** Japan Patent 84,101,775, 1984.
106. **Hirabaru, O., Nakase, T., Hanabusa, K., Shirai, H., Takemoto, K., and Hojo, N.,** Functional metal porphyrazine derivatives and their polymers. X. Secondary fuel cells based on oxygen reduction on platinum electrode modified by polymers and metal phthalocyanines, *Angew. Makromol. Chem.*, 121, 59, 1984.
107. **Hirabaru, O., Nakase, T., Hanabusa, K., Shirai, H., Takemoto, K., and Hojo, N.,** Functional metal porphyrazine derivatives and their polymers. XI. Secondary fuel cells based on oxygen reduction at a platinum electrode modified by metal-2,9,16,23-tetracarboxy phthalocyanine covalently bound to poly(2-vinyl pyridine-styrene), *J. Chem. Soc., Dalton Trans.*, 8, 1485, 1984.
108. **Zagal, J., Herrera, P., Brinck, K., and Ureta-Zanartu, S.,** Electrocatalytic properties of phthalocyanines, *Proc.-Electrochem. Soc.*, 84-12, 602, 1984.
109. **Kobayashi, N. and Nishiyama, Y.,** Catalytic electroreduction of molecular oxygen using iron or cobalt 4,4′,4″,4‴-tetracarboxy phthalocyanine, *J. Phys. Chem.*, 89(7), 1167, 1985.
110. **Goodenough, J. B., Shukla, A. K., Da Silva Paliteiro, C. A., Jamieson, K. R., Hamnett, A., and Manoharan, R. (National Research Development Corp.),** PCT International Application 86 01,642, 1986.
111. **Imada, K., Yamanobe, T., Watabe, M., and Sato, H. (Toshiba Battery Co., Ltd.),** Japan Patent 87,184,769, 1987.
112. **Iliev, I.,** Air electrodes for aluminum-air batteries, *Glas Hem. Drus. Beograd*, 48, 317, 1983.
113. **Zecevic, S. K., Despic, A. R., Drazic, D. M., Kaisheva, A., Gamburzev, S., and Iliev, I.,** Long term performance of carbon gas diffusion air electrodes in saline electrolyte, *J. Appl. Electrochem.*, 17(3), 657, 1987.
114. **Yamana, M., Darby, R., and White, R. E.,** Preparation of iron phthalocyanine catalyzed carbon electrodes by chemical modification, *Electrochim. Acta*, 29(3), 329, 1984.
115. **Kanda, M. and Sato, Y. (Toshiba Corp.),** Europe Patent Application 149,846, 1985.
116. **TDK Corp.,** Japan Patent 84,160,966, 1984.
117. **TDK Corp.,** Japan Patent 84,163,764, 1984.
118. **TDK Corp.,** Japan Patent 84,163,763, 1984.
119. **TDK Corp.,** Japan Patent 85,32,257, 1985.
120. **Fukuzawa, T., Itoh, H., Koyama, T., Masuda, E., Hanabusa, K., Shirai, H., and Hirabaru, O.,** Functional metallomacrocycle derivatives and their polymers. XXI. Iodine secondary cells using metallophthalocyanine polymers modified electrodes, *Kobunshi Ronbunshu*, 44(4), 287, 1987.
121. **Shembel, E. M., Ksenzhek, O. S., Danilova, N. P., and Shustov, V. A.,** Cathodic reduction of sulfur dioxide from nonaqueous electrolytes on porous phthalocyanine containing electrodes, *Elecktrokhimiya*, 23(9), 1262, 1987.
122. **Loutfy, R. O. and Hsiao, C. K.,** Photovoltaic properties of polymeric phthalocyanines, *Polym. Prepr.*, 23(1), 237, 1982.

123. **Kreja, L. and Plewka, A.,** Poland Patent 120,256, 1983.
124. **Dainippon Ink and Chemicals, Inc.,** Japan Patent 85,47,028, 1985.
125. **Kreja, L.,** The decomposition of hydrogen peroxide in the presence of iron polyphthalocyanines, *Monatsh. Chem.*, 118(6—7), 717, 1987.
126. **Doddapaneni, N.,** Low temperature performance of high rate lithium/thionyl chloride cells, *Proc. 30th Power Sources Symp.*, 169, 1982.
127. **Doddapaneni, N., Chua, D. L., and Nelson, J. T.,** Development of a spin activated high rate lithium/thionyl chloride bipolar reserve battery, *Proc. 30th Power Sources Symp.*, 201, 1982.
128. **Doddapaneni, N.,** Safety Studies on Lithium/Sulfur Monoxychloride Reserve Battery, Report 1983, Order No. AD-A132279, National Technical Information Service, Springfield, VA, 108 pp.
129. **Doddapaneni, N.,** Lithium/sulfuryl chloride (SO_2 Cl_2) system: the influence of cathode substrate and electrocatalysts on performance characteristics, *Proc.-Electrochem. Soc.*, 84-1, 122, 1984.
130. **Arakawa, M., Yamaki, J., and Okada, T.,** The decomposition of propylene carbonate in a lithium/metal phthalocyanine cell, *J. Electrochem. Soc.*, 131(11), 2605, 1984.
131. **Dalard, F., Deroo, D., Djellab, H., Addad, J. P. C., and Mercier, J.,** Kinetic studies of the electrochemical reaction between lithium and phthalocyanine, *J. Power Sources,* 14(1—3), 71, 1985.
132. **Woehrle, D., Kirschenmann, M., and Jaeger, N. I.,** Polymeric phthalocyanines and their precursors. IX. Electrocatalytic activity of phthalocyanines in secondary lithium cells, *J. Electrochem. Soc.*, 132(5), 1150, 1985.
133. **Driscoll, J. R., Pollard, R., Smith, J. J., and Szpak, S.,** Selected aspects of high rate lithium-thionyl chloride power sources technology, *Proc. 20th Intersoc. Energy Convers. Eng. Conf.*, (Vol. 2), 2.55, 1985.
134. **Shimizu, Y. (Japan Storage Battery Co., Ltd.),** Japan Patent 85,163,371, 1985.
135. **Shimizu, Y. (Japan Storage Battery Co., Ltd.),** Japan Patent 85,177,555, 1985.
136. **Doddapaneni, N.,** Catalyzed cathodes for lithium/thionyl chloride cells, *Proc. 31st Power Sources Symp.*, 411, 1984.
137. **James, S. D., Smith, P. H., O'Neil, K. M., and Wilson, M. H. (U.S. Department of the Navy),** U.S. Patent Application 870,213, 1986.
138. **Doddapaneni, N.,** The effect of cathode additives on $Li/SOCl_2$ cell performance after high temperature storage, *Proc. 32nd Int. Power Sources Symp.*, 525, 1986.
139. **Doddapaneni, N. (Honeywell, Inc.),** U.S. Patent 4,698,283, 1987.
140. **Maruyama, K., Tamiaki, H., and Kawabata, S.,** Insoluble catalyst for heat gain in a water soluble solar energy storage system, *Chem. Lett.*, 5, 743, 1984.
141. **Yoshida, Z., Miki, S., and Morimoto, M. (Kajima Corp.),** Japan Patent 86,179,287, 1986.
142. **Abdul-Ghani, A. J., Bashi, N. O. T., and Maree, S. N.,** Solar energy storage through catalyzed reversible photovalence isomerization of 9-tert-butylanthracene, *J. Sol. Energy Res.*, 5(2), 53, 1987.

Chapter 22

CONDUCTION, SEMICONDUCTION, AND PHOTOCONDUCTION

I. DI HYDROGEN PCS

H_2 phthalocyanine, pc, (H_2 pc = pc = di hydrogen pc = free base pc) doped to H_2 pc $I_{2.5}$ experiences a resistivity decrease from 10^{15} to 1 Ω-cm while [tetra(9,10-phenanthro)tetraazaporphyrin]$I_{0.26}$ experiences a decrease from 10^{11} to 10^8 Ω-cm. Benzene rings fused to the periphery of the tetraazaporphyrin ring probably have little π-electron interaction.[1]

Oxygen and iodine adsorption on β-H_2 pc films is studied vs. temperature.[2]

That a metal ion is not required for high electrical conductivity is demonstrated by the simplest pc molecular metal: H_2 pcI. The crystal consists of staggered H_2 pc units and parallel chains of I_3^- counter ions. At 300 K, σ_{11} = 700 1/Ω-cm. At 15 K, σ_{11} attains a maximum of ~4000 1/Ω-cm.[3]

Electrical noise occurs in thin α-H_2 pc films and at <8 Hz is of the 1/f type. A generation-recombination noise with two time constants of ~0.004 and 0.00005 s prevails at higher frequencies at room temperature.[4]

At low temperatures noise dependences of 1/f and $1/f^2$ are observed at low and high frequencies. At 200 to 300 K, pc film electrical properties are controlled by a generation-recombination noise ascribed to traps.[5]

The photogeneration mechanism in pc thin films is extrinsic.[6]

Transient photocurrents in α-H_2 pc are recorded.[7]

Action spectra of the photoconductivity of thin films of H_2 pc in surface cells are studied.[8]

Using pulsed photoconduction methods, transient action spectra are recorded on sublimed α-H_2 pc thin films at 350 to 700 nm.[9]

Photoconductive material is composed of modified τ-H_2 pc or of modified η-H_2 pc. High efficiency of energy conversion of modified τ-H_2 pc can be obtained.[10]

An ohmic contact test uses the activation energy of steady-state space charge-limited control. The method is illustrated on the sandwich cells containing H_2 pc.[11]

Nonlinear photoeffects are studied in metal-island-type semiconductor film-electrolyte structures from pc.[12]

The nonlinear dependence of current on voltage is not the result of current-carrier injection from the electrode but of thermal field emission of current carriers from traps.[13]

Antioxidants, neutral pc, and titania are added to linear low-density polyethylene resin and films are then extruded; 0.01 wt % pc increases the dark lifetime five times and increases the conduction activation energy as well.[14,15]

The method of temperature-modulated steady-state space charge-limited currents is reviewed. The applicability of the method to determine the bulk density of electron states is illustrated on thin films of polycrystalline H_2 pc.[16]

Theoretical studies are made of intramolecular electron transfer in bimetallic mixed valency molecules and pc-based linear chain conductors.[17]

Electrically conductive circuit substrates contain a pc homopolymer.[18]

The electrical and electroluminescent properties of Au/pc Langmuir-Blodgett film/GaP diodes are reported. The electroluminescence conversion efficiency is maximum at a 5.6-nm thick film. The maximum power conversion efficiency approaches that of an unencapsulated p-n junction diode.[19]

A photovoltaic device with an improved stability suitable as a light sensor is obtained as follows: a light active film consists of a polyvinylidene compound containing a dispersion of pc sandwiched between a CdS barrier electrode and an ohmic electrode, characterized in that the CdS electrode is an electrochemically formed film which is then heat treated at 200 to 600°C.[20]

Platinum particles are deposited on x-pc. The photoactivity of a methanol dispersion of the platinized particles is increased by one order of magnitude over unreacted pc.[21]

New structural and electrical conductivity results are presented for H_2 pcI and for Ni pc Br.[22]

II. MG PCS

Doping pc with metallic impurities increases its specific conductivity by several orders of magnitude·Mg pc and Cu pc are doped with Ag, Al, and Cu.[23]

Adsorbed Mg pc on zinc oxide increases electrical conductivity of zinc oxide in the presence of photostimulated benzene or naphthalene vapors, due to the triplet-triplet energy transfer from the excited aromatic molecules to the adsorbed Mg pc then to the zinc oxide semiconductor.[24]

The current-voltage characteristics are studied of oxygen-doped Mg pc films with aluminum or silver sandwich electrodes.[25]

III. NEODYMIUM (ND) PCS

An improvement in iodination to enhance the electrical conductivity of Nd(III)pc_2 H relates to the linear chain composed to stacked Nd(III) pc_2 H sandwich molecules.[26]

The electrochemical preparation, electrical conductivity, and magnetic behavior of single crystals of H bis pc neodymate (1-) are reported.[27,28]

Nd pc_2 (ox I) and Nd pc_2 (ox II) are prepared as crystals from bis pc Nd. Both phases behave as metallic conductors with conductivities of 50 and 400 1/Ω-cm, respectively, at room temperature.[29]

IV. LUTETIUM (LU) PCS

Intrinsic conductivies of thin films and of thin crystals of bis pc Lu, Lu pc_2, are reported. Lu pc_2 is more conductive than the usual metal pcs by >6 orders of magnitude.[30]

Thin films of Lu pc_2 derivatives are doped with dichlorodicyanoquinodimethane electron acceptor and tetraphenyldithiapyanylidene electron donor. Electrical conduction properties of Lu pc_2 and of p- and n- type doped materials are correlated with the optical characteristics of the films. Lu pc_2 is the first molecular semiconductor to satisfy the criteria: it is a molecular material; its intrinsic conductivity is 10^{-6} to 0.1 1/Ω-cm; and it can be doped with both electron acceptors and donors.[31]

Crystals of solvate bis pc Lu, $pc_2Lu{\cdot}CH_2Cl_2$, are intrinsic molecular semiconductors of 0.00006 1/Ω-cm at room temperature.[32]

The electrical properties of thin films and of single crystals of bis pc Lu, Lu pc_2, and of Li pc are determined. Undoped single crystals of Lu pc_2 have a room temperature conductivity of 0.00006 1/Ω-cm, > 6 orders of magnitude higher than that for standard divalent (Cu, Ni...) M pcs.[33]

V. FE PCS

Fe pc Cl is characterized by IR absorption, X-ray diffraction, Moessbauer spectra, thermogravimetric analysis and DTA plots, dielectric constant and loss, and thermally stimulated depolarization current curves.[34]

Fe(II) (μ-1,4-diisocyanobenzene) pcs are doped with iodine. Increasing dopant levels result in higher electrical conductivities ≤0.01 1/Ω-cm and in decreasing activation energies.[35]

Inherently conductive, 0.4 1/Ω-cm, Fe(III) cyano pc crystals are prepared by electrocrystallization of Fe(III)K dicyano pc in methyl cyanide.[36]

The electrical conductivity of powdered Fe pc (sym-tetrazine)n is 0.2 1/Ω-cm. The first use of $N = C = N^{2-}$ as a bridging ligand in $[Ge\ pc\ (NCN)]_n$ is reported.[37]

VI. CO PCS

Potassium dicyano cobalt (III) pc is electrolyzed to form cyano cobalt (III) pc, Co pc CN in the form of large single crystals >4 mm long. The electrical conductivity is 57 1/Ω-cm without doping.[38]

During electrolysis of potassium dicyano cobalt (III) pc many platelike crystals grow on the platinum point electrode. They are Co(III) pc CN, in which one pc is overlapped with six adjacent pc planes supporting the three-dimensional overlapping of π-orbitals.[39]

Potassium dicyano cobalt (III) pc in methyl cyanide is electrolyzed to precipitate Co pc CN, the conductivity of which can be increased by doping with either electron acceptors or donors.[40]

A DMF solution of Co $pc(CN)_2$ 0.0005 *M* is electrolyzed. Black needlelike crystals are deposited on the anode surface. They are 3 to 5 mm long and their electrical conductivity is 0.039 1/Ω-cm.[41]

A voltage of 3 V is applied between platinum electrodes in DMF solution of 0.02 *M* Co pc $(CN)_2$ to deposit 10 to 70 μ thick crystals on the anode with conductivity of about 0.00001 1/Ω-cm.[42]

The room temperature conductivity of Co pcI is about 50 1/Ω-cm, comparable with the best platinum-spine conductors, although cobalt is a first transition series element at a Co-Co spacing of 3.12 Å instead of only ~2.95 Å for Pt-Pt.[43]

Dicyano Co pc in DMF is dried on a glass plate and heated to grow black columns 1 to 2 μ wide by 4 to 6 μ long with conductivity ~0.0001 1/Ω-cm.[44]

Co pc CN crystals are prepared with conductivity 57 1/Ω-cm. Three-dimensional conductive paths may be the cause of the high conductivity.[45]

VII. RHODIUM (RH) PCS

Rh (III) μ-cyano pc $(Rh\ pc\ CN)_n$ is formed by splitting off potassium cyanide from $K(CN)_2$ pc Rh(III), $K[Rhpc(CN)_2]$. Monomeric Rh pc $[(C_4H_9NH_2)\ CN]$ is formed upon treatment of $(RhpcCN)_n$ with $C_4H_9NH_2$. $(Rh\ pc\ CN)_n$ has a conductivity of 0.0004 1/Ω-cm which is diminished greatly when the polymeric structure is decomposed.[46]

VIII. NI PCS

The temperature dependence of ohmic and space charge-limited bulk current densities flowing in the c-direction in β-Ni pc crystals in ambient oxygen is studied at 293 to 450 K. The results agree with Barbe and Westgate's 1970 model for H_2 pc in oxygen.[47]

Using an Ar^+ ion beam, conducting patterns are established in otherwise high resistivity thin films including Ni pc.[48,49]

Carbon contains high resistivity pc or Ni pc with 1 Ω-cm resistivity prior to irradiation with Ar^+. The irradiated region has a resistivity as low as 0.1 Ω-cm.[50]

The structural, magnetic, and electrical properties of Ni pc are studied.[51]

Electrical properties of Ni and Co pcs are studied, and the results are related to oxygen adsorption.[52]

The conductivity of the linear chain molecular metal Ni pc I remains metallic down to about 30 K, goes through a maximum, and as T → O, levels off to about 0.5 to 2 times the value at room temperature (~500 1/Ω-cm).[53]

The first pc molecular metal without halogen counterions, $Nipc(BF_4)_{0.33}$ is prepared by high temperature electrocrystallization. Conductivity is metal-like at about 300 K, 1000 1/Ω-cm, and a broad metal to semiconductor transition occurs below ~80 K, to 300 to 600 1/Ω-cm at 1.5K.[54]

A metal pc dicarboxylic acid and its alkyl ester having increased conductivity are described. The formula of the ester is Ni pc $(COOC_{10}H_{21})_x$.[55]

The conductivity of Ni pc I is ~400 1/Ω-cm at room temperature, increasing to almost 4000 1/Ω-cm at 50 K.[56]

A model is presented for the high static pressure conductivity of pc and M pc complexes including Ni pc. The calculated activation energy for the Ni pc agrees with an experiment.[57]

Synthesis and crystal structure of a highly conducting Ni pc hexafluoroantimonate, (Ni $pc)_2$ SbF_6, are described.[58]

Highly conducting (Ni $pc)_2AsF_6$ and (Ni $pc)_2$ SbF_6 are prepared electrochemically. The crystal structures resemble that of Ni pc $(I_3)_{0.33}$, composed of Ni pc molecular columns surrounded by I_3. The electrical conductivity of these salts is metallic down to at least 200 K.[59]

Ni pc $(C10_4)_y$ (y = 0.39 to 0.47) is made by electrocrystallization techniques. At 300 K, σ_{11} = 700 ± 200 1/Ω-cm with metal-like behavior.[60]

The electrical properties of thin films based on two new highly soluble Ni pcs involving four long alkyl amides (Am pc 1 and Am pc 2) are studied. Film conductivities increase two to four times upon iodine vapor exposure and reach $\sim 10^{-6}$ 1/Ω-cm.[61]

Heating trimellitic anhydride, phthalic anhydride, nickel chloride, and urea in nitrobenzene gives a pc ester, which is hydrolyzed to the acid and reesterified to the decyl ester. A vinyl acetate-vinyl chloride polymer-polyurethane blend containing 20% of this ester cast on a polyester film has an electrical conductivity of 1 to 10 1/Ω-cm after doping with sulfur trioxide, and 0.000026 1/Ω-cm before doping.[62]

Ni tetracarboxamido pc is hydrolyzed, treated with thionyl chloride to give the acid chloride, and mixed with natural rubber modified with styrene and 2-vinylpyridine. Aluminum trichloride is added. A sheet is formed, and it is doped with iodine to give a conductivity of 0.00001 1/Ω-cnm.[63]

Ni pc Br is prepared by oxidizing Ni pc with bromine. The structure is the same as Ni pc I. Electrical conductivity of Ni pc Br is ~100 1/Ω-cm at room temperature.[64]

IX. CU PCS

The thermally stimulated discharge of 0.8 mm thick Cu pc pellets is studied. Cu pc electrets are formed at 100, 140, and 60°C by applying 2, 4, and 6 kV/cm electric fields.[65]

Photoconductivity and ellipsometric spectroscopy are used to study Cu pc overlayers on nonpolar zinc oxide surfaces.[66]

A photosensitive composition contains Cu 1,2,3,4,8,9,10,11,15,16,17,18,22,23,24,25-hexadecabromo pc for printed circuit boards.[67]

Preferentially oriented Cu pc films are obtained and a Schottky diode is prepared.[68]

The effect of oxygen dopant on dark and photoconductivity in Mg and Cu pcs is studied.[69]

AC conductivity, dielectric constant, and loss dependence on frequency are studied for Cu pc acetylene carbon black mixtures.[70]

Highly conductive Cu pcarbon black mixtures are generated. Transition between semiconducting and metallic states is observable. The conductivity depends on the carbon black grain size.[71]

Metallic conductivity and magnetic interactions in Ni and Cu pc iodides are studied. Replacement of diamagnetic Ni^{2+} by paramagnetic Cu^{2+} introduces a novel coupling between local Cu^{2+} spins and the mobile charge carriers.[72]

The electrical resistivity of Cu pc-carbon black and pc-carbon black mixtures at 300 to 450 K is determined. Mixtures with 10 and 15 wt % carbon black exhibit reversible semiconductor metal transitions.[73]

Electrical conduction of Cu pc with carbon black is measured.[74] The conductivity of the mixtures increases linearly with the conductivity of the carbon black; the conductivity of the carbon black decreases with increasing grain size.[75]

The current-voltage characterization of copper, lead, and tin electrode materials are studied on 0.9 mm thick pellets of Cu pc.[76]

The electrical properties of thin Cu pc films <10 μ thick are studied.[77]

The conductivity of Cu PcI_y peaks at 150 to 170°C doping temperature and increases as the iodine concentration increases. Optimum Cu pc and iodine ratio is 1:1.5.[78]

The conductivity of Cu pc I_x ($0 < x \leqslant 2.0$) is determined of Cu pc I_x prepared by sublimation of Cu pc and iodine. Cu pc I_x is stable at $x = 2.0$. Cu pc I_x is a solid solution of Cu pc and Cu pc I_2.[79]

Cu pc iodide is a molecular metal, the conducting stacks of which incorporate a one-dimensional array of local moments strongly coupled to conduction electrons. There is little static magnetic order <8 K.[80]

The dielectric properties of pc, Cu pc, and p-terphenyl pellets at 1000 to 10,000 Hz are determined. The dielectric loss increases as the frequency decreases in pc and Cu pc.[81]

The n-p^+-GaAs Langmuir-Blodgett film switching devices using ω-tricosenoic acid and Cu pc are studied. Switching appears to be due to a punch-through mechanism.[82]

An asymmetrically substituted Cu pc has a maximum short circuit photocurrent of the order of 10^{-8} A/cm² for 100 μW/cm² monochromatic light, indicating the superiority of asymmetrical Cu pc as a film forming molecule for photovoltaic Langmuir-Blodgett films.[83]

The variations in mobility and trapping concentration in Cu pc thin films are ascribed to the scattering and trapping effects of incorporated ambient gas molecules.[84]

A photoelectric device capable of distinguishing incident wavelengths comprises a film containing UV- or IR-absorbing photoconductive Cu pc and a pair of electrodes on both sides of the film.[85]

Cells using $Cd_{1-y}Zn_yS$ and Cu pc bilayers between gold and F-doped stannic oxide (SnO_2:F/$Cd_{1-y}Zn_yS$/Cu pc/Au) exhibit a photovoltaic effect with a weak energy conversion efficiency of ≲0.06%.[86]

The photovoltaic effects of Cu pc/CdS and pc/CdS heterojunctions may be able to be developed as light detectors or solar cells under 7 mW/cm²; the pc heterojunctions have a conversion efficiency of ~0.15%.[87]

Epoxy resin coatings contain Cu pc for photovoltaic devices.[88]

A semiconductor organic thin film is prepared by plasma treatment of a Cu pc thin film.[89]

The current-voltage characteristics are found for Au/Cu pc/Au and Ag/Cu pc/Ag structures. The Frenkel-Poole tunneling mode predictions fit the data.[90]

Trap parameters for α-Cu pc films are studied vs. temperature and electric fields by the isothermal decay current method. The field dependence of the isothermal decay current indicates field-assisted detrapping, probably by a Poole-Frenkel mechanism.[91]

Semiconductor electrorheological fluids are described.[92]

A quartz powder and β-Cu pc mixture is heated at 1050°C for 6 h and then cooled after 0.5 h to give a powder with a conductivity of 10 1/Ω-cm.[93]

The increase in the number of sulfonate groups in a polypyrrole film doped with Cu pc causes localization of positive charge carriers in the polypyrrole films, thereby decreasing the film conductivity.[94]

Photoelectric devices are made from dehydrohalogenated KF 1000 containing Cu pc.[95]

Heat-resistant conductive polymers contain Cu pc. The polymers contain Nylon 66 and brass fibers. The systems exhibit 0.001 to 10^{12} Ω-cm volume resistivity.[96]

A photoreceptor or photosemiconductive material is made from a mixture of β-Cu pc, Cu tetra-2,3-pyridinoporphyrazine, acrylic polyol, epoxy resin, and cellosolve acetate.[97]

ϵ-Cu pc, 2,5-bis(4-diethylaminophenyl)-1,3,4-oxadiazole, Cu piperidinomethyl pc, an acrylpolyol, an epoxy resin, xylene, and cellosolve acetate are kneaded to give a photoconductive composition.[98]

X. ZN PCS

The conductivity of oxidized Zn pc single crystals on platinum anodes in organic solutions containing chlorides and bromides is described.[99]

The effect of different substituents on pc and nc (nc = naphthalocyanine) is studied vis-a-vis sublimed films of Zn pc and Zn nc derivatives and their dark conductivity.[100]

The conductivity of α-Zn pc crystals in ammonia decreases abruptly due to the formation of the Zn pc-ammonia complex at 25°C.[101]

Zn pc $(CN)_8$ is ~2 orders of magnitude more photosensitive than Zn pc $(OCH_3)_8$. This difference is discussed in terms of the Onsager mechanism of free carrier generation.[102]

α-Zn pc grows in the form of randomly oriented microcrystallites, but the β-form shows oriented needle-like crystals and whisker growth. The electrical properties are dependent on material purity.[103]

XI. AL AND GA PCS

The electric resistivities of Al pc F and Ga pc F decrease by doping with I^-, BF_4^-, PF_6^- and SbF_6^-. Measurements are also made with Al pc OH and Al pc Cl doped with PF_6^-.[104]

Photocurrent spectroscopy provides evidence for charge trapping at the Au/Ga pc Cl interface using photocurrent spectroscopy.[105]

The electric conductivity is studied of I^--implanted Al pc F films. The conductivity increases as the iodine concentration increases.[106]

Thermally stimulated current is measured in Al pc Cl film at ~150 to 300 K after 600-nm laser irradiation. Trap parameters are evaluated; the trap sites are described by the Poole-Frenkel model.[107]

Ga pc Cl and In pc Cl are thought to be p-type and n-type semiconductors, respectively. The dark and photoconductivity of Al, Ga, and In pcs show ohmic conduction after correction with the value of photovoltage.[108]

XII. INDIUM (IN) PCS

Electric field gradients are studied in In pc Cl by perturbed angular correlation of γ rays. The electric field gradient is large, despite the electron withdrawing properties of the pc ring.[109]

XIII. SI PCS

The Wolfsberg Helmholtz correlation between overlap and exchange integrals is used along with a tight binding band scheme to study the effects of geometric distortions on the calculated bandwidths of $(Si\ pc\ O)_n$.[110]

Schottky barrier structures are made by evaporating aluminum onto Langmuir-Blodgett films of a substituted Si pc. The substituted Si pc appears to be a p-type semiconductor.[111]

XIV. PB PCS

An electronic switching model for polycrystalline films of Pb pc is proposed.[112]

The effect of series resistance is studied on the voltage-capacitance characteristics of a Ni/Pb pc/Au structure. Current-voltage, capacitance-voltage, and the dependence of the photoemf on the intensity of strongly adsorbed light are measured. At the Ni/Pb pc interface, a Schottky barrier forms; the gold electrode, on the other hand, is ohmic.[113]

The electric switching in thin Pb pc films is based on an electrical field-induced structural disorder-order transition.[114] The effect of crystal structure on the properties of the Ni/Pb pc interface is studied. With increasing crystal size, the Schottky barrier height decreases, whereas the space charge region is expanded.[115]

Metallic conduction is observed at 150 to 295 K for most of the sandwiches M/Pb pc/M (M = Al, Au, Cu, Ni, or Pb).[116]

For Pb pc single crystals, the decay of photocurrent is slower than the rise in vacuum and ammonia, but especially in nitrogen dioxide.[117]

Pb pc dark current in ammonia or oxygen obeys Ohm's law. The photoconductivity has a broad maximum at 420 to 1100 nm. Oxygen increases both dark conductivity and photoconductivity.[118]

Iodine doping of Pb pc thin films increases Pb pc conductivity seven to nine orders of magnitude.[119]

A method to make stable and miniaturizable switches and memory cells from M pcs consists of depositing a mixture of pcs in an electric field and heating in an oxidizing atmosphere to passivate the surface. An example is given of Pb pc deposition on a gold electrode.[120]

XV. METAL (M) PCS

The electrical conductivities of M pcs, Fe(III), Co, Cu, and Ni pcs bound to 2-vinylpyridine-styrene copolymer films are studied. The films become highly conductive when doped with iodine: 0.00001 to 10 1/Ω-cm.[121,122]

The electronic structures of one-dimensional M pc and M pc L (L = ligand attached to the central metal atom, M) are studied by the tight binding LCAO method. Calculations are made for M = Co, Cr, Cu, Fe, Mg, Ni, and di hydrogen.[123]

The dark electrical properties of M pc-metal contacts are studied at 0.001 to 100,000 Hz. The standard Schottky model is not applicable to these conductors. The current-voltage relationship seems to show a Frenkel-Poole mechanism of charge detrapping.[124]

The physical properties of a new class of charge-transfer salts based on M pc and M pc I_x ($0.2 \leq x \leq 1.7$) are studied. The electrical conductivity of M pc I_x at 293 K is 8 to 15 times higher than that of M pc. Semiconductivity occurs at 77 to 500 K.[125]

The structure and electronic configuration of molecular and macromolecular pc conductors are studied by high resolution C-13 NMR with cross polarization, H-1 dipolar decoupling, and magic angle spinning.[126]

The effect of pressure is studied on the electrical conductivity of pc and of Co, Cu, Ni, and Zn pcs.[127]

Junctions based on *trans*-polyacetylene and different M pcs are studied. The dark conductivity vs. frequency and vs. applied voltage of M_1/t-$(CH)_x$/M_2 or M_1/M pc/M_2 sandwich cells demonstrate the similarity of their properties. If the films show an apparent Schottky behavior, the standing model is not applicable.[128]

Evaporation deposition of Cu pc films with good crystallinity involves rotating and moving a deposition substrate in the direction of the evaporation-material motion.[129]

Bulk conductivity of pc and of its Co, Ni, and Zn complexes at 293-373 K and ≤3 GPa is studied. Two conduction mechanisms appear to be operating simultaneously.[130]

Langmuir-Blodgett films of tetrakis(cumylphenoxy) pc compounds and stearyl alcohol have changes in electrical conductivity and mass changes caused by iodine vapor doping measured by a 52 M Hz surface acoustic wave device. The conductivity increases by four orders of magnitude and there are two to four iodine atoms per pc ring. The central metal cation, Co, Cu, Ni, Pd, Pt, or Zn, or di hydrogen, has very little influence on either the magnitude of the conductivity increase or on the I-M pc stoichiometry.[131]

The effect of NO_2, NO, F_2, BCl_3, BF_3, ammonia, hydrogen sulfide, sulfur dioxide, and hydrogen chloride at <1 ppm in air on the electrical conductivities of Co, Cu, Fe, Ni, Pb, and Zn pcs is determined. As the metals become heavier, sensitivity to nitrogen dioxide increases. Mg pc is affected only by hydrogen chloride. These pcs are evaluated as gas sensors.[132]

Changes in behavior of electrochemical and photoelectrochemical properties in thin films of tri- and tetravalent M pcs are studied.[133]

An electrorheological fluid comprises a liquid continuous phase and $\geqslant$ 1 dispersed phases. The dispersed phase is preferably an M pc or poly(acene-quinone). The electrorheological fluid can support a current density of $\leqslant$400 A/m^2 at an applied electric field of $\leqslant$1 MV/m.[134]

The pc compounds containing a central metal cation (M = Co, Cu, Ni, Pt, or Pd, or di hydrogen), and any of the following groups or their mixtures aryloxy, arylthio, alkoxy, hydrogen substituted on the four benzene rings are prepared and then made into semiconducting thin films by the Langmuir-Blodgett technique.[135]

M pcs and M C_n pcs [tetra(alkoxy) M pcs] are fabricated into thin films by the Langmuir-Blodgett (LB) method, vacuum deposition (VD), and organic MBE (molecular beam epitaxy) method and their electric conductivities are determined. The relative surface roughness of films prepared by LB, VD, and OMBE methods is 20, 70, and $<$10 Å, respectively, when each sample is 200Å thick.[136]

A thin film of tetraalkoxy M pc where M = Mg or transition metal, and R = $C_{3\text{-}25}$ linear or branched hydrocarbon group, with OR on the 5 or 6 position of the isoindoline groups, is used as the semiconductor in a MSM (metal/semiconductor/metal) or MIS (metal/insulator/semiconductor) device.[137]

Photocurrent spectra of thin film sandwich cells of vacuum deposited di hydrogen and M pcs are compared with their optical absorption spectra. VO pc absorption spectrum depends considerably on the deposition condition of thin films, unlike that of other planar pcs.[138]

Electrical and optical properties of thin films of pcs are studied in ultrahigh vacuum.[139]

A new method to obtain thin films of doped molecular materials in the absence of oxygen is developed, by simultaneously subliming under vacuum the electroactive compound, i.e., an M pc, and the electron acceptor or donor doping agent. A rectifying contact between doped pc thin films and aluminum is observed in the absence of oxygen.[140]

The effect of the metal M is studied on the electrical and photoconductive properties of the M pcs (M = CO, Cu, Fe, Mn, Ni, VO, and Zn). Absorption maxima at 650 to 700 nm is due to charge transfer from metal to nitrogen in the ligands.[141]

Metal complexes of the pc-type compounds shown in Figure 1 are semiconductors.[142] (M = Al, Cd, Co, Cr(III), Cu, Fe(II), Ga, Ge(IV), In, Mg, Ni, Pb, Sn(II), Sn(IV), Zn, and Zr(IV).)

Transition metal pcs of the fourth, fifth, and sixth periods and their catalytic, electrochemical, and semiconducting properties are studied.[143]

XVI. PC POLYMERS

A. AL PC F

Al pc F and Ga pc F polymers doped with BF_4^- and PF_6^- are highly stable to temperature and exposure to ambient air.[144]

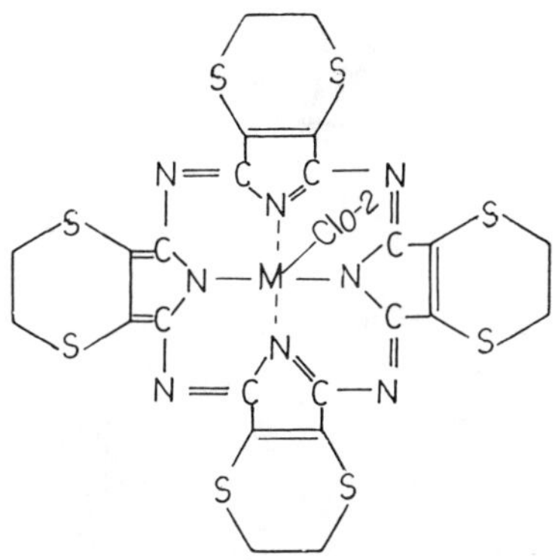

FIGURE 1. Metal complex of tetra (1,4-dithiacyclohexene) porphyrazine.

(Al pc F)$_n$ is deposited in layers several hundred angstroms thick. The electrical conductivity of the layers increases when exposed to an iodine and nitrogen gaseous mixture.[145]

(Al pc F)$_n$ is doped with AsF_5 and AsF_6^- to as many as 4 or 5 AsF_5 per pc molecule. A model attempts to describe their low conductivity.[146]

Al pc F polymer thin films are doped by iodine implantation at 100 to 200 keV.[147]

Thin Al pc F layers are sensitive to ambient nitrogen dioxide and have potential as the sensitive part of gas sensors.[148]

(Al pc F)$_n$ is doped with IF_5. An optimum conductivity of 0.06 1/Ω-cm is attained for 0.5 to 0.8 molecules of IF_5 per each Al pc F molecule.[149]

Doping of (Al pc F)$_n$ by IF_5, MoF_6, and WF_6 is effective to prepare polymers that conduct well such as (Al pc F)-$(IF_5)_{0.43}$, (Al pc F)-$(MoF_6)_{0.38}$, and (Al pc F)-$(WF_6)_{0.42}$.[150]

B. SI PC O

Strong, very conductive fibers that are air stable are obtained from co-spinning I-doped (Si pc O)$_n$ and Kevlar.[151]

Conductive solutions of (Si pc O)$_n$ in F_3CSO_3H containing Kevlar are wet spun and then doped with iodine to give (Si pc O) $(\text{Kevlar}_{n1}\ I_{n2})_n$ where $n_1 = 0.59$ to 4.38 and $n_2 = 0$ to 3.69, stable in air with conductivities at 300 K of 0.0025 to 0.091 1/Ω-cm. Without Kevlar, $[(\text{Si pc O})I_{1.76}]_n$ is obtained with conductivity at 300 K of 1.8 1/Ω-cm.[152]

Spinning a 1:1 Kevlar 29 aramid fiber pulp and Si pc O polymer in CF_3SO_3H into water gives strong, electrically conductive, smooth, purple black fibers.[153]

(Si pc O)$_n$ and Kevlar in solution are wet spun, doped with halogen or electrochemically, to give strong fibers that are stable in air and that have a thermally active conductivity as high as 5 1/Ω-cm. Kevlar and $[(\text{SipcO})I_{1.1}]_n$ are preferentially oriented.[154]

Kevlar 29 aramid pulp and (Si pc O)n in equivalent proportions are dissolved in triflic acid with about 10 wt % solids, and extruded into water. Electrically conducting, strong, purple black smooth fibers are obtained.[155]

(Si pc O)$_n$ is reversibly oxidized to yield $[(\text{Si pc O})(BF_4)_y]_n$ polymers ($y = 0$ to 0.50). Varying y provides a continuous evolution from semiconductor to a low-dimensional metal-like substance.[156]

Oxidative electrochemical methods yield polymers with doping levels as high as $[(\text{Si pc O})(BF_4)_{0.50}]$, $[(\text{Si pc O})(\text{p-toluenesulfonato})_{0.85}]_n$, and $[(\text{Si pc O})(SO_4)_{0.09}]_n$. Partially reduced polymers are also made such as $\{(\text{Si pc O})[(C_2H_5)_4N]_{0.25}\}_n$. (Si pc O)$_n$ can also be doped with 2,3-dichloro-5,6-dicyano-1,4-benzoquinone (DDQ) to give substances with doping levels as high as $y = 0.91$: $[(\text{Si pc O})(\text{DDQ})_y]_n$. As y is increased from zero, the polymer changes from a filled band insulator to a molecular metal ($y \sim 0.30$ to 0.50), then to an integral oxidation state Mott insulator ($y > 0.90$).[157]

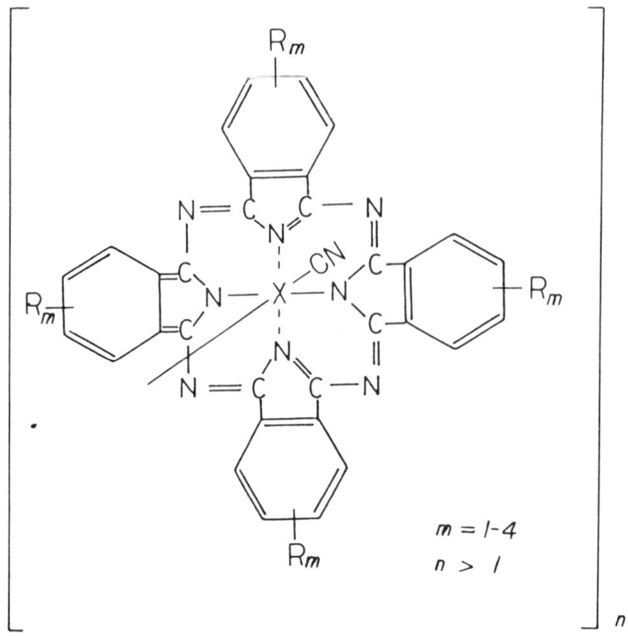

FIGURE 2. An electrically conducting U-cyano pc transition metal polymer.

Electrical properties of (Si pc O)$_n$ are studied vs. dopants added: BF, PF_6, SbF_6, and I.[158]

The photoconductivity of (Si pc O)$_n$ films is much less than that of Si pc $(OH)_2$ films.[159]

Four-probe electrical conductivity studies of polycrystalline $\{[\text{Si pc O}]X_y\}_n$ reveal a sharp increase in conductivity with increasing y. X = BF_4, PF_6, or SbF_6.[160]

The electronic structure of the conductive building block Si pc $(OH)_2$ and HO Si pc O Si pc OH is studied.[161]

The principal charge-transport pathway in (Si pc O)$_n$ is by the pc π systems.[162]

C. FE PCS

Electrically conducting μ-cyano pc transition metal polymers as shown in Figure 2 (X = Co, Cr, Fe, Mn, Rh; R = H, among other entities) may be used as semiconductors or photoconductors. [Fe(III) pc CN]$_n$ has a conductivity of 0.0008 1/Ω-cm. It is made by suspending 0.57 g Fe(III) pc Cl in a solution of 0.26 potassium cyanide in 30 ml water and refluxing.[163]

Various monomeric and polymeric Fe pcs are prepared and treated with atmospheric oxygen. Electrical conductivity tends to increase with polymerization, oxygen inclusion, and as the ratio of the two valencies shifts to one.[164]

The AC conducting mechanisms of films of poly(2-vinylpyridine-costyrene) having Fe(II) pc groups, which may be doped with iodine, are examined.[165] The frequency dependence of Fe pc homopolymer is studied in terms of a frequency exponent and a model dependent factor.[166]

Conductive polymer coatings containing Fe pc are made on catalytic mold surfaces in the moldings of static-free plastic parts.[167]

Tetrasulfonated Fe pcs can be incorporated into electronically conductive polypyrrole by electrochemical polymerization of pyrrole in the presence of the tetrasulfonated Fe pcs.[168]

Iodine-doped [Fe pc(μ-pyrazine)]$_n$ is not highly crystalline. The iodine is reduced and

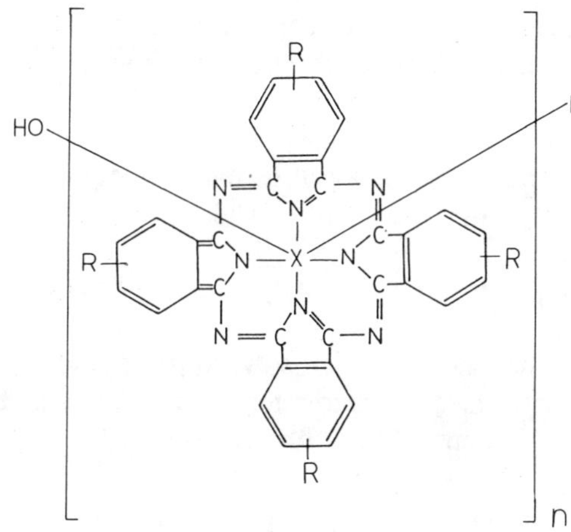

FIGURE 3. A doped M pc conductor.

the [Fe pc(μ-pyrazine)]$_n$ is oxidized. The conductivity of [Fe pc(μ-pyrazine)]$_n$ increases greatly with increasing doping.[169]

D. CU PCS

Cu pc homopolymer is included in a discussion of whether postpolymerization processing of conductive polymers is a viable way to make conductive polymers.[170]

A semiconductive polymer is formed by treating poly(vinylidene fluoride) (Kureha KF 1000) dissolved in DMF by adding $(C_4H_9)_4$ NBr and Cu pc powder and immersing the precipitate in alkaline solution. The resultant film has a short-circuit current of 10^{-8} A under a 150-W xenon lamp.[171]

Poly Cu pc films are made by evaporating tetracyanobenzene and copper in a vacuum followed by a heat treatment.[172]

The room temperature electrical conductivity of Co, Cu, and Ni pc sheet oligomers are 2.22×10^{-8}, 6.54×10^{-7}, and 2.42×10^{-8} 1/Ω-cm, respectively. The planar oligomer increases the π-electron delocalization and accounts for the greater conductivity of the oligomers compared with their monomers.[173]

E. A VARIETY OF M PC POLYMERS

Stacked bridged macrocyclic metal complexes are made from pyrazine and 1,4-diisocyanobenzene as the bridging ligands, pc, and tetrabenzoporphine as the macrocycle, and Fe and Ru as central metal atoms, and doped with iodine. (M pc CN)$_n$ polymers with M = Co, Cr, Mn, and Fe with CN as bridging ligand are described. With Co and Fe these polymers exhibit room temperature conductivities of about 0.01 1/Ω-cm without doping, which is about the same as the above M pc ligand polymers doped with iodine.[174]

A conductor and photoconductor has the structure shown in Figure 3, where R may = H, X = Ge, Si, or Sn, $n < 1000$, and is doped by reacting with NO^+, NO_2^+, and a variety of inorganic salts. As example, X = Si, R = H, $n > 1$, and the dopant is ferric chloride in methyl cyanide to give a film with conductivity 0.011 1/Ω-cm.[175]

Temperature and composition dependence of electrical conductivity of Ni pc/Kevlar/I_2 fibers are similar to the behavior of composites.[176]

A polymerized M pc derivative (M = Mg or transition metal) is used as the semiconductor layer of an MSM (metal/semiconductor/metal) or of an MIS device.[177]

A polymerized pc derivative containing a dopant is prepared by high-frequency irradiation of a mixture containing a dopant (I, SO_3, AsF_5, BF_3, SbF_5, NbF_5, TaF_5, and PF_5), a tetra-R-pc, with R = H, halo, or $C_{\leq 5}$ alkoxy, and M = Mg, transition metal, or di hydrogen. A MSM or MIS device having a semiconductor layer from the polymerized derivative is described.[178]

A semiconductor device involves a doped Cr pc F polymer on a substrate used as an interconnection layer and the other nondoped regions of the polymer as an insulator.[179]

A semiconducting dehydrogenated poly(vinylidene fluoride) polymer is doped with di hydrogen pc derivatives.[180]

A doped polymer has M pc groups on its side chains and is formed into a conductive film.[181]

The photocurrents in polymeric Co(III) μ-cyano pc depend on temperature, intensity, voltage, and wavelength.[182] The same is true for polymeric $(\text{Ge pc S})_x$.[183]

The physical characterization of M pc polymers with variable band filling is the subject of a thesis.[184]

M pc sheet polymers under study give benzene, cyanobenzene, and dicyanobenzene on thermal degradation. Electric conductivity measurements show that these polymers are semiconductive.[185]

Methods to modify polypyrrole with electroactive groups, such as sulfonated Co pc, are discussed.[186]

Polypyrrole-sulfonated Co pc films are a conducting polymer stable to air and moisture.[187]

An electroconductive polypyrrole-PVC film contains Co pc for redox reaction catalysts.[188]

An apparatus is described for molding electrically conductive Co pc homopolymer.[189]

It is attempted to make semiconducting polymers by the high-temperature free radical polymerization of nitriles, such as of di hydrogen phthalonitrile.[190]

Polymeric Mo pc is prepared by heating pyromellitonitrile-$Mo(CO)_6$. Its conductivity is 1.07×10^{-10} 1/Ω-cm. Di hydrogen pc polymers are prepared by heating pyromellitonitrile-urea mixtures with polymer conductivity 7.55×10^{-8} to 8.67×10^{-10} 1/Ω-cm. Polymer conductivity increases 2 to 3 times with iodine doping but monomer conductivity increases 12 times with iodine doping.[191]

Introduction of Cu, Fe, Ni, or Zn into unbranched poly pc and Cu or Zn into branched poly pc does not change their electrical properties.[192]

In a study of polymeric pcs and of their precursors, the electrochemical properties of poly pc-modified titanium electrodes are optimized.[193]

Di hydrogen poly pc in tablet form has a network structure exhibiting electron switching at $\geq 200 \pm 20$°C.[194]

The role of π-orbital interactions in the electronic properties of conductive pc polymers are the subject of a thesis.[195]

XVII. REVIEW

A review with 69 references relates to the electrical properties of pcs, poly pcs, partially oxidized pcs, and doped film and polymer conductors stemming from pcs.[196]

A review with 93 references covers M pcs exhibiting metallic properties, and film and polymeric electrically conducting materials derived from pcs.[197]

A review with 49 references covers application of pcs including electrocatalysts, photovoltaics, photocatalysts, molecular metals, photoconductivity, electrochromism, chemical sensors, batteries, and biological and medical applications.[198]

A review with 33 references covers M pc functions, reactions catalyzed by M pcs, and photo and electronic functions of M pcs.[199]

A review with 15 references emphasizing pc derivatives covers intrinsic semiconductivity, doping, and the extended space charge layer in a junction.[200]

A review with 15 references discusses the synthesis and conducting properties of macrocycles containing axial bridges of nitrogen bases, cyano, or thiocyanato groups. (Co octamethyl pc)$_n$ and (Co tetra-*tert*-butyl pc)$_n$ are prepared and their electrical conductivity determined.[201]

A review with 29 references discusses the work of the author of the review on the synthesis and conducting properties of Co, Fe, and Ru polymers of pc, tetrabenzoporphyrin, and 2,3-naphthalocyanine with pyrazine, 4,4′-bipyridine, CN^-, SCN^-, 1,4-dicyanobenzene, and 1,2,4,5-tetrazine as the bridging ligand. The polymers in which the bridging ligand is bound by two coordination bonds or by a coordinate bond and a sigma bond, have good semiconductive properties.[202]

A review with 49 references relates to the photoconductive properties of M pcs and their application in electrophotography and photovoltaic cells.[203]

REFERENCES

1. **Snow, A. and Price, T. R.,** Tetra(9,10-phenanthro)tetraazaporphyrin-phthalocyanine: comparative electrical conductivities, *Synth. Met.*, 9(2), 329, 1984.
2. **Shimura, M., Kishi, H., and Baba, H.,** Temperature dependence of dark conductivity for β-metal-free phthalocyanine sublimed films, *Denki Kagaku oyobi Kogyo Butsuri Kagaku*, 52(11), 767, 1984.
3. **Inabe, T., Marks, T. J., Burton, R. L., Lyding, J. W., McCarthy, W. J., Kannewurf, C. R., Reisner, G. M., and Herbstein, F. H.,** Highly conductive metallophthalocyanine assemblies. Structure, charge transport, and anisotropy in the metal-free molecular metal hydrogen phthalocyanine iodide (H_2 (Pc)I), *Solid State Commun.*, 54(6), 501, 1985.
4. **Nespurek, S., Cermakova, E., and Sikula, J.,** Electrical noise in metal-free phthalocyanine thin films, *Mater. Sci.*, 10(1-2), 193, 1984.
5. **Cermakova, E., Musilova, V., and Nespurek, S.,** Study of noise in organic substances, *Knicznice Odb. Ved. Spisu Vys. Uceni Tech. Brne, B*, 97, 95, 1984.
6. **Nespurek, S.,** Photoconduction in metal-free phthalocyanine, *Wiss. Ber.-Akad. Wiss. D.D.R., Zentralinst. Festkoerperphys. Werkstofforsch.*, 29, 185, 1984.
7. **Nespurek, S.,** Influence of acceptors on charge carrier photogeneration in phthalocyanine, *Mater. Sci.*, 10(1-2), 189, 1984.
8. **Koropecky, I. and Nespurek, S.,** Surface cell photoconductivity of α-metal-free phthalocyanine, *Mater. Sci.*, 10(1-2), 129, 1984.
9. **Nespurek, S., Hart, R. H. G., Bonham, J. S., and Lyons, L. E.,** Photoconduction in metal-free phthalocyanine films, *Aust. J. Chem.*, 38(7), 1061, 1985.
10. **Toyo Ink Manufacturing Co., Ltd. and Hitachi, Ltd.,** Japan Patent 85,86,552, 1985.
11. **Nespurek, S., Zmeskal, O., and Schauer, F.,** An ohmic contact test using the activation energy of steady state space charge limited current, *Phys. Status Solidi A*, 85(2), 619, 1984.
12. **Prosvirin, K. A., Shumov, Y. S., and Bratukhin, Y. K.,** Nonlinear photoeffect in an island-type semiconductor film, *Poverkhnost*, 9, 68, 1985.
13. **Vidadi, Y. A. and Sadraddinov, S. A.,** Thermal-field ionization of traps in phthalocyanine, *Dokl. Akad. Nauk Az. SSR*, 39(9), 44, 1983.
14. **Perlman, M. M. and Haridoss, S.,** Bulk modification of charge trapping and conductivity in linear low-density polyethylene, *IEEE Trans. Electr. Insul.*, EI-22(1), 9, 1987.
15. **Perlman, M. M. and Haridoss, S.,** Bulk modification of charge trapping and conductivity in linear low density polyethylene, *Proc.-Int. Symp. Electrets*, 88, 1985.
16. **Schauer, F., Nespurek, S., and Zmeskal, O.,** The bulk trap spectroscopy of solids by temperature-modulated space-charge-limited currents (TMSCLC): application to real crystalline and amorphous semiconductors, *J. Phys. C*, 19(36), 7231, 1986.
17. **Hale, P. D.,** Theoretical Studies of Intramolecular Electron Transfer in Bimetallic Mixed-Valency Molecules and Phthalocyanine-based Linear Chain Conductors, Ph.D. thesis, Northwestern University, Evanston, IL, 1987.
18. **Tachibana, K. (Nissha Printing Co., Ltd.),** Japan Patent 87,04,388, 1987.
19. **Batey, J., Petty, M. C., Roberts, G. G., and Wight, D. R.,** Gallium phosphide/phthalocyanine Langmuir-Blodgett film electroluminescent diode, *Electron. Lett.*, 20(12), 489, 1984.

20. **Minami, S., Tanabe, G., Tsuda, K., Yonehara, Y., and Kinoshita, S. (Agency of Industrial Sciences and Technology and Kawamura Physical and Chemical Research Institute)**, Japan Patent 86,202,478, 1986.
21. **Harbour, J. R., Perovic, A., Issler, S. L., and Wolkow, R.**, Inhomogeneous deposition of platinum particles on the organic pigment metal-free, x-phthalocyanine, *J. Phys. Chem.*, 88(5), 880, 1984.
22. **Palmer, S. M., Stanton, J. L., Martinsen, J., Ogawa, M. Y., Heuer, W. B., Van Wallendael, S. E., Hoffman, B. M., and Ibers, J. A.**, Porphyrinic molecular metals, *Mol. Cryst. Liq. Cryst.*, 125(1-4), 1, 1985.
23. **Fedorov, M. I. and Shorin, V. A.**, Study of Doping of Phthalocyanines with Metals, *Deposited Doc. 1984*, VINITI, All Union Institute of Scientific and Technical Information, U.S.S.R., 1679, 11 pp.
24. **Tsivenko, V. I., Myasnikov, I. A., Bol'shun, E. V., and Nikolaeva, O. Y.**, Sensitization by magnesium phthalocyanine for the transport of the electronic excitation energy of benzene and naphthalene to zinc oxide, *Zh. Fiz. Khim.*, 60(5), 1314, 1986.
25. **Vidadi, Y. A., Sadraddinov, S. A., and Khalilov, S. A.**, Properties of film structures with the Schottky barrier based on oxygen-doped magnesium phthalocyanine, *Elektron. Org. Mater.*, 179, 1985.
26. **Ter Haar, L. W., Hatfield, W. E., and Tsutsui, M.**, Effects of Iodination on the Electrical Conductivity of Hydrogenbis(Phthalocyaninato) Neodymium (III), Report 1984, TR-22, Order No. AD-A137561, National Technical Information Service, 13 pp.
27. **Sullivan, B. W., Dominey, R. N., Helms, J. H., Schwartz, M., and Ter Haar, L. W.**, Preparation and Properties of Single Crystals of Hydrogen Bis(Phthalocyaninato) Neodymium (II), Report 1984, TR-23, Order No. AD-A144933, National Technical Information Service, Springfield, VA, 8 pp.
28. **Sullivan, B. W., Dominey, R. N., Helms, J. H., Schwartz, M., Ter Haar, L. W., and Hatfield, W. E.**, Preparation and properties of single crystals of hydrogen bis(phthalocyaninato) neodymium (III), *Mol. Cryst. Liq. Cryst.*, 120(1-4), 433, 1985.
29. **Jullien, J., Mossoyan-Deneux, M., Pierrot, M., Sorbier, J. P., Fournel, A., and Benlian, D.**, On two conductor compounds prepared from bis(phthalocyaninato)-neodymium, *C. R. Acad. Sci., Ser. 2*, 303(8), 669, 1986.
30. **Andre, J. J., Holczer, K., Petit, P., Riou, M. T., Clarisse, C., Even, R., Fourmigue, M., and Simon, J.**, Electrical and magnetic properties of thin films and single crystals of bis(phthalocyaninato) lutetium, *Chem. Phys. lett.*, 115(4-5), 463, 1985.
31. **Maitrot, M., Guillaud, G., Boudjema, B., Andre, J. J., Strzelecka, H., Simon, J., and Even, R.**, Lutetium bisphthalocyanine: the first molecular semiconductor. Conduction properties of thin films of p- and n-doped materials, *Chem. Phys. Lett.*, 133(1), 59, 1987.
32. **Petit, P., Holczer, K., and Andre, J. J.**, Low-dimensional molecular semiconductor: the radicalar bis(phthalocyaninato) lutetium Pc_2 Lu.CH_2 Cl_2, *J. Phys. (Les Ulis, Fr.)*, 48(8), 1363, 1987.
33. **Turek, P., Petit, P., Andre, J. J., Simon, J., Even, R., Boudjema, B., Guillaud, G., and Maitrot, M.**, A new series of molecular semiconductors: phthalocyanine radicals, *J. Am. Chem. Soc.*, 109(17), 5119, 1987.
34. **Nalwa, H. S. and Vasudevan, P.**, Chloro-iron phthalocyanine: characterization and electrical properties, *J. Mater. Sci. Lett.*, 4(8), 943, 1985.
35. **Keppeler, U., Schneider, O., Stoeffler, W., and Hanack, M.**, Conductivity of doped (μ-1,4-diisocyanobenzene)phthalocyaninato iron (II) compounds, *Tetrahedron Lett.*, 25(34), 3679, 1984.
36. **Orihashi, Y., Ohno, H., Tsuchida, E., Matsuda, H., Nakanishi, H., and Kato, M.**, Highly conductive crystals prepared by electrocrystallization of potassium dicyanophthalocyaninato iron (III), *Chem. Lett.*, 4, 601, 1987.
37. **Hanack, M., Datz, A., Fay, R., Fischer, K., Hedtmann-Rein, C., Keppeler, U., Koch, J., and Stoeffler, W.**, Conductivity of doped and nondoped bridged macrocyclic transition metal complexes, *Polym. Prepr. (Am. Chem. Soc., Div. Polym. Chem.)*, 25(2), 232, 1984.
38. **Orihashi, Y., Kobayashi, N., Tsuchida, E., Matsuda, H., Nakanishi, H., and Kato, M.**, Highly conductive single crystals of metallophthalocyanine, *Chem. Lett.*, 11, 1617, 1985.
39. **Tsuchida, E., Orihashi, Y., Kobayashi, N., and Ohno, H.**, Highly conductive crystals of cyanocobalt (III) phthalocyanine prepared by electrolysis, *Synth. Met.*, 15(2-3), 201, 1986.
40. **Matsuda, H., Nakanishi, H., Kato, M., Orihashi, Y., Kobayashi, N., and Tsuchida, H. (Agency of Industrial Sciences and Technology)**, Japan Patent 86,200,996, 1986.
41. **Kuwae, Y., Haruta, M., Yuasa, S., Munakata, H., and Nishimura, Y. (Canon K. K.)**, Japan Patent 86,246,393, 1986.
42. **Kuwae, Y., Haruta, M., and Yuasa, S. (Canon K. K.)**, Japan Patent 86,157,689, 1986.
43. **Martinsen, J., Stanton, J. L., Greene, R. L., Tanaka, J., Hoffman, B. M., and Ibers, J. A.**, Metal-spine conductivity in a partially oxidized metallomacrocycle: (phthalocyaninato) cobalt iodide, *J. Am. Chem. Soc.*, 107(24), 6915, 1985.
44. **Haruta, M., Kuwae, Y., and Yuasa, S. (Canon K. K.)**, Japan Patent 86,157,560, 1986.

45. **Orihashi, Y., Kobayashi, N., Ono, H., Tsuchida, E., Matsuda, H., Nakanishi, H., and Kato, M.,** Inherently conductive crystals of cyanocobalt(III) phthalocyanine, *Synth. Met.*, 19(1-3), 751, 1987.
46. **Hanack, M. and Muenz, X.,** Synthesis and properties of μ-cyanonphthalocyaninato) rhodium (III), *Synth. Met.*, 10(5), 357, 1985.
47. **Abdel-Malik, T. G.,** Effect of oxygen on the electrical properties of β-nickel phthalocyanine single crystals, *Indian J. Phys., A,* 57A(4), 231, 1983.
48. **Kaplan, M. L., Forrest, S. R., Schmidt, P. H., and Venkatesan, T.,** Optical and electrical properties of ion-beam-irradiated films of organic molecular solids and polymers, *J. Appl. Phys.*, 55(3), 732, 1984.
49. **Venkatesan, T., Forrest, S. R., Kaplan, M. L., Schmidt, P. H., Murray, C. A., Brown, W. L., Wilkens, B. J., Roberts, R. F., Rupp, L., Jr., and Schonhorn, H.,** Structural and chemical analysis of ion beam produced conductive regions on highly resistive organic films, *J. Appl. Phys.*, 56(10), 2778, 1984.
50. **Westinghouse Electric Corp.,** Japan Patent 84,12,502, 1984.
51. **Martinsen, J.,** Studies of the Structural, Magnetic, and Charge Transport Properties of a Series of Quasi One-Dimensional Nickel Porphyrin Iodide Systems, Ph.D. thesis, Northwestern University, Evanston, IL, 1984.
52. **Sao, G. D. and Tiwary, H. V.,** Electrical properties of nickel and cobalt phthalocyanines, *Proc. Nucl. Phys. Solid State Phys. Symp. 1980,* 23C, 253, 1983.
53. **Martinsen, J., Palmer, S. M., Tanaka, J., Greene, R. C., and Hoffman, B. M.,** Nickel phthalocyanine iodide: a highly one dimensional low temperature molecular metal, *Phys. Rev. B,* 30(11), 6269, 1984.
54. **Inabe, T., Nakamura, S., Liang, W. B., Marks, T. J., Burton, R. L., Kannewurf, C. R., and Imaeda, K.,** Highly conductive metallomacrocyclic assemblies. Synthesis via electrocrystallization and single crystal properties of a phthalocyanine molecular metal without halogen counterions, *J. Am. Chem. Soc.*, 107(24), 7224, 1985.
55. **TDK Corp.,** Japan Patent 85,94,981, 1985.
56. **Hale, P. D. and Ratner, M. A.,** Conductivity mechanisms in phthalocyanine-based ''molecular metals'': calculation of the temperature-dependent resistivity, *J. Chem. Phys.*, 83(10), 5277, 1985.
57. **Berlin, Y. A., Danielyan, N. G., and Enikolopyan, N. S.,** Model of impurity electric conductivity of organic semiconductors with conjugate bonds in the presence of high static pressures, *Dokl. Akad. Nauk SSSR,* 287(2), 363, 1986.
58. **Yakushi, K., Sakuda, M., Kuroda, H., Kawamoto, A., and Tanaka, J.,** Synthesis and crystal structure of a highly conducting (phthalocyaninato) nickel hexafluoroantimonate, $(NiPc)_2$ SbF_6, *Chem. Lett.*, 7, 1161, 1986.
59. **Yakushi, K., Sakuda, M., Hamada, I., Kuroda, H., Kawamoto, A., Tanaka, J., Sugano, T., and Kinoshita, M.,** Preparation, structure, and properties of metallic (phthalocyaninato) nickel salts, *Synth. Met.*, 19(1-3), 769, 1987.
60. **Almeida, M., Kanatzidis, M. G., Tonge, L. M., Marks, T. J., Marcy, H. O., McCarthy, W. J., and Kannewurf, C. R.,** Unusual band-filling and counterion ordering effects in a phthalocyanine molecular metal. Single crystal studies of nickel phthalocyanine perchlorate $(Ni(Pc)(C10_4)_y)$, *Solid State Commun.*, 63(6), 457, 1987.
61. **Fujiki, M., Tabei, H., and Imamura, S.,** Direct patterning and electrical properties of phthalocyanine thin films prepared by Langmuir-Blodgett and spin cast techniques, *Jpn. J. Appl. Phys., Part 1,* 26(8), 1224, 1987.
62. **Shirai, H., Hanabusa, K., Kitamura, M., Hojo, N., and Hirabaru, O. (TDK Corp.),** German Patent 3,439,846, 1985.
63. **Imoto, T. (NOK Corp.),** Japan Patent 86,62,504, 1986.
64. **Palmer, S. M., Stanton, J. M., Hoffman, B. M., and Ibers, J. A.,** Structure, conductivity, and magnetic properties of (phthalocyaninato) nickel (II) bromide, *Inorg. Chem.*, 25(14), 2296, 1986.
65. **Das, A., Tripathi, A. K., and Srivastava, A. P.,** Thermally stimulated discharge current studies of copper phthalocyanine, *Indian J. Phys., A,* 57A(5), 343, 1983.
66. **Ritz, A., Matz, R., and Lueth, H.,** Copper phthalocyanine overlayers on zinc oxide (1100) surfaces, *Appl. Phys. A,* A33(1), 37, 1984.
67. **Showa Denko, K. K.,** Japan Patent 84,03,878, 1984.
68. **Usui, H., Yamada, I., and Takagi, T.,** Formation of an organic cluster beam and its application to film formation, *Proc. Int. Ion Eng. Congr.*, 2, 1247, 1983.
69. **Fedorov, M. I. and Shorin, V. A.,** Study of Oxygen Doping of Phthalocyanines, Deposited Doc. 1984, VINITI 521, All Union Institute of Scientific and Technical Information, U.S.S.R., 10 pp. Avail. VINITI.
70. **Zabkowska, M., Kulesza, M., and Waclawek, W.,** Dielectric behavior of copper phthalocyanine-carbon black mixtures, *Mater. Sci.*, 10(1-2), 303, 1984.
71. **Kulesza, M., Zabkowska, M., and Waclawek, W.,** Highly conductive copper phthalocyanine-carbon black mixtures, *Mol. Cryst. Liq. Cryst.*, 118(1-4), 371, 1985.

72. **Palmer, S. M., Ogawa, M. Y., Martinsen, J., Stanton, J. L., Hoffman, B. M., Ibers, J. A., and Greene, R. L.,** Metallic conductivity and magnetic interactions in nickel and copper phthalocyanine iodides, *Mol. Cryst. Liq. Cryst.*, 120(1-4), 427, 1985.
73. **Kulesza, M., Koziol, J., and Waclawek, W.,** Electrical resistivity and IR behavior of metal-free and copper phthalocyanine-carbon black mixtures, *Springer Ser. Solid-State Sci.*, 63, 331, 1985.
74. **Waclawek, W. and Zabkowska-Waclawek, M.,** Conducting compositions of phthalocyanines with carbon blacks, *Pr. Nauk. Inst. Technol. Nieorg. Nawozow Miner. Politech. Wroclaw*, 28, 239, 1985.
75. **Waclawek, W. and Zabrowska-Waclawek, M.,** Factors influencing electrical conduction in carbon black-copper phthalocyanine mixtures, *J. Mater. Sci. Lett.*, 6(7), 784, 1987.
76. **Tripathi, A., Tripathi, A. K., and Srivastava, A. P.,** Electrical conduction in copper phthalocyanine with special reference to electrode material, *J. Electrost.*, 16(1), 99, 1984.
77. **Gould, R. D.,** D.C. electrical measurements on evaporated thin films of copper phthalocyanine, *Thin Solid Films*, 125(1-2), 63, 1985.
78. **Zhukhovitskii, V. B., Ezhova, N. N., and Letuchii, Y. A.,** Effects of the temperature of doping copper phthalocyanine with iodine and the ratio of the initial components on the electric conductivity of partially oxidized complexes, *Zh. Fiz. Khim.*, 59(11), 2867, 1985.
79. **Grigoryan, L. S., Simonyan, M. V., and Sharoyan, E. G.,** Study of highly conducting heat resistant compounds of phthalocyanines with iodine, *Elektron. Org. Mater.*, 31, 1985.
80. **Ogawa, M. Y., Hoffman, B. M., Lee, S., Yudkowsky, M., and Halperin, W. P.,** Transition of local moments coupled to itinerant electrons in the quasi one dimensional conductor copper phthalocyanine iodide, *Phys. Rev. Lett.*, 57(9), 1177, 1986.
81. **Bak, G. W.,** Low frequency dielectric propertes of some organic materials, *Phys. Status Solidi A*, 93(2), K195, 1986.
82. **Thomas, N. J., Roberts, G. G., and Petty, M. C.,** Switching characteristics for gallium arsenide/LB film MISS devices, in *Insul. Films Semicondors, Proceedings of the International Conference INFOS 85 1985*, Vol. 71, Simonne, J. J. and Buxo, J., Eds., North-Holland, Amsterdam, 1986.
83. **Yoneyama, M., Sugi, M., Saito, M., Ikegami, K., Kuroda, S., and Iizima, S.,** Photoelectric properties of copper phthalocyanine Langmuir-Blodgett film, *Jpn. J. Appl. Phys., Part 1*, 25(7), 961, 1986.
84. **Gould, R. D.,** Dependence of the mobility and trap concentration in evaporated copper phthalocyanine thin films on background pressure and evaporation rate, *J. Phys. D*, 19(9), 1785, 1986.
85. **Takaoka, K. (Mitsubishi Paper Mills, Ltd.),** Japan Patent 86,102,075, 1986.
86. **Said, M. A. B., Belgacem, S., Dachraoui, M., Bennaceur, R., and Bouchriha, H.,** Structural and optical properties of cadmium zinc sulfide/copper phthalocyanine (Cd_{1-y} Zn_yS/CuPc): evidence of a photovoltaic effect, *Rev. Phys. Appl.*, 21(7), 407, 1986.
87. **Pan, H., Huang, S., and Zhou, C.,** Photovoltaic effects of copper phthalocyanine/cadmium sulfide and phthalocyanine/cadmium sulfide heterojunctions, *Huadong Huagong Xueyuan Xuebao*, 12(5), 583, 1986.
88. **Ishihara, S. (Matsushita Electric Industrial Co., Ltd.),** Japan Patent 86,196,584, 1986.
89. **Miyata, S., Boku, Z., Soeda, Y., and Uno, Y. (Mitsubishi Petrochemical Co., Ltd.),** Japan Patent 86,210,680, 1986.
90. **Vidadi, Y. A. and Sadraddinov, S. A.,** Frenkel-Poole effect in copper phthalocyanine, *Elektron. Org. Mater.*, 208, 1985.
91. **Qian, R., Zhou, S., Jin, X., Wang, Z., and Jiang, S.,** α-Copper phthalocyanine evaporated film trap — an isothermal decay current spectrum, *Yingyong Kexue Xuebao*, 5(1), 37, 1987.
92. **Block, H. and Kelly, J. P. (National Research Development Corp.),** British Patent Application 2,170,510, 1986.
93. **Baebler, F. and Munk, K. (Ciba-Geigy AG),** Europe Patent Application 131,544, 1985.
94. **Kuwabata, S., Okamoto, K., Ikeda, O., and Yoneyama, H.,** Effect of organic dopants on the electrical conductivity of polypyrrole films, *Synth. Met.*, 18(1-3), 101, 1987.
95. **Uno, Y. and Suzuki, S. (Mitsubishi Petrochemical Co., Ltd.),** Japan Patent 86,176,605, 1986.
96. **Shibata, M. and Momoi, Y. (Dainichiseika Color and Chemicals Manufacturing Co., Ltd.),** Japan Patent 87,20,540, 1987.
97. **Toyo Ink Manufacturing Co., Ltd.,** Japan Patent 85,95,441, 1985.
98. **Tamura, S., Seta, T., Mizukami, S., and Ohashi, M. (Toyo Ink Manufacturing Co., Ltd.),** Japan Patent 85,203,957, 1985.
99. **Mossoyan-Deneux, M., Benlian, D., Ley, M., Pierrot, M., Sorbier, J. P., and Fournel, A.,** Structure, spectra and conductivity of oxidized zinc phthalocyanine single crystals, *Mol. Cryst. Liq. Cryst.*, 120(1-4), 437, 1985.
100. **Mokshin, V. M., Postnikova, O. A., Gal'perin, M. G., Goncharova, G. I., Mikhalenko, S. A., and Luk'yanets, E. A.,** Effect of structural factors on the electrophysical properties of derivatives of phthalocyanine and 2,3-naphthalocyanine, *Khim. Fiz.*, 4(3), 325, 1985.
101. **Sakaguchi, M. and Ohta, M.,** Environmental effect on surface conductivity and the α→β phase transition of zinc phthalocyanine, *J. Solid State Chem.*, 61(1), 130, 1986.

102. **Meier, H., Albrecht, W., Woehrle, D., and Jahn, A.,** Correlation of chemical structure to photoconductivity: octacyano- and octamethoxy-substituted zinc phthalocyanine, *J. Phys. Chem.,* 90(23), 6349, 1986.
103. **Collins, R. A. and Mohammed, K. A.,** Electrical, structural and gas sensing properties of zinc phthalocyanine thin films, *Thin Solid Films,* 145(1), 133, 1986.
104. **Webb, A. W., Brant, P., Nohr, R. S., and Weber, D. C.,** Resistivity of doped phthalocyanines to 6.5 GPa, *Polym. Prepr. (Am. Chem. Soc., Div. Polym. Chem.),* 23(1), 127, 1982.
105. **Buttner, W. J., Rieke, P. C., and Armstrong, N. R.,** Evidence for charge trapping at the gold/chlorogallium phthalocyanine interface using photocurrent spectroscopy with one or two illumination sources, *J. Phys. Chem.,* 89(7), 1116, 1985.
106. **Dann, A. J., Fahy, M. R., Jeynes, C., and Willis, M. R.,** Electrical properties of ion implanted phthalocyaninatoaluminum fluoride films, *Phys. Status Solidi A,* 101(1), K45, 1987.
107. **Takeda, Y., Kinjo, N., and Narahara, T.,** Thermally stimulated current in aluminum phthalocyanine chloride films, *J. Mater. Sci. Lett.,* 4(1), 80, 1985.
108. **Tanaka, T. and Hirohashi, R.,** Photoconduction of metal phthalocyanines (II). Photoelectric properties of IIIa metal phthalocyanines, *Denshi Shashin Gakkaishi,* 25(4), 284, 1986.
109. **Mullins, O. C. and Kaplan, M.,** Perturbed angular correlation studies of indium metalloporphyrin complexes, *J. Chem. Phys.,* 79(9), 4475, 1983.
110. **Pietro, W. J., Marks, T. J., and Ratner, M. A.,** Resistivity mechanisms in phthalocyanine-based linear chain and polymeric conductors: variation of bandwidth with geometry, *J. Am. Chem. Soc.,* 107(19), 5387, 1985.
111. **Hua, Y. L., Petty, M. C., Robert, G. G., Ahmad, M. M., Hanack, M., and Rein, M.,** Photoelectric properties of substituted silicon phthalocyanine Langmuir-Blodgett film Schottky barrier and metal/insulator/semiconductor devices, *Thin Solid Films,* 149(2), 163, 1987.
112. **Frauenheim, T.,** Electronic switching model for organic polycrystalline films of lead phthalocyanine (PbPc), *Wiss. Ber.-Akad. Wiss. D.D.R., Zentralinst. Festkoerperphys. Werkstofforsch.,* 29, 197, 1984.
113. **Koval'chuk, A. V.,** Effect of series resistance on the voltage-capacitance characteristics of metal-organic semiconductor-metal barrier structures, *Fiz. Tekh. Poluprovodn. (Leningrad),* 19(2), 307, 1985.
114. **Frauenheim, T., Hamann, C., and Mueller, M.,** Electric field induced disorder-order transition in organic polycrystallization films of quasi-one-dimensional lead phthalocyanine, *Phys. Status. Solidi A,* 86(2), 735, 1984.
115. **Hamann, C., Koval'chuk, A. V., and Kurik, M. V.,** Effect of crystal structure on properties of a Schottky barrier nickel/PbPc, *Mater. Sci.,* 10(1-2), 109, 1984.
116. **Mueller, M., Frauenheim, T., and Hamann, C.,** Peierls transition and fluctuation conductivity in thin lead-phthalocyanine (PbPc) films, *Phys. Status Solidi A,* 89(1), K89, 1985.
117. **Bahra, G. S., Chadwick, A. V., and Wright, J. D.,** Kinetics of photocurrent rise and decay in molecular crystalline materials, *Mol. Cryst. Liq. Cryst.,* 134(1-4), 121, 1986.
118. **Vartanyan, A. T.,** Electric and photoelectric properties of lead phthalocyanine, *Dokl. Akad. Nauk SSSR,* 284(6), 1396, 1985.
119. **Waclawek, W. and Zabkowski-Waclawek, M.,** The influence of iodine on the electrical properties of lead phthalocyanine, *Thin Solid Films,* 146 (1), 1, 1987.
120. **Hamann, C., Mueller, M., Starke, M., Hoehne, H. J., Pryzyboroski, F., Mrwa, W. A., Richter, J., Reinhardt, C., and Hoenig, J.,** German Patent 3,408,741, 1984.
121. **Hirabaru, O., Tonami, N., Higaki, S., Hanabusa, K., Shirai, H., Takemoto, K., and Hojo, N.,** Conducting mechanisms of metal phthalocyanine bound to 2-vinylpyridine-styrene copolymer film, *Oyo Butsuri,* 52(12), 1051, 1983.
123. **Canadell, E. and Alvarez, S.,** Theoretical study of the electrical behavior of one dimensional metallophthalocyanines and related metallomacrocyclic compounds, *Inorg. Chem.,* 23(5), 573, 1984.
124. **Boudjema, B., Guillaud, G., Gamoudi, M., Maitrot, M., Andre, J. J., Martin, M., and Simon, J.,** Characterization of metallophthalocyanine-metal contacts: electrical properties in a large frequency range, *J. Appl. Phys.,* 56(8), 2323, 1984.
125. **Grigoryan, L. S., Simonyan, M. V., and Sharoyan, E. G.,** Electrical, magnetic, and spectral properties of the new class of charge transfer salts based on phthalocyanines and iodine, *Phys. Status Solidi A,* 84(2), 597, 1984.
126. **Toscano, P. J. and Marks, T. J.,** Electrically conductive polymers composed of cofacially joined metallomacrocycles. Studies of architecture and electronic structure in phthalocyanine materials by high resolution solid state NMR, *Mol. Cryst. Liq. Cryst.,* 118(1-4), 337, 1985.
127. **Berlin, Y. A., Danielyan, N. G., Beshenko, S. I., Zhorin, V. A., Mikhalenko, S. A., and Enikolopyan, N. S.,** High pressure sensitivity of conductivity observed for phthalocyanine complexes with divalent metals, *Dokl. Akad. Nauk SSSR,* 282(2), 347, 1985.
128. **Andre, J. J., Simon, J., Guillaud, G., Boudjema, B., and Maitrot, M.,** Role of oxygen in the space charge layer formation in organic-metal junctions: trans-polyacetylene and metallo-phthalocyanine, *Mol. Cryst. Liq. Cryst.,* 121(1-4), 277, 1985.

129. **Ebisawa, F. (Nippon Telegraph and Telephone Public Corp.)**, Japan Patent 86,09,569, 1986.
130. **Berlin, Y. A., Danielyan, N. G., Beshenko, S. I., Zhorin, V. A., and Enikolopyan, N. S.**, Activation energy for conductivity of metal phthalocyanine complexes under static uniaxial compression, *Dokl. Akad. Nauk SSSR*, 286(2), 364, 1986.
131. **Snow, A. W., Barger, W. R., Klusty, M., Wohltjen, H., and Jarvis, N. L.**, Simultaneous electrical conductivity and piezoelectric mass measurements on iodine doped phthalocyanine Langmuir-Blodgett films, *Langmuir*, 2(4), 513, 1986.
132. **Jones, T. A. and Bott, B.**, Gas induced electrical conductivity changes in metal phthalocyanines, *Sens. Actuators*, 9(1), 27, 1986.
133. **Klofta, T. J.**, The Modification of Electrochemical and Photoelectrochemical Properties in Thin Films of Tri- and Tetravalent Metal Phthalocyanines, Ph.D. thesis, University of Arizona, Tucson, 1986.
134. **Block, H. and Kelly, J. P. (National Research Development Corp.)**, Europe Patent Application 191,585, 1986.
135. **Barger, W. R., Jarvis, N. L., Snow, A. W., and Wohltjen, H. (U.S. Department of the Navy)**, U.S. Patent Application 767,925, 1986.
136. **Yamada, A., Shigehara, K., and Hara, M.**, Electronic conductivity of metallophthalocyanine crystal thin films, *Synth. Met.*, 18(1-3), 821, 1987.
137. **Yamada, A. and Shigehara, J. (Institute of Physical and Chemical Research)**, Japan Patent 86,206,259, 1986.
138. **Minami, N. and Asai, M.**, Photocurrent spectra of phthalocyanine films in relation to excited state properties, *Jpn. J. Appl. Phys., Part 1*, 26(10), 1754, 1987.
139. **Laurs, H. and Heiland, G.**, Electrical and optical properties of phthalocyanine films, *Thin Solid Films*, 149(2), 129, 1987.
140. **Maitrot, M., Guillaud, G., Boudjema, B., Andre, J. J., and Simon, J.**, Molecular material based junctions: formation of a Schottky contact with metallophthalocyanine thin films doped by the cosublimation method, *J. Appl. Phys.*, 60(7), 2396, 1986.
141. **Hirohashi, R. and Kasai, T.**, Photoconduction of metal phthalocyanines. I. The effect of coordination metal on electrical and photoconductive properties, *Denshi Shashin Gakkaishi*, 25(1), 24, 1986.
142. **Shaposhnikov, G. P., Kulinich, V. P., Osipov, Y. M., and Smirnov, R. P.**, Optical and electrophysical properties of metal complexes of tetra(1,4-dithiacyclohexene)porphyrazine, *Khim. Geterotsikl Soedin.*, 9, 1276, 1986.
143. **Bazanov, M. I.**, Phthalocyanines of transition metals of periods IV, V, and VI, their catalytic, electrochemical and semiconducting properties, *Soversh. Protsessov Krasheniya i Metodov Sinteza Krasitelei Ivanovo*, 22, 1983.
144. **Nohr, R. S., Brant, P., Weber, D., and Wynne, K. J.**, Stable, conducting boron tetrafluoride- and phosphorus hexafluoride-doped fluoro metal phthalocyanines, *Polym. Prepr. (Am. Chem. Soc., Div. Polym. Chem.)*, 23(1), 129, 1982.
145. **Berthet, G., Djurado, D., Fabre, C., Faury, F., Maleysson, C., and Robert, H.**, Iodine doping-dedoping of thin layers of the polyphthalocyanine $(AlPcF)_n$, *Mol. Cryst. Liq. Cryst.*, 118(1-4), 345, 1985.
146. **Djurado, D., Hamwi, A., Cousseins, J. C., Bidar, H., Fabre, C., and Berthet, G.**, Study of chemically (arsenic pentafluoride) and electrochemically doped aluminum polyfluorophthalocyanine, $(PcAlF)_n$, *Synth. Met.*, 11(3), 109, 1985.
147. **Dann, A. J., Fahy, M. R., Jeynes, C., and Willis, M. R.**, Iodine implantation of polymeric phthalocyanines, *J. Phys. D*, 19(11), L217, 1986.
148. **Berthet, G., Blanc, J. P., Germain, J. P., Larbi, A., Maleysson, C., and Robert, H.**, Electroactive polymers in thin layers: a potential application as a gas sensor, *Synth. Met.*, 18(1-3), 715, 1987.
149. **Djurado, D., Hamwi, A., Fabre, C., Avignant, D., and Cousseins, J. C.**, Study of iodine fluoride (IF_5) chemical doping of aluminum polyfluorophthalocyanine: $(PcAlF)_n$, *Synth. Met.*, 16(2), 227, 1986.
150. **Djurado, D., Fabre, C., Hamwi, A., and Cousseins, J. C.**, Preparation of conducting polymers by chemical doping of aluminum polyfluorophthalocyanine by iodine pentafluoride, molybdenum hexafluoride and tungsten hexafluoride. Magnetic and electric property studies, *Mater. Res. Bull.*, 22(7), 911, 1987.
151. **Inabe, T., Lomax, J. F., Lyding, J. W., Kannewurf, C. R., and Marks, T. J.**, "Metal-like" metallomacrocyclic polymers. Chemical, physical, and processing studies, *Polym. Mater. Sci. Eng.*, 49, 89, 1983.
152. **Inabe, T., Lyding, J. W., and Marks, T. J.**, Spinning fibers and fiber blends of an electrically conductive polymer, *J. Chem. Soc., Chem. Commun.*, 19, 1084, 1983.
153. **Marks, T. J. and Inabe, T. (Northwestern University)**, U.S. Patent 4,563,301, 1986.
154. **Inabe, T., Lomax, J. F., Lyding, J. W., Kannewurf, C. R., and Marks, T. J.**, Electrically conductive, processable polymeric materials constructed from metallophthalocyanines, *Synth. Met.*, 9(2), 303, 1984.
155. **Wynne, K. J., Marks, T. J., and Inabe, T. (Northwestern University)**, U.S. Patent 4,563,300, 1986.

156. **Gaudiello, J. G., Almeida, M., Marks, T. J., McCarthy, W. J., Butler, J. C., and Kannewurf, C. R.,** Widely tunable band filling in a molecular metal. Chemical and physical consequences of electrochemically doping a cofacially joined metallomacrocyclic assembly, *J. Phys. Chem.*, 90(21), 4917, 1986.
157. **Gaudiello, J. G., Marcy, H. O., McCarthy, W. J., Moguel, M. K., Kannewurf, C. R., and Marks, T. J.,** Highly conductive metallomacrocyclic assemblies. Synthetic approaches to molecular metals with variable band-filling, *Synth. Met.*, 15(2-3), 115, 1986.
158. **Inabe, T., Moguel, M. K., Marks, T. J., Burton, R., Lyding, J. W., and Kannewurf, C. R.,** Electronic properties of the conductive polymers $\{[Si(Pc)O]X\}_{yn}$ with different doping agents, *Mol. Cryst. Liq. Cryst.*, 118(1-4), 349, 1985.
159. **Yudasaka, M., Kawai, M., Kurita, S., Nakanishi, K., and Kuwae, Y.,** Formation of silicon phthalocyaninato dihydroxide and polymeric silicon phthalocyaninato oxide films and their optical and electrical properties, *Thin Solid Films,* 151(3), L115, 1987.
160. **Inabe, T., Gaudiello, J. G., Moguel, M. K., Lyding, J. W., Burton, R. L., McCarthy, W. J., Kannewurf, C. R., and Marks, T. J.,** Cofacial assembly of partially oxidized metallomacrocycles as an approach to controlling lattice architecture in low dimensional molecular ''metals''. Probing band structure-counterion interactions in conductive $[M(phthalocyaninato)O]_n$ macromolecules using nitrosonium oxidants, *J. Am. Chem. Soc.*, 108(24), 7595, 1986.
161. **Pietro, W. J., Ellis, D. E., Marks, T. J., and Ratner, M. A.,** Building blocks for molecular and macromolecular ''metals''. Electronic structure of Group IVA phthalocyanine monomers and cofacially joined dimers, *Mol. Cryst. Liq. Cryst.*, 105(1-4), 273, 1984.
162. **Ciliberto, E., Doris, K. A., Pietro, W. J., Reisner, G. M., Ellis, D. E., Fragala, I., Herbstein, F. H., Ratner, M. A., and Marks, T. J.,** The π-π interactions and bandwidths in molecular metals. A chemical, structural, photoelectron spectroscopic, and Hartree-Fock-Slater study of monomeric and cofacially joined dimeric silicon phthalocyanines, *J. Am. Chem. Soc.*, 106(25), 7748, 1984.
163. **Hanack, M., Datz, A., Fay, R., Metz, J., and Schneider, O. (Bayer AG),** German Patent 3,245,750, 1984.
164. **Bijwe, J., Pandey, P. S., Vasudevan, P., Tripathic, A.,** Electrical properties of iron phthalocyanine systems with a mixed valence central iron atom, *Eur. Polym. J.*, 23(2), 167, 1987.
165. **Hirabaru, O., Nakahara, Y., Shirai, H., and Shiokawa, J.,** Functional and cyclic metal complexes and their polymers. XX. The a.c. conducting mechanisms of the films of poly(2-vinylpyridine-co-styrene) having metal phthalocyanine groups, *Kobunshi Ronbunshu,* 44(4), 291, 1987.
166. **Rehwald, W., Kiess, H., and Binggeli, B.,** Frequency dependent conductivity in polymers and other disordered materials, *Z. Phys. B: Condens. Matter,* 68(2), 143, 1987.
167. **Preston, J. R. and Hillenbrand, L. J. (Battelle Development Corp.),** U.S. Patent 4,557,807, 1985.
168. **Bull, R. A., Fan, F. R., and Bard, A. J.,** Polymer films on electrodes. XIII. Incorporation of catalysts into electronically conductive polymers: iron phthalocyanine in polypyrrole, *J. Electrochem. Soc.*, 131(3), 687, 1984.
169. **Diel, B. N., Inabe, T., Jaggi, N. K., Lyding, J. W., Schneider, O., Hanack, M., Kannewurf, C. R., Marks, T. J., and Schwarz, L. H.,** Cofacial assembly of metallomacrocycles as an approach to controlling lattice architecture in low dimensional molecular solids. Chemical, structural, oxidation state, transport, and optical properties of the iron coordination polymer $[Fe(phthalocyaninato)(\mu\text{-pyrazine})]_n$ and the consequences of halogen doping, *J. Am. Chem. Soc.*, 106(11), 3207, 1984.
170. **Wessling, B., and Volk, H.,** Postpolymerization processing of conductive polymers: a way of converting conductive polymers to conductive materials?, *Synth. Met.*, 15(2-3), 183, 1986.
171. **Uno, Y. and Suzuki, S. (Mitsubishi Petrochemical Co., Ltd.),** Japan Patent 86,176,606, 1986.
172. **Yudasaka, M., Nakanishi, K., Hara, T., Tanaka, M., Kurita, S., and Kawai, M.,** Metal phthalocyanine polymer film formation by the double source evaporation if tetracyanobenzene and metal, *Synth. Met.*, 19(1-3), 775, 1987.
173. **Achar, B. N., Fohlen, G. M., and Parker, J. A.,** Conductive phthalocyanine sheet polymers, *J. Polym. Mater.*, 1(3), 176, 1984.
174. **Hanack, M.,** Synthesis and properties of conducting bridged macrocyclic metal complexes, *Mol. Cryst. Liq. Cryst.*, 105(1-4), 133, 1984.
175. **Jonas, F. and Hocker, J. (Bayer AG),** German Patent 3,242,712, 1984.
176. **Inabe, T., Liang, W. B., Lomax, J. F., Nakamura, S., Lyding, J. W., McCarthy, W. J., Carr, S. H., Kannewurf, C. R., and Marks, T. J.,** Phthalocyanine-based electrically conductive, processible molecular/macromolecular hybrid materials, *Synth. Met.*, 13(1-3), 219, 1986.
177. **Yamada, A. and Osada, Y. (Institute of Physical and Chemical Research),** Japan Patent 86,207,431, 1986.
178. **Yamada, A. and Osada, Y. (Institute of Physical and Chemical Research),** Japan Patent 86,207,461, 1986.
179. **Matsushita Electric Industrial Co., Ltd.,** Japan Patent 84,186,365, 1984.
180. **Uno, Y. and Suzuki, S. (Mitsubishi Petrochemical Co., Ltd.),** Japan Patent 86,176,607, 1986.

181. **TDK Corp.**, Japan Patent 84,54,104, 1984.
182. **Meier, H., Albrecht, W., Zimmerhackl, E., Hanack, M., and Metz, J.**, Photoconductivity of μ-cyanophthalocyaninato cobalt (III) $[PcCoCN]_x$, *Synth. Met.*, 11(6), 333, 1985.
183. **Meier, H., Albrecht, W., Zimmerhackl, E., Hanack, M., and Fischer, K.**, Photoconductivity of phthalocyaninato-μ-thiagermanium (IV) $[PcGeS]_x$, *J. Mol. Electron.*, 1(1), 47, 1985.
184. **Moguel, M. K.**, Electrically conductive metallophthalocyanine polymers. Synthetic Approaches and Physical Characterization of Conductive Polymers with Variable Band-Filling, Ph.D. thesis, Northwestern University, Evanston, IL, 1986.
185. **Achar, B. N., Fohlen, G. M., Parker, J. A., and Keshavayya, J.**, Organic semiconductors — metal phthalocyanine sheet polymers, *J. Polym. Sci., Part A*, 25(2), 443, 1987.
186. **Rosenthal, M. V., Skotheim, T. A., Linkous, C., Florit, M. I., and Salmon, M.**, Chemical modification of polypyrrole conducting polymers, *Polym. Prepr. (Am. Chem. Soc., Div. Polym. Chem.)*, 25(2), 258, 1984.
187. **Skotheim, T., Velazquez Rosenthal, M., and Linkous, C. A.**, Polypyrrole complexed with cobalt phthalocyanine. A conducting polymer which is stable to air and moisture, *J. Chem. Soc., Chem. Commun.*, 9, 612, 1985.
188. **Mizutani, F., Iijima, S., Tanabe, G., and Tsuda, K. (Agency of Industrial Sciences and Technology)**, Japan Patent 87,04,898, 1987.
189. **Wessling, B. and Volk, H. (Zipperling Kessler und Co. GmbH)**, German Patent 3,421,993, 1985.
190. **Padias, A. B. and Hall, H. K., Jr.**, Semiconducting polymers via the high temperature free radical polymerization of multinitriles, *J. Polym. Sci., Part A*, 24(7), 1675, 1986.
191. **Venkatachalam, S., Rao, K. V. C., and Mendez, R.**, Electrically conductive partially oxidized polymeric phthalocyanines, *J. Polym. Mater.*, 3(1), 65, 1986.
192. **Dulov, A. A., Abramova, L. A., Sherle, A. I., and Epshtein, V. R.**, Modification of electrical properties of polyphthalocyanines with different metals, *Elektron. Org. Mater.*, 242, 1985.
193. **Woerhle, D., Bannehr, R., Schumann, B., and Jaeger, N.**, Polymeric phthalocyanines and their precursors. IV. Optimization of electrochemical properties of polyphthalocyanine modified titanium electrodes, *Angew. Makromol. Chem.*, 117, 103, 1983.
194. **Dulov, A. A., Sherle, A. I., and Epshtein, V. R.**, Some Characteristics of Electrical Conductivity of Metal Free Polyphthalocyanine, Deposited Doc. 1983, VINITI 437-83, All Union Institute of Scientific and Technical Information, U.S.S.R., 10 pp.
195. **Doris, K. A.**, The Role of Pi-Orbital Interactions in the Electronic Properties of Conductive Phthalocyanine Polymers, Ph.D. thesis, Northwestern University, Evanston, IL, 1984.
196. **Khidekel, M. L., Zhukhovitskii, V. B., and Dyumaev, K. M.**, Organic metals from metal phthalocyanines and porphyrins, *Wiss. Ber.-Akad. Wiss. D.D.R., Zentralinst. Festkoerperphys. Werkstofforsch.*, 29, 157, 1984.
197. **Zhukhovitskii, V. B., Khidekel, M. L., and Dyumaev, K. M.**, Conducting compounds from phthalocyanine derivatives, *Usp. Khim.*, 54(2), 239, 1985.
198. **Lever, A. B. P.**, The other periodic chart, *ChemTech*, 17(8), 506, 1987.
199. **Shirai, H. and Hanabusa, K.**, Advances in functions of metallophthalocyanines, *Kagaku Kogyo*, 38(4), 332, 1987.
200. **Andre, J. J., Simon, J., Even, R., Boudjema, B., Guillaud, G., and Maitrot, M.**, Molecular semiconductors and junction formation: phthalocyanine derivatives, *Synth. Met.*, 18(1-3), 683, 1987.
201. **Hanack, M., Deger, S., Lange, A., and Zipplies, T.**, Axially bridged macrocyclic metal complexes as conductive polymers, *Synth. Met.*, 15(2-3), 207, 1986.
202. **Hanack, M.**, Bridged metal complexes as organic conductors, *GIT Fachz. Lab.*, 31(2), 75, 1987.
203. **Tamura, S.**, Photoconductive application of organic pigments. Phthalocyanine, *Shikizai Kyokaishi*, 58(9), 528, 1985.

Chapter 23

MAGNETIC PHENOMENA

I. FE PCS

Magnetic moments, and ESR and Moessbauer spectra are obtained for Fe(III) pc Cl. The magnetic moment vs. temperature relation is given for 4.2 to 300 K. The Moessbauer spectrum is determined at 4.2 K; the ESR spectrum is determined at 77K.[1]

Magnetic, ESR, and Moessbauer spectral properties are given for μ-oxoiron(III) phthalocyanine (pc) obtained from a dihydroxo iron(III) precursor. The compound contains high-spin Fe(III) centers with strong antiferromagnetic coupling magnitude similar to that displayed by other FeOFe species. These determinations are the first direct proof of the pure high-spin, S = 5/2, state for Fe(III) in pc complexes.[2]

Magnetic susceptibility, MCD, UV-visible absorption, and ESR spectroscopy show that the spin state of Fe in Fe(III) 1,2,8,9,15,16,22,23-octadecyloxycarbonyl pc dissolved in CH_2Cl_2 or $CHCl_3$ changes from a high- to a low-spin state upon addition of tetrabutylammonium chloride and/or tetrabutylammonium bromide.[3]

ESR and absorption spectroscopies are used to study the reduction of Fe(III) tetrasulfo pc by ascorbic acid and cysteine. The monomeric form of Fe(III) tetrasulfo pc gives an ESR spectrum; the dimer does not. The complex is low spin and is rapidly reduced by ascorbic acid and cysteine to Fe(II) tetrasulfo pc.[4]

II. MN AND CU PCS

ESR shows that Mn(II) 3,10,17,24-tetrasulfo pc exists in aqueous solution in high- and low-spin forms with the low-spin form being favored in slightly alkaline solution because of the interaction of Mn(II) with the hydroxyl anion.[5]

ESR at 77 K studies the effect of hydrogen peroxide, DMF, and sodium hydroxide on the spin state of Mn(II) tetrasulfo pc.[6]

The magnetic susceptibility and moment temperature dependence of Mn(II) pc are explained by assuming the existence of a Heisenberg exchange interaction between the Mn cations.[7]

An ESR study is made of substituted Cu pc dimers, and their structural parameters are determined by computer modeling.[8]

Measurements of the proton spin-lattice relaxation rate reveal that Cu pc is a highly one-dimensional Heisenberg system. It is indicated that Cu pc is an excellent one-dimensional Heisenberg magnet.[9]

Cu tetrakis(cumylphenoxy) pc is a building block of Langmuir-Blodgett multilayer films. A structural model of these films is developed by computing the ESR spectra of different orientation distribution of Cu tetrakis(cumylphenoxy) pc molecules with respect to the film surface.[10]

III. VO PCS

The ESR spectra of vanadyl (VO) pc are drawn.[11]

Salts with the composition VO(IV)L/Ni(II)L′, VO(IV)L/Co(II)L′, or Co(II)L/Ni(II)L′ (L = N,N′,N″,N‴-tetramethyl tetra-2,3-pyridineporphyrazine; L′ = 3,10,17,24-tetrasulfo pc) give rise to temperature-dependent, well-resolved ESR spectra from which the nature of the magnetic interaction between the metal ion centers is deduced.[12]

IV. MO PC

EPR measurements are made of the spin Hamiltonian parameters of a molybdenum (Mo) pc and its solution in sulfuric acid. The EPR are analyzed in terms of an axial spin Hamiltonian ($S = 1/2$) and computer spectra simulations using the analysis are made, with suggestions as to the compound structure.[13]

V. MAGNETIC EXCHANGE INTERACTIONS

Magnetic susceptibility and magnetization studies on the linear chain Cr(II) compounds, chromium(III) catena-fluoro pc $(Cr\ pc\ F)_n$, and chromium(II) catena-cyano pc $(Cr\ pc\ CN)_n$ show that the Cr(III) ions in $(Cr\ pc\ F)_n$ and $(Cr\ pc\ CN)_n$ are magnetically exchange coupled.[14]

The magnetic susceptibilities of the dinuclear rare earth pc complexes, $(Rare\ Earth)_2$ pc [(2,2,6,6-tetramethylheptane-3,5-dionato (1-)] where the rare earth is Sm(III), Yb(III), and Y(III), are measured at 77 to 300 K. Er(III) and Tm(III) dinuclear complexes are the first examples having an observable intramolecular spin-spin exchange interaction between tri-positive rare earth ions.[15]

Extreme spin exchange narrowing is observed in a neutral Li pc radical which is a molecular semiconductor of conductivity of about 0.001 $1/\Omega$-cm at 300 K with activation energy of 0.2 eV. There is a complex magnetic behavior attributed to competing ferro-magnetic and antiferromagnetic interactions.[16]

Exchange interactions between unpaired π-electrons in polycrystalline sodium-doped H pc is studied by ESR. At $x = 1.7$ and 2.5, π-electron ferromagnetic interactions are observed.[17]

VI. MAGNETIC POWDERS AND THIN FILMS

The surface of a magnetic powder is coated with pc compounds. The powder, with good wetting and dispersibility, is useful in magnetic-coating paints for magnetic tapes and disks. A Li_2 pc in ethanol is mixed with magnetic iron powder to prepare a paint which, coated on polyester film, gives a magnetic coating with good brightness.[18]

Ferromagnetic Mn pc complexes form Langmuir-Blodgett films to make molecules oriented in one direction. The films resist weathering and oxidation well.[19]

Pc compounds are a coupling agent for magnetic recording disk and tape fabrication.[20]

VII. NMR STUDIES

A model eutectic composed of a solid glass with its pores filled with pc is said to exhibit double nuclear magnetooptical resonance.[21]

VIII. MOLECULAR AND MACROMOLECULAR METALS

The physical and magnetic properties of molecular metals formed by the oxidation of M pcs by iodine and bromine are the subject of a thesis.[22]

High-resolution NMR spectroscopy probes local architecture and electronic structure in pc molecular and macromolecular metals: Ni pc I, H_2 pc I, $[(Si\ pc\ O)\ I_{1.1}]_n$, $[(Si\ pc\ O)\ X_y]_n$ where $X = BF_4$, PF_6, SbF_6, $y \sim 0.35$, $[(Ge\ pc)\ I_{1.1}]_n$ and Ni pc X_2, as well as the precursors Ni pc, H_2 pc, $(Si\ pc\ O)_n$, and $(Ge\ pc\ O)_n$.[23] A review with 5 references is given of studies on the low-temperature magnetic properties of Cu pc I_2. EPR indicates the existence of a transition at 8 K involving a highly one-dimensional array of local moments strongly coupled to conduction electrons in the Cu pc I_2 crystal.[24]

The exchange interactions between Cu^{2+} ions in polycrystalline Cu pc $I_{0.2}$ is studied by ESR. Addition of iodine induces a maximum in the integrated intensity at 20 K.[25]

ESR spectra of Ni pc crystals are determined. There are two different magnetic centers, each with axial symmetry. Center 1 arises from a π-radical cation of the type $(Ni\ pc)^{x+}$ O_2^{x-} analogous to that observed in Ni pc I. Center 2 may arise from a Ni(III) species.[26]

IX. MAGIC-ANGLE SPINNING

A low-temperature magic-angle spinning experiment using liquid nitrogen as the spinning gas is described. The magic-angle spinning spectra of N-15 enriched pc obtained at 153 and 300 K are given.[27]

REFERENCES

1. **Kennedy, B. J., Brain, G., and Murray, K. S.,** Chlorophthalocyanineiron(III). FePc(-2)Cl. A spin admixed (S = 3/2S = 5/2) system, *Inorg. Chim. Acta,* 81(2), L29, 1984.
2. **Kennedy, B. J., Murray, K. S., Zwack, P. R., Homborg, H., and Kalz, W.,** μ-Oxo iron(III) phthalocyanine. Electronic structure of the solid form obtained from a dihydroxyiron(III) precursor, *Inorg. Chem.,* 24(20), 3302, 1985.
3. **Kobayashi, N., Koshiyama, M., Ishikawa, Y., Osa, T., Shirai, H., and Hojo, N.,** Electrolyte induced spin state transition of iron in iron(III) phthalocyanine, *Chem. Lett.,* 10, 1633, 1984.
4. **Wagnerova, D. M., Stopka, P., Lassmann, G., and Ebert, B.,** Reduction and spin transition of some macrocyclic iron(III) complexes, *Collect. Czech. Chem. Commun.,* 49(11), 2579, 1984.
5. **Livorness, J., Smith, T. D., Pilbrow, J. R., and Sinclair, G. R.,** Electron spin resonance study of the manganese(II) chelate of tetrasodium 3,10,17,24-tetrasulfonatophthalocyanine, *J. Chem. Soc., Faraday Trans. 2,* 80(4), 425, 1984.
6. **Stopka, P. and Wagnerova, D. M.,** Spin transition of manganese(II)-tetrasulfophthalocyanine in the presence of hydrogen peroxide, *Stud. Biophys.,* 103(2), 147, 1984.
7. **Labarta, A., Molins, E., and Tejada, J.,** A new interpreation of the magnetic properties and electronic structure of manganese(II) phthalocyanine, *Z. Phys. B,* 58(4), 299, 1985.
8. **Selyutin, G. E., Shklyaev, A. A., and Shul'ga, A. M.,** Structure of porphyrin and phthalocyanine copper(II) dimers based on ESR, *Izv. Akad. Nauk SSSR. Ser. Khim.,* 6, 1331, 1985.
9. **Lee, S., Yudkowsky, M., Halperin, W. P., Ogawa, M. Y., and Hoffman, B. M.,** One dimensional magnetism in copper phthalocyanine, *Phys. Rev. B,* 35(10), 5003, 1987.
10. **Pace, M. D., Barger, W. R., and Snow, A. W.,** EPR of copper tetrakis(cumylphenoxy)phthalocyanine Langmuir-Blodgett films, *J. Magn. Reson.,* 75(1), 73, 1987.
11. **Selyutin, G. E. and Shklyaev, A. A.,** Electron paramagnetic resonance spectra of vanadium(IV) complexes with cation radicals of etioporphyrin and phthalocyanine, *Khim. Fiz.,* 6, 850, 1983.
12. **Skorobogaty, A., Smith, T. D., Pilbrow, J. R., and Rawlings, S. J.,** Electron spin resonance study of the salts formed by combinations of transition-metal ion chelates of tetracationic porphyrazines with tetraanionic phthalocyanines, *J. Chem. Soc., Faraday Trans. 2,* 82(2), 173, 1986.
13. **Galindo, S., Gonzalez-Tovany, L., and Guia-Garcia, M. A.,** An electron paramagnetic resonance (EPR) study of a molybdenum phthalocyanine powder, *Rev. Mex. Fis.,* 32(3), 435, 1986.
14. **Schwartz, M., Hatfield, W. E., Joesten, M. D., Hanack, M., and Datz, A.,** Magnetic exchange interactions in the linear chain chromium(III) compounds, *Inorg. Chem.,* 24(24), 4198, 1985.
15. **Maeda, A. and Sugimoto, H.,** Intramolecular exchange interactions between two rare earth ions, *J. Chem. Soc., Faraday Trans. 2,* 82(11), 2019, 1986.
16. **Turek, P., Andre, J. J., and Simon, J.,** Extreme spin exchange narrowing in a neutral phthalocyanine radical: the lithium phthalocyanine, *Solid State Commun.,* 63(8), 741, 1987.
17. **Arutyunyan, A. R., Grigoryan, L. S., and Sharoyan, E. G.,** Magnetic properties of free base phthalocyanine doped with sodium, *Phys. Status Solidi B,* 142(2), K169, 1987.
18. **Ishihara, S. (Dainippon Ink and Chemicals, Inc.),** Japan Patent 86,104,601, 1986.
19. **Ogawa, K.,** Japan Patent 85,216,515, 1985.
20. **Anon.,** Magnetic recording medium, *Res. Discl.,* 266, 350, 1986.

21. **Chesnokov, A. D., Anisimov, V. V., Kolontaevskaya, L. A., Safronov, V. F., and Volosova, T. L.,** Double nuclear magnetooptical resonance in a solid eutectic, *Sovrem. Metody YaMR EPR Khim. Tverd. Tela, [Mater. Vses. Koord. Sovesch.], 4th,* 149, 1985.
22. **Palmer, S. M.,** Molecular Metals Formed by the Oxidation of Metallophthalocyanines by Iodine and Bromine: Physical and Magnetic Properties, Ph.D. thesis, Northwestern University, Evanston, IL, 1985.
23. **Toscano, P. J. and Marks, T. J.,** Electrically conductive metallomacrocyclic assemblies. High resolution solid state NMR spectrsocopy as a probe of local architecture and electronic structure in phthalocyanine molecular and macromolecular ''metals'', *J. Am. Chem. Soc.,* 108(3), 437, 1986.
24. **Hu, R.,** Principle of magnetooptical disk, *Wuli,* 15(3), 145, 1986.
25. **Grigoryan, L. S.,** Exchange interaction in the low dimensional charge transfer salt based on copper phthalocyanine and iodine, $CuPcI_{0.2,}$ Fiz. Tverd. Tela (Leningrad), *29(9), 2823, 1987.*
26. **Boas, J. F.,** Anomalous ESR spectra of nickel phthalocyanine. A single crystal study, *Aust. J. Chem.,* 40(3), 557, 1987.
27. **Kendrick, R. D., Friedrich, S., Wehrle, B., Limbach, H. H., and Yannoni, C. S.,** A stratagem for low temperature magic-angle spinning using nitrogen spinning gas, *J. Magn. Reson.,* 65(1), 159, 1985.

Chapter 24

PCS IN ELECTROCHEMISTRY

I. AIR BATTERIES

An air battery is obtained by making an air electrode by baking Co, Fe, or Ni complex with ligands having a phthalocyanine (pc) ring and fluorinated carbon.[1]

An air battery is obtained with an air electrode made by heating a mixture of a metal-sulfonated pc made from a sulfonated phthalic acid salt and a Co, Cu, Fe, Ni, or Ru salt, and carbon. The battery has good high-load discharge characteristics.[2] Another air battery is obtained with an air electrode prepared by heating a mixture containing a Co, Fe, or Ni complex of a sulfonated phthalic acid, such as Cu pc 2,9,16,23-tetrasulfonic acid or a phthalic acid derivative and carbon.[3]

A catalyst for an oxygen electrode for an air battery is prepared from a Co, Fe, Mn, or Ni chelate with a pc ring.[4]

Oxygen reduction catalysts for button air batteries are prepared from $MnSO_4$ mixed with Co pc sulfonate, then mixed with activated carbon, acetylene black, and a fluoro resin. The discharge period of an IEC-R-44 button air-zinc cell with 100-Ω resistance is about 30 h.[5]

The discharge characteristics of an air-zinc battery are improved with a polymeric Co pc together with manganese dioxide-carbon cathode.[6]

There are three direct methods to prepare an Fe pc polymer for an air electrode: melt, liquid phase, and gas phase. The discharge test of an aluminum-air battery with these electrodes is given.[7]

The electrochemical activity in potassium hydroxide electrolyte of carbon-based air electrodes containing polymeric Fe pc is studied by three direct synthesis methods: melt, liquid phase, and gas phase. The activity of poly Fe pc is in the order melt > liquid phase > gas phase.[8]

The influence of an intermediate decomposition rate on the electrode potential of an air cathode is studied by observing the catalytic activity of hydrogen peroxide decomposition by vitamin B12, Co and Fe pcs, platinum black, and activated charcoal.[9]

The catalytic effects of Fe pc on types A and B of practical air electrodes are compared with those of Co pc. The improved performance of type B electrode is thought to be due to the presence of many pores as a result of PTFE yarn connecting the carbon electrode particles.[10]

To improve air electrode performance, the dependence of Fe pc catalytic effects on the method of preparation of the electrode is examined.[11]

High surface area carbon is prepared from coconut shell. The carbon is then impregnated with Fe pc or Co tetraphenyl porphyrin and hot pressed into porous air electrodes with a polyethylene binder into a platinum screen. Oxygen reduction on the carbon electrodes impregnated with catalyst is in terms of activity: Fe pc > Co pc ~ Co-TPP > Ni pc >>> Mn pc ~ carbon.[12]

The effect of Co pc as catalyst on the electrochemical activity of oxygen and air electrodes is studied.[13]

II. DI HYDROGEN PCS

Pc (H_2pc) is used in the sensitization of an electrochemically modified roughened aluminum support used to make offset printing plates.[14]

Pc is incorporated in light-sensitive layers on anodized aluminum support for offset printing plates. The plates are flexible and haze free when developed.[15,16]

Photoinduced carrier generation is studied in particle dispersions of β-pc purified by temperature gradient sublimation, providing new evidence for the observation of time-resolved carrier generation first reported by Mort et al., in 1982, in polyvinyl carbazole.[17]

AC impedance and transient photoelectrochemical measurements are made on pc-modified electrodes.[18]

Restricted diffusion impedance is studied theroretically and experimentally vis-a-vis the reaction of oxygen on a pc film.[19]

Pc film electrodes are prepared by dispersing pc particles in poly(vinylidene fluoride) polymer binder.[20]

The photoelectrochemical behavior of a polycrystalline zinc oxide electrode is confirmed and its spectral response is determined with electrodes sensitized with pc and Rose Bengal.[21]

A redox couple in the photoelectrochemical production of oxalates from carbon dioxide involves pc complexes.[22]

Pc catalysts participate in the electrochemical reduction of oxygen on pyrolyzed carbon-supported transition metal chelates.[23]

Several possible ways to prepare electrodes coated by polymeric pc are described.[24]

Reversible stoichiometric reduction and oxidation of pc thin films 2 to 4 μ thick on stainless steel are observed.[25]

The photoreduction of oxygen on pc thin films deposited on gold substrates is studied in potassium hydroxide solutions containing hydrogen peroxide.[26]

The mechanism of reduction of pc in the presence of alkali cations is studied.[27]

A gold/pc/wet poly(vinyl alcohol)/gold sandwich cell shows positive photovoltage for the gold/pc electrode.[28]

III. LI PCS

A high-capacity lithium (Li) secondary battery with superior charging-discharging properties contains a cathode-active material with ultrafine particles of a pc type compound and a lithium anode active material.[29]

IV. MG PCS

Reversible oxidation and reduction of magnesium (Mg) pc thin films (1000 to 2000 Å thick) are observed on gold electrodes at 0.0 to 0.9 V vs. a saturated calomel electrode in various electrolytes. Raman spectra show the movement of anions into and from the electrodes during redox processes.[30]

V. RARE EARTH PCS

A. A VARIETY OF RARE EARTH PCS

Oxygen reduction on rare earth mono and di pcs is studied. Rare earth di pcs have a catalytic activity for oxidation reduction but rare earth mono pcs do not.[31]

A large number of rare earth di pcs (Ln pc_2H) are synthesized and their electrochemical properties are analyzed in neutral and acidic media. The lanthanum and neodymium compounds appear to exist as a super pc complex: Ln_2pc_3.[32]

B. LU DI PC

Mechanisms and kinetics of lutetium (Lu) di pc processes are studied, providing a link in the formulation of a unified redox reaction scheme.[33]

Several redox processes are observed with Lu di pc films on tin oxide electrodes in aqueous chloride electrolytes.[34]

It is proposed that oxygen adds reversibly to the blue reduction product Lu H pc_2 (formed after acidification of Lu pc_2 in DMF) to form the green Lu H pc_2^+ O_2^-.[35] However, the evidence for oxygen addition to Lu di pc may not be clear.[36]

Protonation of the reduced blue form of Lu di pc in dichloromethane solutions leads to the green Lu di pc disproportionation.[37]

The electrochemistry, redox and spectroscopic properties are given of alkyl-substituted Lu di pcs, Lu $(octyl)_8pc_2$ and Lu $(dodecyl)_{12}$ pc_2, in dichloromethane.[38]

Cyclic voltammograms of the green Lu pc_2 solutions exhibit three single electron reversible steps leading to the turquoise blue Lu pc_2^-, dark blue Lu pc_2^{2-}, and violet Lu pc_2^{3-}. No proton is present on the green Lu pc molecule.[39]

An experimental study of the electrochemistry of the blue form of Lu octa-4-*tert*-butyl di pc leads to the proposal of a mechanism of the redox conversion from the blue to the green forms.[40]

Disproportionation of Lu di pc in Lu di pc film occurs in an acidic aqueous medium leading to the blue/orange transition:[41]

$$\text{H Lu pc}_2 - 2e^- \rightarrow \text{Lu}^+ \text{ pc}_2 + \text{H}^+$$

VI. TITANYL (TIO) PCS

The photoelectrochemical properties of vacuum-deposited thin films of TiO pc are described. Continuous films at 60 to 400 nm thickness behave like a highly doped p-type semiconductor. XPS of both TiO pc and VO pc (vanadyl pc) films show the presence of extra oxygen atoms for films vacuum deposited in moderate vacuum vs. stoichiometric oxygen quantities of oxygen in films deposited under ultra-high vacuum.[42]

VII. VANADYL (VO) PCS

Photoelectrochemical studies are made of VO pc thin film electrodes. The photoactive thin films are deposited in a vacuum on gold metallized plastic optically transparent electrode substrates. Two types of crystallites exist at the surface.[43]

Platinum electrodes are first coated with Zn-etioporphyrin-II followed by VO pc. Since the absorption spectra of the two pigment layers at 400 to 580 nm overlap, the efficiency of use of light is greatly increased.[44]

VIII. MO PCS

For a constant potential, macrocyclic catalysts such as Mo nc (nc = naphthalocyanine) effect a current density increase with respect to the graphite electrodes.[45]

IX. FE PCS

The relation between the electrochemical and catalytic activities of Fe(II) polypc in oxygen reduction is studied. The dependence of electrochemical activity on catalytic activity indicates two different mechanisms for hydrogen peroxide decomposition during the reduction of oxygen.[46]

The stability of Fe pc electrodes is studied with respect to their use in oxygen reduction.[47]

Fe-tetrasulfonated pcs can be incorporated into electronically conducting polypyrrole by electrochemical pyrrole polymerization in the presence of Fe tetrasulfonated pcs. Glassy carbon electrodes containing this system catalyze oxygen reduction.[48]

Oxygen is almost exclusively reduced to hydroxyl ion on thick Fe pc layers in an alkaline solution using the rotating ring disk electrode.[49] The reduction kinetics are explained by a redox mechanism.[50]

A battery pellet electrode, 0.5 mm thick, is formed by mixing carbon black, PTFE, and Fe pc. It can be used in a lithium battery.[51]

High-activity stable catalysts for oxygen reduction in acid electrolytes are prepared. They include Co and Fe pcs, including the tetrasulfonated pcs.[52]

Oxygen reduction on Co and Fe tetrasulfo pcs coadsorbed on graphite to see if the water produced on the cobalt sites is decomposed on the iron sites on the adsorbed pcs shows that there is independent action of both catalysts on the surface.[53]

The cathodic and photocathodic behavior of electrodes of pc, Cu pc, or Fe pc is studied in acidic, neutral, and basic solutions. In the dark, the polarization curve of the electrode with Fe pc shows two reduction peaks not observed with the other pcs.[54]

The adsorption of Fe-tetrasulfonated pc is studied on single crystal silver electrodes. Evidence favors perpendicular adsorption of Fe-tetrasulfonated pc by the Ag(100) surface.[55]

When modified with Fe pc, carbon fiber electrodes show a significant catalytic effect on the electrochemical reduction of oxygen in aqueous and methanolic solutions.[56]

The mechanism of oxygen electroreduction catalyzed by Fe tetrasulfo pc is studied.[57]

Fe pc is incorporated in a diffusion membrane of electrochemical measurement cells.[58]

The electrochemical behavior of (μ-oxo)bis Fe(III) pc $(\text{Fe pc})_2\text{O}$ is studied in neat pyridine at a platinum button electrode. An electron transfer pathway is proposed and compared with that observed for the analogous porphyrin complex (μ-oxo)bis Fe(III) tetraphenylporphyrin.[59]

Using the impedance technique, the frequency response character of electrodes modified with Fe polypc is determined. Frequency spectra show that the electrode process is charge transfer controlled.[60]

Fe and Co tetrasulfonated pcs are studied in aqueous solution at pH 2 to 10 spectroelectrochemically. Reducing $[\text{Fe(III) tetrasulfo pc (2-)}]^+$ solution gives green Fe(II) tetrasulfo pc (2-) and pink $[\text{Fe(I) tetrasulfo pc (2-)}]^-$ over the entire pH range.[61]

The reduction of oxygen is studied on vacuum-deposited Co and Fe pc films, in acidic and alkaline solutions. Co pc reduces oxygen to hydrogen peroxide only. Fe pc produces a mixture of hydrogen peroxide and water. The difference between Co and Fe as the central metal atom is discussed.[62]

The redox potentials of Co and Fe, pc and of Co tetraazaannulene, irreversibly adsorbed on pyrolytic graphite, are studied by cyclic voltammetry vs. pH.[63]

Fe pc and its two μ-oxo derivatives obtained by mixing them with Vulcan XC-72C yield a common set of voltammetric peaks. The conductivity increase in Fe induced by its exposure to oxygen is probably due to the formation of μ-oxo-type derivatives.[64]

Oxygen reduction at Fe tetrasulfo pc incorporated in polypyrrole is studied. Dimeric Fe tetrasulfo pc species are present in polypyrrole, and the shift in the reduction onset potential is due to their presence.[65]

Electrochemical reduction of molecular oxygen at a polycarbazole electrode in aqueous solution is studied. The reduction is electrocatalyzed with or without Fe pc but is greater at the surface containing the Fe pc.[66]

Fe pc vacuum deposited and adsorbed on glassy carbon and gold exhibits catalytic activity in the reduction of molecular oxygen in an acidic aqueous electrolyte.[67]

X. CO PCS

Cobalt (Co) pc tetrasulfonate is incorporated into a polypyrrole film, for catalytic reduction of oxygen, using a platinum ring-glassy carbon disk rotating electrode.[68]

Co and Ni pcs deposited on carbon electrodes catalyze electroreduction of carbon dioxide to formic acid at pH 3 to 7.[69]

Adsorption of Co pc by carbon leads to electrocatalytic carbon dioxide reduction in aqueous solution at pH 5. Carbon monoxide is produced for over 100,000 turnovers of Co pc catalyst.[70]

In the cathodic reduction of oxygen in air, the activity of a Co polypc carbon electrode is greater than that of manganese dioxide/carbon.[71]

The electrochemical reduction of carbon dioxide takes place on mercury in the presence of Co pc.[72]

Oxidation of an ionic form of a compound occurs in a membrane electrolyte cell by reducing oxygen to hydrogen peroxide with a reducing agent in the cathode compartment. The reducing agent can be a Co pc.[73]

Cyclic voltammetry indicates cathodic reduction of thionyl chloride is enhanced by Co and Fe pcs but is relatively unaffected by Cu pc.[74]

The electroreduction of nitric oxide is studied by using electrodes modified with Co pc and a redox polymer. Conversion of nitric oxide to ammonia is obtained at a Co pc coated glassy carbon electrode with 92.3% maximum selectivity.[75]

The effect is studied of the compression pressure, and of the amounts of Ftoroplast and Co pc added, on the electrochemical activity and porous structure of a carbon electrode.[76]

Films of M pcs, Co, Cu, Fe(III), and Ni, bound to styrene-2-vinylpyridine copolymer can be used as the positive electrode of secondary cells in metal iodide solution saturated with iodine.[77]

Co pc tetrasulfonic acid is used as an electrochemical reduction catalyst of hydrogen peroxide and oxygen.[78]

Co pc is a photosensitizer in the photoelectrochemical reduction of carbon dioxide.[79]

For the electrooxidation of cysteine the catalytic activity of metal tetrasulfonated pcs decreases in the sequence Co > Fe > Mn > Ni > Cu.[80]

Modification of the carbon paste electrode by incorporating Co pc offers interesting properties to the electrocatalytic capability of the electrode.[81]

A new technique is offered to measure the activity and selectivity of carbon-supported catalysts impregnated with Co pc on a rotating ring disk electrode.[82]

Oxygen reduction is studied with Co pc or Co tetrasulfo pc but prepared by different methods. To determine turnover number, electrodes with an irreversibly adsorbed monolayer are recommended.[83]

A comparative study is made of glucose electrooxidation on anodes made of group VIII metals, Ag, Au, Cu, and glassy carbon as well as of pcs and porphyrins of Co, Fe, and Mn.[84]

A Co pc containing polymer film electrode with immobilized catalytic sites shows catalytic activity in a variety of reactions. Co pc is dispersed in tetrahydrofuran (THF) solution of PVC and cast into films on the surface of a glassy carbon electrode.[85]

Cyclic voltammetric measurements of adsorbed Co- and Fe- tetrasulfonated pc and Co pc are carried out on ordinary pyrolytic graphite and silver electrodes at pH 1 to 13.[86]

N_4-chelates containing Co and Fe in connection with active carbon are the prevalent materials to serve as electrocatalyst for cathodic reduction of oxygen in acid solution. Pc homopolymer metal complexes are prepared as electrocatalysts for oxygen reduction.[87]

The Co macroheterocyclic complexes of bi- (I), tri- (II), and tetradentate (III) structure (Figure 1) are studied by cyclic voltammetry in a neutral solution.[88]

Cyclic voltammetric studies of thin films of Co pc and of Co 4,4′,4″,4‴-tetracarboxy pc in nitrogen-saturated 1 *M* sulfuric acid show that both compounds give irreversibly semistable reduction products at about −0.70 V vs. a saturated calomel electrode.[89]

The conditions for forming monomer, dimer, or 6-coordinate complex are found for Co

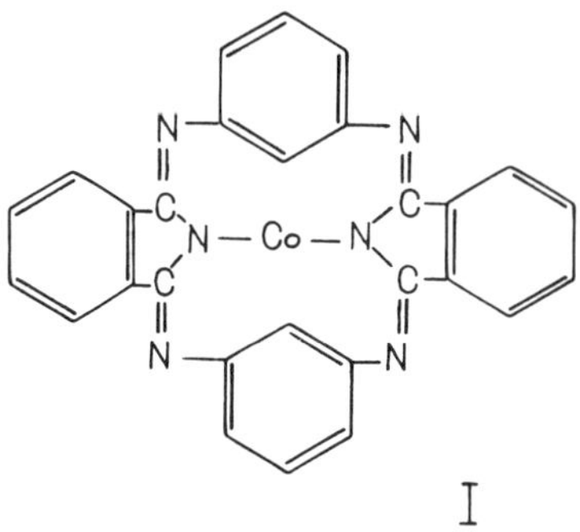

I

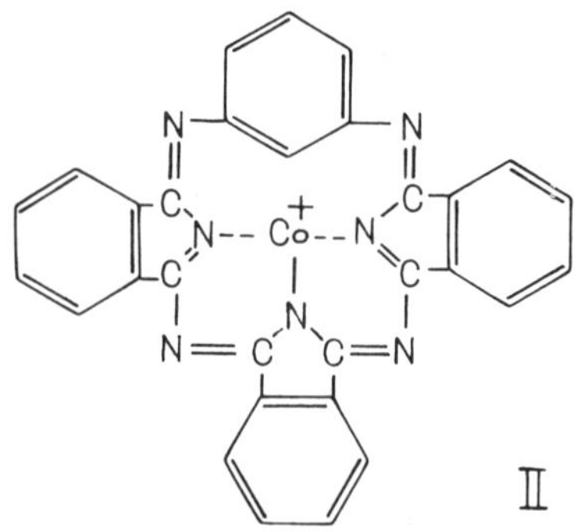

II

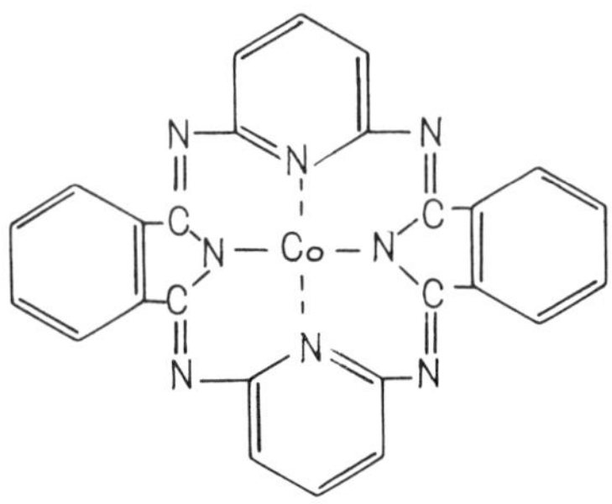

III

FIGURE 1. (I) Bi-, (II) tri-, and (III) tetradentate structures.

tetrasulfo pc, which reacts with cyanide ion to form a 6-coordinate complex. The electrocatalytic activity of Co tetrasulfo pc for oxygen reduction is increased in an acidic electrolyte.[90]

The spectroelectrochemistry of Co and Fe tetrasulfonated pcs is studied in aqueous solution at pH 2 to 10. For Co tetrasulfo pc, two reduced species are seen: yellow Co tetrasulfo pc (1-) and Co pink tetrasulfo pc (2-) over the entire pH range.[91]

The electrochemistry of Co 3,9,16,23-tetra(neopentoxy) pc and some binuclear analogs are studied in dichlorobenzene and DMF. Using an optically thin electrode, the electronic spectra of seven different oxidation states of the mononuclear species are reported.[92,93]

A glassy carbon electrode is modified with a composite consisting of Co pc and a conducting polymer support. Rotating ring-disk experiments indicate that water instead of hydrogen peroxide is the main reduction product of oxygen on the Co pc-modified electrode.[94]

Two main aspects of electrocatalysis of electrogenerated polypyrrole layers are presented. The cathodic reduction of oxygen with Co and Fe pcs is discussed.[95]

A quinone derivative-modified Co pc electrode does not act as a mediator in the catalytic reduction of carbon dioxide since the redox reaction is completely irreversible.[96]

Binuclear Co pcs are more effective electrocatalysts than corresponding mononuclear Co pcs in the reduction of molecular oxygen.[97]

A comparison between vacuum-deposited films and irreversibly adsorbed monolayers of Co pc and Fe pc indicates that with vacuum deposited films only a very small film fraction is electrochemically active.[98]

Conditions that promote the adsorption of tetrasulfo pcs on pyrolytic graphite are given. A model is presented to explain the observed kinetic limitation for electrodes with a low coverage of Co tetrasulfo pc.[99]

A process for the electrocatalytic reduction of carbon dioxide to carbon monoxide comprises immersing a transition metal complex such as a Co pc into a solution acidified to H^+ concentration of ~0.00001 to 0.1 *M*, adding the carbon dioxide, applying ~ −0.8 to ~ −1.5 V vs. a saturated calomel electrode (SCE), and reducing the carbon dioxide to carbon monoxide.[100]

The full resonance Raman spectra of Co(II) and Co(I) pcs are presented. They indicate that Co(I) pc is a reactive intermediate in the electrochemical reduction of carbon dioxide to carbon monoxide at Co(II) pc-impregnated electrodes.[101]

XI. PT PCS

Platinum (Pt) pc is studied as an electroreduction catalyst since Pt pc is much more stable than Fe(III) pc. The catalytic activity of Pt pc to reduce oxygen is not understandable in terms of a classical redox film.[102]

XII. NI PCS

Nickel (Ni) pc catalysts are referenced in the electrochemical reduction of carbon dioxide.[103]

XIII. CU PCS

Cu pcs are referenced as electrocatalysts in the reduction of oxygen.[104]

Cu pc trisulfonic acid is referenced in the doping of electric conductors in electroplating.[105]

Hydrogen photoevolution takes place at p-type silicon electrodes coated with discontinuous metal layers. The silicon electrodes are coated with platinum with an intermediate layer of Cu pc.[106]

XIV. ZN PCS

P-p iso-type junction electrodes providing photocurrents may consist of InSn oxide/Zn pc/Zn 5,10,15,20-tetraphenylporphyrin in potassium chloride solutions.[107]

An IBM 9001 laboratory computer aids the coulometry of Zn pc.[108]

The electrochemical reduction of H_2, Cu, and Zn octacyano pc is studied in DMF at Hg and at solid Au or Pt electrodes. Four reversible single-electron transfer steps are seen for each complex, corresponding to formation of a mono-, di-, tri-, and tetraamine.[109]

XV. AL PCS

A spin coating is used to produce Al pc OH photoelectrochemical cells. The spectral and electrical behavior of SnO_2/Al pc OH/benzoquinone/hydroquinone/Pt cells are studied and compared with cells based on sublimed Al pc OH films.[110]

The effect of phthalic acid treatment on the performance of Al pc OH photoelectrochemical cells is studied. Al pc OH thin films are prepared by vacuum deposition on tin oxide substrates. A phthalic acid treatment induces a dimorphic transformation of α-Al pc OH to a new n-Al pc OH polymorph absorbing over a broader spectrum 550 to 900 nm. The phthalic acid treatment strongly improves Al pc OH cell performance.[111]

XVI. GA PCS

The role of light source and its direction in the photoelectrochemical properties of Ga pc thin film electrodes doped with chlorine is studied.[112] Light-activated electrochemical reactions on Ga pc Cl-modified electrode surfaces are also investigated.[113]

A photocell consists of a thin film of Ga pc Cl photoconductor vacuum deposited onto thin gold or platinum films. A Au/Ga pc Cl electrode and a Pt/Ga pc Cl electrode in contact with a ferricyanide, ferrocyanide solution, and illuminated, develops a photopotential. Under ~100 mW/cm^2 illumination, a power conversion efficiency of about 0.03 to 0.05% is obtained with a fill factor up to 0.55.[114]

A photovoltaic cell, Au/Ga pc Cl/$Fe(CN)_6^{3-/4-}$/Ga pc Cl/Au has an open circuit photovoltage of 0.64 V.[115]

Thin films of vacuum-deposited Ga pc Cl and other trivalent M pcs exhibit widely different variable photoelectrochemical properties depending on vacuum deposition conditions and postdeposition doping with either oxygen or hydrogen.[116]

XVII. SI PCS

Electrochemical and photoelectrochemical studies are made on well-defined Si pc stacked ring oligomers.[117]

Thin films of Si pc with dimethyl-*tert*-butylsilyl caps on stannic oxide and gold substrates are studied as photoelectrodes. As thin solid films, the monomer Si pc proves to yield high photocurrent quantum efficiencies for hydroquinone and ferrocyanide, up to 10% at surface coverages <40 equivalent monolayers.[118]

Spectroelectrochemical data are obtained for monomer, dimer, trimer, and tetramer stacked-ring Si pcs on silver electrodes in aqueous solutions.[119]

XVIII. PB PCS

The electrochemical redox properties of Pb pc are studied by cyclic voltammetry, rotating disk voltammetry, and DC polarography in DMF containing 0.1 *M* $(C_2H_5)_4NClO_4$. An overall redox mechanism is postulated.[120]

XIX. METAL (M) PCS

Electrochemical studies are made of metal pcs.[121]

A method is given to chemically anchor metal pc polymer to semiconducting electrodes.[122]

Thionyl chloride reduction is a function of the cathode material; the rates are enhanced by transition metal pcs irrespective of cathode material.[123]

Electrochemical reduction of oxygen in sulfuric acid is catalyzed by porphyrin-like complexes. After heat treating, their catalytic activity is enchanced. This activity is compared with that of other heat-treated N_4 clusters such as tetraphenylporphyrins and pcs.[124]

The electrocatalytic activity of several metal pcs for the electrooxidation of hydrazine is in the order Mn pc > Mn tetrasulfo pc > Cr tetrasulfo pc > Ni tetrasulfo pc > VO pc > Ni pc ~ Cu pc ~ VO tetrasulfo pc ~ Zn tetrasulfo pc > Co tetrasulfo pc > graphite.[125]

An attempt is made to correlate the energy levels in M pcs with their electrocatalytic activity for oxygen reduction and hydrazine oxidation. pcs and tetrasulfo pcs of Co, Cr, Cu, Fe, Mn, Ni, V, and Zn are used.[126]

Effective electrocatalytic activity and stability for carbon dioxide reduction by M pc thin films is effected by their plasma-assisted deposition on a glassy carbon.[127]

Electrochemical oxidation on platinum in benzonitrile and reduction of complexes of tetra-4-*tert*-butyl pc with groups II and III metal ions Mg, Zn, Ga(III) and In(III) are studied. The difference of potentials of oxidation and reduction of these pcs does not depend on the redox properties of the central metal ion.[128]

The anodic oxidation of Cu tetra-2,9,16,23-*tert*-butyl pc and of Co tetrabromo-tetra-

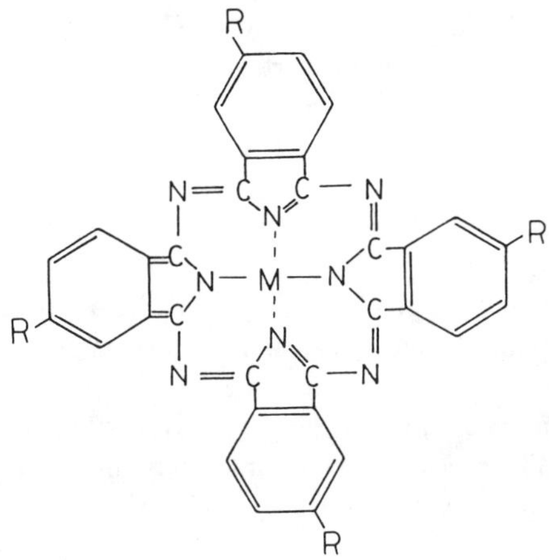

FIGURE 2. Substituted pcs and M pcs for which electrode potentials are recorded in liquid methylnaphthalene.

tert-butyl pc can be principally explained by the association of the redox forms of the pc with anions of the supporting electrolyte.[129]

A correlation is given from electronic absorption spectra of Co, Cu, and Zn tetra-4-*tert*-butyl pc and certain Zn pcs with macro ring substituents of standard redox potentials formed by substituted Zn pcs and by the corresponding products of 1-electron anodic oxidation, on the σ_n substituent constant.[130]

Electrode potentials for 30 pcs and porphyrins (R = H, $-C(CH_3)_3$, $-SO_3^-$; M = H_2, Co, Cu, Fe, Mg, Ni, Zn) are recorded in liquid methylnaphthalene electrolyte at 150°C. The pcs (Figure 2) show two distinct 1-electron reductions.[131]

The photopotential of an M pc electrode increases when it is polarized by a current of constant value. The galvanostatic potential exceeds eight to ten times the photopotential in the open circuit, reaching values sufficient for water photoelectrolysis.[132]

XX. ELECTROCHROMISM

The electrochemical properties of solid films of Cu, Zn, and H_2 2,3,9,10,16,17,23,24-octacyano pc are studied by cyclic voltammetry in aqueous electrolytes. The 10 to 100 -nm film electrodes show markedly stable electrochromic redox behavior.[133]

Cyclic voltammograms are obtained with Lu di pc electrodes in contact with various acids; the green/yellow/orange transition of Lu di pc in various acids is discussed.[134]

The electrochromic properties of film electrodes Er, Gd, Lu, and Yb di pcs in ethylene glycol solutions are studied. Some of the problems of this family of electrochromic materials are discussed.[135]

Polypyrrole films incorporating tetra- and monosulfonated Co pc and tetrasulfonated H_2 pc are strongly electrochromic and exhibit four different color stages at 1.2 to 1.6 V vs. a SCE.[136]

A spectral-electrochemical study is made of the electrochromic properties of films of the blue form *tert*-butyl substituted Lu di pc which is the product of chemical dimerization of the green form of Lu octa-4-*tert*-butyl di pc.[137]

The electrochemical process in electrochromic films of Lu di pc is studied by cyclic voltammetry.[138]

Semiconductor/solid electrolyte junctions are studied for optical information storage. The junctions are studied with solid polymer electrolytes with incorporated electrochromic materials including Lu di pc.[139]

The electrochemical oxidation of Hf(IV) and Zr(IV) pcs is studied on a gold electrode. These compounds have electrochromic properties.[140]

XXI. REVIEW

A review with 75 references relates to the photovoltaic effect of pc electrodes vis-a-vis the mechanism of photosynthesis.[141]

A review with 84 references is given on the activity of pcs in redox processes, electrical conductivity, activity in photovoltaic cells, and electrochemical and electroanalytical behavior.[142]

A review with 44 references surveys the preparation of polynuclear pc complexes, the electrochemistry of Co and Fe pcs, the electronic coupling in H pcs and Co pcs, and the catalytic activity of pc complexes toward oxygen reduction.[143]

REFERENCES

1. **Toshiba Battery Co., Ltd.,** Japan Patent 83,189,965, 1983.
2. **Toshiba Battery Co., Ltd.,** Japan Patent 83,189,964, 1983.
3. **Toshiba Battery Co., Ltd.,** Japan Patent 83,169,775, 1983.
4. **Matsushita Electric Industrial Co., Ltd.,** Japan Patent 84,68,172, 1984.
5. **Matsushita Electric Industrial Co., Ltd.,** Japan Patent 84,73,051, 1984.
6. **Pobedinskii, S. N., Erin, V. A., and Pobedinskaya, A. I.,** Discharge characteristics of an air-zinc battery with a phthalocyanine-carbon cathode, *Zh. Prikl. Khim. (Leningrad),* 57(5), 1031, 1984.
7. **Hirai, T., Yamaki, J., and Yamaji, A.,** Electrochemical characteristics of air electrodes with iron phthalocyanine polymers prepared by direct synthesis methods, *Denki Kagaku Kogyo Butsuri Kagaku,* 52(8), 534, 1984.
8. **Hirai, T. and Yamaki, J.,** Electrochemical activity in potassium hydroxide electrolyte of carbon based air electrodes containing polymeric iron phthalocyanine by direct synthesis methods, *J. Electrochem. Soc.,* 132(9), 2125, 1985.
9. **Hirai, T. and Yamaki, J.,** Influence of intermediate decomposition rate on electrode potential of an air cathode, *Denki Kagaku oyobi Kogyo Butsuri Kagaku,* 53(2), 145, 1985.
10. **Hirai, T., Yamaki, J., and Yamaji, A.,** Effect of method of preparation of iron phthalocyanine oxygen reduction catalyst on the activity of practical air electrodes, *J. Appl. Electrochem.,* 15(1), 77, 1985.
11. **Hirai, T., Yamaki, J., and Yamaji, A.,** The influence of catalyst-supporting methods on electrochemical activity and the resultant stability of air electrodes activated with iron phthalocyanine, *J. Appl. Electrochem.,* 15(3), 441, 1985.
12. **Shukla, A. K., Manoharan, R., Paliteiro, C., Hamnett, A., and Goodenough, J. B.,** High efficiency cathodes for alkaline air electrodes, *J. Appl. Electrochem.,* 15(5), 774, 1985.
13. **Bazanov, M. I., Sokolov, A. V., and Smirnov, R. P.,** Oxygen and air electrodes with polymeric cobalt phthalocyanine in the active layer, *Izv. Vyssh. Uchebn. Zaved., Khim. Khim. Tekhnol.,* 29(6), 36, 1986.
14. **Frass, W. and Dieter, M. (Hoechst AG),** German Patent 3,222,967, 1983.
15. **Mohr, D. (Hoechst AG),** German Patent 3,328,049, 1985.
16. **Mohr, D. (Hoechst AG),** German Patent 3,328,048, 1985.
17. **Popovic, Z. D.,** A study of carrier generation in β-metal free phthalocyanine, *Chem. Phys.,* 86(3), 311, 1984.
18. **Chan, K. W.,** A.C. Impedance and Transient Photoelectrochemical Measurements on Phthalocyanine Modified Electrodes, Ph.D. thesis, University of Arizona, Tucson, 1984.
19. **Contamin, O., Levart, E., Magner, G., Parsons, R., and Savy, M.,** Restricted diffusion impedance theory and application to the reaction of oxygen on a hydrogen phthalocyanine film, *J. Electroanal. Chem. Interfacial Electrochem.,* 179(1-2), 41, 1984.
20. **Minami, N., Sasaki, K., and Tsuda, K.,** Photoelectrochemical behavior of metal free phthalocyanine particulate film electrodes, *Kenkyu Hokoku-Sen'i Kobunshi Zairyo Kenkyusho,* 143, 19, 1984.
21. **Regull Climent, P. and Victori Companys, L.,** Photoelectrochemical behavior of a polycrystalline zinc oxide electrode: construction, photoelectrochemical activity and sensitization by a dye, *Afinidad,* 42(399), 453, 1985.

22. **Ang, P. G. P., Sammells, A. F., and Morduchowitz, A. (Texaco Inc.),** U.S. Patent 4,595,465, 1986.
23. **Van der Putten, A., Elzing, A., Visscher, W., and Barendrecht, E.,** Oxygen reduction on pyrolyzed carbon supported transition metal chelates, *J. Electroanal. Chem. Interfacial Electrochem.*, 205(1-2), 233, 1986.
24. **Woerhle, D., Schumann, B., Schmidt, V., and Jaeger, N.,** Preparation of polymer phthalocyanine electrodes and their electrochemical properties, *Makromol. Chem., Macromol. Symp.*, 8, 195, 1987.
25. **Djellab, H. and Dalard, F.,** Reduction and reoxidation of metal-free phthalocyanine electrodes in non-aqueous media. Electrochemical behavior and spectroscopic analysis, *J. Electroanal. Chem. Interfacial Electrochem.*, 221(1-2), 105, 1987.
26. **Contamin, O., Levart, E., Magner, G., Savy, M., and Scarbeck, G.,** Oxygen photoelectroreduction on hydrogen phthalocyanine films in alkaline solutions, *J. Electroanal. Chem. Interfacial Electrochem.*, 237(1), 39, 1987.
27. **Djellab, H., Dalard, F., Deroo, D., and Armand, M. B.,** Kinetic studies of the electrochemical reaction between lithium and phthalocyanine. II. Reduction reaction mechanism, *J. Power Sources*, 16(1), 45, 1985.
28. **Uehara, K., Iwai, T., and Tanaka, M.,** Photovoltaic effect of gold/metal-free phthalocyanine/wet poly(vinyl alcohol)/gold sandwich cell, *Chem. Express*, 1(3), 137, 1986.
29. **Nippon Telegraph and Telephone Public Corp.,** Japan Patent 83,223,265, 1983.
30. **Kahl, J. L., Faulkner, L. R., Dwarakanath, K., and Tachikawa, H.,** Reversible oxidation and rereduction of magnesium phthalocyanine electrodes. Electrochemical behavior and *in situ* Raman spectroscopy, *J. Am. Chem. Soc.*, 108(18), 5434, 1986.
31. **Wang, B., Shi, Z., and Zhang, Y.,** Electrochemical oxygen reduction on rare earth phthalocyanines, *Huaxue Xuebao*, 42(7), 727, 1984.
32. **M'sadak, M., Roncali, J., and Garnier, F.,** Rare-earth substitution effect on the electrochemical properties of lanthanide diphthalocyanines, *J. Electroanal. Chem. Interfacial Electrochem.*, 189(1), 99, 1985.
33. **Nicholson, M. M., Weismuller, T. P., and Pizzarello, F. A.,** Mechanisms and Kinetics of Diphthalocyanine Electrode Processes, Report 1983, C83-522/201, AFOSR-TR-83-0721, Order No. AD-A131726, National Technical Information Service, Springfield, VA, 31 pp.
34. **Nicholson, M. M. and Weismuller, T. P.,** Electrochemistry of Rare-Earth Diphthalocyanines, Report 1984, SC5383. 1FR, AFOSR-TR-84-0769, Order No. AD-A145813/2/GAR, National Technical Information Service, Springfield, VA, 25 pp.
35. **Nicholson, M. M., and Weismuller, T. P.,** The role of oxygen in the redox chemistry of lutetium diapthalocyanine. Reply to comments, *J. Electronchem. Soc.*, 132(7), 1774, 1985
36. **L'Her, M.,** The role of oxygen in the redox chemistry of lutetium diphthalocyanine. Comments, *J. Electrochem. Soc.*, 132(7), 1774, 1985.
37. **L'Her, M., Cozien, Y., and Courtot-Coupez, J.,** Effect of acidity on the electrochemical and spectrophotometric behavior of lutetium diphthalocyanine in dichloromethane solutions, *C.R. Acad. Sci., Ser. 2*, 300(11), 487, 1985.
38. **Castaneda, F., Piechocki, C., Plichon, V., Simon, J., and Vaxiviere, J.,** Spectroscopic and redox properties of alkyl-substituted lutetium diphthalocyanines in dichloromethane, *Electrochim. Acta*, 31(1), 131, 1986.
39. **L'Her, M., Cozien, Y., and Courtot-Coupez, J.,** Electrochemical reduction of lutetium diphthalocyanine in solution, *C.R. Acad. Sci., Ser. 2*, 302(1), 9, 1986.
40. **Gavrilova, V. I., Vazhnina, V. A., Konstantinov, A. P., and Shelepin, I. V.,** Electrochemistry of the blue form of lutetium diphthalocyanine, *Elektrokhimiya*. 22(8), 1112, 1986.
41. **Castaneda, F., Plichon, V., Clarisse, C., and Riou, M. T.,** Electrochemistry of a lutetium diphthalocyanine film in contact with an acidic aqueous medium, *J. Electroanal. Chem. Interfacial Electrochem.*, 233(1-2), 77, 1987.
42. **Klofta, T. J., Danziger, J., Lee, P., Pankow, J., Nebesny, K. W., and Armstrong, N. R.,** Photoelectrochemical and spectroscopic characterization of thin films of titanyl phthalocyanine: comparisons with vanadyl phthalocyanine, *J. Phys. Chem.*, 91(22), 5646, 1987.
43. **Klofta, T., Linkous, C., and Armstrong, N. R.,** Photoelectrochemical studies of vanadyl phthalocyanine (VOPc) thin film electrodes, *J. Electroanal. Chem. Interfacial Electrochem.*, 185(1), 73, 1985.
44. **Ilatovskii, V. L., Rudakov, V. M., Dmitriev, I. B., and Komissarov, G. G.,** Becquerel effect in two layer pigmented electrodes, *Zh. Fiz. Khim.*, 59(9), 2273, 1985.
45. **Costa, M. and Savy, M.,** Study of Organic Catalysts for Electrochemical Generators, Report 1984, Order No. B85-176519, National Technical Information Service, Springfield, VA, 25 pp.
46. **Kreja, L.,** The relation between electrochemical and catalytic activities of iron(II) polyphthalocyanine in the reduction of oxygen, *Electrochim. Acta*, 28(12), 1807, 1983.
47. **Melendres, C. A. and Xu, S. Q.,** Anodic dissolution of iron phthalocyanine film electrodes, *J. Electrochem. Soc.*, 131(2), 466, 1984.
48. **Bull, R. A., Fan, F. R., and Bard, A. J.,** Polymer Films on Electrodes. XIII. Incorporation of Catalysts into Electronically Conductive Polymers: Iron Phthalocyanine in Polypyrrole, Report 1983, TR-13, Order No. AD-A127700, National Technical Information Service, Springfield, VA, 22 pp.

49. **Van den Brink, F., Visscher, W., and Barendrecht, E.**, Electrocatalysis of cathodic oxygen reduction by metal phthalocyanines. III. Iron phthalocyanine as electrocatalyst: experimental part, *J. Electroanal. Chem. Interfacial Electrochem.*, 172(1 2), 301, 1984.
50. **Van den Brink, F., Visscher, W., and Barendrecht, E.**, Electrocatalysis of cathodic oxygen reduction by metal phthalocyanines. IV. Iron phthalocyanine as electrocatalyst: mechanism, *J. Electroanal. Chem. Interfacial Electrochem.*, 175(1-2), 279, 1984.
51. **Nippon Telegraph and Telephone Public Corp.**, Japan Patent 84,25,167, 1984.
52. **Aldred, W., Fierro, C., Gupta, S., Simic-Glavaski, B., and Yeager, E.**, Transition Metal Macrocycles and Related Complexes as Catalysts for Oxygen Electrodes, Report 1984, LBL-18542, Order No. DE85004439, National Technical Information Service, Springfield, VA, 138 pp.
53. **Zagal, J. H., Paez, M., Sturm, J., and Ureta-Zanartu, S.**, Electroreduction of oxygen on mixtures of phthalocyanines co-adbsorbed on a graphite electrode, *J. Electroanal. Chem. Interfacial Electrochem.*, 181(1-2), 295, 1984.
54. **Marsan, B., Belanger, G., and Piron, D. L.**, Photoelectrochemistry of copper and iron phthalocyanines, *Can. J. Chem.*, 63(7), 1580, 1985.
55. **Adzic, R., Simic-Glavaski, B., and Yeager, E.**, Spectroscopic and electrochemical studies of transition metal tetrasulfonated phthalocyanines. VI. The adsorption of iron tetrasulfonated phthalocyanine on single crystal silver electrodes, *J. Electroanal. Chem. Interfacial Electrochem.*, 194(1), 155, 1985.
56. **Jannakoudakis, A. D., Missaelidis, N., and Theodoridou, E.**, Chemical modification of electrochemically oxidized and partially rereduced carbon fiber electrodes, *Synth. Met.*, 11(2), 101, 1985.
57. **Chen, Z.**, Catalytic activity of chemically modified FeTSPc graphite electrode for oxygen electroreduction, *Xiamen Daxue Xuebao, Ziran Kexueban,* 25(1), 82, 1986.
58. **Hoelscher, U. (Draegerwerk AG)**, German Offenlegungsschrift 3,432,949, 1986.
59. **Bottomley, L. A., Ercolani, C., Gorce, J. N., Pennesi, G., and Rossi, G.**, Spectroelectrochemistry of (μ-oxo)bis[phthalocyaninato)iron(III)], *Inorg. Chem.*, 25(14), 2338, 1986.
60. **Qiu, X., Song, S., and Zhao, T.**, Titanium electrodes modified with metalloporphyrins and phthalocyanines. IV. Frequency response analysis of titanium electrodes modified with metalloporphyrins and phthalocyanines, *Zhongnan Kuangye Xueyuan Xuebao,* 1, 101, 1986.
61. **Nevin, W. A., Liu, W., Melnik, M., and Lever, A. B. P.**, Spectroelectrochemistry of cobalt and iron tetrasulfonated phthalocyanines, *J. Electroanal. Chem. Interfacial Electrochem.*, 213(2), 217, 1986.
62. **Van der Putten, A., Elzing, A., Visscher, W., and Barendrecht, E.**, Oxygen reduction on vacuum deposited and adsorbed transition metal phthalocyanine films, *J. Electroanal. Chem. Interfacial Electrochem.*, 214(1-2), 523, 1986.
63. **Van der Putten, A., Elzing, A., Visscher, W., and Barendrecht, E.**, Redox potential and electrocatalysis of oxygen reduction on transition metal chelates, *J. Electroanal. Chem. Interfacial Electrochem.*, 221(1-2), 95, 1987.
64. **Tanaka, A. A., Fierro, C., Scherson, D., and Yeager, E. B.**, Electrocatalytic aspects of iron phthalocyanine and its μ-oxo derivatives dispersed on high surface area carbon, *J. Phys. Chem.*, 91(14), 3799, 1987.
65. **Elzing, A., Van der Putten, A., Visscher, W., and Barendrecht, E.**, The mechanism of oxygen reduction at iron tetrasulfonatophthalocyanine incorporated in polypyrrole, *J. Electroanal. Chem. Interfacial Electrochem.*, 233(1-2), 113, 1987.
66. **O'Brien, R. N. and Santhanam, K. S. V.**, Electrocatalysis of oxygen reduction at a polycarbazole electrode, *Electrochim. Acta,* 32(8), 1209, 1987.
67. **Van den Ham, D., Hinnen, C., Magner, G., and Savy, M.**, Electrocatalytic oxygen reduction: the role of oxygen bridges as a structural factor in the activity of transition metal phthalocyanines, *J. Phys. Chem.*, 91(18), 4743, 1987.
68. **Osaka, T., Nishikawa, M., and Nakamura, S.**, Electrocatalytic oxygen reduction with metal porphyrins or metal phthalocyanine incorporated in polypyrrole film electrode, *Denki Kagaku oyobi Kogyo Butsuri Kagaku,* 52(6), 370, 1984.
69. **Kapusta, S. and Hackerman, N.**, Carbon dioxide reduction at a metal phthalocyanine catalyzed carbon electrode, *J. Electrochem. Soc.*, 131(7), 1511, 1984.
70. **Lieber, C. M. and Lewis, N. S.**, Catalytic reduction of carbon dioxide at carbon electrodes modified with cobalt phthalocyanine, *J. Am. Chem. Soc.*, 106(17), 5033, 1984.
71. **Pobedinskii, S. N., Erin, V. A., and Pobedinskaya, A. I.**, Effect of metal phthalocyanines and manganese dioxide on the activity of an oxygen carbon electrode in an alkaline solution, *Zh. Prikl. Khim. (Leningrad),* 57(5), 1034, 1984.
72. **Tinnemans, A. H. A., Koster, T. P. M., Thewissen, D. H. M. W., and Mackor, A.**, Tetraazamacrocyclic cobalt(II) and nickel(II) complexes as electron-transfer agents in the photo(electro)chemical and electrochemical reduction of carbon dioxide, *Recl.: J. R. Neth. Chem. Soc.*, 103(10), 288, 1984.
73. **Kuwana, T. (Koppers Co., Inc.)**, U.S. Patent 4,487,669, 1984.
74. **Madou, M., Kinoshita, K., McKubre, M. C. H., and Szpak, S.**, Electrocatalysis studies on the cathodic reduction of thionyl chloride, *Proc.-Electrochem. Soc.*, 84-12, 618, 1984.

75. **Ogura, K. and Yamasaki, S.,** Electroreduction of nitric oxide to ammonia at chemically modified electrodes, *J. Appl. Electrochem.*, 15(2), 279, 1985.
76. **Tret'yak, A. N., Novikov, G. I., Grishaeva, G. A., Butyrin, G. M., and Shulepova, N. Y.,** Selection of conditions for the production of carbon electrodes, *Khim. Khim. Tekhnol. (Minsk)*, 19, 49, 1984.
77. **Hirabaru, O., Nakase, T., Hanabusa, K., Shirai, H., Takemoto, K., and Hojo, N.,** Functional metal porphyrazine derivatives and their polymers. XIII. Secondary cells using metal phthalocyanine bound to poly(2-vinylpyridine-co-styrene) as positive electrodes, *Polym. Commun.*, 25(9), 284, 1984.
78. **Chen, Z.,** Reaction mechanism of cobalt tetrasulfonated phthalocyanine as a catalyst used for electrocatalytic reduction of oxygen and hydrogen peroxide, *Xiamen Daxue Xueban Ziran Kexueban*, 24(2), 209, 1985.
79. **Sammells, A. F. and Pa, A. P. G. (Texaco Inc.),** U.S. Patent 4,545,872, 1985.
80. **Zagal, J. H. and Herrera, P.,** Electrochemistry of cysteine and cystine on metal phthalocyanines adsorbed on a graphite electrode, *Electrochim. Acta*, 30(4), 449, 1985.
81. **Linders, C. R., Patriarche, G. J., Kauffmann, J. M., and Guilbault, G. C.,** Electrochemical study of thiols and disulfides using modified electrodes, *Anal. Lett.*, 19(1-2), 193, 1986.
82. **Van der Putten, A., Visscher, W., and Barendrecht, E.,** A new method of preparing a rotating ring-disk electrode for the study of carbon supported catalysts, *J. Electronanal. Chem. Interfacial Electrochem.*, 195(1), 63, 1985.
83. **Elzing, A., Van der Putten, A., Visscher, W., and Barendrecht, E.,** The cathodic reduction of oxygen at cobalt phthalocyanine. Influence of electrode preparation on electrocatalysis, *J. Electroanal. Chem. Interfacial Electrochem.*, 200(1-2), 313, 1986.
84. **Vasil'ev, Y. B., Khazova, O. A., and Nikolaeva, N. N.,** Kinetics and mechanism of glucose electrooxidation on different electrode-catalysts. II. Effect of the nature of the electrode and the electrooxidation mechanism, *J. Electroanal. Chem. Interfacial Electrochem.*, 196(1), 127, 1985.
85. **Mizutani, F., Iijima, S., Tanabe, Y., and Tsuda, K.,** Conducting polymer film electrodes with immobilized catalytic sites, *J. Chem. Soc., Chem. Commun.*, 23, 1728, 1985.
86. **Zecevic, S., Simic-Glavaski, B., Yaeger, E., Lever, A. B. P., and Minor, P. C.,** Spectroscopic and electrochemical studies of transition metal tetrasulfonated phthalocyanines. V. Voltammetric studies of adsorbed tetrasulfonated phthalocyanines (MTsPc) in aqueous solutions, *J. Electroanal. Chem. Interfacial Electrochem.*, 196(2), 339, 1985.
87. **Wiesener, K.,** N_4-chelates as electrocatalyst for cathodic oxygen reduction, *Electrochim. Acta*, 31(8), 1073, 1986.
88. **Bazanov, M. I., Kokhova, L. V., Bogdanovskaya, V. A., Tarasevich, M. R., Smirnov, R. P., and Kolesnikov, N. A.,** Redox behavior of cobalt macroheterocyclic compounds, *Izv. Vyssh. Uchebn. Zaved., Khim. Khim. Tekhnol.*, 29(7), 43, 1986.
89. **Kusuda, K., Ishihara, R., Yamaguchi, H., and Izumi, I.,** Electrochemical investigation of thin films of cobalt phthalocyanine and cobalt 4,4′,4″,4‴-tetracarboxyphthalocyanine and the reduction of carbon monoxide, formic acid and formaldehyde mediated by the cobalt(I) complexes, *Electrochim. Acta*, 31(6), 657, 1986.
90. **Chen, Z. and Zhou, T.,** Electrocatalytic capability of tetrasulfonated phthalocyanine for oxygen reduction and the EHMO calculation, *Xiamen Daxue Xuebao, Ziran Kexueban*, 25(5), 551, 1986.
91. **Nevin, W. A., Lia, W., Melnik, M., and Lever, A. B. P.,** Spectroelectrochemistry of Cobalt and Iron Tetrasulfonated Phthalocyanines, Report 1986, TR-6, Order No. AD-A169642/6/GAR, National Technical Information Service, Springfield, VA, 32 pp.
92. **Nevin, W. A., Hempstead, M. R., Liu, W., Leznoff, C. C., and Lever, A. B. P.,** Electrochemistry and spectroelectrochemistry of mononuclear and binuclear cobalt phthalocyanines, *Inorg. Chem.*, 26(4), 570, 1987.
93. **Nevin, W. A., Hempstead, M. R., Liu, W., Leznoff, C. C., and Lever, A. B. P.,** Electrochemistry and Spectroelectrochemistry of Mononuclear and Binuclear Cobalt Phthalocyanines, Report 1986, TR-10, Order No. AD-A174289/9/GAR, National Technical Information Service, Springfield, VA, 47 pp.
94. **Mizutani, F., Iijima, S., Tanabe, Y., and Tsuda, K.,** Conducting polymer film electrode with catalytic sites, *Synth. Met.*, 18(1-3), 111, 1987.
95. **Barendrecht, E., Elzing, A., Janssen, L. J. J., Van der Putten, A., Visscher, W., and Vork, F.,** Electrocatalytic aspects of electrogenerated polymer layers, *Makromol. Chem., Macromol. Symp.*, 8, 211, 1987.
96. **Ogura, K. and Fujita, M.,** Electrocatalytic reduction of carbon dioxide to methanol. VII. With quinone derivatives immobilized on platinum and stainless steel, *J. Mol. Catal.*, 41(3), 303, 1987.
97. **Hempstead, M. R., Lever, A. B. P., and Leznoff, C. C.,** Electrocatalytic reduction of molecular oxygen by mononuclear and binuclear cobalt phthalocyanines, *Can. J. Chem.*, 65(11), 2677, 1987.
98. **Van der Putten, A. M. T. P.,** Electrocatalysts of Dioxygen Reduction on Transition Metal Chelates, Report 1986, ETN-87-99584, National Technical Information Service, Springfield, VA, 117 pp.
99. **Elzing, A., Van der Putten, A., Visscher, W., and Barendrecht, E.,** The cathodic reduction of oxygen at metal tetrasulfonatophthalocyanines. Influence of adsorption conditions on electrocatalysis, *J. Electroanal. Chem. Interfacial Electrochem.*, 233(1-2), 99, 1987.

100. **Cuellar, E. A. and Gaylor, V. F. (Standard Oil Co. (Ohio)),** U.S. Patent 4,668,349, 1987.
101. **Masheder, D. and Williams, K. P. J.,** Raman spectroelectrochemistry. II. *In situ* Raman spectroscopic studies of the electrochemical reduction of carbon dioxide at cobalt(II) phthalocyanine-impregnated PTFE-bonded carbon gas diffusion electrodes, *J. Raman Spectrosc.*, 18(6), 391, 1987.
102. **Paliteiro, C., Hamnett, A., and Goodenough, J. B.,** Cathodic reduction of oxygen by platinum phthalocyanine thin films, *J. Electroanal. Chem. Interfacial Electrochem.*, 160(1-2), 359, 1984.
103. **Ang, P. G. P. and Sammells, A. F. (Texaco Inc.),** U.S. Patent 4,523,981, 1985.
104. **Tarasevich, M. R., Bogdanovskaya, V. A., Gavrilova, E. F., and Orlov, S. B.,** Electrocatalysis of cathodic molecular oxygen reduction with biopolymers-enzymes and their models, *J. Electroanal. Chem. Interfacial Electrochem.*, 206(1-2), 217, 1986.
105. **BASF AG,** Japan Patent 87, 47,499, 1987.
106. **Nakato, Y., Yano, H., Nishiura, S., Ueda, T., and Tsubomura, H.,** Hydrogen photoevolution at p-type silicon electrodes coated with discontinuous metal layers, *J. Electroanal. Chem. Interfacial Electrochem.*, 228(1-2), 97, 1987.
107. **Harima, Y. and Yamashita, K.,** Organic photoelectrodes based on p-p iso-type junctions, *J. Phys. Chem.*, 89(25), 5325, 1985.
108. **Nyokong, T. and Stillman, M. J.,** Computer aided chemistry. I. Control of the PAR 273 electrochemical instrument using the IBM 9001 laboratory computer, *J. Autom. Chem.*, 8(3), 122, 1986.
109. **Louati, A., El Meray, M., Andre, J. J., Simon, J., Kadish, K. M., Gross, M., and Giraudeau, A.,** Electrochemical reduction of new, good electron acceptors: the metalloctacyanophthalocyanines, *Inorg. Chem.*, 24(8), 1175, 1985.
110. **Perrier, G. and Dao, L. H.,** Hydroxyaluminum phthalocyanine photoelectrochemical cells deposited by rotation, *Can. J. Chem.*, 64(12), 2431, 1986.
111. **Perrier, G. and Dao, L. H.,** Improvement of the performance of a hydroxyaluminum phthalocyanine photoelectrochemical cell by a phthalic acid treatment, *J. Electrochem. Soc.*, 134(5), 1148, 1987.
112. **Buttner, W., Rieke, P., and Armstrong, N. R.,** Photoelectrochemical response of chlorine-doped gallium phthalocyanine thin film electrodes using two photon sources and two illumination directions, *J. Electrochem. Soc.*, 131(1), 226, 1984.
113. **Linkous, C. A.,** Light Activated Electrochemical Reactions on Chloro-Gallium Phthalocyanine-Modified Electrode Surfaces, Ph.D. thesis, Michigan State University, East Lansing, 1983.
114. **Buttner, W. J., Rieket, P. C., and Armstrong, N. R.,** The gold/GaPc-Cl/ferri,ferrocyanide/GaPc-Cl/platinum photoelectrochemical cell, *J. Am. Chem. Soc.*, 107(12), 3738, 1985.
115. **Klofta, T., Buttner, W. F., and Armstrong, N. R.,** Effect of crystallite size hydrogen and oxygen uptake in the photoelectrochemistries of thin films of chlorogallium phthalocyanine, *J. Electrochem. Soc.*, 133(7), 1531, 1986.
116. **Klofta, T. J., Sims, T. D., Pankow, J. W., Danziger, J., Nebesny, K. W., and Armstrong, N. R.,** Spectroscopic and photoelectrochemical studies of trivalent metal phthalocyanine thin films: the role of gaseous dopants (oxygen and hydrogen) in determining photoelectrochemical response, *J. Phys. Chem.*, 91(22), 5651, 1987.
117. **Mezza, T. M.,** Electrochemical and Photoelectrochemical Studies on Well Defined Silicon Phthalocyanine Stacked-Ring Oligomers, Ph.D. thesis, University of Arizona, Tucson, 1983.
118. **Mezza, T., Armstrong, N. R., and Kenney, M.,** Photoelectrochemical studies of thin films of monomer, dimer, and trimer stacked-ring silicon phthalocyanines, *J. Electroanal. Chem. Interfacial Electrochem.*, 176(1-2), 259, 1984.
119. **Simic-Glavaski, B., Tanaka, A. A., Kenney, M. E., and Yeager, E.,** Spectroscopic and electrochemical studies of transition-metal tetrasulfonated phthalocyanines. VIII. Spectroelectrochemical studies of stacked-ring silicon phthalocyanines, *J. Electroanal. Chem. Interfacial Electrochem.*, 229(1-2), 285, 1987.
120. **El Meray, M., Louati, A., Simon, J., Giraudeau, A., Gross, M., Malinski, T., and Kadish, K. M.,** Redox properties and demetalation of reduced lead phthalocyanine in dimethylformamide, *Inorg. Chem.*, 23(17), 2606, 1984.
121. **Minor, P. C.,** Some Electrochemical Studies of Metal Phthalocyanines, Ph.D. thesis, York University, Downsview, Ontario, Canada 1983.
122. **Frank, A. J. and Honda, K. (U.S. Department of Energy),** U.S. Patent Application 483,040, 1984.
123. **Doddapaneni, N.,** Thionyl chloride ($SOC1_2$) reduction on rotating and stationary disk electrodes, *Proc.-Electrochem. Soc.*, 84-12, 630, 1984.
124. **Johansson, L. Y. and Larsson, R.,** Electrochemical reduction of oxygen in sulfuric acid catalyzed by porphyrin-like complexes, *J. Mol. Catal.*, 38(1-2), 61, 1986.
125. **Zagal, J. H., Lira, S., and Ureta-Zanartu, S.,** A mechanistic study of the electrooxidation of hydrazine on phthalocyanines of vanadyl, chromium, manganese, nickel, copper and zinc attached to graphite electrodes, *J. Electroanal. Chem. Interfacial Electrochem.*, 210(1), 95, 1986.
126. **Zagal, J., Paez, M., and Fierro, C.,** Correlation between electronic and redox properties of metal macrocycles and their electrocatalytic activity for electrochemical reactions, *Proc.-Electrochem. Soc.*, 87-12, 198, 1987.

127. **Tanabe, H. and Ohno, K.,** Electrocatalysis of metal phthalocyanine thin film prepared by the plasma-assisted deposition on a glassy carbon in the reduction of carbon dioxide, *Electrochim. Acta,* 32(7), 1121, 1987.

128. **Platonova, E. P., Skuridin, E. Y., and Degtyarev, L. S.,** Electrochemical oxidation and reduction of complexes of tetra-4-tert-butylphthalocyanine with Group II and III metal ions, *Zh. Obshch. Khim.,* 54(4), 925, 1984.

129. **Shelepin, I. V., Ushakov, O. A., and Gavrilov, V. I.,** Bifurcation of single electron waves of the anodic oxidation of phthalocyanine complexes, *Elektrokhimiya,* 20(3), 343, 1984.

130. **Gavrilov, V. I., Konstantinov, A. P., Derkacheva, V. M., Luk'yanets, E. A., and Shelepin, I. V.,** Cation-radicals of phthalocyanine complexes, *Zh. Fiz. Khim.,* 60(6), 1448, 1986.

131. **Campbell, R. H., Heath, G. A., Hefter, G. T., and McQueen, R. C. S.,** A.C. and D.C. voltammetry in liquid naphthalene solutions: stepwise electrode potentials for metal phthalocyanines and porphyrins in strictly noncoordinating media, *J. Chem. Soc., Chem. Commun.,* 20, 1123, 1983.

132. **Shumov, Y. S. and Komissarov, G. G.,** Galvanostatic photopotential of magnesium phthalocyanine films, *Biofizika,* 30(5), 916, 1985.

133. **Schumann, B., Woehrle, D., and Jaeger, N. I.,** Reversible reduction and reoxidation of entire thin films of octacyanophthalocyanine, *J. Electrochem. Soc.,* 132(9), 2144, 1985.

134. **Riou, M. T., Auregan, M., and Clarisse, C.,** Electrochromic properties of lutetium diphthalocyanine films, *J. Electroanal. Chem. Interfacial Electrochem.,* 187(2), 349, 1985.

135. **Collins, G. C. S. and Schiffrin, D. J.,** The properties of electrochromic film electrodes of lanthanide diphthalocyanines in ethylene glycol, *J. Electrochem. Soc.,* 132,(8), 1835, 1985.

136. **Velazquez, R. M., Skotheim, T. A., and Linkous, C. A.,** Polypyrrole-phthalocyanine, *Synth. Met.,* 15(2-3), 219, 1986.

137. **Gavrilov, V. I., Konstantinov, A. P., Luk'yanets, E. A., and Shelepin, I. V.,** Electrochromism of solid films of the blue form of a phthalocyanine complex of lutetium, *Elektrokhimiya,* 22(12), 1667, 1986.

138. **Castaneda, F. and Plichon, V.,** Electrochemical process in electrochromic films of lutetium diphthalocyanine, *J. Electroanal. Chem. Interfacial Electrochem.* 236(1-2), 163, 1987.

139. **Sammells, A. F.,** Semiconductor/Solid Electrolyte Junctions for Optical Information Storage, Report 1986, Order No. AD-173228/8/GAR, National Technical Information Service, Springfield, VA, 4 pp.

140. **Tomilova, L. G., Ovchinnikova, N. A., and Luk'yanets, E. A.,** Phthalocyanines and related compounds. XXIX. Spectral-electrochemical study of hafnium and zirconium diphthalocyanines, *Zh. Obshch. Khim.,* 57(9), 2100, 1987.

141. **Shumov, Y. S. and Prosvirin, K. A.,** Photovoltaic Effect of Phthalocyanine Electrodes, Deposited Doc. 1984, VINITI 4208-84, All Union Institute of Scientific and Technical Information, U.S.S.R., 20 pp.

142. **Woehrle, D.,** Phthalocyanines — a system with an unusual structure and properties. II, *Kontakte (Darmstadt),* 1, 24, 1986.

143. **Lever, A. B. P., Hempstead, M. R., Leznoff, C. C., Liu, W., Melnik, M., Nevin, W. A., and Seymour, P.,** Recent studies in phthalocyanine chemistry, *Pure Appl. Chem.,* 58(11), 1467, 1986.

Chapter 25

PC AND METAL (M) PC POLYMERS

I. SYNTHESIS

A. POLYIMIDES

Highly thermal stable M pc imide copolymers are made from Co, Cu, Ni, or Zn 4,4′4″,4‴-pc tetraamine, 4,4′ oxydianiline, and pyromellitic dianhydride. They exhibit good thermal and thermooxidative stability at >500°C in air and nitrogen. They show promise for thermally stable films, fibers, and coatings.[1]

They may also be prepared from 3,3′,4,4′- benzophenonetetracarboxylic dianhydride, Co, Cu, or Ni 4,4,′,4″,4‴-pc tetraamine, *p*-phenylenediamine, 4,4′-methylene dianiline, and 9,9-bis(4-aminophenyl)fluorene. They also have high char yields of >60% at 800°C.[2]

Thermally resistant imide copolymers are made by reacting Co, Cu, or Ni 4,4′,4″,4‴-pc tetraamine, a diamine such as *p*-phenylenediamine, and 1,2,4,5-benzenetetracarboxylic dianhydride.[3]

Poly (Cu pc) imide polymers with superior heat stability, fire resistance, and good toughness are made by polymerizing either of their precursors (Figure 1) with metal powder or salt. These thermosetting polymers should be useful to prepare thermally resistant molded objects and laminates.[4]

Polymerization of (Figure 2) with cuprous chloride at 245 to 305°C gives heat-stable Cu pc polymers with ester-imide linkages.[5]

A polyimide from Co pc tetracarbonyl chloride, 3,3′,4,4′-benzophenonetetracarboxylic dianhydride, and 4,4′-methylenedianiline decomposes at 530°C in nitrogen and 484°C in air.[6]

Tetranitrile polyimide oligomers are made from benzhydrol tetracarboxylates and are crosslinked by CN addition to make pcs, useful as adhesives and heat-resistant binders. A 4-aminophthalonitrile dimethyl 3,3′,4,4′–benzhydro tetracarboxylic-4,4′-methylenedianiline copolymer is impregnated in a glass fabric and cured at 300°C to give a heat-stable, hard, and green composite.[7]

A colored polyimide composition contains polyimides, solvents, and pigments. A mixture of reaction product of Cu pc, thionyl chloride, and $(C_2H_5)_2$ NCH_2 CH_2 NH_2,α-Cu pc mixture, the polyimide precursor Semicofine SP510, and dimethylformamide (DMF) has a viscosity of 770 cp.[8]

Polyimides containing Co, Cu, or Ni pc are made by condensation in *N*-methyl-2-pyrrolidone of Co, Cu, or Ni bis(3,4-dicarboxybenzoyl) pc dianhydride with 2,6-diamino pyridine and then thermally imidated. Room temperature conductivity is 10^{-12} to 10^{-9} 1/-ω-cm.[9]

Imide-linked bisphthalonitrile compounds are prepared by reacting a dianhydride with a dicyano aromatic diamine and polymerizing by heating with or without a metal salt to give a pc polymer with increased toughness and fire resistance appropriate for laminates, coatings, and moldings in aircraft and spacecraft.[10]

B. POLYMERS FROM TETRACYANOBENZENE

A planar Cu pc polymer is made by heating 1,2,4,5-tetracyanobenzene in the presence of copper or cuprous chloride in an evacuated sealed tube. The polymer collected on copper film has a ribbon structure.[11]

By the double source evaporation of copper and tetracyanobenzene, copper pc polymer films are made. Their conductivity at room temperature is about 0.0002 1/ω-cm.[12]

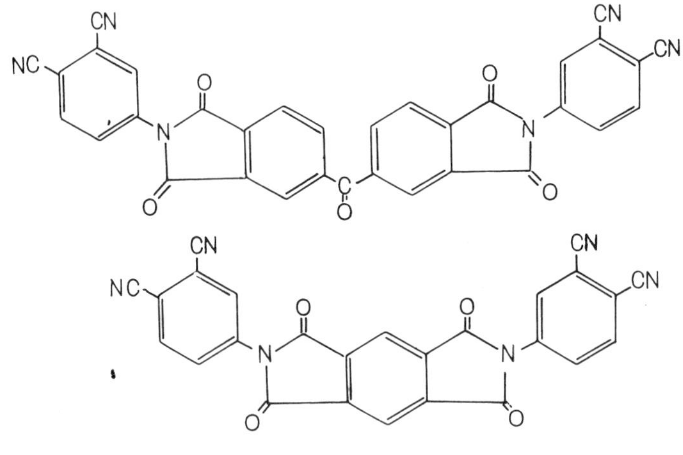

FIGURE 1. Precursors of thermosetting poly (Cu pc) imide polymers.

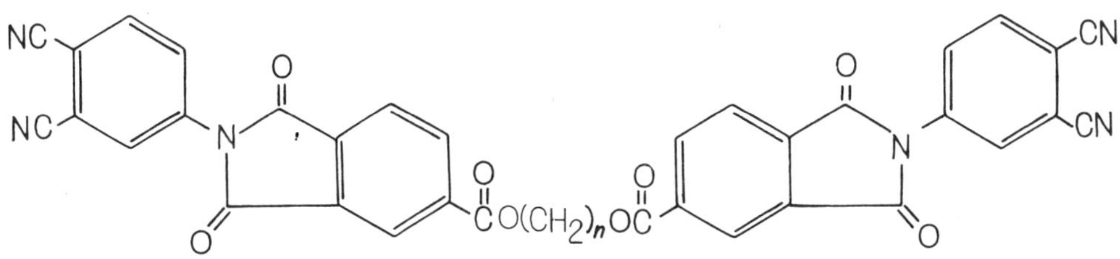

FIGURE 2. Precursor of thermosetting Cu pc polymer.

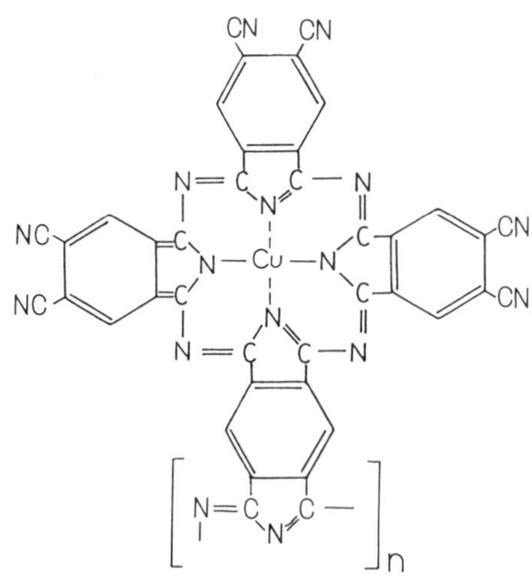

FIGURE 3. Poly (Cu 2,3,9,10,16,17,23,24-octacyano pc).

Reacting tetracyanobenzene and cuprous chloride at 145 to 160°C in *N* - methylpyrrolidone produces a dark blue and heat-stable polymer, as shown in Figure 3, with room temperature conductivity of 6.7×10^{-6} 1/ω-cm in air.[13] A dark blue polymer is prepared from tetracyanobenzene in 1-methyl-2-pyrrolidone at about 150°C.[14]

C. POLYMERS FROM PYROMELLITIC DIANHYDRIDE

Smooth green films of 2,3,9,10,16,17,23,24-pc octacarbonitrile and poly(*N*-vinylcarbazole) or 4,4′-oxydianiline-pyromellitic dianhydride copolymer are prepared on gold substrates by casting from solution. Mixed films of the octacarbonitrile and poly(*N*-vinylcarbazole) exhibit stable electrochromic redox behavior. For mixed films of the pc octacarbonitrile and oxydianiline-pyromellitic dianhydride copolymer, with lesser amount of octacarbonitrile, reduction and reoxidation appears to be diffusion controlled.[15]

D. $[M\ (PC)\ O]_n$

There is evidence of a topotactic mechanism in the solid-state polymerization of M pc $(OH)_2$ (M = Ge, Si, or Sn) into linear high polymers. They are wet spinnable and made into fibers.[16]

Poly Si pc $(OH)_2$ with degree of polymerization ≤ 100 is made by refluxing the Si pc $(OH)_2$ in tertiary amines in the presence of cadmium chloride catalyst.[17]

E. POLYMERS FROM TETRACARBOXY PCS

Co pc -$[CO_2CH_2CH_2O_2CC(CH_3)=CH_2]_4$ is prepared for polymerization in an electron beam or UV light from Co 3,3′,3″,3‴-tetracarboxy pc.[18]

Friedel-Crafts reaction of the acid chlorides of Co, Cu, Fe(III), and Ni 2,9,16,23–tetracarboxy pcs with the benzyl groups of poly (γ-benzyl L-glutamate) gives polypeptides with about 0.2 to 2.8 mol % pc rings covalently linked to the main chain.[19]

Cu pc tetracarboxylate salt, maleic anhydride, hydroquinone, and glycidyl methacrylate react to give a polyester-grafted Cu pc. Such a reaction can form a sheet polymer with 1 million ω-cm resistivity.[20]

Structurally uniform polymeric pcs containing imido end groups are made by reaction of 1,2,4,5-benzenetetracarboxylic acid derivatives with metal salts, urea, and a catalyst. The number of pc units in the polymers is >9.[21]

Polymerizable Co pc -$(COOCH_2CH_2OOCCH=CH_2)_4$ is made from Co pc $(COOH)^4$. The polymerization is effected with 4-vinylpyridine in THF at 60°C to give copolymer at 16.4% yield.[22]

Starting with Co pc $(COOH)_4$, Co pc $[COOCH_2CH_2OOCC(CH_3)=CH_2]_4$ is formed which is polymerized with 4-vinylpyridine in THF at 60°C to give copolymer at 8.4% yield.[23]

Tetrakis[2-(methacryloyloxy)ethyl] Co pc tetracarboxylate is polymerized with 9-vinylcarbazole to give a blue-green polymer.[24]

F. A VARIETY OF PC POLYMERS AND POLYMER-CURING AGENTS

Methods to prepare di hydrogen pc network polymers from oxygen-, sulfur-, and selenium- bridged bis(phthalonitrile) monomers are studied based on pc model compounds derived from phenoxy-, phenylthio-, and phenylseleno-phthalonitrile compounds.[25]

Soluble polyphosphazenes with covalently bound Cu pc side groups are prepared. The electrical conductivity for the iodine-doped trimers and high polymers as pellets and thin films are 0.0001 1/ω-cm for the trimers and 10^{-5} to 10^{-8} 1/ω-cm for the high polymers.[26]

A novel Cu pc polymer containing crown ether moieties is synthesized by polymerizing tetrabromodibenzo-18-crown-6 with CuCN.[27]

Cu 4,4′,4″,4‴-pc tetraamine is useful to cure epoxy resins and in carbon fiber-reinforced composites.[28] A silicon pc is a coupling agent for glass fiber reinforced epoxy resins.[29] A Si pc (OH) OSi $(CH_3)_2CH_2CH_2C_6H_4\ CH_2$ C1 silane coupling agent may form oxygen bridges to and glass fiber surfaces.[30]

Chloromethylated M pcs or di hydrogen pcs containing no or one sulfo group are reacted with 1-alkyl- or with 1,4-dialkylpiperazines in water to give ionene-type polymers with degree of polymerization (d.p.) ~3 to 18.[31]

Reaction of a bound dicarbonitrile with 4-phenoxy-1,2-benzenedicarbonitrile in the presence of zinc acetate yields covalently bound polymeric pc.[32]

Octasubstituted pcs with carboxylic acid, imido, and amido groups are prepared from

2,3,9,10,16,17,23,24-pc octacarbonitrile. They may be considered to be low molecular weight analogs of polymeric pcs with different end groups.[33]

Polymers are prepared by reacting 1,2,4,5-benzenetetracarbonitrile or 2, 3, 9, 10, 16, 17, 23, 24-di hydrogen pc octacarbonitrile with metal salts or metals. The d.p. is >3.[34]

G. PC DEODORIZING POLYMERS

Friedel-Crafts acylation of styrene-vinyl pyridine copolymer with Fe(III) pc tetracarbonyl chloride in nitrobenzene gives a product that catalyzes oxidation of ammonia, amines, hydrogen sulfide, and mercaptans efficiently even after 1 year of use.[35]

Co pc carboxylic acid, terephthalic acid, and ethylene glycol are polymerized to give a polyester that catalyzes oxidation of ammonia, amines, hydrogen sulfide, and mercaptans efficiently, also after 1 year of use.[36]

Fe pc carboxamide is saponified, converted to the acid chloride, and reacted with polystyrene in nitrobenzene in the presence of aluminum chloride to give an acylated polystyrene useful as a deodorant.[37]

A deodorant polymer is prepared by reacting polystyrene in a solution of octyl Fe pc tetracarboxylate. The product decomposes ammonia, amines, hydrogen sulfide, and mercaptans.[38]

Fe(III) pc tetracarboxylic acid, polyallylamine hydrochloride, dipropylene glycol glycidyl ether, and water, mixed and heated, form a film with resistivity 100,000 ω-cm. The product exhibits good antistatic and deodorant properties.[39] Fe(III) pc tetracarboxylic acid, polyallylamine hydrochloride, dipropylene glycol, diglycidyl ether, and water, similarly mixed and heated, show similar good properties.[40]

H. REVIEW

A review with no references includes pc polymers with emphasis on synthesis problems.[41] A review with 38 references relates to the preparation and catalytic properties of pc polymers.[42]

A review with 62 references to June 1985 treats pc polymers emphasizing their preparation, structure, and properties.[43]

A review with 94 references is concerned with the physicochemical properties and coordination compounds of polymeric pcs.[44]

A review with 69 references covers M pc sheet polymers and polymers containing pendant metal pc groups. Their uses in batteries, solar cells, microlithography, and sensors are pointed out.[45]

I. MOLECULAR WEIGHTS

Fe and Co pc polymers are filtered with Sephadex gels. Their molecular weights are, for Fe pc polymer, ~20,000, and for Co pc polymer, ~35,000.[46]

Adding siloxyvinyl pc during radical polymerization of styrene decreases the molecular weight of polystyrene.[47] However, adding Si pc $-[OSi(CH_3)_2OCH_2C(CH_3){=}CH_2]_2$ increases the molecular weight. The styrene polymers formed are highly colored.[48]

J. PCS IN AND PCS INCORPORATED POLYMERICALLY INTO MEMBRANES

Ultrathin membranes, suitable to use in sensors and secondary cell electrodes, are prepared using Fe 4,4′,4″,4‴-tetrachlorocarbonyl pc and styrene-vinylpyridine copolymer. The Fe pc derivative is covalently bonded to the polymer in a water-insoluble solvent.[49]

A semipermeable membrane is based on 1:1 styrene-2-vinylpyridine polymer covalently bonded to Fe 4,4′,4″,4‴-tetracarboxamido pc.[50]

A 50-μ polyoxyphenylene film is coated with a 500-Å layer of Fe pc to give a membrane with an oxygen permeation coefficient of 2.71×10^{-9} ml-cm/cm^2/s/cm Hg and oxygen/nitrogen separation factor of 8.0.[51]

A membrane, useful to separate oxygen from air, is based on casting 10 parts poly(trimethylsilylpropyne) and 1 part Fe tetracarboxylic acid into a dry 35-μ film with permeation coefficient 2.8×10^{-8} and oxygen/nitrogen separation factor 3.4.[52]

A porous membrane suitable to separate dispersed particles from fluids by ultrafiltration is prepared by plasma treatment followed by chemical modification which may include treatment with Cu (aminosulfonyl)[[[3-(dodecyloxy)propyl]amino]sulfonyl] pc disulfonic acid.[53]

A composite membrane for gas enrichment contains organometallic gas carriers dispersed in interpenetrating networks of nitrogen polymers such as acrylonitrile-vinylpyridine copolymer and Co pc cyanic acid derivatives.[54]

REFERENCES

1. **Achar, B. N., Fohlen, G. M., and Parker, J. A.,** Novel high-temperature-resistant metal phthalocyanine imide copolymers, *J. Polym. Sci., Polym. Chem. Ed.*, 21(11), 3063, 1983.
2. **Achar, B. N., Fohlen, G. M., and Parker, J. A.,** Studies on heat-resistant poly(metal phthalocyanine)imide copolymers, *J. Polym. Sci., Polym. Chem. Ed.*, 22(2), 319, 1984.
3. **Achar, B. N., Fohlen, G. M., and Parker, J. A.,** Heat-resistant poly(metal phthalocyanine)imide copolymers using benzenetetracarboxylic dianhydride, *J. Polym. Sci., Polym. Chem. Ed.*, 23(3), 801, 1985.
4. **Achar, B. N., Fohlen, G. M., and Parker, J. A.,** Studies on heat-resistant thermosetting phthalocyanine polymers, *J. Appl. Polym. Sci.*, 29(1), 353, 1984.
5. **Achar, B. N., Fohlen, G. M., and Parker, J. A.,** New thermosetting phthalocyanine polymers, *J. Polym. Mater.*, 2(1), 16, 1985.
6. **Sugiura, Y., Nishimura, S., and Sano, S. (to Nippon Shokubai Kagaku Kogyo Co., Ltd.),** Japan Patent 86,231,029, 1986.
7. **Malinge, J., Rabilloud, G., and Sillion, B. (to Centre d'Étude des Materiaux Organiques pour Technologies Avancées),** Europe Patent Application 170,590, 1986.
8. **Tanaka, T., Hikosaka, M., Sawamura, M., Hoshi, H., and Sugiura, T. (to Toyo Ink Manufacturing Co., Ltd.),** Japan Patent 85,235,860, 1985.
9. **Shirai, H., Takemae, Y., Kobayashi, K., Kondo, Y., Hirabaru, O., and Hojo, N.,** Functional metal-porphyrazine derivatives and their polymers. XII. Synthesis and properties of polyimides containing a metal-phthalocyanine ring and a pyridyl group, *Makromol. Chem.*, 185(7), 1395, 1984.
10. **Achar, B. N., Fohlen, G. M., and Parker, J. A.,** U.S. Patent Application 754,362, 1986.
11. **Fryer, J. R. and Kinnaird, T. A.,** The formation and structure of planar polymeric copper phthalocyanine, *Conf. Ser.-Inst. Phys.*, 68, 27, 1984.
12. **Yudasaka, M., Nakanishi, K., Hara, T., Tanaka, M., and Kurita, S.,** Formation of copper phthalocyanine polymer films by the double source evaporation of tetracyanonbenzene and copper, *Jpn. J. Appl. Phys., Part 2*, 24(11), L887, 1985.
13. **Lin, J. W. P. and Dudek, L. P.,** Poly(copper 2,3,9,10,16,17,23,24-octacyanophthalocyanine) and its properties, *Polym. Prepr. (Am. Chem. Soc., Div. Polym. Chem.)*, 25(2), 236, 1984.
14. **Lin, J. W. P. and Dudek, L. P.,** Preparation and characterization of poly(copper 2,3,9,10,16,17,23,24-octacyanophthalocyanine), *J. Polym. Sci., Polym. Chem. Ed.*, 23(5), 1579, 1985.
15. **Woehrle, D., Kaune, H., Schumann, B., and Jaeger, N.,** Polymeric phthalocyanines and their precursors. XII. Reversible reduction and reoxidation of thin films of octacyanophthalocyanine in polymer matrixes, *Makromol. Chem.*, 187(12), 2947, 1986.
16. **Carr, S. H., Xiao, Z., Inabe, T., Marks, T. J.,** "Metal-like" metallomacrocyclic polymers. Microstructure and morphology of undoped and doped $[M(phthalocyaninato)O]_n$ macromolecules, M = silicon, germanium, and tin, *Polym. Mater. Sci. Eng.*, 49, 94, 1983.
17. **Orthmann, E. and Wegner, G.,** Catalysis of the polycondensation of dihydroxysiliconphthalocyanine, *Makromol. Chem., Rapid Commun.*, 7(5), 243, 1986.
18. **Shirai, H. and Ito, H.,** Japan Patent 85,184,083, 1985.
19. **Hanabusa, K., Kobayashi, C., Koyama, T., Masuda, E., Shirai, H., Kondo, Y., Takemoto, K., Iizuka, E., and Hojo, N.,** Functional metal-porphyrazine derivatives and their polymers. XV. Synthesis and properties of poly (γ-benzyl L-glutamate)s containing covalently bound metal phthalocyanine moieties in the side chain, *Makromol. Chem.*, 187(4), 753, 1986.

20. **Shirasaka, H. and Shirai, H. (to Hokushin K. K.),** Japan Patent 86,159,456, 1986.
21. **Woehrle, D. and Preussner, E.,** Polymeric phthalocyanines and their precursors. VII. Synthesis and analytical characterization of polymers from 1,2,4,5-benzenetetracarboxylic acid derivatives, *Makromol. Chem.*, 186(11), 2189, 1985.
22. **Kondo, S., Ito, H., Hanabusa, K., Shirai, H., and Hojo, N. (to TDK Corp.),** Japan Patent 87,53,990, 1987.
23. **Kondo, S., Ito, H., Hanabusa, K., Shirai, H., and Hojo, N. (to TDK Corp.),** Japan Patent 87,53,989, 1987.
24. **Itoh, H., Kondo, S., Masuda, E., Hanabusa, K., Shirai, H., and Nojo, N.,** Functional metallomacrocycle derivatives and their polymers. XVI. Synthesis of a vinyl monomer containing a covalently bound cobaltphthalocyanine moiety and its copolymerization with 9-vinylcarbazole, *Makromol. Chem., Rapid Commun.*, 7(9), 585, 1986.
25. **Snow, A. W., Griffith, J. R., and Marullo, N. P.,** Synthesis and characterization of heteroatom-bridged metal-free phthalocyanine network polymers and model compounds, *Macromolecules*, 17(8), 1614, 1984.
26. **Allcock, H. R. and Neenan, T. X.,** Synthesis of polyphosphazenes bearing covalently linked copper phthalocyanine units, *Macromolecules*, 19(6), 1495, 1986.
27. **Ahsen, V., Yilmazer, E., Bekaroglu, O., and Gul, A.,** Preparation of a novel polymeric copper phthalocyanine containing crown ether moieties and its alkali metal binding property, *Makromol. Chem., Rapid Commun.*, 8(5), 243, 1987.
28. **Achar, B. N., Fohlen, G. M., and Parker, J. A. (to U.S. National Aeronautics and Space Administration),** U.S. Patent Application 598, 777, 1985.
29. **Shih, G. C. and Ebert, L. J.,** Interface strength effects on the compressive-flexural/shear failure mode transition of composites subjected to four-point bending, *J. Mater. Sci.*, 21(11), 3957, 1986.
30. **Sounik, D. F. and Kenney, M. E.,** New potential silane coupling agents, *Polym. Compos.*, 6(3), 151, 1985.
31. **Anon.** Phthalocyanine products, *Res. Discl.*, 242, 294, 1984.
32. **Woehrle, D. and Krawczyk, G.,** Polymeric bound porphyrins and their precursors. II. Solid phase synthesis of a monosubstituted phthalocyanine, *Polym. Bull. (Berlin)*, 15(3), 193, 1986.
33. **Woehrle, D. and Huendorf, U.,** Polymeric phthalocyanines and their precursors. VI. Synthesis and analytical characterization of some octasubstituted phthalocyanines, *Makromol. Chem.*, 186(11), 2177, 1985.
34. **Woehrle, D., Marose, U., and Knoop, R.,** Polymeric phthalocyanines and their precursors. VIII. Synthesis and analytical characterization of polymers from 1,2,4,5-benzenetetracarbonitrile, *Makromol. Chem.*, 186(11), 2209, 1985.
35. **Fukamachi, K. and Shirai, H. (to Earth Clean K. K.),** Japan Patent 86,258,806, 1986.
36. **Fukamachi, K. and Shirai, H. (to Earth Clean K. K.),** Japan Patent 86,258,815, 1986.
37. **Shirai, H., Kakegawa, E., Ito, Y., Kamimura, M., Yokozeki, T., Shimizu, K., Sakaguchi, I., and Hojo, N. (to Nippon Carbide Industries Co., Inc., Nissetsu K. K., and Earth Clean K. K.),** Japan Patent 87,104,807, 1987.
38. **Fukamachi, K. and Shirai, H. (to Earth Clean K. K.),** Japan Patent 87,01,752, 1987.
39. **Shirai, H., Sugiura, K., Inamura, S., Shirasaka, H., and Ishii, A. (to Hokushin Kogyo, Inc., and Earth Clean K. K.),** Japan Patent 87,101,602, 1987.
40. **Shirai, H., Sugiura, K., Inamura, S., Shirasaka, H., and Ishii, A. (to Hokushin Kogyo, Inc., and Earth Clean K. K.),** Japan Patent 87,101,603, 1987.
41. **Allcock, H. R.,** Inorganic macromolecules, *Chem. Eng. News*, 63(11), 22, 1985.
42. **Hara, T.,** Catalytic activities of metal phthalocyanines. Phthalocyanines as part of a polymer, *Kagaku Kogyo*, 37(5), 390, 1986.
43. **Snow, A. and Griffith, J.,** Phthalocyanine polymers, *Encycl. Polym. Sci. Eng.*, 11, 212, 1987.
44. **Berezin, B. D. and Lomova, T. N.,** Structure and physicochemical properties of coordination compounds of polymeric phthalocyanines, *Izv. Vyssh. Uchebn. Zaved., Khim. Khim. Tekhnol.*, 30(5), 3, 1987.
45. **Shira, H.,** Synthesis of metal phthalocyanine polymers and their function, *Hyomen*, 25(7), 422, 1987.
46. **Johansson, L. Y. and Larsson, R.,** Molecular weight determination of polymeric iron and cobalt phthalocyanines with the method of gel filtration, *Chem. Scr.*, 23(1), 48, 1984.
47. **Bernal-Castillo, J.,** Effect of functionalized phthalocyanine pigments on molecular weight of polystyrene, *Rev. Colomb. Quim.*, 13(2), 5, 1984.
48. **Bernal-Castillo, J.,** Effect of functional phthalocyanine dyes on the molecular weight of polystyrene, *Rev. Ion*, 8(1), 11, 1984.
49. **Imoto, T., Kato, K., and Shirai, H. (to NOK Corp.),** Japan Patent 87,13,412, 1987.
50. **Imoto, T. (to Nippon Oil Seal Industry Co., Ltd.),** Japan Patent 85,172,305, 1985.
51. **Saito, Y., Seki, M., and Asakawa, S. (to Matsushita Electric Industrial Co., Ltd.),** Japan Patent 87,110,730, 1987.
52. **Asakawa, S., Saito, Y., and Kawahito, M. (to Matsushita Electric Industrial Co., Ltd.),** Japan Patent 87,33,526, 1987.

53. **Wolff, J., Koll, J., Ellinghorst, G., and Steinhauser, H. (to Bayer AG),** German Patent 3,509,068, 1986.
54. **Zupancic, J. J. and Swedo, R. J. (to Allied Corp.),** U.S. Patent 4,627,859, 1986.

Chapter 26

PCS IN BIOLOGY

I. TREATMENT OF MUTAGENS

A. METHODS

A method to remove planar mutagens with ≥3 aromatic rings from solution involves absorption into a copper containing phthalocyanine (pc) compound with a reactive terminal group and coupled to an organic solid. The recovery of hydrogen-3 labeled *Trp-P2* from human urine is 91.6% by turquoise blue G on cotton.[1]

A cellulose acetate membrane with immobilized Cu pc is placed on an oxygen electrode to detect mutagens. Cu pc bound to *benzo[a]pyrene* catalyzes hydrogen peroxide decomposition.[2]

A Cu pc membrane biomimetric electrode can be used for rapid preliminary detection of polycyclic mutagens in <30 min. The electrode has selectivity detected polycyclic mutagens such as amino acids pyrolysis products.[3]

Data of *Salmonella* mutagenicity results on 270 coded chemicals, including a Cu pc derivative, relate to >300 tests performed by laboratories under a national toxicological program.[4]

A tobacco smoke filter having a high mutagen- and tar-removing capacity contains pc derivatives. For example, nylon fibers are impregnated with acetic acid, a solution of Glauber salt, and C. I. direct blue 85 in water.[5]

Blue cotton containing a Cu pc trisulfonate is used to detect mutagenicity in human feces excreted after eating fried ground beef.[6]

B. REVIEW

A review with 22 references covers the use of blue cotton with Cu[[[4-[(2-hydroxyethyl)sulfonyl]phenyl]amino]sulfonyl] pc trisulfonate for environmental mutagen removal.[7]

II. PCS AS DETOXICANTS

Buildings or clothing can be coated with Co, Fe, or Ni tetrasulfonated or tetraamino pc to decontaminate or to prevent contaminations of mustard gas by converting the gas to divinylsulfide. Cotetrasulfonated pc is used in military camouflage paint with alkyl paint pigments.[8]

Water-soluble M pcs can be used against cyanide intoxication. The rate of cyanide binding depends on the metal and its oxidation state. The rate constant of Co tetracarboxy pc is about 20 times the rate of binding of cyanide to silated cytochrome oxidase.[9]

III. TOXICITY INDUCED BY PCS

An i.p. administration of Cu pc at 3 g/kg to rats severely affects the liver functions and decreases diuresis. Without its water-soluble salt, Cu pc is a class III moderately toxic substance and its recommended maximum permissible concentration in the working zone is 5 mg/m^3.[10]

Cu pc trisulfonic acid injected into incubated chick embryos induces a highly reproducible caudal malformative syndrome in terms of taillessness and limb anomalies.[11]

The very low solubility values of eleven typical organic pigments including Cu pc indicate that, despite the very high predicted bioaccumulation factors, such pigments cannot build up concentrations in lipids, and thus in fish. It should not be necessary to perform a fish bioaccumulation test to assess the bioaccumulation potential of such compounds.[12]

Four pc dyes, reactive turquoise K, reactive turquoise 2ZT, and Cu pc di- and trisulfonic acids are studied *in vivo* and *in vitro* via their absorption spectra. Following their introduction into the stomach, the dyes are not noticed in the blood but a bluish color of the digestive tract mucosa is observed.[13]

Mechanisms of the oxidation of 1-mmol solutions of Cu pc disulfonic acid in water-acetone solutions containing ferrous sulfate, chelatron III, ascorbic acid, and phosphate are studied. The oxidation mechanisms are discussed vis-a-vis the detoxication and biodegradation of the dyes and wastewater purification from the dye industry.[14]

IV. FORENSIC APPLICATIONS

Automobile paint pigments including Cu pc are identified by IR spectra and matched with the spectra of known pigments. The method may be of use to compare paint samples from accident cases involving hit-and-run situations.[15]

Reflectance spectra of TLC separated paint pigments and dyes are used to identify objects of forensic interest. Pigments of only slight solubility, the pcs, are discussed, among paints and dyes, together with casework applications to paint pigment analysis.[16]

V. PC AND M PC COLOR ADDITIVES

Cu pc is approved under the Food, Drug and Cosmetic Act to color nonabsorbable polybutester sutures for general and ophthalmic surgery.[17]

The tolerance for organic chlorine in Cu pc to color contact lenses and sutures is increased to 0.5% under the Food, Drug and Cosmetic Act.[18]

The prohibition of migration of Cu pc from nonabsorbable sutures to surrounding tissue is deleted under the Food, Drug and Cosmetic Act.[19]

Cationic polymer suspensions of Cu pc are suitable for hair coloring.[20]

Cu pc is a participant as a coloring material in contact lenses.[21]

A dental visualization fluid to monitor occlusion and prosthesis performance contains heliogen blue (Cu pc). The fluid is applied before insertion of prostheses to visualize contact points.[22]

Cu pc is used in dental resin composites to identify extent of curing.[23]

A clear dentrifice containing pigment green 7 (Cu polychloro pc with 15 to 16 atoms per molecule) is dispersed uniformly throughout the dentrifice; the green color is intense and stable with storage.[24]

Cu pc disulfonic acid reaction products with protein hydrolyzates are useful to color cosmetics and hair preparations.[25]

A cationic polymer suspension of Cu pc is suitable to color hair.[26]

A Cu pc pyridinium derivative is suitable as a hair dye.[27]

A Cu methanaminium derivative is determined on a solid biopolymer by laser photoacoustic microscopy.[28]

A Cu benzene methanaminium pc derivative is applied to basophil selective staining.[29]

VI. PCS IN IMMUNOCHEMISTRY

Fe(III) 2,9,16,23-tetrakis(chlorocarbonyl) pc shows a pronounced catalytic activity in the chemiluminescence reaction between luminol and hydrogen peroxide, and albumin is labeled with it. Albumin in the range 50 ng to 50 μg per tube is determined by measuring the chemiluminescence intensity in a competitive immunoassay by using glass beads with an immobilized antibody.[30]

An Al pc homopolymer is used as a semiconductive element in an analytical device for immunoassay.[31]

H_2 pc 1,8,15,22-tetracarboxylic acid metal complexes are studied as labeling reagent catalysts for the chemiluminescence immunoassay of proteins.[32]

Fe(III) 1,8,15,22-tetrakis(chloroformyl) pc is used as labeling agent in an immunoassay known as chemiluminescence complex catalyst immunoassay.[33]

VII. PCS IN FOOD PRESERVATION

Fresh strawberries are hermetically stored in a container along with pc blue sealed in a gas permeable packet for 10 d, with 10 g pc blue, and 30 g L-ascorbic acid in the packet, at 0°C. No changes in the organoleptic qualities of the strawberries are observed.[34]

Active carbon particles are immersed in 15% Co pc tetracarboxylic acid tetrasodium salt, mixed with pulp, binders, and fillers, and made into sheets to preserve apples.[35]

VIII. PCS IN ANALYSIS

Methods to identify 20 inorganic pigments and 20 organic dyes, including H_2 pc and Cu pc, are detailed.[36]

DNA-dependent RNA polymerase purification is accomplished with Cu 2,3-[[(4-chloro-1,3,5-triazin-2-yl)azo]sulfonyl] pc 2,9,16-trisulfonic acid reaction products with Sepharose.[37]

An iron pc is prepared for use in biochemical analysis.[38]

An assay uses amperometric detection at carbon paste electrodes containing Co pc to determine cysteine and glutathione in blood and plasma.[39]

A carbon paste electrode with surface bound Co pc is applied to detect α-ketoacids and oxalic acid in blood plasma and urine.[40]

A carbon paste electrode incorporates Fe pc for the quantitative determination of hydrogen peroxide down to 0.00005 M. This electrode is of interest because of its catalase-like activity.[41]

IX. ANTIVIRAL ACTIVITY

The sulfonated scandium di pc, $Sc[(C_{32}H_{14}N_8)(NH_4SO_3)_2]_2NH_4$ possesses antiviral activity.[42]

X. STAINING TISSUES

Almost 100 acidic, basic, and nonionic dyes including Cu pc disulfonic acid are screened for fluorescent staining of Carnoy fixed rat tissues. The fluorescence/nonfluorescence of a dye is predictable using a conjugated bond number cut-off value.[43]

Ghosts prepared from human erythrocytes stained with direct lightfast turquoise M (Cu pc disulfonic acid) exhibit a morphology similar to those from unstained cells.[44]

Sorption of direct lightfast turquoise M by erythrocytes or their ghosts is studied after processing with trypsin or after incubating them at 37°C in an isotonic phosphate buffer. The absorption spectra indicate that the trypsin treatment increases the monomer/dimer ratio of the whole erythrocytes by 19%.[45]

Pc dyes tend to aggregate in aqueous solutions. Detergents disaggregate them, and the detergents tend to form micelles. The use of detergents is a satisfactory test system for the study of staining interactions between dyes and cell components.[46]

A Cu pc methanaminium derivative (alcian blue GS) stains rat rectum and eyeball sections. Photoacoustic microscopy is applied to the determination of the dye to determine acidic mucopolysaccharide in the rectum and eyeball.[47]

XI. PCS WITH POTENTIAL FOR CANCER THERAPY

The photosensitization of Chinese hamster fibroblast V79-B310H cells by A1 pc C1 is investigated. Increasing light exposure decreases cell survival and increasing A1 pc C1 also decreases cell survival. A1 pc C1 appears to be an efficient photosensitizer for mammalian cells and a possible candidate to replace hematoporphyrin derivatives in cancer phototherapy.[48]

H_2 tetrasulfo-, tetraamino-, and tetranitro pcs are labeled with γ-emitting technetium-99m. They are studied in female Fischer 3441 CRBL rats bearing mammary adenocarcinomas and in healthy Dutch rabbits. All three technetium complexes show *in vivo* stability and nonsignificant activity levels in target organs such as salivary gland, thyroid, and stomach.[49]

Pc complexes of tetrasulfo pc with Tc-99m, Ga-69, and Ga-67 are prepared. In rabbits and tumor-bearing rats, most of the radioactivity accumulates in the kidneys, liver, ovaries, adrenals, and spleen. They are *in vivo* stable.[50]

Tetrasulfo pc and certain among its metal chelates are tested for their ability to photo-inactivate V-79 Chinese hamster cells. The cerium complex is about 5 times more effective for cell killing than the H_2 tetrasulfo pc and 20 times more efficient compared with hema toporphyrin. The potential of sulfonated pcs as photosensitizers for photodynamic cancer therapy (PCT) is discussed.[51]

In PCT of EMT-6 mammary tumors in mice with H_2, A1, Ce, and Ga sulfo pcs, the cerium complex is the most active sensitizer required to induce tumor necrosis and cure but also shows the highest phototoxicity towards healthy skin. These results suggest that sulfonated pcs may offer a new alternative to PCT of light accessible meoplasms.[52]

Pcs are a class of mammalian cell photosensitizers which may be useful in PCT for cancer. The role of molecular oxygen in this PCT effect is discussed for the example of Chinese hamster cells in culture and exposed to white light and different oxygen concentrations.[53]

A1 pc C1 sulfonate in cultured Chinese hamster V79-B310H fibroblasts is compared with those of the unsubstituted water-insoluble aluminum pc.[54]

Using Chinese hamster cells exposed to photoexcited A1 pc C1 tetrasulfonate in PBS solution, cytotoxicity diminishes as the fluency rate is reduced.[55]

Al pc Cl sensitizes the cultured Chinese hamster fibroblast cell line V79-B310H upon exposure to white fluorescent light. Improved spectral properties, low toxicity, and selective retention by the neoplasms make pcs promising candidates for use in PCT.[56]

H_2 pc is included in a comparative study of 28 porphyrins and their abilities to localize in mammary mouse carcinomas are evaluated.[57,58]

Zn pc tetrasulfonate shows high tumor localization and retention in polyvinylalcohol sponge-implanted Swiss-Webster mice.[59]

The PCT of transplantable bladder tumors in rodents after pretreatment with A1 tetrasulfo pc C1 indicate that this M pc may be a useful new agent for PCT.[60]

Since A1 pc C1 photosensitizes cell killing in culture and tumor destruction *in vivo*, and is potentially useful in PCT, its properties as a genotoxic agent are evaluated.[61]

A comparison of nine dyes including A1 pc tetrasulfonate and A1 pc injected into mice with C3H mammary carcinomas is made in terms of their potential as sensitizers for photodynamic cancer therapy.[62]

H_2 and M pc dyes are screened for their ability to photosensitize murine embryonic fibroblasts or fibrosarcoma cells. Mg, Zn, and Zr pc dyes are highly toxic to cells following exposure to red or white light, but A1 pc C1, A1 tetrasulfonated pc, and Sn pc $C1_2$ exhibit phototoxicity only under red light, but not white light, indicating that they may be useful in PCT.[63]

The photosensitizing properties of pcs are compared with hematoporphyrins and are discussed with respect to PCT.[64]

A1 pc C1 sulfonate is studied as a photosensitizer for colonic PCT.[65]

The photosensitizer A1 pc C1 is labeled with tritium and it is shown that A1 pc C1 complexes to some extent with bovine serum albumin.[66]

Factors that affect the photokilling of cultured Chinese hamster cells by A1 pc C1 are studied.[67]

A1 pc C1 and Zn pc are found to be the only active cytotoxic dyes for cultured Chinese hamster cells.[68]

Ga tetrasulfo pc is not phototoxic to cells but Ga pc $(\text{sulfonate})_{2.7}$ is very phototoxic and induces DNA strand breakage.[69]

A review with 54 references discusses pcs as photosensitizers in biological systems and for the photodynamic therapy of tumors.[70]

XII. PC MODIFIED AND ARTIFICIAL CYTOCHROMES

Artificial cytochromesc are formed with VO and Zn tetrasulfonated pc in place of heme.[71]

Studies of Co and Fe tetrasulfonated pc interaction with cytochrome c fragments show complex formation only with the heme fragment apopeptides. The M pc appears to be coordinated axially with histidine-18 and methionine-80 of apocytochrome c.[72]

Fe-tetrasulfonated pc is also incorporated into cytochrome b of the cytochrome bc_1 complex.[73]

XIII. PCS AS HEME-LIKE MOLECULES

Isolated α and β chains of human hemoglobin by Co tetrasulfonated pc and hybrid pc-heme intermediates are studied. Incorporating Co-pc tetrasulfonated pc into α- and β-globins stabilizes the protein structure, shown by a marked increase in its helical content.[74]

Human serum albumin is modified by Co- and Fe-tetrasulfonated pcs. There are two high-affinity sites toward pc on albumin. The strongest one is located in fragment M and that of lower affinity in fragment C. Co pc displaces heme from the heme-M fragment, suggesting the same binding site for heme and pc.[75]

The electrooptical behavior of heme-like Fe tetrasulfonated pc adsorbed on a nerve membrane is studied. The changes in Raman spectra signals on Fe tetrasulfonated pc on the interface of a silver electrode and the nerve membrane vs. the interfacial potential provide a way to optically monitor electrical excitability in nerves.[76]

A monooxygenetically active insoluble carrier has the structure A-B-C. A = a support material, B = Hb, a hemin derivative or a pc derivative, and C = a dehydrogenase.[77]

XIV. PROTEINS AND AMINO ACIDS

The equilibrium distribution of direct lightfast turquoise (Cu pc disulfonic acid), bromothymol blue, bromophenol blue, and phenol red, during contraction and denaturation of actomyosin filaments, between actomyosin and medium, decreases with their hydrophobicity.[78]

Reconstitution of apomyoglobin with Fe tetracarboxy pc and Fe-tetrasulfonated pc gives proteins containing proximal and distal histidine bound to iron.[79]

The effects of poly-L-lysine and polyglycols on the electronic spectral properties of Co tetracarboxy pc are studied to determine the effect of the polymers on the molecular stacking properties of Co tetracarboxy pc.[80]

Sulfonated pc and a series of its metal chelates with red light irradiation lead to degradation of L-tryptophan in oxygenated aqueous solution.[81]

The photooxidation of L-tryptophan by sulfonated Ga pcs is studied.[82]

Zn tetrasulfo pc in an aqueous buffer effectively sensitizes the photooxidation of amino acids and nucleic acid bases and sensitizes the photodynamic inactivation of lysozyme.[83]

XV. ENZYMES

Co pc 2,9,16,23-tetracarboxylic acid reaction products with Amberlite IRA 900 anion exchange resin exhibit peroxidase-like accelerative activity in hydrogen peroxide determination in clinical analysis.[84]

XVI. RETINAL

The isomerization of retinal isomers is studied in the presence of Zn tetra-4-tert-butyl pc in comparison with other porphyrin photosensitizers.[85]

XVII. MICROORGANISMS

The locomotion rate of *Mayorella penardi* in response to a Cu pc methanaminium derivative is studied.[86]

X-ray absorption near edge structure (XANES) of copper compounds with copper in 1+, 2+, and 3+ states is studied. Extended X-ray absorption fine structure (EXAFS) is used to determine bond distance and coordination numbers in model copper compounds. Using XANES and EXAFS, the structure of the copper complex formed by *Pseudomonas aeruginosa* is square planar. EXAFS and XANES are studied vis-a-vis Cu pc.[87]

Finely divided Cu pc is dispersed into a hydrocarbon containing fermentation medium.[88]

XVIII. EPOXIDIZING AGENTS

Transition M pcs are used as epoxidizing agents as illustrated by treating a solution of 3,3-(2,2-dimethyltrimethylenedioxy)-5(10),9(11)-estradien-17-one in methyl cyanide with iodosobenzene and Fe pcs to give 6.7% 3,3-(2,2-dimethyltrimethylenedioxy)-5β,10β-epoxy-9(11)-estren-17-one and 73.7% 5α,10α-epoxy isomer.[89]

XIX. NEUROLOGICAL AND BIOLOGICAL MOLECULAR ELECTROOPTICAL DEVICES AND METHOD

Macrocyclic molecules such as Fe pc are linked to biological or neurological substrates. The state of the substrate can be assessed by detecting the scattered light from directing electromagnetic radiation to the macrocyclic molecule. The procedure is illustrated by detecting changes in macrocyclic compound linked neural cells of a gastropod.[90]

XX. PORPHYRINS

A face-to-face porphyrin-pc dimer bound by electrostatic bonds is obtained from a mixture of a negatively charged metal 4,4′,4″,4‴-tetrasulfonated pc and positively charged metal 5,10,15,20-tetra-(*p*-*N*-methylpyridinium) porphyrin. Treating Zn porphyrin with Cu pc yields a porphyrin-pc complex.[91]

Covalently linked pc-porphyrin dimers (Figure 1) are prepared and studied spectroscopically. There is very efficient intramolecular energy transfer from the Zn porphyrin to the Zn pc moiety from light irradiation.[92]

Cu 2,9,16,23-tetrakis(1,1-dimethylethyl) pc is referenced with respect to the synthesis of tetrabenzoporphyrins.[93]

XXI. CHLOROPHYLL AND PHOTOSYNTHESIS

The spectral and fluorescence parameters of polymer gels containing covalently bound

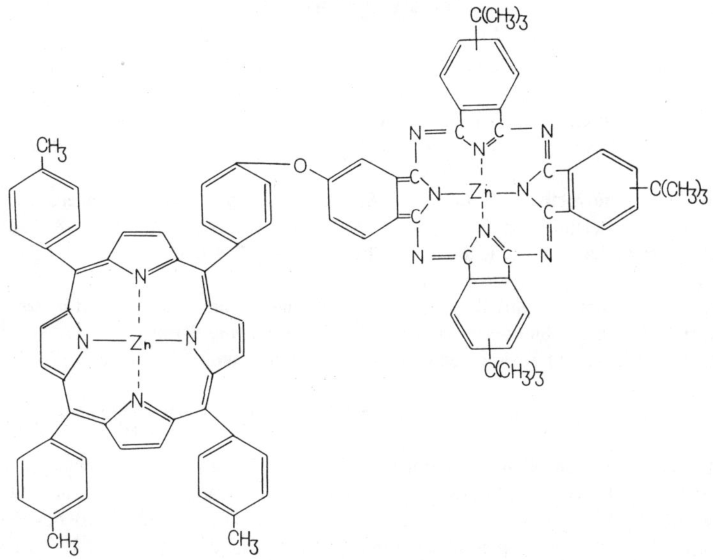

FIGURE 1. A Zn pc-porphyrin dimer.

Zn pc as models of light-harvesting systems in photosynthesis are studied. The gels are usable to model light-harvesting antennae of native photosynthetic systems.[94]

The nonuniform chlorophyll distribution in a protein matrix is modeled. Since the model pigment Zn tetra-*tert*-butyl pc forms coordination bonds with 1-vinyl-1,2,4-triazole, it concentrates in the 1-vinyl-1,2,4-triazole-enriched zones and the local concentration increases with decreasing concentration of 1-vinyl-1,2,4-triazole. Pcs in copolymers based on 1-vinyl-1,2,4-triazole are models of light-harvesting antennae of photosynthetic systems.[95]

Natural pheophorphyrin, chlorophyllide, H_2 pc, and chlorophyll pigments from photosynthetic bacteria and plants are deposited as thin films and the dark DC conductivity is measured in vacuum and in the presence of 1000 ppm nitrogen tetroxide in nitrogen. Photochlorophyllide a has an especially rapid and rapidly reversible response to nitrogen tetroxide.[96]

Zn pc tetrasulfonate is one of four dyes evaluated for their suppression of the rate of oxygen production by the algae *Lyngbya majescula, Anabaena flos-aquae,* and *Oedegonium foveolatum*, which are nuisance plants in several southern states. Zn pc tetrasulfonate and erythrosin are the less effective.[97]

XXII. COATING IMPLANTS

The physical and chemical properties of pcs and their application in coating implants are discussed. The properties of evaporated M pc films and the results *in vitro* and with implants in guinea pigs are covered.[98]

XXIII. LIPOSOMES

Spectroscopic studies on Zn pc show that Zn pc bound to liposomes has potential use for photosensitization studies and for phototherapeutic use *in vivo*.[99]

REFERENCES

1. **Hayatsu, H. and Nakano, M. (Sumitomo Chemical Co., Ltd.),** Europe Patent Application 90,610, 1983.
2. **Matsunaga, T.,** Japan Patent 85,13,255, 1985.
3. **Tomoda, R., Kusunoki, S., Nakashima, K., and Matsunaga, T.,** Use of a copper phthalocyanine membrane electrode for rapid preliminary detection of polycyclic mutagens, *Mutat. Res.*, 164(4), 203, 1986.
4. **Mortelmans, K., Haworth, S., Lawlor, T., Speck, W., Tainer, B., and Zeiger, E.,** Salmonella mutagenicity tests, II. Results from the testing of 270 chemicals, *Environ. Mutagen.*, 8, 1, 1986.
5. **Ohizumi, I. and Okayasu, H. (Sumitomo Chemical Co., Ltd.),** Europe Patent Application 121,436, 1984.
6. **Hayatsu, H., Hayatsu, T., and Wataya, Y.,** Use of blue cotton for detection of mutagenicity in human feces excreted after ingestion of cooked meat, EHP, *Environ. Health Perspect.*, 67, 31, 1986.
7. **Hayatsu, H.,** Detection of environmental mutagens by blue cotton, *Nippon Yakuzaishikai Zasshi,* 39(5), 417, 1987.
8. **Seiders, R. P. and Ward, J. R. (U.S. Department of the Army),** Statutory Invention Register U.S. 223, 1987.
9. **Chan, Y. W. and Wilson, R. B., Jr.,** Metal Complexes for Defense against Cyanide Intoxication, Report 1986, Order No. AD-A174680/9/GAR, National Technical Information Service, Springfield, VA, 61 pp.
10. **Kurlyandskii, B. A., Braude, E. V., Klyachkina, A. M., Torshina, N. L., Khokhlova, S. B., and Zasorina, E. V.,** Study on copper phthalocyanine induced toxicity, *Gig. Sanit.*, 1, 92, 1985.
11. **Sandor, S., Prelipceanu, O., and Checiu, I.,** Sulfonated phthalocyanine induced caudal malformative syndrome in the chick embryo, *Rev. Roum. Morphol., Embryol. Physiol., Morphol. Embryol.*, 31(3), 173, 1985.
12. **Anliker, R. and Moser, P.,** The limits of bioaccumulation of organic pigments in fish: their relation to the partition coefficient and the solubility in water and octanol, *Ecotoxicol. Environ. Saf.*, 13(1), 43, 1987.
13. **Kurlyandskii, B. A., Torshina, N. L., and Braude, E. V.,** Effect of phthalocyanine dye structure on their behavior in the organism, *Gig. Tr. Prof. Zabol.*, 4, 49, 1987.
14. **Matrka, M. and Pipalova, J.,** Chemical models of biotransformation. I. Study of the degradation of organic dyes, *Chem. Prum.*, 35(2), 86, 1985.
15. **Nandagopal, K., Srinivasan, G. J., Sundararajan, K., and Sekharan, P. C.,** The study of pigments in automobile paints by infrared spectrophotometer, *J. Forensic Sci. Soc. India,* 1(2), 55, 1985.
16. **Fuller, N. A.,** Analysis of thin layer chromatograms of paint pigments by direct microspectrophotometry, *Forensic Sci. Int.*, 27(3), 189, 1985.
17. U.S. Food and Drug Administration, [Phthalocyaninato(2-)] copper; listing as a color additive for coloring polybutester nonabsorbable sutures, *Fed. Regist.*, 50(80), 16227, 25 April 1985.
18. U.S. Food and Drug Administration, [Phthalocyaninato(2-)] copper; change in organic chlorine content specification for the color additive for coloring sutures and contact lenses, *Fed. Regist.*, 51(121), 22928, 24 June 1986.
19. U.S. Food and Drug Administration, [Phthalocyaninato(2-)] copper; migration from nonabsorbable sutures, *Fed. Regist.*, 51(156), 28929, 13 August 1986.
20. **Bogaty, H., Brown, K. C., Loveless, N. P., and Wolfram, L. J. (Bristol-Myers Co.),** Europe Patent Application 132,960, 1985.
21. **Loshaek, S. (Schering Corp.),** U.S. Patent 4,668,240, 1987.
22. **Hansen, J. M. (Prodent Gesellschaft fur Zahnmed. Bedarfsartikel mbH),** German Patent 3,403,118, 1985.
23. **De Gee, A. J., Ten Harkel-Hagenaar, E., and Davidson, C. L.,** Color dye for identification of incompletely cured composite resins, *J. Prosthet. Dent.*, 52(5), 626, 1984.
24. **Hayes, H. and Harvey, K. (Colgate-Palmolive Co.),** U.S. Patent 4,456,585, 1984.
25. **Forestier, S., Maille, G., Junino, A., and Lang, G. (Oreal, SA),** German Patent 3,437,360, 1985.
26. **Bogaty, H., Brown, K. C., Loveless, N. P., and Wolfram, L. J. (Bristol-Myers Co.),** Europe Patent Application 132,960, 1985.
27. **Omura, T., Takeshita, A., Kenmochi, H., and Matsuo, Y. (Sumitoma Chemical Co., Ltd.),** Japan Patent 86,27,915, 1986.
28. **Wada, K., Masujima, T., Yoshida, H., Murakami, T., Yata, N., and Imai, H.,** Laser photoacoustic microscopy for the determination of dye on a solid biopolymer, *Chem. Pharm. Bull.*, 33(3), 1316, 1985.
29. **Mann, M. J., Loev, B., and Hwang, D. R. (Technicon Instruments Corp.),** Europe Patent Application 142,369, 1985.
30. **Hara, T., Toriyama, M., and Tsukagoshi, K.,** Immunoassay using a metal complex compound as a chemiluminescent catalyst. II. An improvement of the analytical method for practical use, *Bull. Chem. Soc. Jpn.*, 56(10), 2965, 1983.

31. **Malmroa, M. K.,** U.S. Patent 4,444,892, 1984.
32. **Hara, T., Toriyama, M., Miyoshi, H., and Syogase, S.,** Immunoassay using a metal complex compound as a chemiluminescent catalyst. IV. The investigation of a metal porphine complex as a labeling reagent, *Bull. Chem. Soc. Jpn.,* 57(10), 3009, 1984.
33. **Hara, T., Toriyama, M., and Imaki, M.,** Immunoassay using a metal complex compound as a chemiluminescent catalyst. V. Continuous immunoassay by the use of CLCCIA, *Bull. Chem. Soc. Jpn.,* 58(4), 1299, 1985.
34. **Ota, T.,** Japan Patent 87,146,177, 1987.
35. **Shirai, H. and Yokozeki, T. (Earth Clean K. K.),** Japan Patent 87,186,778, 1987.
36. **Lehmann, G. and Binkle, B.,** Identification of pigments in cosmetic products, *Seifen, Oele, Fette, Wachse,* 110(5), 125, 1984.
37. **Skripal, I. G.,** Ligand chromatography of DNA-dependent RNA polymerases as a convenient method for their purification and concentration, *Mol. Biol. (Kiev),* 37, 33, 1984.
38. **Hu, K. W. J. and Vogelhut, P. O.,** (Miles Laboratories), Europe Patent Application 140,322, 1985.
39. **Halbert, M. K. and Baldwin, R. P.,** Determination of cysteine and glutathione in plasma and blood by liquid chromatography with electrochemical detection using a chemically modified electrode containing cobalt phthalocyanine, *J. Chromatogr.,* 345(1), 43, 1985.
40. **Santos, L. M. and Baldwin, R. P.,** Profiling of oxalic acid and α-keto acids in blood and urine by liquid chromatography with electrochemical detection at a chemically modified electrode, *J. Chromatogr.,* 414(1), 161, 1987.
41. **Linders, C. R., Vincke, B. J., and Patriarche, G. J.,** Catalase like activity of iron phthalocyanine incorporated in a carbon paste electrode, *Anal. Lett.,* 19(17-18), 1831, 1986.
42. **Moskalev, P. N., Komarov, E. V., Shneider, M. A., Rachovskaya, L. A., and Stilbans, E.,** U.S.S.R. Patent 1,153,536, 1985.
43. **Juarranz., A., Horobin, R. W., and Proctor, G. B.,** Prediction of *in situ* fluorescence of histochemical reagents using a structure staining correlation procedure, *Histochemistry,* 84(4-6), 426, 1986.
44. **Brudnaya, M. S., Kirpichnikova, K. M., Komissarchik, Y. Y., and Levin, S. V.,** Changes in the shape and ultrastructure of white ghosts of erythrocytes upon their sealing and phthalocyanine dye sorption, *Tsitologiya,* 26(2), 184, 1984.
45. **Kirpichnikova, K. M. and Shuvalova, E. V.,** Spectrophotometric study of intact and trypsin treated erythrocyte membranes stained with a phthalocyanine dye, *Tsitologiya,* 26(8), 914, 1984.
46. **Juarranz, A., Stockert, J. C., Canete, M., and Villanueva, A.,** A micellar model for the disaggregating effect of detergents on phthalocyanine dyes, *Cell. Mol. Biol.,* 31(5), 379, 1985.
47. **Wada, K., Masujima, T., Yoshida, H., Murakami, T., Yata, N., and Imai, H.,** Application of photoacoustic microscopy to analysis of biological components in tissue sections, *Chem. Pharm. Bull.,* 34(4), 1688, 1986.
48. **Ben-Hur, E. and Rosenthal, I.,** The phthalocyanines: a new class of mammalian cells photosensitizers with a potential for cancer phototherapy, *Int. J. Radiat. Biol. Relat. Stud. Phys., Chem. Med.,* 47(2), 145, 1985.
49. **Van Lier, J. E., Ali, H., and Rousseau, J.,** Phthalocyanines labeled with gamma-emitting radionuclides as possible tumor scanning agents, *Progr. Clin. Biol. Res.,* 170, 315, 1984.
50. **Rousseau, J., Ali, H., Lamoureux, G., Lebel, E., and Van Lier, J. E.,** Synthesis, tissue distribution and tumor uptake of technetium-99m and gallium-67 tetrasulfophthalocyanine, *Int. J. Appl. Radiat. Isot.,* 36(9), 709, 1985.
51. **Brasseur, N., Ali, H., Autenrieth, D., Langlois, R., and Van Lier, J. E.,** Biological activities of phthalocyanines. III. Photoinactivation of V-79 Chinese hamster cells by tetrasulfophthalocyanines, *Photochem. Photobiol.,* 42(5), 515, 1985.
52. **Brasseur, N., Ali, H., Langlois, R., Wagner, J. R., Rousseau, J., and Van Lier, J. E.,** Biological activities of phthalocyanines. V. Photodynamic therapy of EMT-6 mammary tumors in mice with sulfonated phthalocyanines, *Photochem. Photobiol.,* 45(5), 581, 1987.
53. **Rosenthal, I., Krishna, C. M., Riesz, P., and Ben-Hur, E.,** The role of molecular oxygen in the photodynamic effect of phthalocyanines, *Radiat. Res.,* 107(1), 136, 1986.
54. **Ben-Hur, E. and Rosenthal, I.,** Photosensitization of Chinese hamster cells by water soluble phthalocyanines, *Photochem. Photobiol.,* 43(6), 615, 1986.
55. **Ben-Hur, E. Kol, R., Riklis, E., Marko, R., and Rosenthal, I.,** Effect of light fluence rate on mammalian cells photosensitization by chloraluminum phthalocyanine tetrasulfonate, *Int. J. Radiat. Biol. Relat. Stud. Phys., Chem. Med.,* 51(3), 467, 1987.
56. **Ben-Hur, E. and Rosenthal, I.,** Photosensitized inactivation of Chinese hamster cells by phthalocyanines, *Photochem. Photobiol.,* 42(2), 129, 1985.
57. **El-Far, M. and Pimstone, N.,** A comparative study of 28 porphyrins and their abilities to localize in mammary mouse carcinoma: uroporphyrin I superior to hematoporphyrin derivative, *Prog. Clin. Biol. Res.,* 170, 661, 1984.

58. **Anon.**, Selective *in vivo* tumor localization of uroporphyrin isomer I in mouse mammary carcinoma: superiority over other porphyrins in a comparative study, *Cancer Res.*, 46(9), 4390, 1986.
59. **Straight, R. C. and Spikes, J. D.**, Preliminary studies with implanted polyvinyl alcohol sponges as a model for studying the role of neointerstitial and neovascular compartments of tumors in the localization, retention and photodynamic effects of photosensitizers, *Adv. Exp. Med. Biol.*, 193, 77, 1985.
60. **Selman, S. H., Kreimar-Birnbaum, M., Chaudhuri, K., Garbo, G. M., Seaman, D. A., Keck, R. W., Ben-Hur, E., and Rosenthal, I.**, Photodynamic treatment of transplantable bladder tumors in rodents after pretreatment with chloroaluminum tetrasulfophthalocyanine, *J. Urol. (Baltimore)*, 136(1), 141, 1986.
61. **Ben-Hur, E., Fujihara, T., Suzuki, F., and Elkind, M. M.**, Genetic toxicology of the photosensitization of Chinese hamster cells by phthalocyanines, *Photochem. Photobiol.*, 45(2), 227, 1987.
62. **Qian, P., Evensen, J. F., Rimington, C., and Moan, J.**, A comparison of different photosensitizing dyes with respect to uptake C3H-tumors and tissues of mice, *Cancer Lett.(Shannon, Irel.)*, 36(1), 1, 1987.
63. **Chan, W. S., Marshall, J. F., Svensen, R., Phillips, D., and Hart, I. R.**, Photosensitizing activity of phthalocyanine dyes screened against tissue culture cells, *Photochem. Photobiol.*, 45(6), 757, 1987.
64. **Barel, A., Beltramini, M., Jori, G., Menegazzo, E., Perin, A., Reddi, E., Ricchelli, F., Romandini, P., and Valduga, G.**, Photosensitizing and phototherapeutic properties of porphyrins and phthalocyanines, *Med., Biol., Environ.*, 14(1), 35, 1986.
65. **Barr, H, Tralau, C. J., MacRobert, A. J., Krasner, N., Boulos, P. B., Clark, C. G., and Bown, S. G.**, Photodynamic therapy in the normal rat colon with phthalocyanine sensitization, *Br. J. Cancer*, 56(2), 111, 1987.
66. **Shimoni, M., Rosenthal, I., and Ben-Hur, E.**, Tritium labeled chloroaluminum phthalocyanine, *J. Labelled Compd. Radiopharm.*, 22(8), 863, 1985.
67. **Ben-Hur, E. and Rosenthal, I.**, Factors affecting the photokilling of cultured Chinese hamster cells by phthalocyanines, *Radiat. Res.*, 103(3), 403, 1985.
68. **Ben-Hur, E., Carmichael, A., Riesz, P., and Rosenthals, I.**, Photochemical generation of superoxide radical and the cytotoxicity of phthalocyanines, *Int. J. Radiat. Biol. Relat. Stud. Phys., Chem. Med.*, 48(5), 837, 1985.
69. **Hunting, D. J., Gowans, B. J., Brasseur, N., and Van Lier, J. E.**, DNA damage and repair following treatment of V-79 cells with sulfonated phthalocyanines, *Photochem. Photobiol.*, 45(6), 769, 1987.
70. **Spikes, J. D.**, Phthalocyanines as photosensitizers in biological systems and for the photodynamic therapy of tumors, *Photochem. Photobiol.*, 43(6), 691, 1986.
71. **Przywarska-Boniecka, H., Ostropolska, L., and Swirska, H.**, Interaction between some metal tetrasulfonated phthalocyanines and apocytochrome c, *Polyhedron*, 3(1), 129, 1984.
72. **Przywarska-Boniecka, H. and Ostropolska, L.**, Structural studies of iron and cobalt phthalocyanine substituted cytochromes c. Carboxymethylation and peptide fragments investigations, *J. Inorg. Biochem.*, 20(2), 103, 1984.
73. **Ostropolska, L. and Przywarska-Boniecka, H.**, Modification of cytochrome chain proteins with metal tetrasulfonated phthalocyanines, *Proc.-Sch.-Symp. Inorg. Biochem. Mol. Biophys.*, 45, 1985.
74. **Przywarska-Boniecka, H., Trynda, L., and Kosciukiewicz, T.**, Modifications of isolated α and β chains of human hemoglobin by the metal tetrasulfonated phthalocyanines. Studies on hybrid phthalocyanine-heme intermediates, *Inorg. Chim. Acta*, 93(4), 147, 1984.
75. **Trynda, L., Przywarska-Boniecka, H., and Kosciukiewicz, T.**, Modification of the human serum albumin-heme system with metal tetrasulfonated phthalocyanines, *Inorg. Chim. Acta*, 135(1), 55, 1987.
76. **Simic-Glavaski, B.**, Electrooptical behavior of heme-like molecules adsorbed on nerve membrane, *Cell Biophys.*, 7(3), 205, 1985.
77. **Kuehn, M. and Benes, M. (Academie der Wissenschaften der DDR)**, East German Patent 216,958, 1985.
78. **Troshina, V. P. and Kruchinina, N. G.**, Changes in the parameters of equilibrium sorption of acidic dyes by contractile proteins during their functional activity and heat denaturation, *Tsitologiya*, 25(12), 1393, 1983.
79. **Stynes, D. V., Liu, S., and Marcus, H.**, Distal histidine coordination to iron in phthalocyanine reconstituted myoglobin, *Inorg. Chem.*, 24(25), 4335, 1985.
80. **Kornguth, S. E., Kalinke, T., and Pietro, W.**, The electronic spectra of porphyrin-like cobalt tetracarboxy phthalocyanines as modified by polylysine and polyglycols, *Biochim. Biophys. Acta*, 924(1), 19, 1987.
81. **Langlois, R., Ali, H., Brasseur, N., Wagner, J. R., and Van Lier, J. E.**, Biological activities of the phthalocyanines. IV. Type II sensitized photooxidation of L-tryptophan and cholesterol by sulfonated metallo phthalocyanines, *Photochem. Photobiol.*, 44(2), 117, 1986.
82. **Wagner, J. R., Ali, H., Langlois, R., Brasseur, N., and Van Lier, J. E.**, Biological activities of the phthalocyanines. VI. Photooxidation of L-tryptophan by selectively sulfonated gallium phthalocyanines: singlet oxygen yields and effect of aggregation, *Photochem. Photobiol.*, 45(5), 587, 1987.

83. **Spikes, J. D. and Bommer, J. C.,** Zinc tetrasulfophthalocyanine as a photodynamic sensitizer for biomolecules, *Int. J. Radiat. Biol. Relat. Stud. Phys., Chem. Med.*, 50(1), 41, 1986.
84. **Saito, Y., Mifune, M., Nakashima, S., Tanaka, Y., Chikuma, M., and Tanaka, H.,** Peroxidase-like accelerative activity of ion exchange resins modified with metalloporphyrins in the dye formation reaction of phenol with 4-aminoantipyrine by hydrogen peroxide, *Chem. Pharm. Bull.*, 34(12), 5016, 1986.
85. **Jensen, N. H., Wilbrandt, R., Houee-Levin, C., and Bensasson, R. V.,** Isomerization and structure of retinals in the triplet state, *NATO ASI Ser., Ser. A,* 85, 105, 1985.
86. **Oshima, N., Takeda, F., and Ishii, K.,** Effects of chemical substances on rate of locomotion in the amoeba *Mayorella penardi, J. Protozool.*, 32(3), 509, 1985.
87. **Sarode, P. R., Sankar, G., and Rao, C. N. R.,** XANES and EXAFS of copper compounds: studies of copper carboxylates with metal-metal bonds and of the complex formed by *Pseudomonas aeruginosa, Proc. Indian Acad. Sci., Chem. Sci.*, 92(4-5), 527, 1983.
88. **Inoue, S. and Moriyama, N. (Kao Corp.,),** German Patent 3,329,727, 1984.
89. **Rohde, R. and Neef, G. (Schering AG),** German Patent 3,438,484, 1986.
90. **Simic-Glavaski, B. (H.S.G. Venture),** PCT International Application 87,01,807, 1987.
91. **Gaspard, S.,** "Face to face" porphyrin-phthalocyanine dimers, *C.R. Acad. Sci., Ser. 3,* 298(14), 379, 1984.
92. **Gaspard, S., Giannotti, C., Maillard, P., Schaeffer, C., and Tran-Thi, T. H.,** The first synthesis of covalently linked mixed dimer complexes containing phthalocyanine and porphyrin, *J. Chem. Soc., Chem. Commun.*, 16, 1239, 1986.
93. **Kopranenkov, V. N., Makarova, E. A., and Dashkevich, S. N.,** Novel isoindologens in tetrabenzoporphyrin synthesis, *Khim. Geterotsikl. Soedin,* 10, 1372, 1985.
94. **Pshezhetskii, V. S., Lebedeva, T. S., Yaroslavov, A. A., Lukyanets, E. A., and Gal'pern, M. G.,** Spectroscopy of polymer gels containing covalently bound zinc phthalocyanine as models of light harvesting systems in photosynthesis, *Biofizika,* 30(4), 593, 1985.
95. **Rakhnyanskaya, A. A., Yaroslavov, A. A., Pshezhetskii, V. S., Tatarova, L. A., and Lopyrev, V. A.,** Distribution of phthalocyanines in copolymers based on 1-vinyl-1,2,4-triazole — models of light harvesting antennae of the photosynthetic apparatus, *Vysokomol. Soedin., Ser. B,* 27(11), 861, 1985.
96. **Honeybourne, C. L., Houghton, J. D., and Jones, O. T. G.,** Dark d.c. conductivity of clean and gas doped natural tetrapyrrole pigments, *Chemtronics,* 1(2), 84, 1986.
97. **Martin, B. B. and Martin, D. F.,** Effects of four dyes on rates of oxygen production by three filamentous algae, *Microbios Lett.*, 35(139-140), 151, 1987.
98. **Starke, M., Hampel, H., Hamann, C., Knoefler, W., and Bethmann, W.,** Evaporated phthalocyanine films: preparation, properties, and suitability as a coating material for implants, *Wiss. Z.-Karl-Marx-Univ. Leipzig, Math. Naturwiss. Reihe,* 32(5), 479, 1983.
99. **Valduga, G., Reddi, E., and Jori, G.,** Spectroscopic studies on zinc(II)-phthalocyanine in homogeneous and microheterogeneous systems, *J. Inorg. Biochem.*, 29(1), 59, 1987.

Chapter 27

THERMAL, THERMAL TRANSFER, AND OPTICAL RECORDING MEDIA

A number of Japanese patents describe the compositions of thermal transfer recording media.

A medium suitable for heat mode recording is formed from an acrylic disk support coated with polymethyl methacrylate 1 wt % and Cu phthalocyanine (pc) 19 wt %. Recording with a He/Ne laser shows high sensitivity and durability.[1] An optical recording layer for heat-mode recording comprises nitrocellulose 1 part by weight and Cu pc 100 parts by weight dissolved in organic solvents.[2]

A thermal transfer recording material contains a layer containing Cu pc and a binder, and a transfer layer containing a low melting point compound that can dissolve or disperse the coloring agent when melting. [3] A laser heat recording medium includes a vanadyl pc dye film.[4] An optical recording medium contains a pc pigment which is bonded at $\geqslant$5 carbon atoms of the pc macrocycle selected from among the 1, 4, 8, 11, 15, 18, 22, and 25 carbon atoms to an aqueous group by a group VA or group VI metal atom and absorbs the optical signal received by the recording layer.[5]

A best sensitive transferable ink layer contains a urethane binder and Cu pc for thermal recording materials.[6] A thermal transfer recording material contains a metal-crosslinked oxidized polymer binder and Cu pc.[7] A dispersion containing Cu pc and oligostyrene in toluene is a component of a heat transfer recording sheet.[8] A thermal transfer recording with correction sheets contains Cu pc.[9] Thermal transfer recording materials contain Cu pc for images with improved light stability.[10]

A thermal transfer recording contains the colorant Cu pc.[11] Thermal transfer recording materials with an image transferring sheet contains Cu pc.[12] Thermal recording materials contain thermoplastic resin particles, polyvinyl alcohol, and Cu pc.[13] A myristic acid coating for thermal transfer recording material with an ink layer contains Cu pc for improved image density and no background stains.[14] An electrothermal recording material includes Cu pc to give clear blue images.[15]

A thermal recording ink material composition contains Cu pc for an electrothermal transfer sheet.[16] A thermal transfer recording medium contains Cu pc as a filler.[17] A thermal transfer recording ink sheet contains Cu pc disulfonic acid for clear transparent images.[18] A thermal recording ink sheet contains Cu pc.[19] A meltable transparent ink layer contains waxes and Cu pc for thermal transfer recording materials for multicolor print production without color mixing.[20]

A coating process for hot melt ink sheets for thermal transfer recording contains Cu pc.[21] An electrolyte solution contains Cu pc for the electrochemical development of thermal recording material containing a basic polymer-acidic polymer complex.[22] Thermal transfer recording materials have an ink layer that contains Cu pc for high-density images with superior gradation.[23]

Heat-sensitive compositions contain thermoplastic resin particles and fluororesin and water-soluble resin and Cu pc for thermal recording materials.[24] A thermal recording material with transfer sheet contains Cu pc tetrasulfonic acid.[25]

A dispersion contains Cu pc and oligostyrene in toluene to assist formation of a heat transferring layer and give a heat transfer recording sheet.[26] A heat fusible ink layer contains waxes and Cu pc for thermal transfer recording materials.[27] Thermal transfer recording materials contain Cu pc for improved sensitivity and fixability.[28]

A thermal transfer recording material contains Cu pc.[29] Optical printing image transferring sheets contain a photoconductive layer containing Cu pc.[30] A colored thermographic recording sheet contains Cu pc.[31] A thermal transfer recording material has a heat-meltable ink layer containing Cu pc.[32] A thermal transfer recording material contains sulfonated Cu pc to give clear blue images on paper.[33] A thermal transfer recording material contains Cu pc; high-density well fixed images are obtained.[34] Powdered ink contains Cu pc for a transfer recording medium.[35]

Several European and one U.S. patent also address the subject.

A thermal color transfer recording material contains Cu c.[36] A heat-sensitive color optical recording material with a light-absorbing layer contains sulfonated A1 pc.[37] An electrothermal transfer recording material contains pc derivatives.[38] A heat-melting ink layer contains inorganic particles, wax, and Cu pc for thermal transfer recording materials.[39] Thermal transfer recording materials possess an ink layer containing Cu pc for high-density images with superior gradation.[40]

A thermographic imaging system contains the ferric salt of an organic acid, a phenolic compound, and Cu pc with improved image density.[41] An ablative IR sensitive recording composition contains a dispersion of a resinous binder and an absorbing IR sensitive tertiary butyl substituted VO pc.[42] A process and material relate to forming a multicolored image by using a heat-transferable photosensitive material. Cu pc is used to produce a cyan image.[43] A polysubstituted di hydrogen pc, M pc, or MO pc is a radiation absorber in a laser transfer printing medium.[44]

A new optical disk medium for hole-forming laser recording is developed, consisting of a multilayer film of tellurium and Cu pc, as described in a technical journal. The tellurium and Cu pc multilayer medium is promising for large-capacity optical disks with high data transfer rate and long-term durability.[45]

REFERENCES

1. **TDK Corp.,** Japan Patent 84,05,096, 1984.
2. **TDK Corp.,** Japan Patent 84,05,095, 1984.
3. **Konishiroku Photo Industry Co., Ltd.,** Japan Patent 84,109,389, 1984.
4. **Ito, M., Morimoto, A., and Edokoro, S. (NEC Corp.),** Japan Patent 87,86,551, 1987.
5. **TDK Corp.,** Japan Patent 86,154,888, 1986.
6. **Dainippon Printing Co., Ltd.,** Japan Patent 83,199,192, 1983.
7. **Canon K. K.,** Japan Patent 84,224,388, 1984.
8. **Tomoegawa Paper Manufacturing Co., Ltd., and Matsushita Electric Industrial Co., Ltd.,** Japan Patent 85,68,994, 1985.
9. **Watanabe, H., Kubo, T., Takigawa, N., Hakiri, M., and Kawasaki, K. (Ricoh Co., Ltd.),** Japan Patent 85,255,489, 1985.
10. **Sekine, M., Matsushita, T., and Morishita, S. (Mitsubishi Paper Mills, Ltd.),** Japan Patent 86,08,387, 1986.
11. **Kobashi, T. (Matsushita Electric Industrial Co., Ltd.,),** Japan Patent 85,198,294, 1985.
12. **Kobashi, T., Onishi, H., and Ezaki, H. (Matsushita Electric Industrial Co., Ltd.),** Japan Patent 85,159,089, 1985.
13. **Shiraishi, S. (Ricoh Co., Ltd.),** Japan Patent 86,10,488, 1986.
14. **Kawanishi, T. and Tabata, Y. (Ricoh Co., Ltd.),** Japan Patent 86,37,492, 1986.
15. **Fukue, T.,** Japan Patent 85,178,089, 1985.
16. **Hayashi, S., Yamaguchi, Y., Takei, K., Taniguchi, M., and Usui, T. (Epson Corp. and Suwa Seikosha Co., Ltd),** Japan Patent 86,51,392, 1986.
17. **Suzuki, A., Mochizuki, N., Kawanishi, T., Tatsumi, S., Watanabe, H., and Kubo, T. (Ricoh Co., Ltd.),** Japan Patent 86,114,881, 1986.
18. **Minowa, S. (Nippon Kayaku Co., Ltd.),** Japan Patent 86,89,885, 1986.
19. **Onishi, H., Ezaki, H., and Kobashi, T.,** Japan Patent 86,125,890, 1986.
20. **Yokoi, N. (Fuji Xerox Co., Ltd.),** Japan Patent 86,108,586, 1986.
21. **Kobashi, T., Onishi, H., and Ezaki, H. (Matsushita Electric Industrial Co., Ltd.),** Japan Patent 86,125,883, 1986.

22. **Haruta, M., Kuwae, Y., Yuasa, S., and Miyazaki, T. (Canon K. K.)**, Japan Patent 86,169,871, 1986.
23. **Suzuki, A. and Mochizuki, N. (Ricoh Co., Ltd.)**, Japan Patent 86, 230,988, 1986.
24. **Shiraishi, S. (Ricoh Co., Ltd.)**, Japan Patent 86,193,888, 1986.
25. **Kawanishi, T., Mochizuki, N., Suzuki, A., Tatsumi, S., Watanabe, H., and Kubo, T. (Ricoh Co., Ltd.)**, Japan Patent 86,143,184, 1986.
26. **Tomoegawa Paper Manufacturing Co., Ltd.**, Japan Patent 85,68,993, 1985.
27. **Canon K. K.**, Japan Patent 85,25,787, 1985.
28. **Yubagami, K., Matsuda, H., Imai, A., and Taguchi, N. (Matsushita Electric Industrial Co., Ltd.)**, Japan Patent 87,109,689, 1987.
29. **Tezuka, T., Watanabe, H., and Abe, T. (Konishiroku Photo Industry Co., Ltd.)**, Japan Patent 87,80,091, 1987.
30. **Uehara, T., Tomura, S., and Okuyama, T. (Toshiba Corp.)**, Japan Patent 86,295,555, 1986.
31. **Igarashi, A. (Fuji Photo Film Co., Ltd.)**, Japan Patent 86,130,085, 1986.
32. **Suzuki, A., Mochizuki, N. (Ricoh Co., Ltd.)**, Japan Patent 86,297,178, 1986.
33. **Haruta, M., Yoshinaga, Y., Yuasa, S., Munakata, H., and Nishimura, Y. (Canon K. K.)**, Japan Patent 87,05,889, 1987.
34. **Taguchi, N., Yubagami, K., Matsuda, H., and Imai, A. (Matsushita Electric Industrial Co., Ltd.)**, Japan Patent 87,83,192, 1987.
35. **Tamura, Y. and Kaneko, S. (Canon K. K.)**, Japan Patent 87,60,694, 1987.
36. **Seto, T. and Shimazaki, Y. (Fuji Kagakushi Kogyo Co., Ltd.)**, German Patent 3,304,584, 1983.
37. **Morinaka, A., Oikawa, S., and Sato, H. (Nippon Telegraph and Telephone Public Corp.)**, German Patent 3,319,738, 1983.
38. **Kawanishi, T. and Tabata, Y. (Ricoh Co., Ltd.)**, German Patent 3,347,337, 1984.
39. **Kohashi, T., Onishi, H., and Esaki, H. (Matsushita Electric Industrial Co., Ltd.)**, Europe Patent Application 163,297, 1985.
40. **Suzuki, A., Mochizuki, N., Tasaka, M., and Hashimoto, M. (Ricoh Co., Ltd.)**, German Patent 3,606,757, 1986.
41. **Sagawa, B. K. (Minnesota Mining and Manufacturing Co.)**, Europe Patent Application 114,111, 1984.
42. **Law, K. Y. and Johnson, G. E. (Xerox Corp.)**, U.S. Patent 4,529,688, 1985.
43. **Mino, H., Iijima, T., and Monden, K. (Sanyo-Kokusaku Pulp Co., Ltd.)**, German Patent 3,427,648, 1985.
44. **Barlow, W. A. (Imperial Chemical Industries PLC)**, Europe Patent Application 157,568, 1985.
45. **Goto, Y., Koshino, N., and Itoh, K.**, Tellurium and copper phthalocyanine multilayer films for optical disk media, *Fujitsu Sci. Tech. J.*, 20(3), 411, 1984.

Chapter 28

LASER AND OTHER OPTICAL RECORDING APPLICATION

Over 100 Japanese patents relate to the specifics of optical recording media including M phthalocyanines (pcs) and dihydrogen pcs as dye or pigment component. They are briefly summarized.

Cu pc 4,4′,4″,4‴-tetracarboxylic acid is converted to its chloride. It is then Friedel-Crafts reacted with a 1:1 styrene-(4-vinylpyridine) copolymer and coated on a glass disk.[1] Cu pc is a component in a poly(sodium methacrylate) film bonded to a glass disk.[2] Cu pc is found in an organic dye layer.[3] Sn hexadecafluoro dichloro pc is vacuum deposited on an acrylic resin support to give a laser recording sheet.[4] A light reflecting layer for laser recording material contains Pb pc.[5] Di hydrogen pc is a component in laser recording material.[6]

Cu pc is found in a substrate in laser recording material.[7] Cu pc is formed on a substrate of a laser optical disk.[8] Cu pc is in a transparent layer in laser recording material.[9] Cu pc is in a recording layer for laser recording material.[10,11] Cu pc is coated with poly(vinyl acetate) on a glass support to give a laser recording disk.[12]

Pb pc 3,3′-dicarboxylic acid chloride polymerizes with hexamethylenedicarboxylic acid chloride to form a recording layer for laser recording material.[13] Cu pc is subbing for a laser recording disk.[14] A titanium pc is used to absorb laser light at 600 to 750 nm for laser recording material.[15]

Cu pc is in an organic polymer film by vapor deposition in a laser recording material.[16] Titanyl pc is in a recording layer for laser recording material.[17] Cu pc is a sublimable organic substance for laser recording material.[18]

The reaction products of V bis(5-carboxypentyl)ester pc 1,8 dicarboxylic acid with chelates and acetyl cellulose are in a recording layer of laser recording material.[19] Pb pc is in a light reflecting layer of a laser recording medium.[20]

Ni pc finds itself in a recording layer of optical recording material by reaction of 1:1 styrene-4-vinylpyridine copolymer with Ni pc chloride with three carboxyl groups on the benzene nucleus.[21] Al pc Cl is a component in an absorption layer in an optical recording medium.[22] Mn(II) octa(4,5-cyano) pc finds itself in a recording layer.[23] Cu pc is vapor deposited on a Pyrex substrate.[24] A titanium pc complex is an organic thin film recording layer.[25] A recording layer is formed by copolymerization of Co 2,9-[bis(3,4-dicarboxy benzoyl)] pc dianydride with *p*-phenylene diamine.[26] The polymer is shown in Figure 1.

A vanadium pc *N,N′*-dihexyl-1,8-disulfonamide finds itself in laser recording material.[27] The reaction products of a vanadium pc-1,8-carboxylic acid bis[6-(acetyloxy) hexyl] ester with poly(methyl methacrylate) provide a recording layer.[28] A laser recording material contains the bis[6-(acetyloxy) hexyl] ester of the polymer of vanadium pc-1,8-dicarboxylic acid.[29]

Laser disks are made from compositions containing a vanadium pc-1,8-dicarboxylic acid bis(5-carboxypentyl) ester.[30,31]

An oxidized vanadium pc is in a recording layer of an optical recording medium.[32,33] The oxidized vanadium pc complex recording layer may also contain Cu pc.[34] A vanadium pc complex finds use in a multilayered optical recording material.[35]

Di hydrogen pc is a light-absorbing electrode layer for an optical recording medium.[36,37]

Laser recording materials contain vanadyl pc.[38]

An optical laser recording material has a layer containing celluloid and Cu pc-1,8,15,22-tetrasulfonic acid.[39]

In pc Br with substituents finds itself in a dichloroethane solution in spin coating on an acrylic polymer disk for optical recording material.[40]

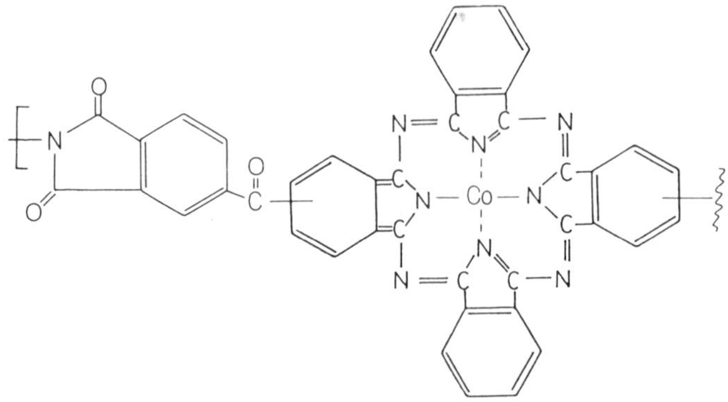

FIGURE 1. A polymer with Co pc in the principal chain.

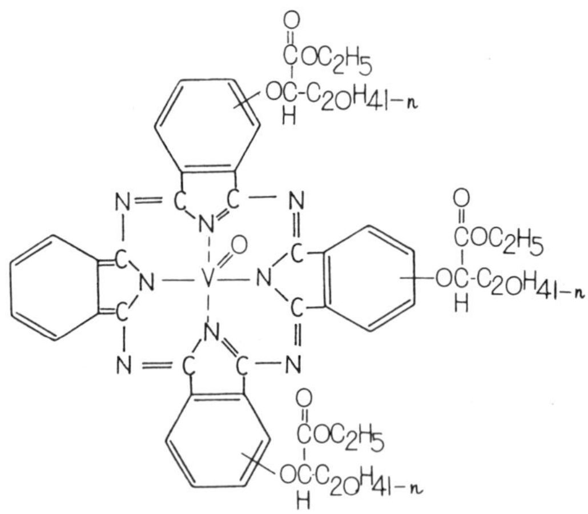

FIGURE 2. A vanadyl pc complex coated as a Langmuir-Blodgett film in a laser recording disk.

Dihydrogen, Co, Cu pc, or A1 pc C1 absorbs a high-energy light beam coated on a glass substrate by vacuum deposition.[41]

VO 3,3′-di(6-hydroxyhexyloxycarbonyl) pc is coated on a poly(methyl methacrylate) laser optical recording disk.[42,43]

Cu pc is a component in a layer along with thermoplastic resin for laser recording material.[44] Vanadyl pc is in a light-absorbing electrode layer in laser recording media.[45]

A Mn pc complex is in a recording layer in a laser recording medium.[46] Di hydrogen pc is a recording layer in a laser-sensitive optical recording medium.[47]

Vanadyl is recording layer in a laser-sensitive optical recording medium.[48] Cu pc is a light absorber in laser optical recording material.[49] Vanadyl pc is in a light-absorption layer for optical recording material.[50] A vanadyl pc complex (Figure 2) is coated as a Langmuir-Blodgett film on a glass substrate in a laser recording disk.[51,52]

A vanadyl pc-poly(vinyl alcohol) mixture acts as a coating in a high-quality laser recording disk.[53,55,56] Vanadyl pc is a recording layer in an optical recording material.[54]

Mn pc forms a layer in optical recording material.[57] Vanadyl pc forms a recording layer in optical recording material.[58] Vanadyl pc and poly(vinyl alcohol) are coated on a glass support as part of a high-quality laser recording material.[59]

VO pc finds itself in laser recording material.[60-62] VO pc finds itself in a laser recording

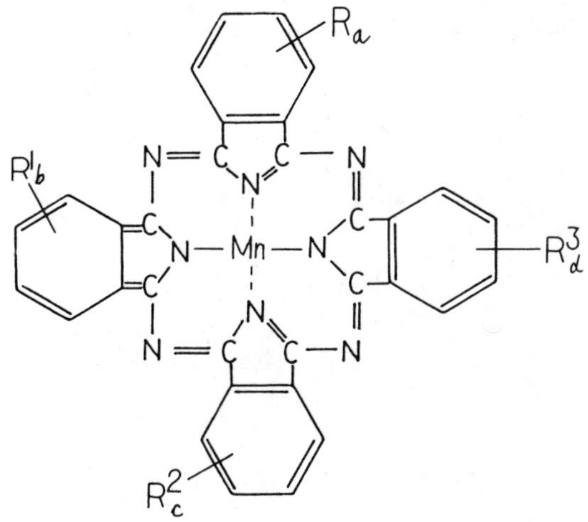

FIGURE 3. A Mn pc derivative for a recording layer.

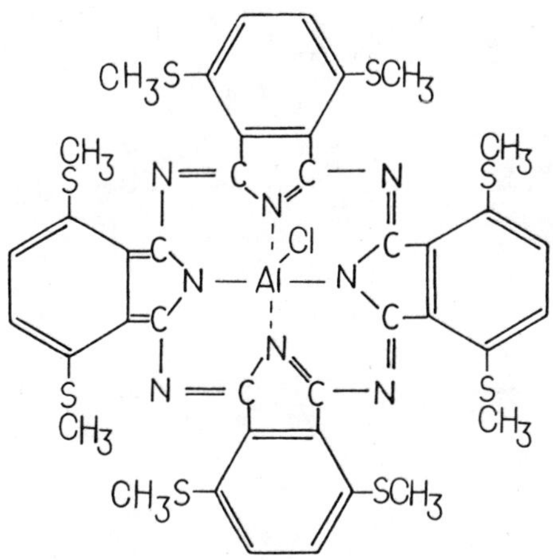

FIGURE 4. An Al pc Cl derivative as component of a recording layer.

medium with thermal coloration.[63] VO pc is in a light absorber layer of a laser recording disk.[64] A VO derivative similar to that in Reference 51 is mixed with poly(vinyl alcohol) to give a laser recording disk[65-68] or to give a high-quality recording medium.[69]

Cu pc is a component in a dye reflection layer.[70] Vanadyl pc docosanoic acid derivative is a light absorber in an optical recording medium.[71] In a composite recording layer, first polyethylene then A1 pc C1 are deposited on a glass plate.[72] In a composite recording layer, first polypropylene then A1 pc C1 are deposited on a glass plate.[73]

Mn pc finds itself in a recording layer for optical information recording media, where R-R^3 = H, halo, and a, b, c, d = 1 to 4 (Figure 3).[74]

A M nc (nc = naphthalocyanine) containing alkyl groups attached to the outer benzene rings forms a laser absorption layer for optical recording readout materials.[75]

Di hydrogen pc is a component in a polymerizable vinyl monomer dispersion used in a photothermographic process to record an image.[76]

An A1 pc C1 derivative (Figure 4) is added to a resin to form a recording layer in an optical recording medium.[77]

A M pc or di hydrogen pc derivative (Figures 5 and 6) is a component in an optical recording medium.

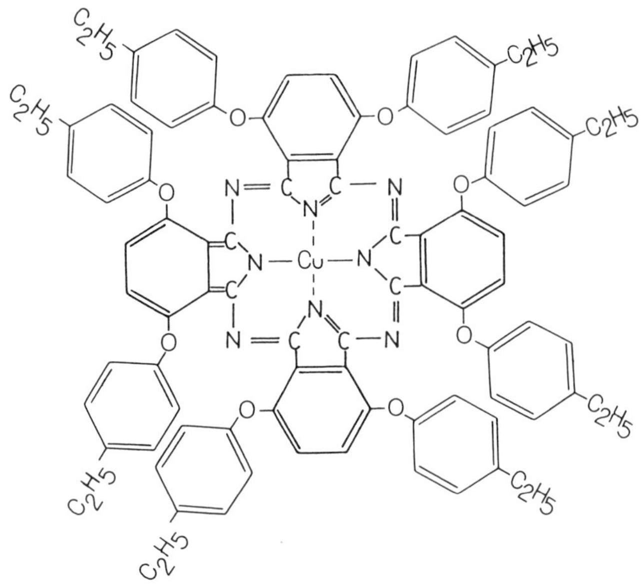

FIGURE 5. A Cu pc derivative that is a component in an optical recording medium.

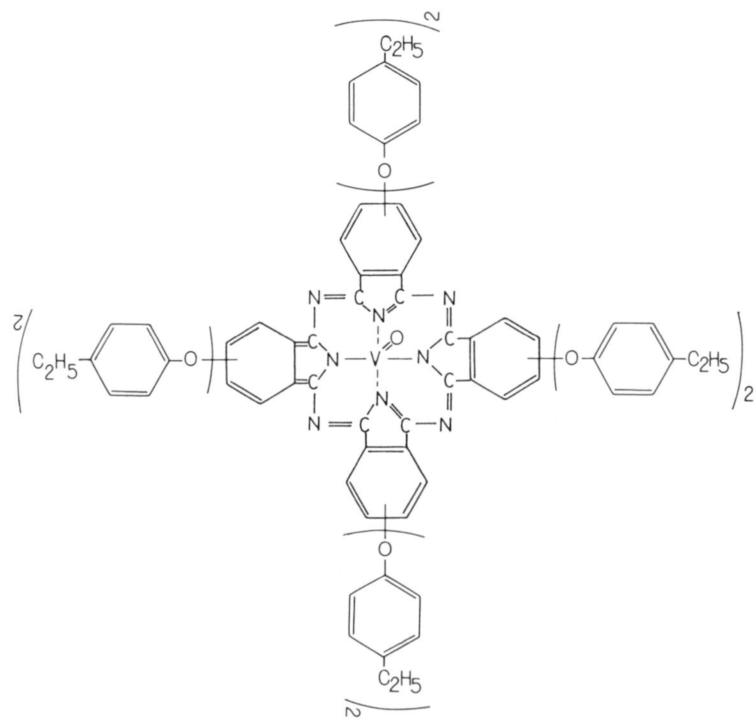

FIGURE 6. A vanadyl pc derivative that is a component in an optical recording medium.

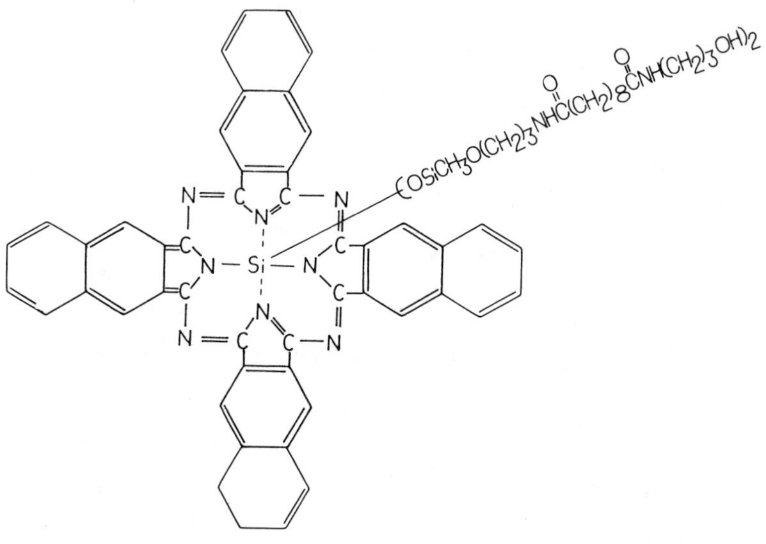

FIGURE 7. A Si pc derivative chromophore in an optical recording medium.

A Si pc derivative (Figure 7) is a chromophore in an optical recording medium.[80]

Di hydrogen pc is a light-absorbing dye in an optical recording medium.[81]

An optical recording material is made from a crosslinked polymer and di hydrogen pc.[82]

An optical disk with an electrochromic layer and a photovoltaic layer contains di hydrogen pc.[83]

Cu pc is a component in a high sensitivity, high stability optical recording medium.[84]

Cu pc tetrasulfonic acid is involved in the electrolysis of an aqueous solution containing pyrrole and Cu pc tetrasulfonic acid in preparation of a colored polymer layer for optical information recording material.[85]

Cu pc tetrasulfonic acid is in an electrolyte composition involved in the deposition of a polymer membrane for optical recording materials.[86] Pyrrole anodic polymerization takes place in a solution of Cu pc tetrasulfonic acid to form a colored polymer layer for recording materials.[87]

Mn(III) pc Cl is vacuum deposited on a polycarbonate support to obtain an optical recording disk.[88]

Cu tetra-*tert*-butyl pc in a Langmuir-Blodgett film that also contains poly-L-Phe and tetracyano quinodimethane is part of a highly efficient recording medium.[89]

M pc derivatives form a component of a recording layer of an optical recording medium.[90]

Ti nc is a component of an optical recording layer coated on a plastic film to produce an optical card.[91]

Cu pc is in a plasma polymerization dielectric layer for optical recording materials.[92]

Vanadyl pc is a component in an optical recording disk containing a polymer liquid crystal layer.[93] Vanadium pc is a component in a light absorbing layer in an optical disk.[94-96]

An optical disk that contains Cu hexadecaphenyl thio pc shows the same quality as a conventional compact disk.[97]

V pc coated on a glass disk is a light absorber in a laser recording disk.[98]

Cu pc is a protective layer for laser recording materials.[99]

A laser recording material contains a vacuum-deposited oxidizable thin film of Al pc Cl.[100]

Te pc and Cu pc are simultaneously vacuum deposited on a support to give a laser recording disk with excellent storage stability.[101]

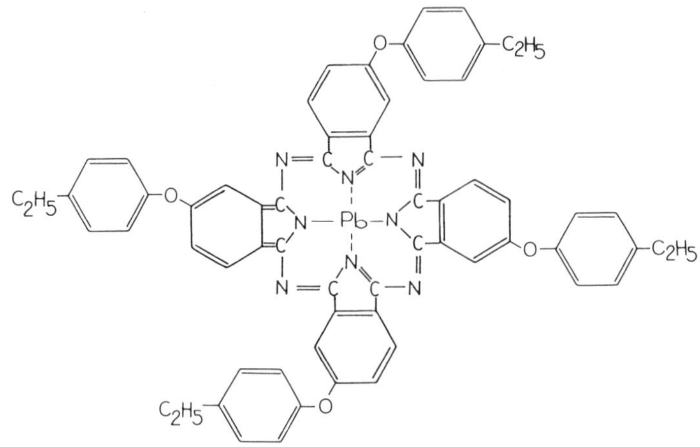

FIGURE 8. A Pb pc complex used in an optical recording disk.

Boats for vacuum deposition of VO pc for laser recording disk preparation are provided with thermally conductive perforated plates.[102] Boats for vacuum deposition of TiO pc for laser recording disks with improved durability are equipped with mesh-structured thermal conductors.[103]

Optical memory materials contain W pcs for improved photosensitivity.[104] Two-dimensional optical patterns are formed by laser beam irradiation of vacuum-deposited Cu pc layers.[105]

An optical recording layer contains titanium pc and >1 shifting agent.[106] A laser optical recording material contains Pb pc.[107] Di hydrogen pc finds itself in a laser recording material.[108,109]

Pb pc is a component in a laser recording material.[110] Di hydrogen pc is also a component in a laser recording material.[111] Zn hexadecafluoro pc is a component in an optical information recording material.[112] Titanium pc is a component in a laser recording medium.[113] Vanadyl pc finds itself in a multiset sandwich laminated recording material with a light-absorbing layer of VO pc.[114]

Several U.S. and European patents also address themselves to the topic of optical recording materials that contain M pcs.

An optical information recording material is made by preparing a coordination compound of Cu pc and Empol-1010 dimer acid-ethylenediamine-*N*-[2-(4-pyridyl)ethyl]ethylene diamine copolymer.[115]

An optical recording material includes an amorphous oxide layer in the presence of Cu pc.[116]

A laser-sensitive optical recording material contains VO 2,9,16,23-tetrakis(4-ethyl-1-piperazinyl). pc.[117]

An optical recording disk containing a Pb pc complex (Figure 8) shows a carrier-to-noise ratio of 52:1 dB before and after a 10-d aging test.[118]

A layer in a recording medium suitable to prepare optical disks contains Cu pc.[119] Pb pc or Cu pc is a component in an information recording medium with a modulated laser beam.[120] Cu pc is a component in a recording element exposed at 220 nm and developed in a water bath to give a sharp image.[121]

An optical recording using di hydrogen pc thin films is the subject of a journal contribution.[122]

REFERENCES

1. **TDK Corp.**, Japan Patent 84,71,894, 1984.
2. **TDK Corp.**, Japan Patent 84,124,893, 1984.
3. **Ricoh Co., Ltd.**, Japan Patent 83,36,490, 1983.
4. **Ricoh Co., Ltd.**, Japan Patent 83,56,892, 1983.
5. **Ricoh Co., Ltd.**, Japan Patent 83,224,448, 1983.
6. **Konishiroku Photo Industry Co., Ltd.**, Japan Patent 83,100,246, 1983.
7. **Fujitsu Ltd.**, Japan Patent 83,62,093, 1983.
8. **NEC Home Electronics, Ltd.**, Japan Patent 84,119,551, 1984.
9. **Fujitsu Ltd.**, Japan Patent 83,128,035, 1983.
10. **Fujitsu Ltd.**, Japan Patent 84,117,744, 1984.
11. **Fujitsu Ltd.**, Japan Patent 83,62,094, 1983.
12. **TDK Electronics Co., Ltd.**, Japan Patent 83,56,895, 1983.
13. **TDK Corp.**, Japan Patent 84,185,694, 1984.
14. **Fujitsu Ltd.**, Japan Patent 84,217,245, 1984.
15. **Nippon Electric Co., Ltd.**, Japan Patent 84,73,996, 1984.
16. **Hoya Corp.**, Japan Patent 85,15,843, 1985.
17. **Sumitomo Chemical Co., Ltd.**, Japan Patent 85,35,345, 1985.
18. **NEC Kansai, Ltd.**, Japan Patent 85,21,295, 1985.
19. **TDK Corp.**, Japan Patent 84,201,244, 1984.
20. **Ricoh Co., Ltd.**, Japan Patent 84,78,891, 1984.
21. **TDK Corp.**, Japan Patent 84,62,188, 1984.
22. **Canon K. K.**, Japan Patent 84,177,743, 1984.
23. **TDK Corp.**, Japan Patent 84,67,093, 1984.
24. **Nippon Telegraph and Telephone Public Corp.**, Japan Patent 84,144,056, 1984.
25. **Sumitomo Chemical Co., Ltd.**, Japan Patent 84,11,292, 1984.
26. **TDK Corp.**, Japan Patent 84,73,994, 1984.
27. **TDK Corp.**, Japan Patent 84,201,241, 1984.
28. **TDK Corp.**, Japan Patent 84,232,894, 1984.
29. **TDK Corp.**, Japan Patent 84,229,396, 1984.
30. **TDK Corp.**, Japan Patent 84,190,895, 1984.
31. **TDK Corp**, Japan Patent 84,198,191, 1984.
32. **Iyo, M. (NEC Corp.)**, Japan Patent 85,161,194, 1985.
33. **Ito, M. (NEC Corp.)**, Japan Patent 85,161,192, 1985.
34. **Ito, M. (NEC Corp.)**, Japan Patent 85,161,193, 1985.
35. **Morinka, A. and Oikawa, S. (Nippon Telegraph and Telephone Public Corp.**, Japan Patent 85,174,688, 1985.
36. **Kuroiwa, A., Asami, S., Aoi, T., Takahashi, K., and Nanba, N. (TDK Corp.)**, Japan Patent 85,178,092, 1985.
37. **Kuroiwa, A., Asami, S., Aoi, T., Takahashi, K., and Nanba, N. (TDK Corp.)**, Japan Patent 85,179,294, 1985.
38. **Morinaka, A. and Oikawa, S. (Nippon Telegraph and Telephone Public Corp.)**, Japan Patent 85,219,647, 1985.
39. **Omori, Y. and Yoneda, M. (Daicel Chemical Industries, Ltd.)**, Japan Patent 85,183,191, 1985.
40. **Shirai, H. and Ito, H.**, Japan Patent 85,184,887, 1985.
41. **Takeda, Y., Kaneshiro, T., and Numata, S. (Hitachi Ltd.)**, Japan Patent 85,154,098, 1985.
42. **TDK Corp.**, Japan Patent 85,124,292, 1985.
43. **TDK Corp.**, Japan Patent 85,124,291, 1985.
44. **Shiraishi, S. (Ricoh Co., Ltd.)**, Japan Patent 85,225,796, 1985.
45. **Kuroiwa, A., Asami, S., Aoi, T., Takahashi, K., and Nanba, N. (TDK Corp.,)**, Japan Patent 85,159,088, 1985.
46. **Hasegawa, M., Umezaki, T., and Muto, M. (Dainichiseika Color and Chemical Manufacturing Co., Ltd.)**, Japan Patent 86,68,293, 1986.
47. **Yamanaka, Y. (NEC Corp.)**, Japan Patent 86,74,148, 1986.
48. **Yamanaka, Y. (NEC Corp.)**, Japan Patent 86,94,249, 1986.
49. **Etsuno, N., Goto, Y., Maeda, M., Shioda, A., and Ito, K. (Fujitsu Ltd.)**, Japan Patent 86,120,797, 1986.
50. **Nishimura, Y., Kawada, H., Haruta, M., Hirai, Y., Mochizuki, N., Nakagiri, T. (Canon K. K.)**, Japan Patent 86,37,475, 1986.
51. **Nishimura, Y., Kawada, H., Haruta, M., Hirai, Y., Mochizuki, N., and Nakagiri, T. (Canon K. K.)**, Japan Patent 86,37,478, 1986.

52. **Nishimura, Y., Kawada, H., Haruta, M., Hirai, Y., Mochizuki, N., and Nakagiri, T. (Canon K. K.)**, Japan Patent 86,63,939, 1986.
53. **Nishimura, Y., Kawada, H., Haruta, M., Hirai, Y., Mochizuki, N., and Nakagiri, T. (Canon K. K.)**, Japan Patent 86,63,479, 1986.
54. **Ueno, T. and Tani, K. (NEC Corp.)**, Japan Patent 86,99,953, 1986.
55. **Nishimura, Y., Kawada, H., Haruta, M., Hirai, Y., Mochizuki, N., and Nakagiri, T. (Canon K. K.)**, Japan Patent 86,63,486, 1986.
56. **Nishimura, Y., Kawada, H., Haruta, M., Hirai, Y., Mochizuki, N., and Nakagiri, T. (Canon K. K.)**, Japan Patent 86,63,487, 1986.
57. **Nishizawa, A., Sueyoshi, N., and Maruta, H. (Victor Co. of Japan, Ltd.)**, Japan Patent 86,125,882, 1986.
58. **Etsuno, N., Goto, Y., Maeda, M., Shioda, A., and Ito, K. (Fujitsu Ltd.)**, Japan Patent 86,114,887, 1986.
59. **Nishimura, Y., Kawada, H., Haruta, M., Hirai, Y., Mochizuki, N., and Nakagiri, T. (Canon K. K.)**, Japan Patent 86,63,481,1986.
60. **Nishimura, Y., Kawada, H., Haruta, M., Hirai, Y., Mochizuki, N., and Nakagiri, T. (Canon K. K.)**, Japan Patent 86,37,476, 1986.
61. **Nishimura, Y., Kawada, H., Haruta, M., Hirai, Y., Mochizuki, N., and Nakagiri, T. (Canon K. K.)**, Japan Patent 86,37,481, 1986.
62. **Nishimura, Y., Kawada, H., Haruta, M., Hirai, Y., Mochizuki, N., and Nakagiri, T. (Canon K. K.)**, Japan Patent 86,37,479, 1986.
63. **Morinaka, A. and Oikawa, S.**, Recording process on an optical recording medium with thermal coloration, *J. Appl. Phys.*, 60(6), 1919, 1986.
64. **Nishimura, Y., Kawada, H., Haruta, M., Hirai, Y., Mochizuki, N., and Nakagiri, T. (Canon K. K.)**, Japan Patent 86,63,483, 1986.
65. **Nishimura, Y., Kawada, H., Haruta, M., Hirai, Y., Mochizuki, N., and Nakagiri, T. (Canon K. K.)**, Japan Patent 86,37,477, 1986.
66. **Nishimura, Y., Kawada, H., Haruta, M., Hirai, Y., Mochizuki, N., and Nakagiri, T.**, Japan Patent 86,63,491, 1986.
67. **Nishimura, Y., Kawada, H., Haruta, M., Hirai, Y., Mochizuki, N., and Nakagiri, T. (Canon K. K.)**, Japan Patent 86,63,482, 1986.
68. **Nishimura, Y., Kawada, H., Haruta, M., Hirai, Y., Mochizuki, N., and Nakagiri, T. (Canon K. K.)**, Japan Patent 86,63,490, 1986.
69. **Nishimura, Y., Kawada, H., Haruta, M., Hirai, Y., Mochizuki, N., and Nakagiri, T. (Canon K. K.)**, Japan Patent 86,63,938, 1986.
70. **Ueda, Y., Umehara, M., Abe, M., Sato, T., and Oba, H. (Ricoh Co., Ltd.)**, Japan Patent 86,22,988, 1986.
71. **Nishimura, Y., Kawada, H., Haruta, M., Hirai, Y., Mochizuki, N., and Nakagiri, T. (Canon K. K.)**, Japan Patent 86,63,489, 1986.
72. **Hashimoto, K. and Saito, K. (Kanegafuchi Chemical Industry Co., Ltd.)**, Japan Patent 86,254,392, 1986.
73. **Hashimoto, K. and Saito, K. (Kanegafuchi Chemical Industry Co., Ltd.)**, Japan Patent 86,254,391, 1986.
74. **Sato, T. (Ricoh Co., Ltd.)**, Japan Patent 86,162,385, 1986.
75. **Hirose, S., Ozawa, H., Abe, K., and Hosono, Y. (Mitsui Toatsu Chemicals, Inc.)**, Japan Patent 86,163,892, 1986.
76. **Takeda, T. and Hayakawa, Y. (Fuji Photo Film Co., Ltd.)**, Japan Patent 86,228,441, 1986.
77. **Kuroiwa, A., Nanba, N., and Kamijo, T. (TDK Corp.)**, Japan Patent 86,197,281, 1986.
78. **Kuroiwa, A., Nanba, N., and Kamijo, T. (TDK Corp.)**, Japan Patent 86,197,280, 1986.
79. **Kuroiwa, A., Nanba, N., and Kamijo, T. (TDK Corp.)**, Japan Patent 86,246,091, 1986.
80. **Celanese Corp.**, Japan Patent 86,235,188, 1986.
81. **Arahara, K. and Oguchi, Y. (Canon K. K.)**, Japan Patent 86,241,190, 1986.
82. **Hiratsuka, H., Nishi, T., and Arishima, K. (Nippon Telegraph and Telephone Public Corp.)**, Japan Patent 86,184,535, 1986.
83. **Yoshike, N. and Kondo, S. (Matsushita Electric Industrial Co., Ltd.)**, Japan Patent 86,211,849, 1986.
84. **Ricoh Co., Ltd.**, Japan Patent 83,169,152, 1983.
85. **Haruta, M., Kuwae, Y., Munakata, H., Yuasa, S., and Sugawa, E. (Canon K. K.)**, Japan Patent 86,143,192, 1986.
86. **Haruta, M., Kuwae, Y., Munakata, H., Yuasa, S., and Sugawa, E. (Canon K. K.)**, Japan Patent 86,144,391, 1986.
87. **Haruta, M., Kuwae, Y., Munakata, H., Yuasa, S., Sugawa, E. (Canon K. K.)**, Japan Patent 86,143,193, 1986.

88. **Hasegawa, M., Umezaki, T., and Muto, M. (Dainichiseika Color and Chemicals Manufacturing Co., Ltd.)**, Japan Patent 86,167,690, 1986.
89. **Mizushima, K., Nakanishi, H., Okamoto, M., and Miura, A. (Toshiba Corp.)**, Japan Patent 87,74,688, 1987.
90. **Ozawa, T., Maeda, S., Kurose, Y., and Iwanami, J. (Mitsubishi Chemical Industries Co., Ltd.)**, Japan Patent 87,124,987, 1987.
91. **Kanno, T., Watanabe, H., and Hamanishi, K. (Olympus Optical Co., Ltd.)**, Japan Patent 87,75,948, 1987.
92. **Fujimori, S., Chiba, R., and Yamazaki, H. (Nippon Telegraph and Telephone Public Corp.)**, Japan Patent 87,102,438, 1987.
93. **Nakamura, T. and Ueno, T. (NEC Corp.)**, Japan Patent 87,12,937, 1987.
94. **Nishimura, Y., Kawada, H., Haruta, M., Hirai, Y., Mochizuki, N., and Nakagiri, T. (Canon K. K.)**, Japan Patent 86,37,486, 1986.
95. **Nishimura, Y., Kawada, H., Haruta, M., Hirai, Y., Mochizuki, N., and Nakagiri, T. (Canon K. K.)**, Japan Patent 86,37,487, 1986.
96. **Nishimura, Y., Kawada, H., Haruta, M., Hirai, Y., Mochizuki, N., and Nakagiri, T. (Canon K. K.)**, Japan Patent 86,37,489, 1986.
97. **Aoi, T., Kamijo, T., and Nanba, N. (TDK Corp.)**, Japan Patent 86,237,239, 1986.
98. **Morinaka, A., Yoshida, T., and Oikawa, S. (Nippon Telegraph and Telephone Co.)**, Japan Patent 86,239,444, 1986.
99. **Horigome, S., Terao, M., Miyauchi, Y., and Nishida, T. (Hitachi, Ltd.)**, Japan Patent 85,135,293, 1985.
100. **Nippon Telegraph and Telephone Public Corp.**, Japan Patent 85,48,396, 1985.
101. **Nippon Columbia Co., Ltd.**, Japan Patent 85,131,649, 1985.
102. **Ito, M. and Edokoro, S. (NEC Corp.)**, Japan Patent 85,149,767, 1985.
103. **Ito, M. and Edokoro, S. (NEC Corp.)**, Japan Patent 85,149,766, 1985.
104. **Nishi, T., Arishima, K., and Hiratsuka, H. (Nippon Telegraph and Telephone Public Corp.)**, Japan Patent 86,32,051, 1986.
105. **Morinaka, A. and Oikawa, S. (Nippon Telegraph and Telephone Public Corp.)**, Japan Patent 86,29,592, 1986.
106. **Sumitomo Chemical Co., Ltd.**, Japan 84,16,785, 1984.
107. **Ricoh Co., Ltd.**, Japan Patent 82,186,242, 1982.
108. **Nippon Telegraph and Telephone Public Corp.**, Japan Patent 83,08,695, 1983.
109. **Ricoh Co., Ltd.**, Japan Patent 82,212,636, 1982.
110. **Ricoh Co., Ltd.**, Japan Patent 82,212,634, 1982.
111. **Ricoh Co., Ltd.**, Japan Patent 82,212,639, 1982.
112. **Ricoh Co., Ltd.**, Japan Patent 83,112,794, 1983.
113. **Sumitomo Chemical Co., Ltd.**, Japan Patent 84,16,153, 1984.
114. **Nippon Telegraph and Telephone Public Corp.**, Japan Patent 84,210,543, 1984.
115. **Nikles, D. E., Jones, R. S., Jr., and Kuder, J. E. (Celanese Corp.)**, U.S. Patent 4,605,607, 1986.
116. **Masumoto, T., Suzuki, K., Okubo, M., Matsumoto, A., Ota, N., and Masuda, S. (Research and Development Corp. of Japan)**, British Patent Application 2,164,462, 1986.
117. **Albert, B. and Neumann, P. (BASF AG)**, German Patent 3,446,418, 1986.
118. **Ozawa, T., Maeda, S., and Kurose, Y. (Mitsubishi Chemical Industries Co., Ltd.)**, PCT International Application 87,01,076, 1987.
119. **Goto, Y., Koshino, N., Ogawa, S., Goto, H., Ogawa, K., Nichimo, H. (Fujitsu Ltd.)**, Europe Patent Application 84,729, 1983.
120. **Sony Corp.**, Netherlands Application 83,03,893, 1984.
121. **Munakata, H., Tomida, Y., Haruta, M., Hirai, Y., Nishimura, Y., and Hamamoto, T. (Canon K. K.)**, German Patent 3,446,878, 1985.
122. **Dautartas, M. F., Suh, S. Y., Forrest, S. R., Kaplan, M. L., Lovinger, A. J., and Schmidt, P. H.**, Optical recording using hydrogen phthalocyanine thin films, *Appl. Phys. A*, A36(2), 71, 1985.

Chapter 29

STUDIES AND TOPICS IN PHOTOCHEMISTRY

The relaxation kinetics of Al, Si, and di hydrogen phthalocyanines (pcs) are studied by conventional and laser flash photolysis. Second-order deviations observed in the relaxation kinetics of the pcs are discussed in terms of an excimer-mediated mechanism.[1]

Triplet-triplet annihilation of the lowest-lying excited state of porphyrins and pcs are studied by laser flash photolysis. It is indicated that the lowest-lying triplet dimerizes into excimers at a rate less than diffusion controlled.[2]

In diffuse reflectance flash photolysis, the decay of transient absorption and time-resolved spectra are reported for triplet states in benzil microcrystals adsorbed as a function of an acridine monolayer on powdered silica and dyed with Al sulfo pc on cotton fabric.[3]

Diffuse reflectance laser flash photolysis is used to study primary photoprocesses in anisotropic media. The use of photolysis and diffuse reflectance techniques is discussed in terms of transients in a cotton fabric dyed with Al sulfo pc.[4]

Visible irradiations at 640 and 590 nm of Rh(III) pcs with high photonic fluxes induce the same abstraction of hydrogen from solvent molecules as in monophotonic UV photolysis. A mechanism for the sequential biphotonic photochemistry of the pcs is discussed.[5]

Sequential biphotonic processes in pcs are studied at 640 and 490 nm.[6]

The photochemical properties of the pc radicals Al pc Cl^+, Rh pc $(CH_3OH)Br^+$, and Zn pc^+ are studied by a time-resolved sequential biphotonic technique.[7]

Relief holographic gratings are formed on the surface of Cu pc films by using ruby laser radiation. The same recording mechanism is observed using Pb pc films and ruby and YAG laser radiation.[8]

Photoacoustic studies of thermally thin Cu pc sublimates on quartz are made in a wide optical absorption range to verify the validity of the Rosencwaig-Gersho (RG) theory.[9]

An ablative optical recording material with sensitivity in the near IR region contains a layer of a naphthalocyanine (Figure 1), where M = di hydrogen, Al(OH), Cd, Co, Cu, Mn, Ni, Pd, $SnCl_2$, Zn, VO . . .,R, R^1 = H, C_{4-20} alkyl; n = 1, 2. The reflectivity of such a film at 800 nm is 12%.[10]

An IR absorbing pc such as octa-3,6-(4-methylphenylthio)-di hydrogen pc absorbing at 778 nm in chloroform is useful to make the IR energy available as heat energy and to make IR inks, liquid crystals, sun visors, etc.[11]

Laser excitation of Lu pc_2 thin films at 670 nm (Q band) and at 1423 nm (intermolecular charge transfer band) leads to the same transient species.[12]

Stable Rh(III) pc (CH_3OH) X^+ (X = Cl or Br) radicals are studied by continuous, laser, and flash photolysis. UV irradiation (250 or 320 nm) induces photoreduction of the cations back to the parent pc and photolysis in the visible region (525 nm) induces photodecomposition of the macrocycle.[13]

Triplet quenching of Al(III), In(III), Mg, and Zn is studied by nanosecond laser photolysis kinetic spectroscopy.[14]

An EPR-flash photolysis system is used to study transient changes of dielectric loss of di hydrogen pc.[15]

The ability of some M pcs to photogenerate singlet oxygen is studied. Their activities decrease in the order Zn $>$ Mg $>$ Cu $>$ Mn $>$ Co $>$ Fe.[16]

The quenching of phosphorescence of Al tetra-*tert*-butyl pc Cl and the formation of triplet exciplexes are studied.[17]

Upon Q band excitation at $>$580 nm, Ru(II) pc complexes can be quantitatively photooxidized to stable π-cation radical species.[18]

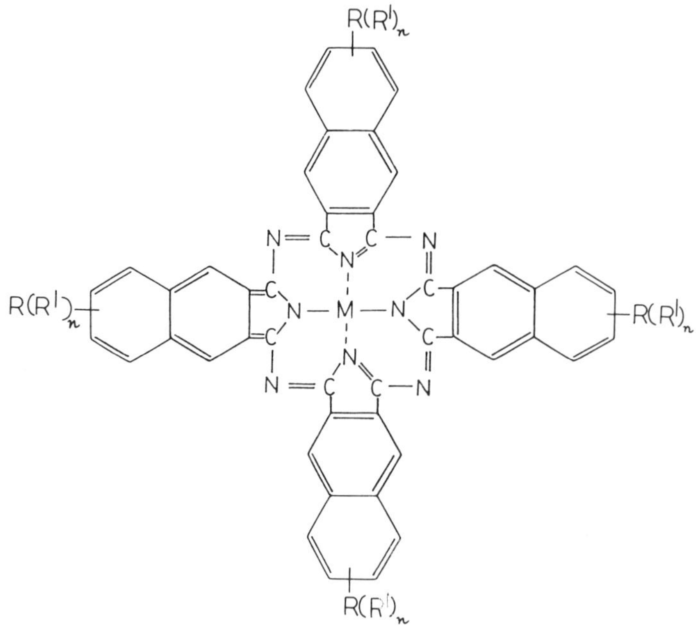

FIGURE 1. A naphthalocyanine for dispersion in a resinous binder.

The kinetics of fluorescence quenching of Al tetra-*tert*-butyl pc Cl are studied in various solvents.[19]

A pc (lanthanum) pc radical is generated from pc La pc H (lanthanum = La, Lu, Nd, Y) in a solvent mixture of methylene chloride and methyl cyanide upon irradiation with light at >320 nm.[20]

The π-cation-radical species of Zn pc, Zn $(pc^-)^+$, is formed quantitatively as a stable product in solution following photochemical reactions with visible light in the presence of electron acceptors.[21]

The spectral sensitization of hydrogenated amorphous silicon by surface layers of In pc Br and 1,1-bis (4-di-*p*-tolylaminophenyl)cyclohexane is studied.[22]

Triplet energy transfer from Al pc Cl to bis(dibutylphenyl)perylene bis(dicarboxamide) is studied.[23]

Cu pc 2,9,16,23-tetracarboxylic acid reaction product with chloromethylstyrene-ethylene graftco polymer is a catalyst for the reversed *cis-trans* dark isomerization of azobenzene in a solar energy conversion system.[24]

Porphyrins and pcs can act as highly efficient electron donor molecules in linked donor-acceptor complexes that may be used in molecular solar cells.[25]

Photocatalysis by Zn pc on zeolites and poisoning of fluid cracking catalysts are the subject of a thesis.[26]

The relating of the thermal and photochemical reactivity of the pcs to the existence of empty and filled π-orbitals at energies which allow the oxidation and reduction of the macrocycle at moderate potentials is the subject of a discussion.[27]

The photo reduction-oxidation behavior of copolymers containing covalently bound pc moieties and/or tetraphenylporphyrin moieties is studied under visible light in the presence of a donor and an acceptor.[28]

The excited state reduction-oxidation properties of Ru pcs are studied under the influence of the axial ligand on the rates of relaxation and electron transfer quenching of the lowest excited state.[29]

Co 4.4′,4″,4‴-tetrasulfo pc covalently linked to the surface of titania particles is an effective photocatalyst to oxidize sulfur dioxide to S^{+6} in aqueous solutions.[30]

M pc photosensitizers, with respect to photosensitization in solar energy conversion systems involved in the photodecomposition of water for hydrogen production, are soluble in ammonia and aliphatic amine solutions.[31] Cr pc dissolves as Cr^{+2} and has interesting photochemical properties, e.g., for new photoreducing systems.

A new photocatalytic system to produce hydrogen using Zn tetracarboxylic pc is studied.[32]

The effects of titania, di hydrogen pc, and perovskite-type compounds on the photooxidation of formic acid are studied. When a minute amount of di hydrogen pc is present in the acid solution, titania catalytic activity increases up to 25%.[33]

M pc-sensitized photoreduction of dimethyl-4-nitrophthalate by ascorbic acid is studied.[34]

Illuminating solutions of Mg pc in methylamine with visible light produces a radical anion. Dimethylhydrazine is formed in the presence of hydrogen peroxide.[35]

A sensitive variation of frequency-modulated spectroscopy is used to detect photochemical holes burned in di hydrogen pc-doped polyethylene at times as low as 100 ns.[36]

The photochemical hole-burning properties of protonated di hydrogen pc in concentrated sulfuric acid glasses at 8300 Å are measured using Ga Al As diode lasers.[37] Line-shape theory is described. The theory links Debye-Weller factors, Debye temperatures, and photochemical quantum yields to hole-burning parameters. The model system is di hydrogen pc in polyethylene, polystyrene, and polymethyl methacrylate.[38]

Photomemory materials using the photochemical hole burning effect contain di lithium pc film dispersed in matrices. The film is placed into liquid helium and irradiated with a pulse color laser beam to form holes with optical density 0.02.[39]

A review with 27 references discusses the photochromic reaction, photochemical hole burning, and compounds used for photochemical hole burning such as di hydrogen pc.[40]

The electrophotographic process is studied using di hydrogen pc sensitized film of polyvinylidene fluoride. The contribution of photoconductivity and the bulk photovoltaic effect to latent image formation are discussed.[41]

Photogeneration of the charge carrier, in an electrophotographic organic photoreceptor is studied on two types of photoreceptor structures consisting of anthanthrone/Ag and anthanthrone/Cu pc heterojunction.[42]

The UV photochemistry of Mn(III) tetrasulfo pc (H_2O) $(OH)^{4-}$ and (H_2O) (tetrasulfo pc) Mn(III)-O-Mn(III) (tetrasulfo pc) $(OH)^{9-}$ is studied by continuous and flash photolysis. There is a significant charge transfer photochemistry, that is, oxidation of the hydroxide or μ-oxo bridge and reduction of the metal center in both complexes.[43]

The fluorescence lifetime of Zn 2,9,16,23-tetramethoxy pc, Zn pc, 2,9,16,23-tetraphenoxy pc, and *N,N′,N″,N‴*-tetrabutyl-Zn pc 2,9,16,23-tetrasulfonamide is determined by the nontransient method.[44]

The photochemistry of di hydrogen pc tetrasulfonic acid is studied along with hematoporphyrin.[45]

Di hydrogen pc is a spectral sensitizer in aqueous compositions for hydrogen production by radiation.[46]

Colloidal suspensions of organic pigments including Cu pc are developed by dispersant addition and ultrasound treatment. They are characterized photometrically.[47]

REFERENCES

1. **Frink, M. E., Geiger, D. K., and Ferraudi, G. J.,** Excimer formation from triplet-triplet annihilation reactions of the lowest lying triplet excited state in aluminum(III), silicon(IV), and metal-free phthalocyanines: medium and magnetic field effects on the rate of reaction, *J. Phys. Chem.*, 90(9), 1924, 1986.
2. **Frink, M. E. and Ferraudi, G.,** Excimer formation in triplet-triplet annihilation reactions of metal ($^3\pi\pi^*$) porphyrins and ($^3\pi\pi^*$) phthalocyanines, *Chem. Phys. Lett.*, 124(6), 576, 1986.
3. **Wilkinson, F.,** Diffuse reflectance flash photolysis, *J. Chem. Soc., Faraday Trans. 2,* 82(12), 2073, 1986.
4. **Wilkinson, F. and Willsher, C. J.,** The use of diffuse reflectance laser flash photolysis to study primary photoprocesses in anisotropic media, *Tetrahedron,* 43(7), 1197, 1987.
5. **Muralidharan, S. and Ferraudi, G.,** Sequential biphotonic processes in rhodium(III) phthalocyanines, *J. Phys. Chem.*, 87(24), 4877, 1983.
6. **Ferraudi, G.,** Sequential dichromatic-biphotonic photochemistry: a simple relationship of the intermediate lifetime with the delay between synthesizing and photolyzing pulses, *J. Phys. Chem.*, 88(18), 3938, 1984.
7. **Van Vlierberge, B. and Ferraudi, G.,** Sequential biphotonic processes: photochemical reactivity of phthalocyanine radicals, *Inorg. Chem.*, 26(2), 337, 1987.
8. **Vertsimakha, Y. I., Grodzinskaya, M. D., Savchuk, A. V., Sal'kova, E. N., and Hamann, C.,** Holographic features of organic thin films, *PotsdamerForsch. Paedagog. Hochsch. ''Karl Liebknecht'' Potsdam, Naturwiss. Reihe,* 37, 136, 1983.
9. **Holter, A. and Perkampus, H. H.,** Photoacoustic investigations of thermally thin sublimates, *Appl. Spectrosc.*, 39(2), 278, 1985.
10. **Law, K. Y., Johnson, G. E., and Lin, J. W. P., (Xerox Corp.),** U.S. Patent 4,492,750, 1985.
11. **Duggan, P. J. and Gordon, P. F. (Imperial Chemical Industries PLC),** Europe Patent Application 155, 780, 1985.
12. **Tran Thi, T. H., Markovitsi, D., Even, R., and Simon, J.,** Laser induced intramolecular charge transfer in a lutetium bis-phthalocyanine thin film, *Chem. Phys. Lett.*, 139(2), 207, 1987.
13. **Ferraudi, G., Oishi, S., and Muraldiharan, S.,** The photochemical properties of rhodium(III) phthalocyanine cation radicals, *J. Phys. Chem.*, 88(22), 5261, 1984.
14. **Ohno, T. and Kato, S.,** Electron transfer reactions of excited phthalocyanines: spin restriction on reaction rate of electron transfer and energy transfer to cobalt compounds, *J. Phys. Chem.*, 88(8), 1670, 1984.
15. **Dobbin, C. J., McIntosh, A. R., Bolton, J. R., Popovic, Z. D., and Harbour, J. R.,** The utilization of time resolved dielectric loss to probe the role of the surface in heterogeneous photochemistry, *J. Chem. Soc., Faraday Trans 1,* 82(12), 3625, 1986.
16. **Wu, S., Zhang, H., Cui, G., Xu, D., and Xu, H.,** Ability of some phthalocyanine compounds for singlet oxygen photogeneration, *Huaxue Xuebao,* 43(1), 10, 1985.
17. **Kapinus, E. I., Aleksankina, M. M., Staryi, V. P., and Dilung, I. I.,** Dependence of lifetimes of triplet exciplexes of porphyrins with electron acceptors on the nature of the reagents, *Dokl. Akad. Nauk SSSR,* 278(5), 1165, 1984.
18. **Nyokong, T., Gasyna, Z., and Stillman, M. J.,** Photochemical formation of ruthenium phthalocyanine π-cation radical species, *Inorg. Chim. Acta,* 112(1), 11, 1986.
19. **Kapinus, E. I., Kucherova, I. Y., and Dilung, I. I.,** Solvent isotope effects on the quenching of porphyrin fluorescence by electron acceptors in alcohols or water, *J. Photochem.*, 34(3), 289, 1986.
20. **Kasuga, K., Morimoto, H., and Ando, M.,** Photoinduced radical formation of bis(phthalocyaninato)lanthanoid(III) in a solvent mixture of dichloromethane and acetonitrile, *Inorg. Chem.*, 25(14), 2478, 1986.
21. **Nyokong, T., Gasyna, Z., and Stillman, M. J.,** Phthalocyanine π-cation radical species: photochemical and electrochemical preparation of $ZnPc(-1)^{\cdot +}$ in solution, *Inorg. Chem.*, 26(4), 548, 1987.
22. **Borsenberger, P. M.,** Spectral sensitization of amorphous silicon by bromoindium phthalocyanine, *J. Appl. Phys.*, 62(7), 2942, 1987.
23. **Ford, W. E. and Kamat, P. V.,** Photochemistry of 3,4,9,10-perylenetetracarboxylic dianhydride dyes. III. Singlet and triplet excited state properties of the bis(2,5-di-*tert*-butylphenyl)imide derivative, *J. Phys. Chem.*, 91(25), 6373, 1987.
24. **Taoda, H., Hayakawa, K., Kawese, K., and Yamakita, H.,** Photochemical conversion and storage of solar energy by azobenzene, *J. Chem. Eng. Jpn.*, 20(3), 265, 1987.
25. **Gasyna, T. N. and Stillman, M. J.,** Photooxidation of phthalocyanines. Photoinduced reactivity of the triplet state, *ACS Symp. Ser.*, 321, 309, 1986.
26. **Psaras, D.,** Photocatalysis by Zinc Phthalocyanine on Zeolites and Poisoning Studies of Fluid Cracking Catalysts, Ph.D. thesis, University of Connecticut, Storrs, 1985.
27. **Ferraudi, G.,** Photochemistry of coordination complexes: the phthalocyanines, *Simp. Quim. Inorg. ''Met. Transicion'', 1st,* 28, 1985.

28. **Woehrle, D. and Krawczyk, G.,** Polymeric bound porphyrins and their precursors. III. Photoredox properties of combined moieties of porphyrin and phthalocyanine, covalently bound to polystyrene, *Markromol. Chem.,* 187(11), 2535, 1986.
29. **Ferraudi, G. J. and Prasad, D. R.,** Excited state redox properties of phthalocyanines: influence of the axial ligand on the rates of relaxation and electron transfer quenching of the lowest $^3\pi\pi^*$ excited state, *J. Chem. Soc., Dalton Trans.,* 10, 2137, 1984.
30. **Hong, A. P., Bahnemann, D. W., and Hoffmann, M. R.,** Cobalt(II) tetrasulfophthalocyanine on titanium dioxide. II. Kinetics and mechanisms of the photocatalytic oxidation of aqueous sulfur dioxide, *J. Phys. Chem.,* 91(24), 6245, 1987.
31. **Sauvage, F. X., De Backer, M., Jacquot, P., and Van Vlierberge, B.,** Photochemical oxidation of amines by metal phthalocyanines, *INTERSOL 85, Proc. Bienn. Congr. Int. Sol. Energy Soc., 9th,* 3, 1892, 1986.
32. **Zhu, Y., Xiao, X., and Xu, H.,** Photocatalytic cleavage of water by zinc tetracarboxylphthalocyanine, *Ganguang Kexue Yu Kuang Huaxue,* 2, 42, 1985.
33. **Yan, Z.,** Photocatalytic oxidation of formic acid, *Huadong Huagong Xeuyuan Xuebuo,* 11(4), 459, 1985.
34. **Xu, H. and Xue, N.,** Metal phthalocyanine sensitized photoreduction of nitro compound, *Huaxue Xuebao,* 45(6), 551, 1987.
35. **De Backer, M., Jacquot, P., Sauvage, F. X., and Van Vlierberge, B., and Lepoutre, G.,** Photoreduction of magnesium phthalocyanine in methylamine solutions — production of hydrazine, *J. Chim. Phys.-Chim. Biol.,* 84(3), 429, 1987.
36. **Romagnoli, M., Moerner, W. E., Schellenberg, F. M., Levenson, M. D., and Bjorklund, G. C.,** Beyond the bottleneck: submicrosecond hole burning in phthalocyanine, *J. Opt. Soc. Am. B,* 1(3), 341, 1984.
37. **Lee, H. W. H., Huston, A. L., Gehrtz, M., and Moerner, W. E.,** Photochemical hole burning in a protonated phthalocyanine with gallium arsenide diode lasers, *Chem. Phys. Lett.,* 114(5-6), 491, 1985.
38. **Kador, L., Schulte, G., and Haarer, D.,** Relation between hole burning parameters and molecular parameters: free-base phthalocyanine in polymer hosts, *J. Phys. Chem.,* 90(7), 1264, 1986.
39. **Tsucha, J. (Idemitsu Kosan Co., Ltd.),** Japan Patent 87,41,083, 1987.
40. **Horie, K.,** Development of functional materials by solid state chemistry and solid state physics, *Kino Zairyo,* 6(9), 41, 1986.
41. **Fridkin, V. M., Shlenskii, A. L., Verchovskaya, K. A., Bilke, W. D., Sydow, M., Markiewitz, N. N., and Pietsch, H.,** Pyroelectric and photovoltaic effects in the ferroelectric polymer poly(vinylidene fluoride) as fundamentals of electrophotographic process, *J. Inf. Rec. Mater.,* 13(6), 421, 1985.
42. **Aizawa, Y., Kuriyagawa, S., Takahashi, Y., and Sakata, T.,** Photocarrier generation in anthanthrone and copper phthalocyanine/anthanthrone photoreceptors, *Denshi Shashin Gakkaishi,* 25(3), 223, 1986.
43. **Ferraudi, G. and Granifo, J.,** Charge transfer and ligand centered photochemistry of manganese(III) phthalocyanines, *J. Phys. Chem.,* 89(7), 1206, 1985.
44. **Shen, T., Yuan, Z., and Xu, H.,** Evaluation of the fluorescence lifetime of zinc phthalocyanines by nontransient experiments, *Wuli Huazue Xeubao,* 1(3), 265, 1985.
45. **Truscott. T. G.,** The photochemistry of hematoporphyrin and some related species, *J. Chem. Soc., Faraday Trans. 2,* 82(12), 2177, 1986.
46. **Herbst, R. L., Jr.,** U.S. Patent 4,637,867, 1987.
47. **Jelinek, Z.,** Czech Patent 215,773, 1984.

Chapter 30

CHROMISM

I. ELECTROCHROMISM

The electrochromism of an Er H(phthalocyanine)$_2$(pc)$_2$, erbium di pc, is studied with (glass)/In Sn oxide (ITO)/Er H (pc)$_2$ + LiF/LiF + $CaCl_2$/Ag) solid cells. The color changes rapidly from green to purple-red when +2 Volts are applied to the ITO electrode.[1]

A di hydrogen film prepared by the Langmuir-Blodgett technique deposited on an ITO/glass shows electrochromic changes in aqueous potassium chloride solution changing from blue-green to purple-red in an oxidative process.[2]

A review of electrochromic materials with many references includes di pc rare earth compounds.[3]

A solid-state electrochromic cell using lutetium di pc Lu H (pc)$_2$ color changes from green to blue to violet vs. the stannic oxide counter electrode.[4]

A solid electrochromic cell using Lu H (pc)$_2$ and lead difluoride experiences color changes from green to orange.[5]

References 6 to 14 relate to the patent literature.

Electrochromic display devices with improved durability of ⩾50,000 cycles are prepared by using Y and Si di pc complexes.[6]

An electrochromic display device includes a Lu di pc complex.[7]

Vanadium oxide impregnated with lutetium pc is a constituent of electrochromic display devices.[8]

An electrochromic display device contains M pc-coated display electrodes.[9]

An electrochromic display device contains M pc complexes as the electrochromic substance.[10]

The color change response of an M di pc complex display material is improved by treating the material with a surfactant. Green Lu di pc films on stannic oxide are so treated with Dow-Corning 193 silicone glycol and Witcamino AL 42-12 imidazoline/tall oil derivative.[11]

Co pc tetracarboxylic acid chloride reacts with poly(γ-benzyl L-glutamate) to form a polymeric liquid crystal which may be of use in color display devices.[12]

A Lu derivative is made from 2,3,9,10,16,17,23,24-octa(dodecycloxy-1-methyl) Cu pc. The product has a metastable electrochromic liquid crystal phase.[13]

Lanthanide di pcs (pc$_2$s) and Y di pc useful for electrochromic purposes are prepared by heating lanthanide or Y derivatives with phthalonitrile above the critical endothermic temperature and then at a temperature less than that of di pc decomposition.[14]

II. THERMOCHROMISM

A thermochromic composition contains Cu hexachlorodecakis [(4-chlorophenyl) thio] pc.[15]

REFERENCES

1. **Yamamoto, H., Noguchi, M., and Tanaka, M.,** Electrochromism of erbium-diphthalocyanine complex films in all solid display cells, *Jpn. J. Appl. Phys., Part 2,* 23(4), 221, 1984.
2. **Yamamoto, H., Sugiyama, T., and Tanaka, M.,** Electrochromism of metal-free phthalocyanine Langmuir-Blodgett films, *Jpn. J. Appl. Phys., Part 2,* 24(5), 305, 1985.
3. **Matsuhiro, K. and Masuda, Y.,** Electrochromic materials, *Kagaku, Zokan (Kyoto),* 104, 107, 1984.
4. **Sammells, A. F. and Pujare, N. U.,** Solid-state electrochromic cell using lutetium diphthalocyanine, *J. Electrochem. Soc.,* 133(5), 1065, 1986.
5. **Egashira, N. and Kokado, H.,** A solid electrochromic cell using lutetium diphthalocyanine and lead fluoride, *Jpn. J. Appl. Phys., Part 2,* 25(6), L462, 1986.
6. **Rockwell International Corp.,** Japan Patent 83,57,113, 1983.
7. **Japan Optics Co., Ltd.,** Electrochromic display devices, Japan Patent 83,78,126, 1983.
8. **Masumoto, T., Suzuki, K., Masuda, S., Ota, Y., Matsumoto, A., and Okubo, M. (Research Development Corp. of Japan),** Japan Patent 86,204,677, 1986.
9. **Kawakami, K. and Matsuhiro, K. (Asahi Glass Co., Ltd.),** Japan Patent 85,182,422, 1985.
10. **Kawakami, K. and Matsuhiro, K. (Asahi Glass Co., Ltd.),** Japan Patent 85,173,523, 1985.
11. **Nicholson, M. M.,** U.S. Patent Application 632, 179, 1985.
12. **TDK Corp.,** Japan Patent 85,55,077, 1985.
13. **Clarisse, C., Riou, M. T., Meston, N., Simon, J., and Piechocki, C.,** French Patent 2,574,806, 1986.
14. **Clarisse, C., Riou, M. T., and Auregan, M.,** Europe Patent Application 150,152, 1985.
15. **Kuroda, Y. (Sakura Color Products Corp.),** Japan Patent 86,233,583, 1986.

Chapter 31

PHOTOREDUCTION OF VIOLOGENS

The mechanism of Cd, Fe, Mg, Mn, or Zn sensitized reduction of methylviologen (MV^{2+}) in dimethylsulfoxide (DMSO) or DMSO-water solution is studied by nanosecond spectroscopy. The fluorescence lifetimes of Mg phthalocyanine (pc), Zn pc, and Cd pc are 7.5, 4.5, and 0.6 ns, respectively.[1]

The quantum yields of the photoreduction of methylviologen (MV^{2+}) photosensitized by the M pcs (Figure 1) where M = Mg, Zn, R =

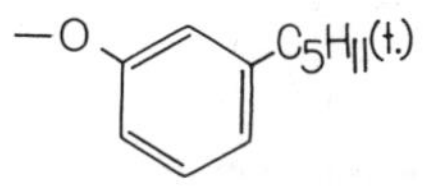

or M = Zn, and R = $-SO_2NHC_4H_9$, or R = $-SO_2NHC_{14}H_{29}$ in aqueous solution containing an electron donor are determined using a potassium ferrioxalate actinometer. From the measured quantum yield values, a photoreduction mechanism is proposed.[2]

The effect of electrostatic interactions between Zn tetrasulfo pc and the electron acceptors methylviologen, propylviologen sulfonate, and 1,4-benzoquinone is studied.[3]

The photoreduction of propylviologen sulfonate by Mg pc in anionic micelles in aqueous solutions is studied.[4]

The fluorescence quantum yield for Mg, Zn, and Co pcs are 0.7, 0.3, and 0.025, respectively. Photosensitized viologen reduction is studied together with the fluorescence quenching. The photosensitization activities of pcs are in the order Mg > Zn > Co, whereas the quenching activity of butylviologen is greater than that of methylviologen.[5]

A remarkably high quantum yield of photoreduction of methylviologen is a notable property of the Zn pc derivatives used in the reaction system to photoreduce methylviologen in the presence of EDTA.[6]

The photoreduction of methylviologen sensitized by Co pc in aqueous dimethylsulfoxide solutions under visible light is studied.[7]

The photoreduction of methylviologen in methanol is catalyzed by Lanthanoid(III) di pc (or pc or bis pc) at >440 nm.[8]

Polymeric (di hydrogen, Cu, Mg, Ni, or Zn) pc thin films prepared from the gas phase by low-temperature plasma polymerization reduce methylviologen under visible light.[9]

Polystyrene with pendant Ni 4,4′,4″,4‴-tetracarboxy pc is synthesized by reacting Ni 4,4′,4″,4‴-tetracarboxy pc with partially chloromethylated polystyrene. Photoreduction of methylviologen by visible light is accomplished with the modified polymer. The photocatalytic activity is enhanced by bonding the Ni 4,4′,4″,4‴-tetracarboxy pc units to the polymer. The relationship between the catalytic activity and the modified polymer structure is discussed.[10]

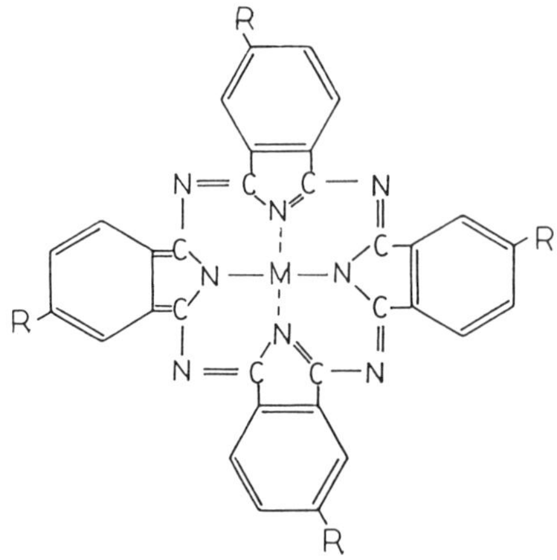

FIGURE 1. M pcs that photosensitize the photoreduction of methylviologen.

REFERENCES

1. **Ohtani, H., Kobayashi, T., Ohno, T., Kato, S., Tanno, T., and Yamada, A.,** Nanosecond spectroscopy on the mechanism of the reduction of methylviologen sensitized by metallophthalocyanine, *J. Phys. Chem.*, 88(19), 4431, 1984.
2. **Yuan, Z., Shou, H., and Xu, H.,** Quantum yield determinations of photosensitized reduction of methyl viologen by metallic phthalocyanines, *Huaxue Tongbao,* 8, 15, 1983.
3. **Sauvage, F. X., De Backer, M., Lepoutre, G., Demortier, A., and Van Vlierberge, B.,** Photoreduction of viologens by zinc phthalocyanine, *J. Chim. Phys. Phys.-Chim. Biol.*, 81(2), 77, 1984.
4. **De Backer, M., Sauvage, F. X., Demortier, A., Vlierberge, B. V., and Lepoutre, G.,** Photoreduction of sulfonated propylviologen by magnesium phthalocyanine, *C.R. Acad. Sci., Ser. 2,* 299(15), 1035, 1984.
5. **Xu, H., Zhou, Q., and Shen, S.,** Photoinduced redox reactions sensitized by metal phthalocyanines. II. Fluorescence spectra characteristics and photosensitized reduction of viologen, *Ganguang Kexue Yu Kuang Huaxue,* 1, 40, 1984.
6. **Zhou, Q., Shen, S., and Xu, H.,** Photoinduced redox reactions sensitized by metal phthalocyanines. III. Photoreduction of methylviologen in micelllar solution, *Ganguang Kexue Yu Kuang Huaxue,* 3, 23, 1984.
7. **Wang, X., Chen, Y., and Liang, X.,** Sensitized photoredox reaction — tetraphenylporphin and metal tetraphenylporphin-induced photoreduction of methyl viologen, *Ganguang Kexue Yu Kuang Huaxue,* 1, 51, 1985.
8. **Kasuga, K. and Atarashi, H.,** Bis(phthalocyaninato) lanthanoid(III)-catalyzed photoreduction of methyviologen, *Chem. Lett.*, 10, 1679, 1986.
9. **Osada, Y., Mizumoti, A., and Tsuruta, H.,** Photovoltaic and catalytic activity of plasma-polymerized phthalocyanine films, *J. Macromol. Sci., Chem.*, A24(3-4), 403, 1987.
10. **Yamaguchi, H., Fujiwara, R., and Kusuda, K.,** Photoinduced charge separation of polystyrene-bound nickel(II)-4,4′,4″,4‴-tetracarboxyphthalocyanine, *J. Macromol. Sci. Chem.*, A24(3-4), 367, 1987.

Chapter 32

COATINGS, FILMS, AND PHOTOGRAPHIC PRODUCTS

Radiation hardenable compositions contain Cu phthalocyanine (pc) for water-resistant multilayer photographic paper supports.[1]

A Cu pc water-soluble derivative is a component in a photographic diffusion transfer film polymeric diffusion control layer.[2]

A copying material is described to prepare duplication of originals such as prints. For example, a thin translucent paper support is coated with a layer containing 4-methoxy-1-naphthol sensitized with erythrosine, Mg pc, trichlrodiethoxyisovalanthrene in an ethyl cellulose binder, and overcoated with a layer of dimethylsiloxane adhesive.[3]

A photographic multicolor element for tanning development contains Cu pc.[4]

Cu pc is vapor deposited to a transmission optical density of 2.0 on a polyester support, then by an alumina/Al containing layer with an optical density of 4.5. The material is then exposed to produce a blue step tablet image on a black background. The material can be used for color proofing and color slide preparation.[5]

A photothermographic copying sheet uses microencapsulated Cu pc.[6]

Wash-off films contain microencapsulated Cu pc.[7]

Cu pc trisulfonic acid is a stain inhibitor for color photographic paper containing an arylthiopyrazolone derivative.[8]

A photosensitive positive acting color proofing sheet contains a thermal adhesive layer and a layer containing Cu pc.[9]

A photoimaging element for the preparation of laminated paper with a photosensitive layer contains silver halide particles, polymeric coupler, and Cu pc.[10]

A photosensitive layer contains a diazo resin and Cu pc for multilayer photoimaging materials for color proofing.[11]

Photosensitive materials contain Cu pc for prepress color proofs.[12]

An electrolyte contains Cu pc tetrasulfonic acid for electrochemical development of photoimaging compositions containing an acidic polymer-basic complex and photosensitizers.[13]

Cu pc is used to form a cyan image in multicolor image formation in a method to produce prepress color proofs.[14]

An imaging material that can perform the function of a material for the proof work in a photomechanical process and also function as a lithographic film includes Cu pc.[15]

An ink jet recording paper has a surface layer that contains Cu pc trisulfonic acid for improved resistance to fading by water.[16]

Color photographic photosensitive material contains Cu pc acylaminonaphthol derivatives. They are oxidizable or reducible dye-releasing compounds in photothermographic materials.[17]

A photoimaging material contains a metallic layer, a photoresist layer, Cu pc layer, and abrasion protective layer vapor deposited on a metal layer.[18]

Photosensitive mixtures for the production of colored relief images contain pc green coated on an aluminum support, imagewise exposed, and developed with an aqueous developer to give a colored relief image.[19]

A photographic element contains a temporary polymer mordant and Cu *N,N′,N″,N‴*-tetrakis [2-(2,5-dihydroxyphenyl)-1-methylethyl] pc-1,8,15,22-tetrasulfonamide, as shown in Figure 1. The mordant can convert to nonmordanting species in an alkaline environment.[20]

An electrooptical color image sensor device contains an In pc halo or Al pc halo derivative layer, to detect in the blue, green, and red regions.[21]

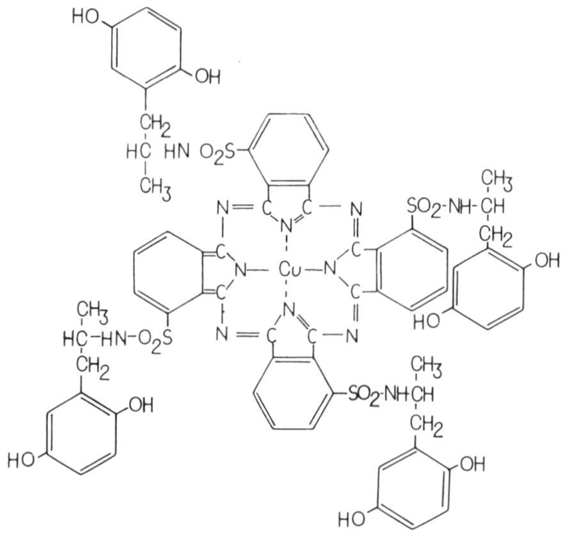

FIGURE 1. A Cu pc tetrasulfonamide complex.

REFERENCES

1. **Anthonsen, R., Sack, W., Trentmann, H., and Woellmer, H. G. (Felix Schoeller Jr, GmbH & Co. KG),** German Patent 3,300,025, 1984.
2. **Jasne, S. J., Schwarzel, W. C., Sullivan, C. I., and Taylor, L. D. (Polaroid Co.),** Europe Patent Application 127,787, 1984.
3. **Wartman, T. G. (Minnesota Mining and Manufacturing Co.),** U.S. Patent 4,508,810, 1985.
4. **Macioch, M. T. (Minnesota Mining and Manufacturing Co.),** U.S. Patent 4,544,620, 1985.
5. **Fisch, R. S. (Minnesota Mining and Manufacturing Co.),** PCT International Application 86 00,575, 1986.
6. **Tamura, Y. and Kaneko, S. (Canon K. K.),** Japan Patent 87,94,383, 1987.
7. **E. I. du Pont de Nemours & Co., Inc.,** Japan Patent 87,141,535, 1987.
8. **Fuji Photo Film Co., Ltd.,** Japan Patent 83,14,832, 1983.
9. **Koelsch, P. M. and Krech, R. I. (Minnesota Mining and Manufacturing Co.),** Europe Patent Application 115,899, 1984.
10. **Chambers, W. J. (E. I. du Pont de Nemours & Co., Inc.),** U.S. Patent 4,520,093, 1985.
11. **Sanyo-Kokusaku Pulp Co., Ltd.,** Japan Patent 85,32,045, 1985.
12. **Mino, H., Iijima, T., and Kadota, K. (Sanyo-Kokusaku Pulp Co., Ltd.),** Japan Patent 86,286,858, 1986.
13. **Haruta, M., Kuwae, Y., Yuasa, S., and Miyazaki, T. (Canon K. K.),** Japan Patent 86,169,868, 1986.
14. **Mino, H., Iijima, T., and Kadota, K. (Sanyo-Kokusaku Pulp Co., Ltd.),** Japan Patent 86,286,850, 1986.
15. **Mino, H. and Iizima, T. (Sanyo-Kokusaku Pulp Co., Ltd.),** U.S. Patent 4,551,408, 1985.
16. **Ricoh Co., Ltd.,** Japan Patent 84,96,988, 1984.
17. **Sato, K., Fujita, S., Naito, H., and Hara, H. (Fuji Photo Film Co., Ltd.),** German Patent 3,248,387, 1983.
18. **Fisch, R. S. and Barjesteh, H. (Minnesota Mining and Manufacturing Co.),** Europe Patent Application 125,086, 1984.
19. **Liu, S. (American Hoechst Corp.),** German Patent 3,639,858, 1987.
20. **Taylor, L. D. (Polaroid Corp.),** U.S. Patent 4,424,272, 1984.
21. **Oki Electric Industry Co., Ltd.,** Japan Patent 84,227,171, 1984.

Chapter 33

PHOTOPOLYMER COMPOSITIONS

The following citations from the Japanese patent literature describe the use of M phthalocyanines (pcs) and of di hydrogen pcs in photopolymerizable compositions.

A resin, a maleic anhydride-2-hydroxyethyl acrylate adduct, 2-hydroxyethyl acrylate, benzoin ethyl ether, talc, and Cu pc are mixed to give a photoresist composition.[1] A photopolymer photoimaging multilayer assembly with imaging layer contains Cu pc.[2] A photopolymerizable imaging composition contains Cu pc.[3] A photopolymerizable imaging composition contains Cu pc.[4] A UV-hardenable resin composition is kneaded with barium sulfate, talc, Cu pc, Aerosil 200, and Modaflow to give an etching resistant ink.[5] A photoimaging composition contains a photosensitive polymer dissolved in cyclohexanone, and mixed with 2-benzoylmethylene-1-methyl-β-naphthothiazoline and di hydrogen pc, it is coated on an aluminum sheet, and dried to give a photosensitive layer.[6] A UV-hardening photoresist composition contains Cu pc.[7] A UV-hardening pattern-forming composition is made into an ink for a screen process by adding talc, barium sulfate, Aerosil, and Cu pc; a Cu pc plate is pattern-printed with the ink, and the ink is hardened by UV irradiation.[8] A UV-hardenable composition for image formation is mixed with ethanol, barium sulfate, Aerosil, and Cu pc to give an ink used to screen print on a copper plate.[9] Pc Green 2GN is kneaded with other compounds into a photohardening polymer composition, then UV irradiated to give a hardened protective film.[10] A photopolymerization initiator for peel-developed photoresist contains Cu pc.[11] A composition containing yellow iron oxide, epoxy resin, and Cu pc is molded into test pieces and subjected to UV radiation by a 308-nm laser, producing a very clear marking. The marks are in clear red, making use of the color change of the yellow iron oxide, useful with respect to products such as condensers and diodes.[12] A photoresist composition contains monoazaindole, acryl group-containing phosphoric acid derivatives, photopolymerizing monomer, themoplastic copolymer, and pc Green SAX, in soldering mask preparation.[13] Cu pc is a component in an image-forming composition that also includes a photopolymerizing unsaturated monomer and a photopolymerization initiator.[14] A method of image recording includes a polymerizable vinyl monomer and di hydrogen pc.[15] Di hydrogen pc is a component in a polymerizable vinyl monomer emulsion to form an image recording material.[16] Di hydrogen pc is a component in a polymerizable vinyl monomer elusion for a method of image recording.[17] Di hydrogen pc is a component in a mixture of a monomer emulsion containing a polymerizable vinyl monomer among other components.[18]

European and U.S. patents also relate to photopolymer preparations that reference M pcs and di hydrogen pcs.

Opaque image formation occurs in photopolymer layers toned with Cu pc using leaching solution.[19] A photopolymer composition contains Cu pc for a peel apart photosensitive element for a positive overlay color proof.[20] A photopolymer coating composition contains Cu pc in a peel-apart element containing a polymeric tacky layer and Cu pc to prepare a color surprint proof for printing.[21] A dry nonelectroscopic toner containing Cu pc for color developing (+) and (−) working photosensitive elements is surface coated with a nonanionic siloxane polyoxyalkylene copolymer.[22] A photopolymerizable dual transfer negative working color proofing system has a photosensitive composition containing Cu pc.[23] A photopolymerizable composition contains acrylated or methoxylated polyurethane having terminal alkane unsaturation and di hydrogen pc for solder masks.[24]

Cu pc is in a layer of a photopolymer composition coated on a polyethylene terephthalate film.[25] A greenish-blue chlorinated pc with "U" pigment with 5% chlorine content is used as a vacuum UV photoresist.[26] A photosensitive composition contains a photoresist layer and a thermally transferable layer of Cu pc for dye transfer imaging.[27] A photopolymeric imaging system with a negative acting photoresist layer contains Cu pc.[28] A photoimaging assembly contains Cu pc for fabrication of lithographic plates and printed circuits.[29] A polymer image that includes di hydrogen pc is formed with photosensitizing comparable to a conventional photographic material.[30]

The following three research articles relate to photopolymers containing M pcs and di hydrogen pcs.

Photoconductive di hydrogen pc is dispersed in a (+) working photopolymer, coated on an aluminum support, and UV-hardened in a high-speed plate making process.[31]

Photoconductive Cu pc is dispersed in a (−) working photopolymer which is coated on an In Sn oxide transparent layer and UV hardened in a process to make a photomasking film.[32]

A camera speed photopolymer imaging emulsion composition contains a coupler polymer, silver halide, and Cu pc.[33]

REFERENCES

1. **Mitsubishi Petrochemical Co., Ltd.,** Japan Patent 82,186,747, 1982.
2. **Ricoh Co., Ltd.,** Japan Patent 84,184,334, 1984.
3. **Sekisui Chemical Co., Ltd.,** Japan Patent 85,103,346, 1985.
4. **Kobayashi, K., Toyama, T., and Yoshida, S. (Fuji Photo Film Co., Ltd.),** Japan Patent 85,162,250, 1985.
5. **Okada, K. and Watanabe, N. (Gooh Chemical Industry Co., Ltd.),** Japan Patent 85,225,150, 1985.
6. **Nakamura, C., Koe, K., and Sasaki, T. (Dainippon Ink and Chemicals, Inc.),** Japan Patent 85,221,748, 1985.
7. **Nippon Shokubai Kagaku Kogyu Co., Ltd.,** Japan Patent 84,128,536, 1984.
8. **Mitsui Toatsu Chemicals, Inc.,** Japan Patent 84,29,246, 1984.
9. **Mitsui Toatsu Chemicals, Inc.,** Japan Patent 84,31,947, 1984.
10. **Toshiba Chemical Corp.,** Japan Patent 85,103,348, 1985.
11. **Fuji Photo Film Co., Ltd.,** Japan Patent 85,22,341, 1985.
12. **Shinozuka, H., Akutagawa, I., and Tajima, H. (Somar Manufacturing Co., Ltd.),** Japan Patent 85,155,493, 1985.
13. **Hitachi Chemical Co., Ltd.,** Japan Patent 84,46,642, 1984.
14. **Nippon Synthetic Chemical Industry Co., Ltd.,** Japan Patent 84,17,548, 1984.
15. **Takeda, T. and Hayakawa, Y. (Fuji Photo Film Co., Ltd.),** Japan Patent 86,188,535, 1986.
16. **Takeda, T. and Hayakawa, Y. (Fuji Photo Film Co., Ltd.),** Japan Patent 86,183,640, 1986.
17. **Takeda, T. (Fuji Photo Film Co., Ltd.),** Japan Patent 86,183,641, 1986.
18. **Takeda, T. (Fuji Photo Film Co., Ltd.),** Japan Patent 86,243,449, 1986.
19. **Chu, V. F.,** U.S. Patent 4,454,217, 1984.
20. **Ashcraft, R. W. and Moehlmann, J. G. (E. I. du Pont de Nemours & Co., Inc.),** U.S. Patent 4,489,153, 1984.
21. **Taylor, H. W., Jr. (E. I. du Pont de Nemours & Co., Inc.),** U.S. Patent 4,489,154, 1984.
22. **Matrick, H. (E. I. du Pont De Nemours & Co., Inc.),** U.S. Patent 4,565,773, 1986.
23. **Barton, O. A. and Wright, J. D. (American Hoechst Corp.),** U.S. Patent 4,596,757, 1986.
24. **Hein, P. R., Miller, H. C., and Brewster, W. D. (W. R. Grace and Co.),** U.S. Patent 4,666,821, 1987.
25. **Quinn, J. A. and O'Neill, J. W. (E. I. du Pont de Nemours & Co., Inc.),** Europe Patent Application 130,581, 1985.
26. **Vasil'ev, A. I., Balabanov, E. I., Bunakov, Y. M., Ignasheva, A. V., Ivashchenko, A. V., Kaloshkin, E. P., Titov., V. V., and Tochitskii, Y. I.,** U.S.S.R. Patent 1,126,581, 1984.
27. **Fisch, R. S. (Minnetosa Mining and Manufacturing Co.),** Europe Patent Application 169,049, 1986.
28. **Kausch, W. L. (Minnesota Mining and Manufacturing Co.),** U.S. Patent 4,544,622, 1985.
29. **Herbert, A. J. (Minnesota Mining and Manufacturing Co.),** U.S. Patent 4,521,503, 1985.
30. **Takeda, K. (Fuji Photo Film Co., Ltd.),** Europe Patent Application 202,490, 1986.

31. **Kitamura, T. and Kokado, H.,** High speed plate making by coupling of a non-conventional presensitized plate and an electrophotographic layer. II. Phthalocyanine pigment. Photopolymer layer, *Denshi Shashin Gakkaishi,* 21(2), 62, 1983.
32. **Kitamura, T., Imamure, S., and Yamaji, K.,** Photomasking film using phthalocyanine electrophotographic layer, *J. Photogr. Sci.,* 32(2), 63, 1984.
33. **Plambeck, L., Chambers, W. J., Donald, D. S., Eaton, D. F., and Cairncross, A.,** Camera-speed photopolymer compositions. III. Applications to litho plates and films, *J. Imaging Sci.,* 30(5), 228, 1986.

Chapter 34

RESISTS

A multilayer peel-development photoresist contains Cu phthalocyanine (pc). The photoresist film is used to form high-quality resist patterns on copper laminate for printed circuits.[1]

A layer of Ni pc is deposited on a resist mask for solid-state imaging device fabrication.[2]

Pinhole defect-free patterns are printed with a resist ink containing Cu pc. The resist ink is printed on a printed circuit board in two steps and cured by UV irradiation to form resist patterns with no defects.[3]

A resist ink for lithographic fabrication of printed circuits contains Cu pc.[4]

A resist composition contains a phenolic composition modified by Cu pc for pattern formation in the fabrication of printed circuits.[5]

A resist composition contains Cu pc for pattern formation in fabrication of printed circuits.[6,7]

A resist film from a Ni pc nonadecanamide derivative for direct pattern fabrication shows good sensitivity to electron beams.[8]

A resin is mixed with trimethylolpropane triacrylate, Irgacure 651, and copper pc to give a resist ink for a resist pattern formation process.[9]

The use of di hydrogen pc improves the resistance to etching in negative-working radiation sensitive resists.[10]

REFERENCES

1. **Dainippon Ink and Chemicals, Inc.,** Japan Patent 83,203,431, 1983.
2. **Canon, K. K.,** Japan Patent 83,144,803, 1983.
3. **Dainippon Ink and Chemicals, Inc.,** Japan Patent 85,24,990, 1985.
4. **Yasui, T., Matsumoto, T., and Akaike, A.(Dainippon Ink Chemical Industry Co.),** U.K. Patent Application 2,141,131, 1984.
5. **Dainippon Ink and Chemicals, Inc.,** Japan Patent 85,39,888. 1985.
6. **Dainippon Ink and Chemicals, Inc.,** Japan Patent 85,39,889, 1985.
7. **Dainippon Ink and Chemicals, Inc.,** Japan Patent 85,39,890, 1985.
8. **Tabei, H., Fujiki, M., and Imamura, S.,** Direct pattern fabrication of substituted phthalocyanine films, *Jpn. J. Appl. Phys., Part 2,* 24(9), L685, 1985.
9. **Yasui, T., Yamazaki, M., and Yamashita, S. Dainippon Ink Chemical Industry Co.),** Japan Patent 86,123,837, 1986.
10. **Fujiki, M., Tabei, H., Imamura, S., and Kurihara, T. (Nippon Telegraph and Telephone Public Corp.),** Japan Patent 86,141,443, 1986.

Chapter 35

CREDIT AND ID CARDS

The following five patents relate to the use of Cu phthalocyanine (pc) and Fe pc in credit and ID cards.

Photopolymerizable imaging materials for ID and credit cards contain Cu pc.[1] A color-image containing card for use as a credit or ID card includes a coloring pigment such as Cu pc in a photosensitive polymer layer.[2] A metallic luster plastic card useful as ID, credit, bank, or magnetic information card contains Cu pc in polyvinyl chloride. The composite structure is heated at 150°C at 30 kg/cm^2 to provide a plastic card with a metallic luster.[3] An optical card with an optical recording unit contains a recording layer of Fe pc.[4] Color photographs prepared by the diffusion transfer process from image receiving elements carrying a preprinted security pattern are used in the production of information-bearing documents such as ID cards. The preparation involves a blue ink that contains Cu pc.[5]

REFERENCES

1. **Toppan Printing Co., Ltd.,** Japan Patent 85,21,047, 1985.
2. **Toppan Printing Co., Ltd. and Osaka Printing Ink Manufacturing Co. Ltd.,** Japan Patent 84,44,616, 1984.
3. **Suzuki, N. and Hayakawa, F. (Dainippon Printing Co. Ltd.),** U.S. Patent 4,479,995, 1984.
4. **Takita, M. (Dainippon Printing Co. Ltd.),** Japan Patent 86,220,148, 1986.
5. **Raphael, T. and Shulman, J. (Polaroid Corp.),** U.S. Patent 4,653,775, 1987.

Chapter 36

COLOR AND OPTICAL FILTERS

A color filter which can be used in a TV camera contains a glass substrate with a layer of Cu phthalocyanine (pc) coated with an Odur 1013 photoresist.[1]

A photoresist containing Cu pc is used during the fabrication of color filters for color TV cameras.[2]

An optical color separation filter which can be used in a TV camera contains methyl cellulose and Cu pc.[3]

Cu pc is a component in an optical color separation filter for use in a solid-state TV camera.[4]

Cu pc tetrasulfonate dye is a filter element in a solid-state imaging device.[5]

A positive working resist Odur 1013 is used to form color filters from Ni pc for a color solid-state image pickup device.[6]

Cu pc disulfonic acid is used in an optical color filter fabrication process.[7]

Dyes for micro color filters for solid-state color imaging devices include Cu pc-$(SO_3pyridinium)_4$.[8]

Cu pc tetrasulfonic acid is involved in color microfilter production for an image pickup device.[9]

An optical filter layer contains Cu pc for a one-shot color electrophotographic photosensitive sheet.[10]

A photosensitive composition contains Cu pc for color separation filters for imaging devices.[11]

Pb pc is vapor deposited on a resist pattern in a process for the photolithographic fabrication of color filters.[12]

A glass plate is coated by electrodeposition with a composition including Cu pc in preparation of an electrically conductive filter for optical display devices.[13]

The reaction products of Cu pc disulfonic acid with ammonium carbonate are used in color filters for electronic devices.[14]

An optical filter contains an Al pc Cl sulfonic acid dye. The filter is useful for solid-state imaging devices such as color TV cameras.[15]

A photoelectric image converter has a blue coloring layer as color separation filter containing Cu pc.[16]

Photosensitive heat-resistant color pastes for color filters contain Cu pc.[17]

Dye compositions for relief patterns for microcolor filters contain Cu pc C. I. reactive blue 14 dye.[18]

The optical filter of a color TV camera contains Cu pc.[19]

Cu pc is involved in the lithographic fabrication of optical filters.[20]

Cu pc, Indenthrene Yellow GCN and di-*p*-xylene are used to prepare an optical filter having cyan and yellow patterns, useful in a solid-state imaging camera.[21]

Ni pc is involved in forming a monochromatic filter pattern on a glass plate.[22]

REFERENCES

1. **Canon K. K.,** Japan Patent 82,148,706, 1982.
2. **Canon K. K.,** Japan Patent 82,178,206, 1982.
3. **Canon K. K.,** Japan Patent 82,172,304, 1982.

4. **Canon K. K.,** Japan Patent 83,09,108, 1983.
5. **Fujita, S., Harada, T., and Matsumoto, K. (Fuji Photo Film Co., Ltd.),** German Patent 3,329,219, 1984.
6. **Canon, K. K.,** Japan Patent 83,147,704, 1983.
7. **Matsushita Electric Industrial Co., Ltd.,** Japan Patent 84,220,706, 1984.
8. **Fuji Photo Film Co., Ltd.,** Japan Patent 84,101,607, 1984.
9. **Fuji Photo Film Co., Ltd.,** Japan Patent 85,10,206, 1985.
10. **Olympus Optical Co., Ltd.,** Japan Patent 85,115,958, 1985.
11. **Agency of Industrial Sciences and Technology Morohoshi Printing Ink Co., Ltd., Dainippon Printing Co., Ltd.,** Japan Patent 85,129,739, 1985.
12. **Yokono, K. and Shindo, H. (Canon K. K.),** German Patent 3,509,198, 1985.
13. **Iwasa, K., Kamamori, H., Suginoya, M., Sano, Y., Terada, Y., Kato, N., Suzuki, T., Yasukawa, J., Nomura, Y., et al. (Seiko Instruments and Electronics, Ltd.; Shinto Paint Co., Ltd.),** German Patent 3,507,309, 1985.
14. **Shimizu, Y., Kojima, K., and Kuwana, K. (Sumitomo Chemical Co., Ltd.),** Japan Patent 86,162,557, 1986.
15. **Miura, K., Ochiai, T., and Makishima, H. (Mitsubishi Chemical Industries Co., Ltd.),** Japan Patent 85,249,102, 1985.
16. **Fukaya, M., Komatsu, T., Shoji, T., Kamio, M., and Sekimura, N. (Canon K. K.),** Japan Patent 86,38,906, 1986.
17. **Miura, Y., Hiramoto, Y., and Eguchi, M.,** Japan Patent 86,254,906, 1986.
18. **Matsumoto, K. (Fuji Photo Film Co., Ltd.),** Japan Patent 87,25,171, 1987.
19. **Dainippon Printing Co., Ltd.,** Japan Patent 83,34,961, 1983.
20. **Canon K. K.,** Japan Patent 85,52,571, 1985.
21. **Dainippon Printing Co., Ltd.,** Japan Patent 84,61,833, 1984.
22. **Ogura, Y., Iijima, S., Kasamori, E., and Kamio, M. (Canon K. K.),** Japan Patent 85,195,543, 1985.

Chapter 37

ELECTROPHORETIC DEVICES

An electrophoretic device is prepared by using dispersions containing τ-, modified τ-, or modified η-di hydrogen phthalocyanines (pcs). α-H_2 pc pigment is converted to τ-H_2 pc by milling and dispersing in Isopar H to give a dispersion for electrophoretic display devices.[1]

Al pc Cl is blended with a polymer matrix containing a condensation polymer or copolymer containing recurring units derived from a bis[4-*N*-(2-hydroxyethyl)piperidyl] alkane and milled with methylene chloride to produce a pc complex that can be used as a cyan colorant in a photoelectrophoretic imaging device.[2]

A photoelectrophoretic imaging material contains electropolymerizable monomer and Cu pc.[3]

REFERENCES

1. **Hitachi Ltd. and Toyo Ink Manufacturing Co. Ltd.,** Japan Patent 85,87,317, 1985.
2. **Ksaacson, H. V. and Wright, H. E. (Eastman Kodak Co.),** U.S. Patent 4,535,046. 1985.
3. **Molaire, M. F., Nielsen, P. L., and Rodenberg, O. C. (Eastman Kodak Co.),** U.S. Patent 4,473,626, 1984.

Chapter 38

ELECTRORADIOGRAPHIC AND ELECTRORECORDING MATERIAL

A layer of Bi_{12} Ge O_{20} particles, zinc oxide, and β-cu phthalocyanine (pc) is dispersed in a polyester resin and formed on an aluminum substrate by evaporation of their suspension in tetrahydrofuran, and heated 3 h at 200°C to give an electroradiographic plate of improved sensitivity.[1]

An ink transfer type color electrorecording material contains Cu pc to give cyan dots.[2]

An electric discharge recording material contains Cu pc.[3]

An electron beam recording material contains an electric conduction layer and a layer of europium or terbium pc complexes.[4]

REFERENCES

1. **Fuji Photo Film Co., Ltd.,** Japan Patent 83,117,552, 1983.
2. **Sekisui Chemical Co., Ltd.,** Japan Patent 84,31,190, 1984.
3. **Sekisui Chemical Co., Ltd.,** Japan Patent 85,49,992, 1985.
4. **Comberg, A. and Welker, T. (Philips Patentverwaltung G.m.b.H.),** German Patent 3,525,994, 1987.

Chapter 39

TONERS

I. ELECTROPHOTOGRAPHIC (ELECTROSTATOGRAPHIC) TONERS

Over 100 Japanese patents detail methods of preparation of toners. They are summarized briefly in the following pages: phthalocyanine (pc) in an electrophotographic toner,[1,2] a suspension polymerization involves a dispersion containing Cu pc for electrostatographic (−)-charging toner preparation,[3] Cu pc in an electrophotographic toner,[4] pc in an electrostatic toner,[5] Cu pc in a positively charged toner useful for color development in electrophotographic and electrostatic recording,[6] Cu pc in a positively charged toner for electrophotography,[7] Cu pc in a toner for development of electrostatic images,[8-10] thiazine derivatives as charge controllers in electrophotographic toner containing Cu pc,[11] Cu pc in an electrostatic toner,[12] Cu pc in a positively charged electrophotographic toner,[13] Cu pc in an electrophotographic toner,[14,15] Cu pc in an electrophotographic toner for the development of electrostatic images,[16] Cu pc in an electrophotographic toner,[17] and Cu pc in a toner to develop electrostatic latent images of electrophotographic or electrostatic recordings.[18]

Electrostatographic toner particles are made by suspension polymerization using an anionic dispersing agent by mixing styrene, paraffin wax, Cu pc, azobisisobutyronitrile, and colloidal silica.[19] A positive chargeable electrophotographic toner contains Cu pc.[20] A styrene-containing Cu pc is added to a mixture of a latex and water to obtain a composite polymer latex and spray dried to give a toner.[21] Cu pc is a component in a toner with narrow particle size distribution which gives high-quality images.[22] Cu pc is a component in a toner mixed with silica to give brilliant blue images.[23] Cu pc is part of a toner for electrostatic image development.[24] Cu pc is a component in a toner which forms an electrostatic developer prepared by mixing the toner with iron powder.[25]

An electrostatographic positive chargeable toner contains bis(dimethyl aminophenyl)methane charge controlling agent and Cu pc.[26] Cu pc is a component in a toner for the development of electrostatic images.[27] Cu pc is a component in a capsule toner for electrostatic latent image development.[28] Cu pc is a component in an electrostatographic alkali treated cationic polymer toner.[29] Cu pc is a component in an electrostatographic toner.[30,31] Cu pc is a component in a positive chargeable electrophotographic toner. The toner is mixed with an iron carrier to obtain a developer. Operation is stable during 10,000 consecutive copying cycles.[32] Cu pc is a component in a toner for electrostatography prepared by suspension polymerization.[33] Electrophotographic toners containing Cu pc are involved in jumping development.[34]

A positive charging electrophotographic developer contains aminobis(dimethylamino)pyrimidine and Cu pc.[35] A toner for electrostatic image development contains Cu pc.[36] Cu pc is a component in an electrostatographic toner.[37]

Cu pc is a component in a positive chargeable toner for electrostatographic jumping development.[38] Cu pc is in an electrostatographic toner in polyethylene microencapsulation.[39] Cu pc is a component in an electrostatographic toner.[40,41] Cu pc is a component in a toner for electrostatic image development.[42]

Cu pc is a component in a polymerizable composition for preparation of electrostatographic toners with controlled sphericity.[43] Positive chargeable toners contain a resin binder, an ethylamine derivative, and Cu pc for electrostatographic developers.[44] Electrostatographic (+)-charging toners contain Cu pc.[45] An electrostatographic toner with spherical particles contains Cu pc.[46]

An electrostatographic toner contains Cu pc.[47] Cu pc is a component in a toner for electrostatic image development.[48-50] Cu pc is a component in an electrostatographic toner of 9.0 μ average toner particle diameter.[51]

Cu pc is a component in a toner for the development of electrostatic images.[52] Cu pc is a component in a toner prepared by suspension polymerization for electrostatic image development.[53] A dispersion containing styrene, azobisisobutyronitrile, divinyl benzene, Cu pc, and helio fast blue is polymerized in a poly(vinyl alcohol) solution to give a toner whose number average particle size, gel content, and Tg are 9.8 μ, 60 wt %, and 58°C, respectively.[54]

Cu pc is a component in a toner for electrostatic image development.[55,56] ϵ-Cu pc is a component in a (+)-charging toner for electrophotography, the main component of which is a resin.[57] An electrophotographic toner contains a binder with hydrocarbon chains, an ionic crosslink, and Cu pc.[58]

Pc is a component in an electrostatographic toner.[59] Cu pc is part of a toner for electrostatic image development containing a resin binder, a lubricant, and an abrasive.[60] Cu pc is in a toner for electrostatic image development containing a naphthol sulfonic acid salt.[61] Pc is a component in an electrophotographic toner. The high quality of images produced therefrom does not change even after 10,000 copies.[62]

A developer for electrostatic images is prepared by mixing the toner containing Cu pc with an iron powder. Electrophotography using the developer gives high-quality images and the developer shows improved durability.[63]

Cu pc is a component in a toner for the development of electrostatic images.[64] Cu pc is a component in an electrophotographic toner.[65-66] Cu pc is a component in an electrostatographic toner.[67-70]

A pressure fixable toner contains wax, magnetite, an amine, silica powder, and Cu pc for electrostatographic developers.[71] Cu pc is a component in an electrophotographic toner.[72]

An electrostatographic toner contains polar and nonpolar polymers, and Cu pc with nonuniform size distribution for electrophotographic developers.[73] Pc is a component in a toner for electrostatographic developers.[74] Cu pc is part of a toner for electrostatic image development.[75]

An electrostatographic developer with toner contains quaternary ammonium hexafluorophosphate and Cu pc mixed with improved charging characteristics.[76] An electrostatographic toner contains pyrrolidine or piperidine derivatives and Cu pc.[77] An electrostatographic toner has a uniform and stable triboelectric charge and gives clear color images; it contains phthalonitrile or its derivative and Cu pc.[78] A toner for electrostatic image development contains a quaternary phosphonium salt charge controlling agent and Cu pc.[79]

An electrostatographic developer with toner contains a cerium compound to give a stable and adequate triboelectric charge and Cu pc for improved charging characteristics.[80] A toner contains a resin binder, quinaldic acid derivative metal complex, and Cu pc for electrostatographic developers with improved triboelectric properties.[81] A toner contains a resin and metal complex charge controlling agent and Cu pc for electrostatographic developers.[82] A toner contains a binder resin and zircoaluminium modifying agent-treated Cu pc for electrostatographic developers with improved durability.[83] An electrostatographic toner contains benzothiazolylguanidine derivative charge-controlling agent and Cu pc.[84] An electrostatographic toner contains a binder resin, a coloring agent, Cu pc, and a dicyclohexylamine nitrite to give stable, adequate triboelectric charges.[85] An electrophotographic toner contains arylantimony halide, Cu pc, and wax.[86] An electrophotographic toner contains triarylstibine oxide and Cu pc.[87]

An electrophotographic toner contains aminoalkylbenzenecarboxylic acid metal complex and Cu pc with stable and controllable triboelectric properties.[88] An electrophotographic toner contains hexahydroanthranilic acid derivative metal complex and Cu pc for improved properties.[89] An electrophotographic toner contains a metal complex of an aliphatic amino acid and Cu pc with improved properties.[90] Electrostatic image development takes place with a toner containing triarylstibine derivative and Cu pc.[91]

Metal chelate compounds with anthranilic acid derivatives are charge-controlling agents for an electrostatographic toner composition containing Cu pc.[92] Toners for electrostatic image development contain bissalicylaldehydediethylenediimine or bissalicylaldehydetrimethylenediimine metal complex and Cu pc with improved charging properties.[93] An electrostatographic toner contains Cu pc; it has an (alkyl)salicylaldehyde chelate charge-controlling agent.[94] An electrostatographic toner contains (alkyl)salicylaldoxime chelate compound charge-controlling agent and Cu pc.[95]

An electrostatographic toner contains hydroxyamine chelate as charge controlling agent and Cu pc.[96] Toners contain (alkyl)salicylamine-metal chelate and Cu pc for electrostatographic developers with improved triboelectric charging properties.[97] Toners contain salicylaldimine-metal chelate and Cu pc for electrostatographic developers with improved triboelectric charging properties.[98] An electrostatographic toner contains a binder resin, Cu pc, and a metal picolinate derivative.[99] An electrophotographic toner is composed of a binder resin, Cu pc, and a metal complex of 2,2′-bipyridine or of a 2.2′-bipyridine derivative.[100]

An electrophotographic toner contains Cu pc.[101] An electrostatographic toner contains nitroquinoline charge controller and Cu pc.[102] An electrostatographic toner contains a triarylphosphine derivative and Cu pc.[103] An electrostatographic developer with a toner contains quaternary ammonium hexafluorophosphate and Cu pc with improved charge characteristics.[104] A toner contains carboxylic acid and low triboelectric charging resin and di hydrogen pc for electrostatographic developers.[105] An electrophotographic nonmagnetic toner for forming uniform layers on toner-transport sleeves contains Cu pc.[106]

An electrophotographic toner contains silicon carbide and Cu pc with good environmental stability.[107] A toner for electrostatic latent image development contains an acrylic copolymer and

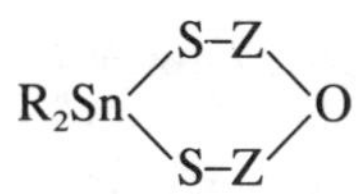

where R = $C_{1\text{-}24}$ alkyl and Z = $C_{1\text{-}12}$ alkylene, charge-controlling agent, and Cu pc for electrostatographic developers.[108]

A toner contains catechol boron compound charge-controlling agent and Cu pc for electrostatographic developers.[109] Positive charging toners contain naphthamidoalkylammonium salt charge-controlling agent and Cu pc for electrophotographic developers.[110]

A toner contains tetraarylstibonium compound charge-controlling agent and Cu pc for electrostatographic developers.[111] Toners contain an organic tin compound charge-controlling agent and Cu pc for electrostatographic developers.[112,113] Toners for developing electrostatic images contain polyolefins and Cu pc.[114] An electrostatic developing toner contains composite metal oxides and Cu pc.[115]

A toner for electrostatic image development contains 1,2-cyclohexanediamine metal complex as charge-controlling agent and Cu pc.[116] Toners contain hexamminecobalt(III) salt charge-controlling agent and Cu pc for electrophotographic developers.[117] Toners for electrostatic image development contain a binder resin, C. I. pigment yellow 17, and C. I. pigment blue 15 (Cu pc) to provide 20,000 copies with good high-quality green images.[118] An electrophotographic blue toner contains di hydrogen pc particles dispersed uniformly in a binder resin.[119]

An electrophotographic toner contains a chlorinated and/or brominated Cu pc treated with rosin. The toner is mixed with a powdered iron carrier to obtain a developer. Imaging

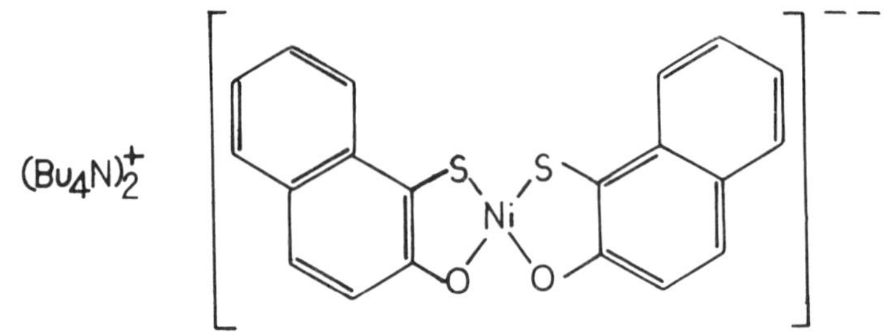

FIGURE 1. A nickel complex component of a blue electrophotographic toner.

with a commercial copier using an organic photoconductive photoreceptor and the developer gives 10,000 clear images without fog.[120]

Epoxy resin, Cu pc, and bis(1,4-dichloro-5,6-dithiophenylate)Ni(II)-tetrabutyl ammonium complex are mixed, kneaded, and pulverized to give an electrophotographic toner.[121] Epoxy resin, Cu pc, and bis(4-chloro-o-phenylenediamino) Ni are mixed, kneaded, and pulverized to give a blue toner.[122] A mixture of epoxy resin, Cu pc, and a metal complex (Figure 1) is kneaded and pulverized to give a blue toner.[123]

β-Cu pc and polystyrene are mixed in 1,2-dichloroethane and spray dried to give an electrophotographic toner.[124] A protective coating for electrophotographic copies is prepared with a toner containing Cu pc.[125] A photoconductor toner dispersion contains Cu pc for electrophotographic imaging.[126] A thermographic copying method involves toners that include Cu pc.[127]

Several U.S. and European patents also relate to electrophotographic toners.

Cu *N,N′,N″,N‴*-tetraoctadecyl pc 2,9,16,23 tetrasulfonamide is a component in an electrostatographic negatively charged toner containing a polyanhydride resin as charge-enhancing agent.[128]

Cu *N,N′,N″,N‴*-tetraoctadecyl pc tetrasulfonamide is in an electrophotographic positively charged toner containing cetyl pyridinium tetrafluoroborate.[129]

A positively charged toner contains a phosphonium compound charge-enhancing agent for an electrophotographic photoconductive compound containing VO pc.[130] A toner composition for an electrophotographic photoreceptor with a photogenerating layer of VO pc contains ammonium sulfate as charge-enhancing additive.[131]

A negatively charged electrophotographic blue toner contains a resin binder, indanthrone dye, and Cu pc Cl.[132]

Charge-controlling agents for electrophotographic positively charging toner compositions contain Cu pc.[133] Cu pc is a photoconductive toner charge-generating agent.[134]

II. COLOR TONERS

Cyan toners for development of electrostatic images contain a resin with a Cu pc derivative (Figure 2),[135] where R to R^3 = H, II, or III, where Z, Z′ = $C_{1\text{-}5}$ alkylene.

A cyan toner for electrophotography contains Cu pc.[136]

Coloring material for an electrophotographic toner is prepared from a mixture containing Cu pc.[137] A cyan toner contains a Cu pc derivative cited in Reference 135 above.[138]

A color electrophotographic toner contains Cu pc with reduced edge effect and fog.[139] An electrophotographic color toner contains butyl acrylate-styrene copolymer and Cu pc coated with an anionic surfactant for reduced surface resistance.[140] An electrophotographic colored toner contains a polyester resin and negatively chargeable Cu pc.[141] An electrophotographic blue toner contains Cu pc.[142] An electrophotographic color toner composition

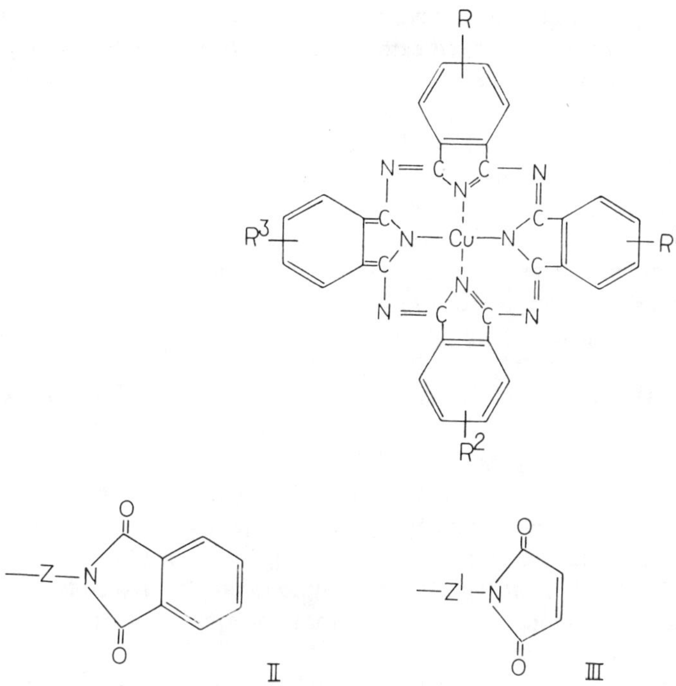

II III

FIGURE 2. A Cu pc cyan toner for electrostatic image development.

includes Cu pc.[143] Electrophotographic blue color toners contain a Cu pc pigment photoconductor.[144] Electrophotographic color separation filter type toners contain Cu pc.[145]

An electrostatographic color toner contains nigrosine base positive charge controlling agent and Cu pc for improved image quality.[146] An electrophotographic toner contains a binder, UV absorber, and Cu pc to give clear diazo images.[147] A color electrophotographic process includes Cu pc as a cyan toner.[148] Color electrophotography takes place by using light-transmitting green particles from compositions that include β-Cu pc.[149]

A color electrophotographic toner contains Cu pc for production of projection slides.[150] An electrophotographic charging color toner contains zinc stearate surface-treated colorant of Cu pc for improved durability.[151] A color toner for development of electrostatic images contains Cu pc. When the toner is mixed with iron powder to give a developer, there is good development performance in 20,000 successive electrophotographic runs.[152]

An electrophotographic color toner contains Cu pc.[153] A coloring agent for electrophotography contains Cu pc.[154] A color toner kit contains three color microencapsulated toners, cyan (Cu pc), magenta, and yellow.[155] An electrophotographic color toner contains Cu pc.[156] A color toner to provide clear transparent images useful in overhead projectors contains β-Cu pc.[157]

An electrophotographic green toner contains Cu pc.[158] Electrophotographic developer toners contain Cu pc.[159] Blue toners for electrophotography contain a chloro-substituted Cu pc added with a red or violet colored pigment. The toners show a high and durable negative charging capacity.[160]

An inorganic pigment treated with an organic material, optionally surface coated with a resin, is added to a composition containing a thermoplastic resin and Cu pc to give an electrophotographic toner.[161]

Cu pc is a component in an electrophotographic color toner.[162] A toner for a two color electronic printer contains Cu pc.[163] Butyl methacrylate-styrene copolymer, Cu pc, and zinc stearate are kneaded and pulverized to give a cyan electrophotographic toner.[164] A green electrophotographic toner includes Sudan Yellow 150 and Heliogen Blue D7072D (Cu pc).[165] Colored toners contain crosslinked carboxylic acid-styrene-acrylate copolymer and Cu pc for electrophotographic developers.[166,167] A color toner for noncontact heat forming contains Cu pc and a sublimable dye.[168] Color decomposition-type electrophotographic toners contain a Cu pc dye to form color images.[169] Electrophotographic color separation filter-type toners contain Cu pc.[170] A color toner contains Cu pc for a transparent electrophotographic plate for multicolor image formation.[171]

Cu pc is a photoconductor in a two color electrophotographic process.[172] Electrophotographic toners in a color process contain di hydrogen pc.[173]

A Netherlands patent awarded to Sony Corporation is referenced for a color electrophotographic toner that contains Cu pc.[174]

III. MICROCAPSULE AND PRESSURE-FIXING TONERS

An electrophotographic microcapsule toner contains Cu pc, Hiwax 220P, and a paraffin wax, the particles of which are coated with polystyrene by spraying.[175]

Polyethylene wax and Cu pc are kneaded to obtain a toner of 10 to 20 μ in diameter. A positive electrostatic latent image is developed, transferred to a paper, and fixed at 125 mm/s and 15 kg/cm line pressure, showing almost complete fixability.[176]

A pressure fixing toner containing hydrocarbon wax and Cu pc is fixed at 125 mm/s and 15 kg/cm line pressure.[177]

Cores containing waxes and Cu pc are microencapsulated with styrene copolymer to give a microencapsulated toner with good pressure fixability for single component-type electrostatographic developers.[178] Cores contain waxes, magnetite, an Cu pc for microencapsulated toners containing vinylpyridine copolymer shells for electrostatographic developers.[179]

Cu pc is in an electrophotographic pressure fixable microencapsulated toner that is spray dried to form a microencapsulated toner with a polystyrene shell.[180]

Cores are prepared from waxes and Cu pc for microencapsulated toners with shell material containing aminophenyl methacrylate-vinyl compound copolymer for electrophotographic developers.[181] Electrophotographic pressure-fixable microencapsulated toners core material includes Cu pc.[182] An electrophotographic pressure fixing toner contains graft copolymers and Cu pc. Complete fixing is carried out at 15 kg/cm line pressure at 125 mm/s line speed.[183]

Binders for electrophotographic pressure fixing toners containing Cu pc are obtained by grafting long hydrocarbon chain compounds with vinyl monomer.[184]

An electrophotographic pressure-fixable toner contains an amine, linear hydrocarbons, and Cu pc.[185] An electrophotographic toner capsule contains a core that includes Cu pc. The core is coated with dimethylaminoethyl methacrylate-styrene copolymer, then the capsule particles are dispersed in a methanol water solution of iodine to give a capsule toner.[186]

A pressure-fixable capsule toner contains diallylammonium halide cyclized polymer walls and cores containing waxes and Cu pc for an electrostatographic developer.[187] An electrophotographic microencapsulated toner contains hydrophobic titania and Cu pc with improved performance under high humidity.[188] An electrostatographic toner core material contains Cu pc encapsulated with diallyl phthalate polymer.[189]

Electrophotographic pressure fixing microcapsule toners have a core that includes Cu pc and a shell made with poly(amino acid) resin.[190]

A pressure-fixable capsule toner contains Cu pc for electrophotographic or electrostatic printing. When used to develop an electrostatic image, a speed of 125 mm/s at 10 kg/cm gives a completely fixed clear and fog-free image with a 1.3 density.[191]

A pressure-fixable microencapsulated toner is composed of a shell and a core that includes Cu pc.[192] An electrostatographic capsule toner contains cores with air bubbles, polyethylene, paraffin wax, and Cu pc, and a shell of 2-diethylaminoethyl methacrylate-styrene copolymer.[193] An electrophotographic toner contains Cu pc for transparency preparation having high image density.[194]

Cu pc is a component in soft resin cores and hard copolymer shells, of insulating nonmagnetic microcapsule toners for electrostatographic developers.[195] A pressure fixable toner contains polyethylene, di-*p*-chlorobenzylidenesorbitol, and Cu pc. Mixed with iron powder, it forms a developer to prepare an image the density of which is greater than that of a toner with no sorbitol derivatives.[196]

A pressure-fixable microencapsulated color electrostatographic toner contains a Cu pc complex with alkyl radicals on the four benzene rings of the macrocycle.[197]

An emulsion-polymerizing composition contains Cu pc for preparation of pressure-fixing toners for electrostatic developers.[198] An electrophotographic pressure-fixable microcapsule toner has a core that includes abrasive particles and Cu pc for improved durability.[199] An electrostatographic capsule toner has a core of paraffin, polyethylene wax, and Cu pc encapsulated in a styrene-acrylate resin.[200]

A microcapsule toner contains fluorinated polymer powder, Hiwax 200P, and Cu pc microencapsulated in 10:90 dimethylaminoethyl methacrylate-styrene copolymer.[201]

Reactive compositions contain safflower oil, an isocyanate, and Cu pc to prepare pressure fixable toners containing polyurea for electrostatographic developers.[202]

IV. MAGNETIC TONERS

The following Japanese patents relate to magnetic toners.

A magnetic toner contains δ-FeOOH, Fe_3O_4, resin, and di hydrogen pc for recording purposes.[203] An electrophotographic toner contains butylacrylate-dimethylaminoethyl methacrylate-styrene copolymer, magnetite, and Cu pc.[204]

A magnetic toner useful for an electrophotographic, electrostatic, or magnetic recording contains an electroless-plated magnetic material. Such a toner is prepared from a blend containing butyl acrylate-maleic acid anhydride-styrene copolymer, a magnetic material and electroless-plated with Ni pc and Cu pc.[205]

A magnetic toner contains butyl acrylate-styrene copolymer, magnetite, Cu pc, colloidal silica, and zinc stearate.[206] An electrophotographic color developer with magnetic toner contains magnetic particles with specified particle size distribution and Cu pc.[207]

A magnetic toner for a protective layer for an electrophotographic photoreceptor contains Cu pc with a protective layer containing orotic acid.[208] An electrophotographic magnetic toner contains magnetite, a binder resin, and Cu pc.[209]

An electrostatographic toner contains styrene, dimethylaminoethyl methacrylate-styrene copolymer, oxidized wax, Cu pc, magnetite, and azobisbutyronitrile, dispersed in water-containing silica, and the styrene polymerized to give a magnetic toner.[210]

A magnetic toner contains a binding resin, a magnetic powder, and Cu pc.[211] An electrostatographic magnetic toner contains Cu pc.[212] A magnetic microcapsule toner contains Cu pc.[213]

A German patent also relates to magnetic toners for electrophotography. Fe_3O_4 powder, styrene-butyl acrylate-dimethylaminoethyl methacrylate polymer, polyethylene, and Cu pc are kneaded and ground to make a blue magnetic toner for electrophotography.[214]

V. MAGNETIC COLOR TONERS

Japanese patents relating to magnetic color toners include the following.

A mixture of titania-coated magnetite, butyl acrylate-styrene copolymer, Cu pc, and nigrosine is kneaded, pulverized, and mixed to give a blue magnetic toner.[215] Reference 216 is similar to Reference 215.

A color magnetic toner contains Cu pc.[217] A color magnetic toner contains Cu pc in a coating layer over a white pigment-hiding layer to give toners providing clear, cyan color images.[218] Reference 219 is similar to Reference 218.

An electroconductive magnetic color toner for electrostatic images contains Cu pc.[220]

An electrophotographic magnetic color toner contains garnet, powder, binder, and Co pc.[221]

A magnetic color toner contains a powdered magnetic material 79 wt % of which is a mixture of iron, Cu pc, and titania.[222]

European patents relating to the use of M pcs in color magnetic toner compositions include the following: 1,2,3,4,8,9,10,11,15,16,17,18,22,23,24,25-hexadecachloro Cu pc is a green pigment in a color magnetic color powder.[223]

Colored magnetic toner particles contain Cu pc for electrophotographic copiers.[224] A blue magnetic toner contains platelike particles of α-iron oxide, Bayertitan R-KB-2 (white pigment), and Heliogen Blue K7080 (Cu pc).[225]

A colored magnetic toner contains an iron powder, magnetic pigment, and Cu pc to give a blue toner.[226]

REFERENCES

1. **Canon K. K.,** Japan Patent 84,62,870, 1984.
2. **Canon K. K.,** Japan Patent 84,62,871, 1984.
3. **Canon K. K.,** Japan Patent 84,53,856, 1984.
4. **Nippon Aerosil Co., Ltd.,** Japan Patent 83,216,252, 1983.
5. **Canon K. K.,** Japan Patent 84,62,868, 1984.
6. **Hitachi Chemical Co., Ltd.,** Japan Patent 83,82,254 and 83,82,256, 1983.
7. **Canon K. K.,** Japan Patent 84,152,451, 1984.
8. **Canon K. K.,** Japan Patent 84,152,450, 1984.
9. **Canon K. K.,** Japan Patent 84,152,449, 1984.
10. **Canon K. K.,** Japan Patent 84,152,448, 1984.
11. **Canon K. K.,** Japan Patent 84,197,053, 1984.
12. **Canon K. K.,** Japan Patent 84,123,853, 1984.
13. **Canon K. K.,** Japan Patent 84,152,453, 1984.
14. **Canon K. K.,** Japan Patent 84,152,454, 1984.
15. **Canon K. K.,** Japan Patent 84,152,452, 1984.
16. **Canon K. K.,** Japan Patent 84,152,447, 1984.
17. **Canon K. K.,** Japan Patent 84,187,356, 1984.
18. **Sumitomo Chemical Co., Ltd.,** Japan Patent 84,172,655, 1984.
19. **Canon K. K.,** Japan Patent 85,121,455, 1985.
20. **Canon K. K.,** JapanPatent 84,195,664, 1984.
21. **Canon K. K.,** Japan Patent 84,78,368, 1984.
22. **Canon K. K.,** Japan Patent 84,123,852, 1984.
23. **Canon K. K.,** Japan Patent 85,04,950, 1985.
24. **Sumitomo Chemical Co., Ltd.,** Japan Patent 84,177,571, 1984.
25. **Sumitomo Chemical Co., Ltd.,** Japan Patent 84,204,850, 1984.
26. **Canon K. K.,** Japan Patent 84,218,461, 1984.
27. **Ricoh Co., Ltd.,** Japan Patent 84,79,256, 1984.
28. **Canon K. K.,** Japan Patent 84,187,350, 1984.
29. **Canon K. K.,** Japan Patent 84,123,851, 1984.
30. **Canon K. K.,** Japan Patent 84,200,251, 1984.
31. **Canon K. K.,** Japan Patent 84,57,252, 1984.
32. **Canon K. K.,** Japan Patent 84,195,662, 1984.
33. **Canon K. K.,** Japan Patent 84,219,755, 1984.
34. **Canon K. K.,** Japan Patent 84,200,264, 1984.
35. **Fujitsu Ltd.,** Japan Patent 84,200,253, 1984.

36. **Sumitomo Chemical Co., Ltd.**, Japan Patent 84,204,851, 1984.
37. **Canon K. K.**, Japan Patent 84,197,052, 1984.
38. **Canon K. K.**, Japan Patent 84,187,348, 1984.
39. **Canon K. K.**, Japan Patent 84,185,353, 1984.
40. **Saruwatari, N., Yamagishi, Y., Ko, K., Narisawa, T., and Okuyama, H. (Fujitsu Ltd.)**, Japan Patent 85,133,460, 1985.
41. **Canon K. K.**, Japan Patent 85,10,258, 1985.
42. **Canon K. K.**, Japan Patent 85,46,565, 1985.
43. **Canon K. K.**, Japan Patent 85,28,665, 1985.
44. **Canon K. K.**, Japan Patent 85,26,952, 1985.
45. **Maruyama, M. and Murai, K. (Nippon Carbide Industries Co., Inc.)**, Japan Patent 85,142,352, 1985.
46. **Canon K. K.**, Japan Patent 84,218,460, 1984.
47. **Canon K. K.**, Japan Patent 84,218,459, 1984.
48. **Canon K. K.**, Japan Patent 85,117,254, 1985.
49. **Canon K. K.**, Japan Patent 85,103,354, 1985.
50. **Canon K. K.**, Japan Patent 85,103,355, 1985.
51. **Canon K. K.**, Japan Patent 85,117,256, 1985.
52. **Canon K. K.**, Japan Patent 84,152,446, 1984.
53. **Canon K. K.**, Japan Patent 85,57,855, 1985.
54. **Canon K. K.**, Japan Patent 85,45,260, 1985.
55. **Canon K. K.**, Japan Patent 85,57,856, 1985.
56. **Canon K. K.**, Japan Patent 85,57,854, 1985.
57. **Tomita, M., Kawabata, T., Minamitani, T., and Murayama, H. (Ricoh Co., Ltd.)**, Japan Patent 85,258,560, 1985.
58. **Matsumoto, T., Yamazaki, M., Wakamya, K., Ushama, N., Osaki, I., Nakahara, T. (Canon K. K.)**, Japan Patent 85,144,757, 1985.
59. **Canon K. K.**, Japan Patent 85,104,956, 1985.
60. **Murayama, H., Kawabata, T., Minamitani, T., and Tomita, M. (Ricoh Co., Ltd.)**, Japan Patent 85,179,748, 1985.
61. **Kawagishi, Y. and Kiryu, T. (Orient Watch Co., Ltd.)**, Japan Patent 85,169,857, 1985.
62. **Canon K. K.**, Japan Patent 85,87,341, 1985.
63. **Shimizu, Y. and Yago, S. (Sumitomo Chemical Co., Ltd.)**, Japan Patent 85,188,958, 1985.
64. **Shimizu, Y. and Yago, S. (Sumitomo Chemical Co., Ltd.)**, Japan Patent 85,188,959, 1985.
65. **Urawa, M., Doi, S., and Mihashi, Y. (Canon K. K.)**, Japan Patent 86,59,348, 1986.
66. **Urawa, M., Doi, S., and Mihashi, Y. (Canon K. K.)**, Japan Patent 86,59,349, 1986.
67. **Hyozu, Y., Takagi, S., and Mori, H. (Canon K. K.)**, Japan Patent 86,35,454, 1986.
68. **Urawa, M., Doi, S., Mihashi, Y. (Canon K. K.)**, Japan Patent 86,59,347, 1986.
69. **Suematsu, H., Ikeda, T., Imai, E., Goto, K., and Yamamoto, A. (Canon K. K.)**, Japan Patent 85,192,957, 1985.
70. **Hyozu, Y., Takagi, S., and Mori, H. (Canon K. K.)**, Japan Patent 86,28,957, 1986.
71. **Wakamiya, K., Koshio, T., Ushiyama, N., Yamazaki, M., Osaki, I., Nakahara, T., Goseki, Y., Kurimoto, J., and Matsumoto, T. (Canon K. K.)**, Japan Patent 86,59,350, 1986.
72. **Urawa, M., Doi, S., and Mihashi, Y. (Canon K. K.)**, Japan Patent 86,65,258, 1986.
73. **Mori, H., Hyozu, Y., and Takagi, S. (Canon K. K.)**, Japan Patent 86,28,956, 1986.
74. **Oosaki, I., Matsumoto, T., Koshio, T., Goseki, Y., Kurimoto, J., Nakahara, T., Ochi, T., Ushama, N., Yamazaki, M., and Wakamya, K. (Canon K. K.)**, Japan Patent 86,99,152, 1986.
75. **Tanaka, K. and Fukumoto, H. (Canon K. K.)**, Japan Patent 86,194,451, 1986.
76. **Tanaka, K. and Fukumoto, H. (Canon K. K.)**, Japan Patent 86,194,450, 1986.
77. **Tanaka, K., Fukumoto, H., Imai, E., and Ikeda, T. (Canon K. K.)**, Japan Patent 86,143,769, 1986.
78. **Takagi, S., Hyozu, Y., and Mori, H. (Canon K. K.)**, Japan Patent 86,156,138, 1986.
79. **Yamamoto, A., Fukumoto, H., and Tanaka, K. (Canon K. K.,)**, Japan Patent 86,217,064, 1986.
80. **Karami, Y., Fukumoto, H., Tanaka, K., and Kasai, M. (Canon K. K.)**, Japan Patent 86,194,453, 1986.
81. **Fukumoto, H. and Tanaka, K. (Canon K. K.,)**, Japan Patent 86,160,758, 1986.
82. **Kawagishi, Y., Narita, S., Kiryu, T., and Uomoto, A. (Orient Chemical Industries, Ltd.)**, Japan Patent 86,91,667, 1986.
83. **Yamazaki, M., Ochi, T., Ushama, N., Nakahara, T., Kurimoto, J., Oosaki, I., Matsumoto, T., Goseki, Y., Koshio, T., and Wakamya, K. (Canon K. K.)**, Japan Patent 86,99,155, 1986.
84. **Fukumoto, H., Tanaka, K., Karami, Y., and Uchida, M. (Canon K. K.)**, Japan Patent 86,217,065, 1986.
85. **Karami, Y., Fukumoto, H., Tanaka, K., and Kasai, M. (Canon K. K.)**, Japan Patent 86,196,255, 1986.

86. **Tanaka, K. and Fukumoto, H. (Canon K. K.),** Japan Patent 86,236,557, 1986.
87. **Tanaka, K. and Fukumoto, H. (Canon K. K.),** Japan Patent 86,236,556, 1986.
88. **Fukumoto, H., Tanaka, K., and Kawagishi, Y. (Canon K. K. and Orient Chemical Industries, Ltd.),** Japan Patent 86,162,056, 1986.
89. **Fukumoto, H., Tanaka, K., and Kawagishi, Y. (Canon K. K. and Orient Chemical Industries, Ltd.),** Japan Patent 86,162,055, 1986.
90. **Fukumoto, H., Tanaka, K., and Kawagishi, Y. (Canon K. K. and Orient Chemical Industries, Ltd.),** Japan Patent 86,162,057, 1986.
91. **Tanaka, K., Fukumoto, H., Imai, E., and Ikeda, T. (Canon K. K.),** Japan Patent 86,143,770, 1986.
92. **Fukumoto, H., Tanaka, K., and Kawagishi, Y. (Canon K. K. and Orient Chemical Industries, Ltd.),** Japan Patent 86,141,450, 1986.
93. **Fukumoto, H., Ikeda, T., and Uchida, M. (Canon K. K.),** Japan Patent 86,149,967, 1986.
94. **Fukumoto, H. and Tanaka, K. (Canon K. K.),** Japan Patent 86,176,944, 1986.
95. **Fukumoto, H. and Tanaka, K. (Canon K. K.),** Japan Patent 86,176,945, 1986.
96. **Fukumoto, H. and Tanaka, K. (Canon K. K.),** Japan Patent 86,176,942, 1986.
97. **Fukumoto, H., Tanaka, K., and Kawagishi, Y. (Canon K. K. and Orient Chemical Industries, Ltd.),** Japan Patent 86,172,153, 1986.
98. **Fukumoto, H., Tanaka, K., and Kawagishi, Y. (Canon K. K.),** Japan Patent 86,172,152, 1986.
99. **Fukumoto, H. and Tanaka, K. (Canon K. K.),** Japan Patent 86,160,757, 1986.
100. **Fukumoto, H., Tanaka, K., Takagi, S., and Yamamoto, A. (Canon K. K.),** Japan Patent 86,160,756, 1986.
101. **Suematsu, H. (Canon K. K.),** Japan Patent 86,130,962, 1986.
102. **Takagi, S., Mori, H., and Huozu, Y. (Canon K. K.),** Japan Patent 86,156,137, 1986.
103. **Tanaka, K., Fukumoto, H., Imai, E., and Ikeda, T. (Canon K. K.),** Japan Patent 86,143,771, 1986.
104. **Tanaka, K. and Fukumoto, H. (Canon K. K.),** Japan Patent 86,194,452, 1986.
105. **Suematsu, H. and Uchiyama, M. (Canon K. K.),** Japan Patent 86,198,247, 1986.
106. **Urawa, M. (Canon K. K.),** Japan Patent 86,248,062, 1986.
107. **Aoki, M. (Ricoh Co., Ltd.),** Japan Patent 86,217,066, 1986.
108. **Fukumoto, H., Tanaka, K., and Uchida, M. (Canon K. K.),** Japan Patent 86,217,062, 1986.
109. **Tanaka, K., Fukumoto, H., and Uchida, M. (Canon K. K.),** Japan Patent 86,217,061, 1986.
110. **Suzuki, N., Sakai, T., Sugiyama, G., and Suzuka, S. (Hodogaya Chemical Co., Ltd.),** Japan Patent 86,267,059, 1986.
111. **Tanaka, K. and Fukumoto, H. (Canon K. K.),** Japan Patent 86,217,059, 1986.
112. **Tanaka, K. and Fukumoto, H. (Canon K. K.),** Japan Patent 86,217,058, 1986.
113. **Tanaka, K., Fukumoto, H., Kasai, M., and Ikeda, T. (Canon K. K.),** Japan Patent 86,217,054, 1986.
114. **Chiba, S., Inoue, S., Asahina, Y., and Ochiai, H. (Ricoh Co., Ltd.),** Japan Patent 87,23,053, 1987.
115. **Fujiki, M., Yushina, H., and Sato, Y. (Mitsubishi Chemical Industries Co., Ltd.),** Japan Patent 87,02,272, 1987.
116. **Fukumoto, H. and Tanaka, K. (Canon K. K.),** Japan Patent 86,176,943, 1986.
117. **Tanaka, K. and Fukumoto, H. (Canon K. K.),** Japan Patent 86,217,060, 1986.
118. **Sasaki, F., Inoue, S., Matsui, A., Minamitani, T., and Tomita, M. (Ricoh Co., Ltd.),** Japan Patent 87,54,275, 1987.
119. **Fujii, M. (Mita Industrial Co., Ltd),** Japan Patent 87,127,847, 1987.
120. **Arikawa, A., Hayashi, M., and Namatame, Y. (Toyo Ink Manufacturing Co., Ltd.),** Japan Patent 87,100,768, 1987.
121. **Fujitsu Ltd.,** Japan Patent 85,57,858, 1985.
122. **Fujitsu Ltd.,** Japan Patent 85,63,546, 1985.
123. **Fujitsu Ltd.,** Japan Patent 85,63,545, 1985.
124. **Fujiwara, S., Suzuki, Y., Okuyama, T., and Uehara, T. (Toshiba Corp.),** Japan Patent 85,138,566, 1985.
125. **Canon K. K.,** Japan Patent 84,160,162, 1984.
126. **Otsuka, S. (Fuji Photo Film Co., Ltd.),** Japan Patent 85,46,708, 1985.
127. **Ookuma, A. (Matsushita Electric Industrial Co. Ltd.),** Japan Patent 86,213,192, 1986.
128. **Lu, C. H. and Auclair, C. J. (Xerox Corp),** U.S. Patent 4,442,189, 1984.
129. **Gruber, R. J., Julien, P. C., and Yourd, R. A., III (Xerox Corp.),** U.S. Patent 4,454,214, 1984.
130. **Yourd, R. A., III, Majumdar, D., and Gruber, R. J. (Xerox Corp.),** U.S. Patent 4,537,848, 1985.
131. **Hoffend, T. R. and Barbetta, A. J. (Xerox Corp.),** U.S. Patent 4,560,635, 1985.
132. **Fujii, M., Honda, N., Horiuchi, A., and Kubo, M. (Mita Industrial Co., Ltd.),** Europe Patent Application 159,166, 1985.
133. **Ikeda, T., Imai, E., Fukumoto, H., Tanaka, K., Suematsu, K., Urawa, M., and Takenouchi, M. (Canon K. K.),** Europe Patent Application 178,952, 1986.

134. **Sano, Y. and Morimoto, T. (Mita Industrial Co., Ltd.),** Europe Patent Application 166,576, 1986.
135. **Canon K. K.,** Japan Patent 83,203,455, 1983.
136. **Tomoegawa Paper Manufacturing Co., Ltd.,** Japan Patent 83,171,060, 1983.
137. **Ricoh Co., Ltd.,** Japan Patent 82,186,758, 1982.
138. **Canon K. K.,** Japan Patent 83,211,161, 1983.
139. **Mita Industrial Co., Ltd.,** Japan Patent 84,200,250, 1984.
140. **Toshiba Corp.,** Japan Patent 84,135,480, 1984.
141. **Mita Industrial Co., Ltd.,** Japan Patent 84,100,453, 1984.
142. **Mita Industrial Co., Ltd.,** Japan Patent 84,181,360, 1984.
143. **Toshiba Corp.,** Japan Patent 84,135,479, 1984.
144. **Mita Industrial Co., Ltd.,** Japan Patent 84,181,361, 1984.
145. **Matsushita Electric Industrial Co., Ltd.,** Japan Patent 85,17,453, 1985.
146. **Mita Industrial Co., Ltd.,** Japan Patent 85,118,851, 1985.
147. **Toyo Printing Ink Manufacturing Co., Ltd.,** Japan Patent 85,93,453, 1985.
148. **Canon K. K.,** Japan Patent 85,15,066, 1985.
149. **Matsushita Electric Industrial Co., Ltd.,** Japan Patent 85,125,855, 1985.
150. **Canon K. K.,** Japan Patent 85,67,959, 1985.
151. **Fujiki, M., Yushina, H., Kosugi, H., and Sato, Y. (Mitsubishi Chemical Industries Co., Ltd.),** Japan Patent 85,205,460, 1985.
152. **Konishiroku Photo Industry Co., Ltd.,** Japan Patent 85,129,756, 1985.
153. **Yamauchi, T. (Toshiba Corp.),** Japan Patent 85,163,050, 1985.
154. **Tsubushi, K. (Ricoh Co., Ltd.),** Japan Patent 85,162,262, 1985.
155. **Ushiyama, N., Ochi, T., Nakahara, T., Kurimoto, J., Osaki, I., Matsumoto, T., Goseki, Y., Koshio, T., Yamazaki, M., and Wakamiya, K. (Canon K. K.),** Japan Patent 86,83,546, 1986.
156. **Koizumi, F., Nakahara, S., and Watanabe, N.,** Japan Patent 86,77,862, 1986.
157. **Hyozu, Y., Takagi, S., and Mori, H. (Canon K. K.),** Japan Patent 86,46,955, 1986.
158. **Koizumi, F., Nakahara, S., and Watanabe, N. (Konishiroku Photo Industry Co., Ltd.),** Japan Patent 86,77,864, 1986.
159. **Koori, S. (Minolta Camera Co., Ltd.),** Japan Patent 86,126,565, 1986.
160. **Fujii, M., Horiuchi, A., and Kubo, M. (Mita Industrial Co., Ltd.),** Japan Patent 85,230,145, 1985.
161. **Nakajo, T. (Toshiba Corp.),** Japan Patent 86,94,064, 1986.
162. **Koizumi, F., Nakahara, S., and Watanabe, N. (Konishiroku Photo Industry Co., Ltd.),** Japan Patent 86,118,760, 1986.
163. **Saruwatari, N., Ko, K., Yamagishi, Y., Katagiri, Y., and Narisawa, T. (Fujitsu Ltd.),** Japan Patent 86,118,761, 1986.
164. **Takayama, H. and Matsumura, Y. (Fuji Xerox Co., Ltd.),** Japan Patent 86,112,158, 1986.
165. **Inagaki, S., Yamamoto, M. (Minolta Camera Co., Ltd.),** Japan Patent 86,245,169, 1986.
166. **Kawabe, N. (Toshiba Corp.),** Japan Patent 86,94,059, 1986.
167. **Kawabe, N. (Toshiba Corp.),** Japan Patent 86,94,060, 1986.
168. **Inagaki, S. and Yamamoto, M. (Minolta Camera Co., Ltd.),** Japan Patent 86,245,167, 1986.
169. **Matsushita Electric Industrial Co., Ltd.,** Japan Patent 82,185,447, 1982.
170. **Matsushita Electric Industrial Co., Ltd.,** Japan Patent 85,17,767, 1985.
171. **Dainippon Printing Co., Ltd.,** Japan Patent 85,23,865, 1985.
172. **Ricoh Co., Ltd.,** Japan Patent 57,167,035, 1982.
173. **Matsushita Electric Industrial Co., Ltd.,** Japan Patent 85,118,852, 1985.
174. **Sony Corp.,** Netherlands Application 84,02,383, 1985.
175. **Canon K. K.,** Japan Patent 84,127,060, 1984.
176. **Canon K. K.,** Japan Patent 84,188,657, 1984.
177. **Canon K. K.,** Japan Patent 84,188,658, 1984.
178. **Canon K. K.,** Japan Patent 84,187,352, 1984.
179. **Canon K. K.,** Japan Patent 84,189,354, 1984.
180. **Canon K. K.,** Japan Patent 84,187,349, 1984.
181. **Canon K. K.,** Japan Patent 84,189,355, 1984.
182. **Canon K. K.,** Japan Patent 84,170,856, 1984.
183. **Canon K. K.,** Japan Patent 85,15,645, 1985.
184. **Canon K. K.,** Japan Patent 85,15,646, 1985.
185. **Canon K. K.,** Japan Patent 84,189,348, 1984.
186. **Canon K. K.,** Japan Patent 84,187,346, 1984.
187. **Canon K. K.,** Japan Patent 85,26,346, 1985.
188. **Canon K. K.,** Japan Patent 85,88,956, 1985.
189. **Canon K. K.,** Japan Patent 85,69,658, 1985.
190. **Canon K. K.,** Japan Patent 85,121,456, 1985.

191. **Nakahara, T., Ohsaki, I., Yamazaki, M., Wakimiya, K., Matsumoto, T., Ushiyama, H. (Canon K. K.)**, German Patent 3,433,191, 1985.
192. **Canon K. K.**, Japan Patent 85,103,356, 1985.
193. **Canon K. K.**, Japan Patent 85,104,957, 1985.
194. **Canon K. K.**, Japan Patent 85,88,958, 1985.
195. **Canon K. K.**, Japan Patent 85,39,656, 1985.
196. **Canon K. K.**, Japan Patent 85,88,957, 1985.
197. **Canon K. K.**, Japan Patent 85,26,951, 1985.
198. **Canon K. K.**, Japan Patent 85,101,547, 1985.
199. **Kurimoto, J., Nakahara, T., Ochi, T., Ushama, N., Matsumoto, T., Oosaki, I., Goseki, Y., Koshio, T., Wakamya, K., and Yamazaki, M. (Canon K. K.)**, Japan Patent 86,99,154, 1986.
200. **Matsumoto, T., Koshio, T., Goseki, Y., Ushiyama, N., Yamazaki, M., Osaki, I., Nakahara, T., Kurimoto, J., and Wakamiya, K. (Canon K. K.)**, Japan Patent 86,59,352, 1986.
201. **Nakahara, T., Matsumoto, T., Koshio, T., Yamazaki, M., Osaki, I., Goseki, Y., Wakamiya, K., Kurimoto, J., and Ushiyama, N. (Canon K. K.)**, Japan Patent 86,59,351, 1986.
202. **Matkin, J., Trebeaven, R. J. (Minnesota Mining and Manufacturing Co.)**, U.S. Patent 4,636,151, 1987.
203. **Canon K. K.**, Japan Patent 84,15,257, 1984.
204. **Canon K. K.**, Japan Patent 83,189,654, 1983.
205. **Canon K. K.**, Japan Patent 83,85,445, 1983.
206. **Canon K. K.**, Japan Patent 84,187,347, 1984.
207. **Canon K. K.**, Japan Patent 85,06,951, 1985.
208. **Hitachi Chemical Co., Ltd.**, Japan Patent 84,185,347, 1984.
209. **Canon K. K.**, Japan Patent 84,170,847, 1984.
210. **Canon K. K.**, Japan Patent 85,117,255, 1985.
211. **Kasuya, R., Koizumi, F., Daidoji, T., and Nakahara, S. (Konishiroku Photo Industry Co., Ltd.)**, Japan Patent 87,11,863, 1987.
212. **Mori, H. and Hyozu, Y. (Canon K. K.)**, Japan Patent 86,292,160, 1986.
213. **Ricoh Co., Ltd.**, Japan Patent 85,18,059, 1985.
214. **Murakawa, K., Mitsuhashi, Y., Uchide, H., Takagi, S., Uchiyama, M., Ochi, H., Urawa, M., and Imai, E. (Canon K. K.)**, German Patent 3,315,005, 1983.
215. **Canon K. K.**, Japan Patent 85,26,350, 1985.
216. **Canon K. K.**, Japan Patent 85,26,349, 1985.
217. **Seiko Instruments and Electronics, Ltd.**, Japan Patent 85,73,549, 1985.
218. **Ricoh Co., Ltd.**, Japan Patent 85,86,554, 1985.
219. **Suzuki, A. (Ricoh Co., Ltd.)**, Japan Patent 85,173,553, 1985.
220. **Okuyama, T. and Koizumi, F. (Konishiroku Photo Industry Co., Ltd.)**, Japan Patent 86,29,854, 1986.
221. **Naruse, T. and Yamaguchi, S. (Victor Co. of Japan, Ltd)**, Japan Patent 86,177,469, 1986.
222. **Koizumi, F., Kasuya, R., Daidoji, T., and Nakahara, S. (Konishiroku Photo Industry Co., Ltd.)**, Japan Patent 87,70,863, 1987.
223. **Oce-Nederland BV**, Netherlands Application 84,00,368, 1985.
224. **Kitabaake, Y. (Sharp Corp)**, German Patent 3,240,091, 1986.
225. **Wiese, J. and Buxbaum, G. (Bayer AG)**, German Patent 3,444,869, 1986.
226. **Ostertag, W., Czech, E., Schmitt, F. U., and Schulze-Hagenest, D. (BASF AG)**, German Patent 3,542,834, 1987.

Chapter 40

LITHOGRAPHIC PRINTING PLATES AND PLANOGRAPHIC PLATES

I. LITHOGRAPHIC PRINTING PLATES

The development of presensitized lithographic plates that contain photoconductive M phthalocyanines (pcs) and di hydrogen pcs is detailed in the Japanese patent literature. The central atom of predominant use is copper.

A water-developable presensitized plate has a photosensitive layer containing a diazo resin, an azide polymer, a polymer that can be softened with water, and Cu pc, from which a high-quality lithographic plate with excellent durability is obtained.[1]

A water-developed presensitized plate containing Cu pc can be used to prepare a lithographic plate showing excellent image quality and durability.[2]

Cu pc is an electrophotographic photoconductor in lithographic plate fabrication. The electrophotographic plate is charged, imagewise exposed, developed, fixed, and etched to give a high-quality lithographic plate.[3]

A lithographic plate with an ink image is coated with a lacquer emulsion containing Cu pc. The plate gives 110,000 good copies.[4] A UV-curable coating composition contains Cu pc for lithographic plate preparation.[5] An electrophotographic plate to be used to prepare a printing plate by removing the nonimage areas by dissolution in an alkaline solution contains ε-Cu pc as photoconductor.[6] An electrophotographic plate for lithographic printing plate fabrication includes a photoconductive composition containing Cu pc.[7]

Lithographic printing plates are prepared from electrophotographic plates containing ε-Cu pc and a pc derivative prepared by acid pasting; the derivative has an electron attracting group.[8] A photoconductive composition contains Cu pc and Zn tetranitro pc, zinc oxide, and zinc sulfide as photosensitive layer in lithographic plate preparation.[9]

A photosensitive lithographic plate material contains an electrophotographic photosensitive layer that incorporates photosensitive β-Cu pc dispersed in thermoplastic polymer.[10]

An electrophotographic photoreceptor contains Cu pc for lithographic plate preparation.[11]

A presensitized color correction sheet contains Cu pc. The sheet is useful to correct color separation negatives for printing plate fabrication processes.[12] A presensitized lithographic plate contains Cu pc.[13] A waterless photosensitive lithographic plate contains Cu pc as a component.[14] A photosensitive plate for lithographic plate preparation contains photoconductive ε-Cu pc.[15] A photosensitive composition contains vinyl polymer with pendant unsaturated water groups and di hydrogen pc for lithographic plate fabrication. The plate gives >50,000 prints without blemishes.[16]

An electrophotographic plate for printing plate fabrication contains β-Cu pc as a photoconductive pigment.[17] A heat-sensitive composition contains Cu pc for laser recording preparation of printing plates.[18] A wash-off development of an electrophotographic lithographic plate that contains Cu pc is described.[19]

An electrophotographic plate for lithographic plate preparation contains Al pc Cl in the charge-generating layer.[20] A lithographic plate preparation method includes Cu pc in the presensitized plate.[21] A presensitized lithographic plate contains Cu pc.[22] A laser-sensitive composition contains thermoplastic particles and Cu pc for direct preparation of screen printing plates.[23]

A number of European patents and one U.S. patent relate to the use of Cu pcs to make lithographic printing plates. One patent relates to di hydrogen pc.

A presensitized lithographic plate with an electrophotographic photoconductive layer contains Cu pc.[24] Cu pc is a photosensitive material in preparing lithographic printing plates.[25] An electrophotographic plate for lithographic printing plate fabrication has a coating that contains Cu pc.[26] A composite electrophotographic photoreceptor with a charge carrier generating layer contains Cu pc for lithographic plate fabrication.[27] The photosensitive layer of a lithographic plate contains Cu pc.[28] A lithographic printing plate contains Cu pc. It provides 220,000 good quality prints.[29]

A water-developable lithographic positive printing plate has an oleophilic layer containing Cu pc.[30] A photoimaging composition contains a halomethyl oxadiazole derivative as a radical generator and Cu pc for lithographic plate fabrication.[31] A presensitized lithographic plate contains a diazo resin sulfopolyester adduct and Cu pc with improved anti-blocking properites and storage stability.[32]

A method to prepare lithographic printing plates requiring no dampening solution has a photosensitive layer containing Cu pc.[33] A aqueous alkaline solution-developable photosensitive composition contains Cu pc for lithographic plate fabrication.[34] An electrophotographic lithographic printing plate has a photoconductive layer including Cu pc, Cu pc $(NO_2)_4$, zinc oxide, and zinc sulfide in a binder resin. The printing plate can produce 10,000 prints.[35]

A photosensitive composition contains di hydrogen pc for production of a rough surface layer on a lithographic plate for improved vacuum contact.[36] A (−)-working photosensitive composition contains a diazo compound condensate and Cu pc for lithographic plate fabrication.[37]

A photoconductive layer of an electrophotographic plate used to prepare a lithographic plate contains Cu pc, Cu pc $(NO_2)_4$, zinc oxide, and zinc sulfide in a binder resin.[38]

A lithographic plate has a photosensitive and a photoconductive layer of Cu pc.[39]

A dry presensitized printing plate for lithographic printing contains Cu pc.[40] A dry lithographic plate contains a photosensitive layer containing Cu pc.[41]

II. PLANOGRAPHIC PLATES

The Toppan Printing Co., Ltd. has patented a method to fabricate planographic plates.[42,43] An electrically conductive substrate (polyester film carrying an aluminum layer) is coated with a photoconductive layer containing ϵ-Cu pc, a mixture of silicone and vinyl acetate resins, and zinc oxide, electrically charged, imagewise exposed to a semiconductor layer, developed with a magnetic brush, and fixed to give a planographic plate with excellent characteristics.

REFERENCES

1. **Konishiroku Photo Industry Co., Ltd,** Japan Patent 83,02,834, 1983.
2. **Konishiroku Photo Industry Co., Ltd.,** Japan Patent 83,02,847, 1983.
3. **Mitsubishi Paper Mills, Ltd.,** Japan Patent 82,161,863, 1982.
4. **Konishiroku Photo Industry Co., Ltd.,** Japan Patent 82,185,152, 1982.
5. **Toyo Ink Manufacturing Co., Ltd.,** Japan Patent 83,62,051, 1983.
6. **Konishiroku Photo Industry Co., Ltd.,** Japan Patent 84,125,751, 1984.
7. **Konishiroku Photo Industry Co., Ltd.,** Japan Patent 84,176,756, 1984.
8. **Toppan Printing Co., Ltd. and Toyo Ink Manufacturing Co., Ltd.,** Japan Patent 84,220,753, 1984.
9. **Toppan Printing Co., Ltd. and Toyo Ink Manufacturing Co., Ltd.,** Japan Patent 84,220,755, 1984.
10. **Fuji Photo Film Co., Ltd.,** Japan Patent 84,22,218, 1984.
11. **Mitsubishi Paper Mills, Ltd.,** Japan Patent 84,166,969, 1984.
12. **Minnesota Mining and Manufacturing Co.,** Japan Patent 85,263,936, 1985.

13. **Imai, M. and Kanzaki, T. (Fuji Photo Film Co., Ltd.)**, Japan Patent 85,138,551, 1985.
14. **Yoshida, S., Takahashi, H., and Shiba, K. (Fuji Photo Film Co., Ltd.)**, Japan Patent 85,229,031, 1985.
15. **Iijima, A. (Konishiroku Photo Industry Co., Ltd.)**, Japan Patent 85,194,467, 1985.
16. **Ichimura, K., Nishio, Y., and Oe, K. (Agency of Industrial Sciences and Technology and Dainippon Ink and Chemicals, Inc.)**, Japan Patent 86,80,241, 1986.
17. **Iijima, A., Kita, N., and Shimizu, Y. (Konishiroku Photo Industry Co., Ltd.)**, Japan Patent 86,32,861, 1986.
18. **Shiraishi, S. and Yamaoka, T. (Ricoh Co., Ltd.)**, Japan Patent 86,24,451, 1986.
19. **Nogami, A. and Kita, N. (Konishiroku Photo Industry Co., Ltd.)**, Japan Patent 86,230,149, 1986.
20. **Kamisaka, T., Katagiri, Y., and Miyagawa, M. (Fujitsu Ltd)**, Japan Patent 87,95,550, 1987.
21. **Akiyama, K. and Kita, N. (Fuji Photo Film Co., Ltd.)**, Japan Patent 87,19,859, 1987.
22. **Akiyama, K., Kobayashi, K., and Kita, N. (Fuji Photo Film Co., Ltd.)**, Japan Patent 86,285,449, 1986.
23. **Shiraishi, S. and Yamaoka, T. (Ricoh Co., Ltd.)**, Japan Patent 86,229,595, 1986.
24. **Nakayama, T. (Fuji Photo Film Co., Ltd.)**, German Patent 3,307,613, 1983.
25. **Nakayama, T. (Fuji Photo Film Co., Ltd.)**, German Patent 3,310,804, 1983.
26. **Yamamoto, T., Suzuki, N., Aoki, T., and Iwaki, A. (Konishiroku Photo Industry Co., Ltd.)**, Europe Patent Application 97,503, 1984.
27. **Yamada, J., Senga, T., and Suzuki, S. (Mitsubishi Paper Mills, Ltd.)**, German Patent 3,300,244, 1983.
28. **Nakayama, T. (Fuji Photo Film Co., Ltd.)**, German Patent 3,310,805, 1983.
29. **Suzuki, N., Aoki, T., Iwaki, A., and Yamamoto, T. (Konishiroku Photo Industry Co., Ltd.)**, Europe Patent Application 98,166, 1984.
30. **Rousseau, A. D., Fohrenkamm, E. A., and Kausch, W. L. (Minnesota Mining and Manufacturing Co.)**, U.S. Patent 4,507,382, 1985.
31. **Kita, N. and Goto, K. (Konishiroku Photo Industry Co., Ltd.)**, Europe Patent Application 118,766, 1984.
32. **Lorenz, W. J. and Larson, W. K. (Minnesota Mining and Manufacturing Co.)**, Europe Patent Application 136,903, 1985.
33. **Takahashi, H., Tamaki, H., Ohishi, C., and Shiba, K. (Fuji Photo Film Co., Ltd.)**, Europe Patent Application 154,891, 1985.
34. **Pawlowski, G. (Hoechst AG)**, German Patent 3,404,366, 1985.
35. **Naganuma, T., Hirayama, S., Kumagai, H., Sawada, M., Tanaka, T., and Kumano, I. (Toppan Printing Co., Ltd. and Toyo Ink Manufacturing Co., Ltd.)**, German Patent 3,423,141, 1986.
36. **Seibel, M. and Kaempf. G. (Hoechst AG)**, German Patent 3,433,247, 1986.
37. **Frommeld, H. D., Lutz, W., and Steppan, H. (Hoechst AG)**, German Patent 3,425,328, 1986.
38. **Naganuma, T., Hirayama, S., Kumagai, H., Sawada, M., Tanaka, T., and Kumano, I. (Toppan Printing Co., Ltd. and Toyo Ink Manufacturing Co., Ltd.)**, British Patent Application 2,160,670, 1985.
39. **Inukai, Y. (Fuji Photo Film Co., Ltd.)**, German Patent 3,522,367, 1986.
40. **Naritomi, Y., Takahashi, H., and Shiba, K. (Fuji Photo Film Co., Ltd.)**, German Patent 3,543,961, 1986.
41. **Yoshida, S., Takahashi, H., Aokia, H., Kondow, K., and Mayuzumi, T. (Fuji Photo Film Co., Ltd. and Shin-Etsu Chemical Co., Ltd.)**, German Patent 3,546,024, 1986.
42. **Toppan Printing Co., Ltd. and Toyo Ink Manufacturing Co., Ltd.**, Japan Patent 84,116,759, 1984.
43. **Toppan Printing Co., Ltd.**, Japan Patent 84,116,760, 1984.

Chapter 41

ELECTROPHOTOGRAPHY AND ELECTROPHOTOGRAPHIC PLATES: ELECTROPHOTOGRAPHIC PHOTORECEPTORS AND PHOTOSENSITIZERS

Over 400 Japanese patents relate to the use of phthalocyanines (pcs) as electrophotographic photoreceptors and photosensitizers. The pcs are also referred to as charge carrier-generating substances. The patent citations will be briefly identified. In sum, they heavily emphasize the use of copper as the central metal atom in the pc molecule: Cu pc as charge carrier generating substance,[1] Cu pc as electrophotographic photosensitive material,[2] Ti pc in charge transfer layer,[3] Cu pc in photoconductor layer,[4] Cu pc as photoconductor dye in charge carrier-generating layer,[5] Cu pc in middle layer,[6] β-Cu pc as charge carrier-generating pigment,[7] Cu pc as photoconductive material,[8] pc as photoconductor,[9] VO pc as charge carrier generating agent,[10,11] p-type Cu pc as photoconductor,[12] pc Green SAX in a photosensitive polymer composition in a heat-resistant soldering mask,[13] pc Green SAX in a photosensitive resin composition for soldering resistant pattern formation,[14] Cu pc in charge carrier-generating layer,[15] Cu pc as an electrophotographic photosensitive material,[16] Cu pc in charge carrier generating layer,[17] electrophotgraphic photoreceptor containing Cu pc,[18] β-Cu pc in charge carrier-generating layer,[19] Cu pc in photoconductor layer,[20] Cu pc in charge generating layer,[21] di hydrogen pc as charge-generating pigment,[22] β-pc in charge generating layer,[23] β-Cu pc as photoconductive pigment,[24] ϵ-Cu pc as sensitizer,[25] M pc in charge carrier generating layer (Figure 1) where M = Al, Cr, Ga, Sb, In, Si, Ti, Ge, Sn, Te; R = H or halogen, $R^1 = R^2 =$ halogen, alkyl, alkoxy, phenyl when M is tetravalent;[26] $R^2 = O$ when M is trivalent,[26] Cu pc in photosensitive resin composition for screen printing plates.[27] β-Cu pc in charge carrier-generating layer,[28] Cu pc in the middle layer of a composite dichromatic photoreceptor,[29] Cu pc in photosensitive resin composition for presensitized lithographic plates,[30] Cu pc as charge-generating pigment,[31] Cu pc as charge-generating pigment,[32] Cu pc in charge carrier-generating layer,[33] Cu pc as photoconductor,[34] Cu pc in charge-generating layer,[35] Cu pc in photoconductive powder,[36] Cu pc in charge carrier generating layer,[37-39] Cu pc as photoreceptor,[40] Cu pc as photosensitive material,[41] di hydrogen pc in charge carrier-generating layer,[42] Cu pc in charge carrier-generating layer,[43,44] Cu pc in photoconductive layer,[45] Cu pc as charge transfer agent,[46] Cu pc in charge generating layer,[47] di hydrogen pc in a photosensitive composition for presensitized printing plate fabrication,[48] Cu pc in charge generating layer,[49] Cu pc in photoreceptor,[50] Cu pc in charge carrier generating layer,[51] β-Cu pc in photoconductive layer,[52] β-Cu pc with NO_2, CN, or COOH on the pc ring as photoconductor,[53] Cu pc as photosensitive material,[54] Cu pc as photosensitive material,[55] β-Cu pc as photoconductor in photoconductive material,[56] Cu pc containing >1 sulfonic acid group as an intermediate layer,[57] Cu pc in photosensitive resin for lithographic or relief printing plates and photoresists,[58] ϵ-Cu pc in charge generating layer,[59] Al pc Cl in photoconductive layer,[61] TiO pc in charge generating layer,[62] Al (pc Cl) Cl in photosensitive material,[63] ϵ-Cu pc as photoconductor,[64] In pc Cl in charge-generating layer,[65] α-Cu pc in photosensitive layer,[66] Al (pc Cl) Cl as photoconductor in photosensitive layer,[67] Al (pc Cl) Cl as photoconductor in charge generating layer,[68] Ge pc in photosensitive material,[68] β-Cu pc in photosensitive material,[70] Cu pc in a photoconductive composition,[71] pc Green SAX in photosensitive compositions for solder resistant pattern formation,[72] Cu pc in an electrophotographic plate,[73] di hydrogen pc in a photosensitive polyester-urethane resin composition,[74] Al pc Cl in a charge carrier-generating layer,[75] di hydrogen pc in a charge-generating layer,[76] Cu pc in electrophotographic plate,[77] Cu pc in charge carrier-generating layer,[78-80] ϵ-Cu pc in photoconductive layer,[81] Cu pc as photoreceptor,[82] Cu pc in photoreceptor charge generator layer,[83] Cu pc in charge generator,[84] Cu pc as photocon-

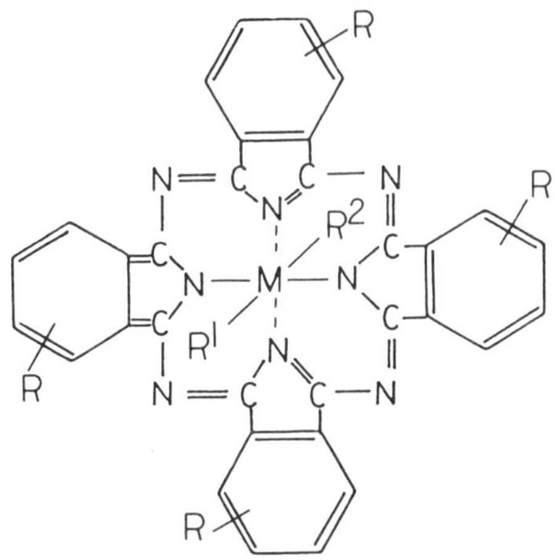

FIGURE 1. A M pc in a charge carrier generating layer.

ductor,[85,86] Cu pc in charge generating layer,[87] β-Cu pc in charge carrier-generating layer,[88] di hydrogen pc in a photosensitive resin for a laser printing plate,[89] di hydrogen pc in a photosensitive composition for a presensitized plate,[90] Cu pc in a photosensitive resin useful as a photoresist,[91] τ′-pc in photoconductors,[92] pc and $pc(NO_2)_{1\text{-}8}$ or $pc(CN)_{1\text{-}8}$ as photoconductor,[93] Cu pc in charge-generating layer,[94,95] Cu pc in photosensitive material,[96] Cu pc in charge carrier-generating layer,[97] Cu pc in photoreceptor,[98-100] α-Cu pc in photoreceptor,[101] Ti pc in charge carrier-generating layer,[102] β-Cu pc in charge generating layer,[103] Cu pc + Cu pc NO_2 in photosensitive layer,[104] In (pc Cl) Cl in photoreceptor layer,[105] Cu pc $(SO_2NHCH_2CH_2OCH_3)_4$ + α-Cu pc in photoreceptor,[106] β-Cu pc as charge-generating agent,[107] τ,τ′,η,η′-pc in charge-generating layer,[108] ε-Cu pc and other pcs in photoconductive layer,[109] ε-Cu pc in photoreceptor,[110] Al pc Cl in photosensitive material,[111] ε-Cu pc as photoconductor,[112] ε-Cu pc in charge-generating layer,[113] τ-pc as photoreceptor,[114] τ,τ′,η,η′-pc in photoconductors,[115] ε-Cu pc as photoconductive agent,[116] Cu pc as photoconductive material,[117] Cu pc in charge-generating layer,[118] TiO pc in charge-generating layer,[119] Cu pc in charge-generating layer,[120] β-Cu pc in charge-generating layer,[121] β-Cu pc in photoconductive layer,[122] pc in photosensitive composition for a lithographic plate,[123] τ,τ′,η,η′-pc as a charge-generating pigment,[124] β-Cu pc in a charge carrier-generating layer,[125] Mn pc, Ba pc, fluorinated Cr pc, halogenated pc derivatives in photosensitive material,[126] pc, pc derivative with NO_2, CN, halo, sulfo, and COOH electron withdrawing groups on the pc ring as a photoconductive composition,[127] Cu pc and di hydrogen $pc(NO_2)_2$ as photoreceptor,[128] partially chlorinated Al pc Cl in a photosensitive material,[129] TiO pc in a charge carrier-generating layer,[130] Cu pc and Cu pc NO_2 in photosensitive material,[131] Cu pc and di hydrogen pc $(NO_2)_4$ as photoreceptor,[132] ε-Cu pc in a photoconductive composition,[133] β-Cu pc in a photosensitive material,[134] a di hydrogen pc as photoreceptor,[135] Ti pc in a charge-generating layer,[136] β-Cu pc in a photosensitive material,[137] Cu pc in charge carrier-generating layer,[138] pc in a charge-generating composition,[139] Cu pc $(NO_2)_2$ as photoreceptor,[140] Cu pc $(NO_2)_2$ as photoconductive material,[141] Cu pc in a photoconductive composition,[142] Al pc Cl in a charge-generating layer,[143] ε-Cu pc as photoconductor,[144] Cu pc in a masking layer in a photoimaging assembly,[145] Al pc Cl in a charge-generating layer,[146] Al (pc Cl) Cl or TiO pc as charge-generating agent,[147] Ti pc as photoreceptor,[148] Cu pc in charge-generator layer,[149] Cu pc in alkali developing photosensitive polymer composition,[150] Cu pc in charge-generating layer,[151,152] pc in photoreceptor,[153] Cu pc in charge-generating layer,[154-156] Cu pc in photoreceptor,[157] Cu pc as photoconductor,[158] ε-pc as photoreceptor,[159] τ-pc in charge carrier-generating layer,[160] Cu pc in charge-generating layer,[161] M pc azo

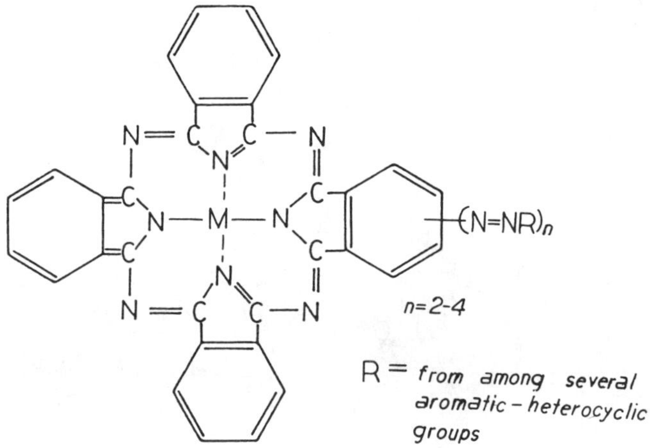

FIGURE 2. An azo M pc in a photosensitive layer in an electrophotographic photoreceptor.

derivative in a photosensitive layer (Figure 2), where M = di hydrogen, Cr, Fe, Co, Cu, Ni, Pd, Pt, Zn, n = 2 to 4[162] Al(pc Cl)Cl in a charge carrier-generated layer,[163] ϵ-Cu pc as a photoreceptor,[164] τ,τ′,η,η′-pc in charge-generating layer,[165] τ,τ′,η,η′-pc in charge-generating layer,[166] Cu pc in charge carrier-generating layer,[167] τ,τ′,η,η′-pc in charge carrier-generating layer,[168] β-Cu pc in charge-generating layer,[169] α-pc in charge carrier-generating layer,[170] τ,τ′,η,η′-pc in charge-generating layer,[171] Cu pc in photoreceptor,[172] ϵ-Cu pc in photoconductive layer,[173] β-Cu pc and Cu pc (CF_3) in photoconductor,[174] τ,τ′,η,η′-pc in charge-generating layer,[175] τ,η-pc in photoreceptor,[176] τ,τ′,η,η′-pc in charge-generating layer,[177,178] Cu pc in charge-generating layer,[179] In pc Cl or In (pc Cl) Cl_n as photoreceptor,[180] τ,τ′,η,η′-pc in charge-generating layer,[181] ϵ-Cu pc as charge generator,[182] τ,τ′,η,η′-pc in charge-generating layer,[183,184] Cu pc in charge generating layer,[185] Cu pc and Cu polychloro pc in charge-generating layer,[186] Ga pc I in charge generating layer,[187] Cu pc and Cu pc $(NO_2)_4$ in photosensitive layer,[188] Cu pc in charge carrier-generating layer,[189] α-pc in charge-generating layer,[190] τ,τ′,η,η′-pc in charge-generating agent,[191] Al pc in photoreceptor,[192] pc in photoreceptor,[193] Cu pc as charge carrier-generating pigment,[194] VO pc in charge carrier-generating layer,[195] pc in charge carrier-generating layer,[196] Cu pc in electrophotographic color toner,[197] Cu pc in charge carrier-generating layer,[198-200] M and di hydrogen pcs as charge generators, where M includes Co, Cr, Cu, Fe, Mn, Pd, Ru, Ti . . .,[201] Cu pc and Cu pc $(NO_2)_3$ in photoconductive layer,[202] Cu pc and Cu pc $(NO_2)_4$ photoreceptors,[203] Cu pc in charge carrier-generating layer,[204] Cu pc photoreceptor,[205] M pc or dihydrogen pc derivative as charge generator (Figure 3) where R = H, halo, CN, NO; M = Co, Cu, Ni, V . . . R^1 = halo, R^2 = halo, R^1 and R^2 = oxygen when M = V,[206] Cu pc from among numerous M pcs and M pc derivatives, similar to Reference 206, as charge generator,[207] Cu pc in photoconductive layer[208] TiO pc in charge carrier-generating layer,[209] di hydrogen or M pc derivative similar to Reference 206 as charge generator,[210] β-pc in a charge-generating layer,[211] β-Cu pc in a photosensitive layer,[212] α-Cu pc as a photoconductive pigment,[213] Cu pc in a charge carrier-generating layer,[214,215] ϵ-Cu pc in a charge-generating layer,[216] Cu pc in a charge generation layer,[217,218] τ,τ′,η,η′-pc as charge generation agents,[219] Cu pc in charge generation layer,[220] Cu pc in a charge-generating layer,[221] ϵ-Cu pc in charge-generating layer,[222] a pc colorant in a photosensitive composition,[223] a pc dye as photoreceptor, Cu pc and Cupc$(NO_2)_4$ as photoreceptor,[225] ϵ-Cu pc as photoreceptor,[226] Ga pc Br in charge-generating layer of the photoreceptor,[227] Cu pc and Cu pc NO_2 in a photosensitive layer,[228] β-Cu pc in a charge carrier generation layer,[229] ϵ-Cu pc in a photoreceptor,[230,231]

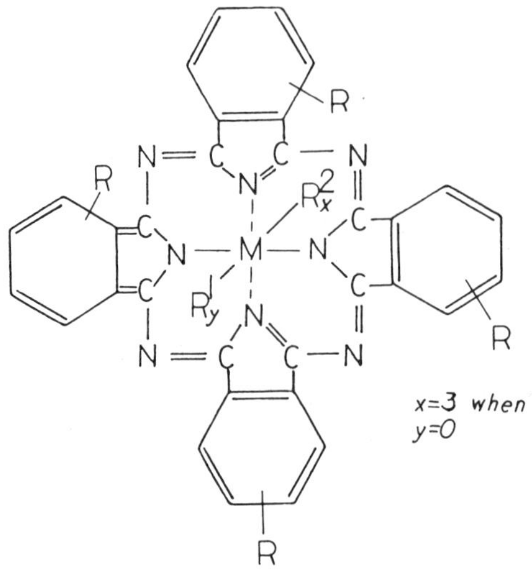

FIGURE 3. M pc charge generator in an electrophotographic photoreceptor.

ε-Cu pc in a charge carrier-generating layer,[232,233] Cu pc as photoconductive agent,[234] α,β, or ε-Cu pc in a photoreceptor,[235,236] α-pc in a photoconductive layer,[237] Cu pc in a one-shot color electrophotographic photoreceptor,[238] Cu pc in a charge carrier-generating layer in a one-shot color electrophotographic photoreceptor,[239] Cu pc in a charge carrier-generating layer,[240,241] Cu pc in photoimaging systems (photosensitive material optimum exposure determining method and apparatus),[242] Cu pc in charge carrier-generating layer,[243,244] Cu pc in photosensitive layer,[245] Cu pc in photoconductive layer,[246] Cu pc in charge carrier generating layer,[247] Cu pc in photoconductive film,[248] Cu pc in charge carrier-generating layer,[249-253] Cu pc in photoconductive composition,[254] Cu pc in charge-generating layer,[255,256] pc in photosensitive resin for alkali-developable photoimaging compositions,[257] pc in a charge-generating layer,[258] β-pc in a charge generating layer,[259] ε-pc in a charge generating layer,[260] η′-pc or Cu pc in a charge-generating layer,[261] Al pc Cl in a charge-generating layer,[262] Cu pc in a charge-generating layer,[263] ε-Cu pc in a charge carrier-generating layer,[264] β-Cu pc photoconductor,[265] Cu pc in a charge carrier-generating layer,[266] Cu pc blue electrophotographic toner,[267] Cu pc as charge-generating agent,[268] pc in a photoconductor,[269] Cu pc photoconductor,[270,271] pc in photoreceptor,[272] Cu pc in charge-generating layer,[273] Cu pc as electrophotographic toner,[274] Cu as photoconductor,[275] Cu pc in charge-generating layer,[276-278] τ,τ′-pc and η,η′-pc in a charge-generating layer,[279] β-Cu pc in a charge-generating layer,[280] ε-Cu pc and sulfonated Cu pc in a photoconductive layer,[281] τ,τ′,η,η′-pc in a charge-generating layer,[282] Cu pc and Cu pc $(NO_2)_4$ in a photoconductive layer,[283] ε-Cu pc as photoconductive material,[284] Cu pc and Cu pc $(NO_2)_4$ as photoconductor,[285] In pc Cl in charge carrier-generating layer,[286] ε-Cu pc in charge-generating layer,[287] Cu pc polymer in charge generating layer (Figure 4),[288] Cu pc and Cu pc $(NO_2)_4$ as photoreceptor,[289] ε-Cu pc as photoconductive material,[290] Cu pc as photoreceptor,[291] ε-Cu pc as photoreceptor,[292] ε-Cu pc in a photoconductive layer,[293] τ-pc in a charge-generating layer,[294] Cu pc as a photoconductive interlayer in a color photoreceptor,[295] Cu pc as a photoreceptor in a two color photoreceptor,[296] Cu pc as charge-generating layer,[297] Cu pc as photoconductive toner,[298] Cu pc in a photosensitive layer,[299] Cu pc in a charge carrier-generating layer,[300] 4,4′,4″,4‴-tetrabutoxychloro Al pc in charge-generating layer,[301] Cu pc in a photosensitive resin composition for a presensitized lithographic plate,[302] x-pc as photoconductor,[303] in pc Cl in photoreceptor,[304] a pc derivative in photoreceptor,[305] pc and pc with >1 (NO_2), halo, SO_2H, CO_2H, and CF_3; and >1 alkoxy, aryloxy, alkyl thio, and aralkyl thio in photoconductor,[306]

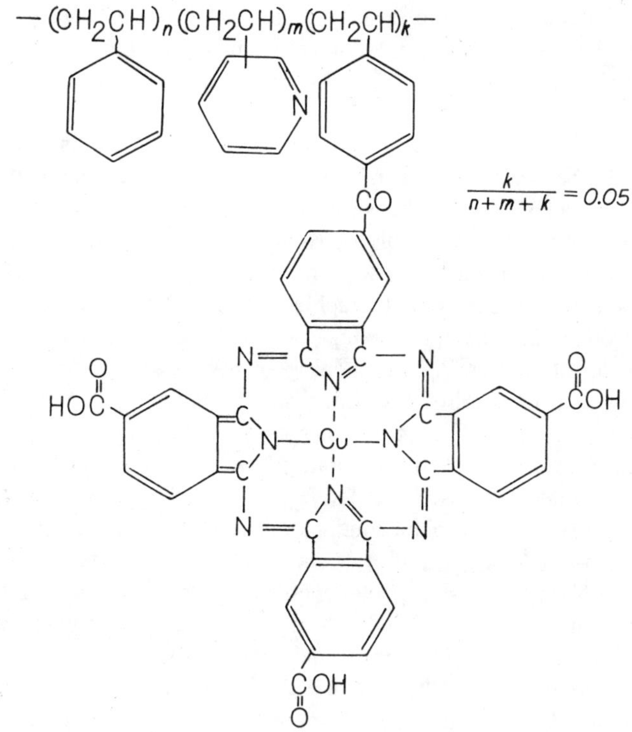

FIGURE 4. A Cu pc polymer binder charge generator in an electrophotographic photoreceptor.

Cu pc and Cu pc NO_2 in photoreceptor,[307] ϵ-Cu pc photoconductive material,[308] pc in photosensitive composition,[309] Cu pc in charge-generating layer,[310] Al pc Cl $(SCH_3)_{5\text{-}16}$ in carrier-generating layer,[311] Cu pc and Cu pc $(NO_2)_4$ in photoreceptor,[312] β-Cu pc in charge-generating layer,[313-315] Cu pc in one of two photoconductive layers,[316] β-Cu pc in photoreceptor,[317] x-pc in photosensitive layer,[318] Cu pc in charge-generating layer,[319] pc in photoreceptor,[320] M pc similar to Reference 206 in charge-generating layer,[321] τ-pc in charge-generating layer,[322] Cu pc in charge-generating layer,[323-327] α-TiO pc in photosensitive layer,[328] α-TiO pc in photosensitive layer,[329] Cu pc in charge-transporting layer,[330] pc in charge carrier generator,[331] Cu pc as photoreceptor,[332] Cu pc as photoconductive toner,[333] pc or M pc derivative in charge carrier-generating layer,[334] TiO pc in photosensitive layer,[335] pc or M pc derivative in charge-generating layer,[336] Al pc derivative in a photoreceptor,[337] α-pc in photosensitive layer,[338] TiO pc in photosensitive layer,[339] ϵ-Cu pc in a photosensitive dye layer,[340] pc in a photoreceptor,[341-342] Cu pc and Cu pc $(NO_2)_4$ in a charge-generating layer,[343] Li pc in a photoreceptor,[344] Mg in a photosensitive layer,[345] Cu pc in a photoreceptor,[346-347] Cu pc (p-type semiconductor) in a photosensitive layer,[348] TiO pc as charge generator,[349] pc as photosensitive material,[350] pc as photoconductor,[351] Cu pc in a photoconductor layer,[352] Al (pc Cl) Cl in charge-generating layer,[353] pc as a charge carrier generating agent,[354] pc in a photoreceptor,[355] ϵ-Cu pc in a charge-generating layer,[356] ϵ-Cu pc in a photoreceptor,[357] ϵ-Cu pc in photoreceptor,[358,359] Cu pc $(NO_2)_4$ in photoreceptor,[360] Al (pc Cl)Cl in charge-generating layer,[361] Cu pc in light-sensitive composition for presensitized lithographic plate,[362] $SnCl_2$ pc as photoconductor,[363] ϵ-Cu pc in lithographic plate,[364] ϵ-Cu pc in photoconductive layer on an electrophotographic plate,[365,366] Cu pc in an electrophotographic toner,[367,368] pc in a photosensitive layer in a lithographic plate,[369] ϵ-Cu pc in a photosensitive layer in a printing plate,[370] Cu pc in an electrostatic image development toner,[371] electrophotographic photoreceptors contain photoconductive pc and hydrazine coatings,[372] pc as a photoconductive material in a photosensitive plate for electrophotography,[373] ϵ-Cu pc in a charge-generating layer of an electrophotographic material,[374] Al (pc Cl) Cl in a charge generating layer of an electrophotographic material,[375] γ-pc in a charge-generating layer of an electrophotographic material,[376] photosensitive material for laser electrophotography includes Cu pc,[377] Ti pc in a charge transfer layer of an electrophotographic material,[378] Cu pc and poly(β-*N*-carbazolylethyl acrylate) are coated on an aluminum support to give an electrophotographic material, highly sensitive to 760 nm monochromatic light,[379] α-, β-, or γ-Cu pc in a photoconductive layer of α-, β-, or γ-Cu pc and a photohardening acrylic resin,[380] ϵ-Cu pc as photoconductive powder to be placed on an aluminum support to give, with other components, an electrophotographic material,[381] a color electrographic process involves an electrophotographic plate having photoconductor layer areas (Cu pc areas) with different spectral sensitivities,[382] Cu pc in charge-generating layer,[383,384] ϵ-Cu pc and a Cu nitro pc derivative as photosensitive element,[385] Cu pc in charge-generating layer,[386] Al pc Cl in charge-generating layer,[387] Cu pc in charge-generating layer,[388] pc derivative in photoconductive layer,[389] Cu pc in charge-generating layer,[390] Cu pc in a photosensitive screen in an electrophotographic process,[391] pc derivatives in a photoconductor layer for an electrographic process,[392] Cu pc in photoconductor layer,[393-395] Cu pc and Cu pc $(NO_2)_2$ as photoconductive powder,[396] Cu pc $(NO_2)_4$, Cu pc, and ϵ-Cu pc as photoconductor,[397] Cu pc as photoconductive pigment,[398] pc as photoconductor,[399] Cu pc in charge-generating layer,[400] α-Cu pc in photoconductive layer,[401] Cu pc as charge-generating pigment,[402] Cu pc in a layer containing Cu pc, trinitrofluorenone, and a polycarbonate resin in a two-color electrophotographic composite plate,[403] Cu pc in a charge-generating layer of a composite electrophotographic plate,[404,405] β-Cu pc in a composite electrophotographic plate,[406] Cu pc in a charge-generating layer of an electrophotographic plate,[407] Cu pc as a coating with a polyester resin as a layer in a composite electrophotographic plate,[408] Cu pc in a charge carrier-generating layer to prepare a composite electrophotographic plate,[409] Cu pc as charge-generating material in a composite electrophotographic plate,[410] ϵ-Cu pc in a photosensitive layer to prepare an electrostatic printing plate,[411] an electrophotographic plate has a charge carrier-generating layer containing Cu pc,[412] an electrophotographic plate photoconductor layer contains alumina and Cu pc in a binder resin with increased mechanical strength,[413] ϵ-Cu pc is a photoconductor in an electrophotographic printing plate,[414] Cu pc is a photo material in an electrophotographic printing plate,[415] a composite photosensitive plate for electrophotography contains a layer with microencapsulated Cu pc for black and white and color copies with a single exposure,[416] a printing plate for electrophotographic fabrication has a recording layer containing β-Cu pc,[417] an electrophotographic composite photoreceptor has a charge carrier-generating layer containing VO pc,[418] an aluminum substrate is coated with a substrate containing ϵ-Cu pc and polyester polyol,[419] a photoconductive composition for an electrostatic printing plate includes Cu pc, Cu pc $(NO_2)_4$, and ϵ-Cu pc,[420-422] composite particles for electrostatic image formation include Cu pc and pc,[423] a photoconductive element for an electrophotographic plate includes Cu pc, Cu pc $(NO_2)_4$, and ϵ-Cu pc,[424] a 3-μ charge-generating layer for a composite electrophotographic plate contains α-pc, acrylic resin, and a xylene tetrahydrofuran mixture,[425] photoconductor particles for an electrophotographic imaging process are made from β-Cu pc, benzoylleucomethylene blue, and a polyamide mixed in a methyl benzene methanol mixture, then dried and pulverized,[426] a laser electrophotographic plate charge carrier-transporting layer contains a light absorbing substance, i.e., Cu pc. to reduce or prevent stripe image formation,[427] a mixture of phenoxy resin, 4,6,8-trimethylazulene, and Al pc Cl is coated on an aluminum plate to give an electrophotographic plate,[428] a composite electrophotographic plate contains pc,[429] color image-forming particles contain Cu pc for electrophotography,[430] electrophotographic composite photoconductors with a charge carrier-generating layer contain β-Cu pc,[431] photoconductive α-Cu pc and ϵ-Cu pc are components of an electrophotographic printing plate,[432] a 15-μ photoconductive layer in an electrophotographic plate contains ϵ-Cu pc, hexyl methacrylate-methyl methacrylate copolymer, and methyl ethyl ketone,[433] a lithographic printing plate for electrophotographic processing contains a photoconductive composition including Cu pc, Cu pc $(NO_2)_4$ and ϵ-Cu pc,[434] polyester-coated layers contain Cu pc $(SO_3Na)_{1.8}$ for electrophotographic uses,[435] a charge-generating layer contains β-Cu pc to form a flexible photosensitive material to give 10,000 good quality images,[436] and di hydrogen, Al, Cu, Fe, Mg, or Ni pcs are in a photoconductor layer of a 1:1 image sensor,[437]

About 50 European and U.S. patents also relate to M pcs or di hydrogen pcs as a component in photoconductive, charge generating layers in electrophotography: τ,τ′,η,η′-pc as electrophotographic photoreceptor,[438] Cu pc in charge carrier-generating layer of electrophotographic photoreceptor,[439] di hydrgen pc in photoconductor layer,[440] VO pc in photoconductor layer,[441] di hydrogen pc in charge-generating layer,[442] τ-pc in charge-generating layer with a butyral resin,[443] Cu pc in charge-generator layer,[444] an electrophotographic photoreceptor has a charge carrier-generating layer containing a binder and di hydrogen pc,[445] Cu pc in a charge carrier-generating layer in a photoreceptor,[446] di hydrogen pc in a photosensitive layer,[447] Cu pc in charge carrier-injecting layer in electrophotographic composite photoreceptor,[448] Al pc Cl or Ga pc Cl or In pc Cl in a charge generation layer of a photoconductor,[449] Cu pc NO_2 and Cu pc/Cu pc NO_2 mixture in photoconductive layer,[450] VO pc in photoconductive layer,[451] β-Cu pc in charge generation layer of photoconductive material,[452] VO pc in photoconductive layer sensitive to IR,[453] VO pc in a charge-photogenerating layer,[454] di hydrogen pc in charge-generating layer,[455] In pc Cl in charge carrier generating layer of photoreceptor with high sensitivity to light of about 800 nm,[456] VO pc in photoconductor,[457] VO pc in photoreceptor with overcoat layer containing VO pc,[458] Cu pc in a photoconductive double layer,[459] Cu pc in a charge carrier-forming layer,[460] Cu pc as photoconductive film with Langmuir-Blodgett film structure,[461] VO pc as photoconductor in charge-generating layer,[462] Ge pc $(OH)_2$ or Sn pc $(OH)_2$ or Si pc $(OH)_2$ in charge-generating layer,[463] β-Cu pc as charge carrier-generating layer with poly(vinyl butyral),[464] ϵ-Cu pc in charge carrier-generating layer,[465] di hydrogen pc in photoconductive layer containing selenium and di hydrogen pc,[466] In pc Cl in photoreceptor sensitive to visible and IR light,[467] VO pc in photoreceptor exposed with 400- to 700-nm light,[468] VO pc in photoreceptor photogenerating layer,[469] Cu pc in charge carrier-generating layer,[470] TiO pc $Cl_{0\text{-}1}$ in the charge carrier-generating layer of a photoreceptor,[471] VO pc in charge-generating layer of a composite photoreceptor,[472] Cu pc in charge carrier-generating layer,[473] VO pc in photoreceptor charge-generating layer,[474] Cu pc in photoconductive layer,[475] di hydrogen pc in composite photoreceptor charge carrier generating layer,[476] Cu pc in photosensitive layer for photosensitive film for electrophotographic use,[477] Al pc $Cl_{1.4}$ as charge-generating layer in photoconductor with maximum absorption at 774 nm,[478] TiO pc in charge-generating layer for photoreceptor,[479] In pc Br in charge generating layer of a photoreceptor sensitive to IR radiation and exposed to 830 nm,[480] an electrophotographic photoreceptor has a charge carrier-generating layer containing di hydrogen pc,[481] a mixture of Cu pc and Cu pc $(NO_2)_4$ is a photoconductor mixed with resin for an electrophotographic plate,[482] VO pc is a photoconductor in an electrophotographic material,[483] β-Cu pc is a photoconductor in a color electrophotographic process,[484] an electrophotographic plate for a laser xerographic copier is made of a charge-generating layer including di hydrogen pc,[485] an electron beam-curable composition contains Cu pc for printing plate fabrication by electrophotography,[486] and an aluminum plate coated with a solution of τ-pc and KR-255 in methanol gives a charge-generating layer in preparation of an electrophotographic plate.[487]

Electrophotographic photoreceptors involving M pcs and pcs are the subject of about 2 dozen research articles worldwide.

The electrophotographic characteristics of Cu pc poly(vinyl cinnamate) binder layers are studied.[488] Di hydrogen pc as a photoreceptor for electrophotography is the subject of a review with 25 references.[489] The energy level and density of traps in electrophotographic photoreceptors containing di hydrogen pc, a polymer binder, and an electron acceptor as dopant are studied.[490] An electrophotographic photoconductor for the near IR comprises ϵ-Cu pc 40, poly(*N*-vinyl carbazole) 10, and a polyester binder 50 wt %.[491]

The reciprocity law is measured in laser-exposed electrophotographic photoreceptors including Cu pc.[492] The electrophotographic properties of di hydrogen pc/poly(*N*-vinylcarbazole) multilayered receptors are studied.[493] The charge carrier transport properteis of Cu pc in poly(vinyl carbazole) binder are studied.[494] The pyroelectric and photovoltaic effects in relation to the electrophotographic process in polyvinylidene fluoride films sensitized by Cu pc are studied.[495] The electrophotographic properties of photoreceptors prepared by using large needle-shaped crystals and powdered surface-treated pigmentized particles of β-Cu pc and acrylic polymers and novolak phenolic resins as binders are studied.[496]

Layered electrophotographic photoconductors using In (pc Cl) Cl are studied. Their good sensitivity at 800 to 900 nm makes them suitable for a laser printer with a Al Ga As laser diode.[497] A nontoxic organic photoconductor for near IR consists of poly(*N*-vinylcarbazole) dispersed in a Cu pc binder resin.[498] A novel thermally metastable τ-pc is made by milling α-pc. A duble-layered photoreceptor using τ-pc as the charge-generation layer is made and tested.[499] A 40:60 weight ratio of Cu pc/PMMA [(poly(methyl methacrylate)] is suitable for electrophotography.[500] The thermal stability of charging an electrophotographic photoconductor containing a ferrocene derivative hole transport agent and Cu pc is studied.[501]

Electrophotographic properties are improved without affecting the charge acceptance in a photoreceptor with a large carrier-generating layer containing pc.[502] The influence of space charges in the carrier generation layer on electrophotographic characteristics of layered photoreceptors, Al/pc/hydrazone derivative is studied.[503] Electrophotographic photoreceptors incorporating aggregated pcs are prepared. Both Mg pc and Al pc Cl are studied as the photogeneration pigment.[504] In pc Cl, chloroindium pc, is synthesized and purified. The xerographic properties of negatively charged dual layer photoreceptors incorporating the crude and purified In pc Cl pigments are determined.[505] Al pc Cl and TiO pc, titanyl pc, are studied as photoreceptors for electrophotography highly sensitive to IR light.[506]

A new morphological structure of Mg pc is observed when Mg pc is recrystallized from morpholine. It absorbs IR strongly (800 ~ 850 nm) and is a candidate as an electrophotographic receptor available for laser diode recording.[507]

An electrophotographic receptor film with photosensitive layer contains Cu pc for projection slides.[508]

α-TiO pc photoreceptors are studied for devices for solid-state diode laser beam printers. The α-TiO pc absorbs strongly in the IR (maximum absorption at 830 nm).[509] Near-IR photoreceptor devices incorporate evaporated In pc Cl, for excellent visible and near IR sensitivity. The In pc Cl is a photogenerator.[510] Electrophotographic properties of pc and of a triboelectric magnetic toner for a new diode laser printer are studied for a nonimpact printer with a diode laser.[511]

An instant, dry, electrophotographic near-IR migration imaging film with high sensitivity and resolution contains di hydrogen pc.[512] A new xerographic photoreceptor uses a 1:4 by weight mixture of Cu pc and acrylate resin.[513] A double-layered photoreceptor, the charge generation layer of which is τ-pc, is tested with a proven life of >100,000 prints.[514]

REFERENCES

1. **Canon K. K.,** Japan Patent 83,17,448, 1983.
2. **Ricoh Co., Ltd..** Japan Patent 82,201,239, 1982.

3. **Kimoto and Co., Ltd.**, Japan Patent 84,30,541, 1984.
4. **Fujitsu Ltd.**, Japan Patent 83,09,152, 1983.
5. **Canon K. K.**, Japan Patent 82,208,558, 1982.
6. **Ricoh Co., Ltd.**. Japan Patent 82,146,256, 1982.
7. **Canon K. K.**, Japan Patent 83,181,049, 1983.
8. **Fujitsu Ltd.**, Japan Patent 83,181,053, 1983.
9. **Mita Industrial Co., Ltd.**, Japan Patent 82,132,153, 1982.
10. **Fuji-Xerox Co., Ltd.**. Japan Patent 82,148,747, 1982.
11. **Fuji-Xerox Co., Ltd.**, Japan Patent 82,146,255, 1982.
12. **Konishiroku Photo Industry Co., Ltd.**, Japan Patent 82,192,960, 1982.
13. **Hitachi Chemical Co., Ltd.**, Japan Patent 83,09,138, 1983.
14. **Hitachi Chemical Co., Ltd.**, Japan Patent 83,04,138, 1983.
15. **Canon K. K. Copyer Co., Ltd.**, Japan Patent 83,10,747, 1983.
16. **Canon K. K.**, Japan Patent 83,108,541, 1983.
17. **Asahi-Dow Ltd.**, Japan Patent 83,209,749, 1983.
18. **Canon K. K.**, Japan Patent 84,24,853, 1984.
19. **Nippon Telegraph and Telephone Public Corp.**, Japan Patent 83,09,148, 1983.
20. **Dainippon Ink and Chemicals, Inc.**, Japan Patent 82,176,050, 1982.
21. **Hitachi Chemical Co., Ltd.**, Japan Patent 83,120,256, 1983.
22. **Asahi Glass Co., Ltd.**, Japan Patent 82,165,841, 1982.
23. **Toray Industries, Inc.**, Japan Patent 83,203,447, 1983.
24. **Canon K. K.**, Japan Patent 83,17,447, 1983.
25. **Hitachi, Ltd.**, Japan Patent 83,184,948, 1983.
26. **Nippon Telegraph and Telephone Public Corp.**, Japan Patent 82,148,745.
27. **Sericol Group Ltd.**, Japan Patent 83,80,301, 1983.
28. **Canon K. K.**, Japan Patent 82,212,453, 1982.
29. **Ricoh Co., Ltd.**, Japan Patent 82,188,044, 1982.
30. **Tokyo Ohka Kogyo Co., Ltd.**, Japan Patent 82,185,034, 1982.
31. **Hitachi Chemical Co., Ltd.**, Japan Patent 82,185,045, 1982.
32. **Ricoh Co., Ltd.**, Japan Patent 82,189,142, 1982.
33. **Copyer Co., Ltd.**, Japan Patent 82,169,753, 1982.
34. **Ricoh Co., Ltd.**, Japan Patent 82,157,256, 1982.
35. **Mitsubishi Paper Mills, Ltd.**, Japan Patent 83,166,352, 1983.
36. **Hitachi Koki Co., Ltd.**, Japan Patent 82,177,157, 1982.
37. **Canon K. K.**, Japan Patent 83,111,044, 1983.
38. **Asahi-Dow Ltd.**, Japan Patent 83,62,654, 1983.
39. **Canon K. K.**, Japan Patent 83,194,036, 1983.
40. **Canon K. K.**, Japan Patent 83,95,351, 1983.
41. **Mitsubishi Paper Mills, Ltd.**, Japan Patent 83,40,555, 1983.
42. **Toray Industries, Inc.**, Japan Patent 83,211,157, 1983.
43. **Canon K. K. Copyer Co., Ltd.**, Japan Patent 83,68,750, 1983.
44. **Canon K. K.**, Japan Patent 83,181,050, 1983.
45. **Hitachi Chemical Co., Ltd.**, Japan Patent 84,46,652, 1984.
46. **Ricoh Co., Ltd.**, Japan Patent 83,178,362, 1983.
47. **Canon K. K.**, Japan Patent 83,156,941, 1983.
48. **Dainippon Ink and Chemicals, Inc.**, Japan Patent 84,116,650, 1984.
49. **Fuji Photo Film Co., Ltd.**, Japan Patent 83,107,545, 1983.
50. **Canon K. K.**, Japan Patent 83,126,538, 1983.
51. **Asahi Chemical Industry Co., Ltd.**, Japan Patent 84,07,958, 1984.
52. **Minolta Camera Co., Ltd.**, Japan Patent 84,15,250, 1984.
53. **Toyo Ink Manufacturing Co., Ltd.**, Japan Patent 84,105,649, 1984.
54. **Canon K. K.**, Japan Patent 83,199,353, 1983.
55. **Canon K. K.**, Japan Patent 83,199,352, 1983.
56. **Toyo Ink Manufacturing Co., Ltd.**, Japan Patent 84,105,649, 1984.
57. **Ricoh Co., Ltd.**, Japan Patent 83,65,441, 1983.
58. **Konishiroku Photo Industry Co., Ltd.**, Japan Patent 83,49,940, 1983.
59. **Canon K. K.**, Japan Patent 83,82,249, 1983.
60. **Canon K. K.**, Japan Patent 84,07,956, 1984.
61. **Toshiba Corp.**, Japan Patent 83,194,038, 1983.
62. **Asahi Chemical Industry Co., Ltd.**, Japan Patent 84,49,544, 1984.
63. **Nippon Telegraph and Telephone Public Corp.**, Japan Patent 84,71,057, 1984.
64. **Toyo Ink Manufacturing Co., Ltd.**, Japan Patent 84,21,019, 1984.

65. **Borsenberger, P. M., Groner, C. F., and Regan, M. T. (Eastman Kodak Co.),** Europe Patent Application 109,842, 1984.
66. **Minolta Camera Co., Ltd.,** Japan Patent 84,135,476, 1984.
67. **Nippon Telegraph and Telephone Public Corp. and Fuji Electric Co., Ltd.,** Japan Patent 84,15,249, 1984.
68. **Nippon Telegraph and Telephone Public Corp. and Fuji Electric Co., Ltd.,** Japan Patent 84,15,253, 1984.
69. **Konishiroku Photo Industry Co., Ltd.,** Japan Patent 83,215,655, 1983.
70. **Fujitsu Ltd.,** Japan Patent 83,54,344, 1983.
71. **Minolta Camera Co., Ltd.,** Japan Patent 84,12,443, 1984.
72. **Hitachi Chemical Co., Ltd.,** Japan Patent 84,48,752, 1984.
73. **Canon K. K.,** Japan Patent 83,200,239, 1983.
74. **Dainippon Ink and Chemicals, Inc.,** Japan Patent 84,115,328, 1984.
75. **Sumitomo Chemical Co., Ltd.,** Japan Patent 84,162,556, 1984.
76. **Dainippon Ink and Chemicals, Inc.,** Japan Patent 84,33,454, 1984.
77. **Canon K. K.,** Japan Patent 84,149, 369, 1984.
78. **Hitachi Chemical Co., Ltd.,** Japan Patent 84,229,563, 1984.
79. **Canon K. K.,** Japan Patent 84,182,455, 1984.
80. **Canon K. K.,** Japan Patent 84,182,456, 1984.
81. **Minolta Camera Co., Ltd.,** Japan Patent 84,142,556, 1984.
82. **Canon K. K.,** Japan Patent 84,182,460, 1984.
83. **Mitsubishi Paper Mills, Ltd.,** Japan Patent 84,223,434, 1984.
84. **Canon K. K.,** Japan Patent 84,182,457, 1984.
85. **Canon K. K.,** Japan Patent 84,185,341, 1984.
86. **Canon K. K.,** Japan Patent 84,185,342, 1984.
87. **Hitachi Chemical Co., Ltd.,** Japan Patent 84,177,560, 1984.
88. **Canon K. K.,** Japan Patent 84,191,058, 1984.
89. **Dainippon Ink and Chemicals, Inc.,** Japan Patent 84,206,431, 1984.
90. **Dainippon Ink and Chemicals, Inc.,** Japan Patent 84,206,432, 1984.
91. **Ono, T. and Ishii, G.,** Japan Patent 85,03,624, 1985.
92. **Hitachi, Ltd. and Toyo Ink Manufacturing Co., Ltd.,** Japan Patent 85,19,153, 1985.
93. **Toyo Ink Manufacturing Co., Ltd.,** Japan Patent 84,191,061, 1984.
94. **Canon K. K.,** Japan Patent 84,195,246, 1984.
95. **Canon K. K.,** Japan Patent 84,195,247, 1984.
96. **Canon K. K.,** Japan Patent 84,149,368, 1984.
97. **Dainippon Ink and Chemicals, Inc.,** Japan Patent 84,113,445, 1984.
98. **Minolta Camera Co., Ltd.,** Japan Patent 84,195,245, 1984.
99. **Canon K. K.,** Japan Patent 84,182,459, 1984.
100. **Canon K. K.,** Japan Patent 84,182,458, 1984.
101. **Minolta Camera Co., Ltd.,** Japan Patent 84,135,474, 1984.
102. **Sumitomo Chemical Co., Ltd.,** Japan Patent 84,214,034, 1984.
103. **Canon K. K.,** Japan Patent 84,29,249, 1984.
104. **Minolta Camera Co., Ltd.,** Japan Patent 84,136,744, 1984.
105. **Oki Electric Industry Co., Ltd.,** Japan Patent 84,226,350, 1984.
106. **Dainippon Ink and Chemicals, Inc.,** Japan Patent 84,200,243, 1984.
107. **Hitachi, Ltd.,** Japan Patent 85,12,551, 1985.
108. **Hitachi, Ltd. and Toyo Ink Manufacturing Co., Ltd.,** Japan Patent 85,19,144, 1985.
109. **Toyo Ink Manufacturing Co., Ltd. and Toppan Printing Co., Ltd.,** Japan Patent 84,220,741, 1984.
110. **Toyo Ink Manufacturing Co., Ltd.,** Japan Patent 84,220,739, 1984.
111. **Asahi Chemical Industry Co., Ltd.,** Japan Patent 84,230,021, 1984.
112. **Minolta Camera Co., Ltd.,** Japan Patent 84,231,546, 1984.
113. **Hitachi, Ltd.,** Japan Patent 85,11,847, 1985.
114. **Hitachi, Ltd. and Toyo Ink Manufacturing Co., Ltd.,** Japan Patent 85,19,149, 1985.
115. **Hitachi, Ltd. and Toyo Ink Manufacturing Co., Ltd.,** Japan Patent 85,19,152, 1985.
116. **Minolta Camera Co., Ltd.,** Japan Patent 84,226,351, 1984.
117. **Fuji Photo Film Co., Ltd.,** Japan Patent 85,15,644, 1985.
118. **Toshiba Corp.,** Japan Patent 84,123,845, 1984.
119. **Sumitomo Chemical Co., Ltd.,** Japan Patent 84,204,045, 1984.
120. **Toshiba Corp.,** Japan Patent 84,211,048, 1984.
121. **Toshiba Corp.,** Japan Patent 84,211,049, 1984.
122. **Konishiroku Photo Industry Co., Ltd.,** Japan Patent 84,214,858, 1984.
123. **Dainippon Ink and Chemicals, Inc.,** Japan Patent 84,114,534, 1984.

124. **Hitachi, Ltd. and Toyo Ink Manufacturing Co., Ltd.,** Japan Patent 85,19,154, 1985.
125. **Canon K. K.,** Japan Patent 84,133,552, 1984.
126. **Asahi Chemical Industry Co., Ltd.,** Japan Patent 84,162,553, 1984.
127. **Minolta Camera Co., Ltd., Toyo Ink Manufacturing Co., Ltd.,** Japan Patent 84,155,850, 1984.
128. **Minolta Camera Co., Ltd.,** Japan Patent 84,135,477, 1984.
129. **Toyo Ink Manufacturing Co., Ltd.,** Japan Patent 84,155,847, 1984.
130. **Nippon Telegraph and Telephone Public Corp.,** Japan Patent 84,166,959, 1984.
131. **Minolta Camera Co., Ltd. and Toyo Ink Manufacturing Co., Ltd.,** Japan Patent 84,155,846, 1984.
132. **Minolta Camera Co., Ltd. and Toyo Ink Manufacturing Co., Ltd.,** Japan Patent 84,157,649, 1984.
133. **Minolta Camera Co., Ltd.,** Japan Patent 84,142,556, 1984.
134. **Canon K. K.,** Japan Patent 84,168,456, 1984.
135. **Minolta Camera Co., Ltd.,** Japan Patent 84,151,158, 1984.
136. **Sumitomo Chemical Co., Ltd.,** Japan Patent 84,31,965, 1984.
137. **Canon K. K.,** Japan Patent 84,168,452, 1984.
138. **Canon K. K.,** Japan Patent 84,170,843, 1984.
139. **Mita Industrial Co., Ltd.,** Japan Patent 84,184,347, 1984.
140. **Minolta Camera Co., Ltd.,** Japan Patent 84,168,461, 1984.
141. **Minolta Camera Co., Ltd.,** Japan Patent 84,180,566, 1984.
142. **Fuji Xerox Co., Ltd.,** Japan Patent 84,225,771, 1984.
143. **Nippon Telegraph and Telephone Public Corp.,** Japan Patent 84,171,959, 1984.
144. **Ueda, H. (Minolta Camera Co., Ltd.),** German Patent 3,421,969, 1984.
145. **Konishiroku Photo Industry Co., Ltd.,** Japan Patent 84,31,943, 1984.
146. **Asahi Chemical Industry Co., Ltd.,** Japan Patent 85,52,853, 1985.
147. **Asahi Chemical Industry Co., Ltd.,** Japan Patent 85,28,663, 1985.
148. **Sumitomo Chemical Co., Ltd.,** Japan Patent 85,63,538, 1985.
149. **Ricoh Co., Ltd.,** Japan Patent 85,98,437, 1985.
150. **Nippon Soda Co., Ltd.,** Japan Patent 85,121,444, 1985.
151. **Mitsubishi Paper Mills, Ltd.,** Japan Patent 85,39,653, 1985.
152. **Toshiba Corp.,** Japan Patent 85,69,654, 1985.
153. **Toray Industries, Inc.,** Japan Patent 85,57,849, 1985.
154. **Ricoh Co., Ltd.,** Japan Patent 85,37,560, 1985.
155. **Mitsubishi Paper Mills, Ltd.,** Japan Patent 85,24,553, 1985.
156. **Hitachi Chemical Co., Ltd.,** Japan Patent 85,118,842, 1985.
157. **Canon K. K.,** Japan Patent 84,185,340, 1984.
158. **Okuyama, T., Suzuki, Y., Fujiwara, S., and Uehara, T. (Toshiba Corp.),** Japan Patent 85,133,459, 1985.
159. **Hitachi, Ltd.,** Japan Patent 85,87,333, 1985.
160. **Kokado, H., Suzuki, S., Tsunoda, A., Amo, S., Mori, Y., and Narahara, T. (Hitachi, Ltd.),** Japan Patent 85,142,341, 1985.
161. **Hitachi, Ltd.,** Japan Patent 85,87,331, 1985.
162. **Dainichiseika Color and Chemicals Manufacturing Co., Ltd.,** Japan Patent 85,67,949, 1985.
163. **Asahi Chemical Industry Co., Ltd.,** Japan Patent 85,26,947, 1985.
164. **Mitsubishi Paper Mills, Ltd.,** Japan Patent 85,39,654, 1985.
165. **Hitachi, Ltd. and Toyo Ink Manufacturing Co., Ltd.,** Japan Patent 85,19,146, 1985.
166. **Hitachi, Ltd. and Toyo Ink Manufacturing Co., Ltd.,** Japan Patent 85,19,151, 1985.
167. **Toshiba Corp.,** Japan Patent 85,70,453, 1985.
168. **Hitachi, Ltd. and Toyo Ink Manufacturing Co., Ltd.,** Japan Patent 85,19,147, 1985.
169. **Canon K. K.,** Japan Patent 85,32,056, 1985.
170. **Hitachi Chemical Co., Ltd.,** Japan Patent 85,52,855, 1985.
171. **Hitachi, Ltd. and Toyo Ink Manufacturing Co., Ltd.,** Japan Patent 85,19,150, 1985.
172. **Toshiba Corp.,** Japan Patent 85,59,353, 1985.
173. **Minolta Camera Co., Ltd.,** Japan Patent 84,219,752, 1984.
174. **Toyo Ink Manufacturing Co., Ltd.,** Japan Patent 85,73,541, 1985.
175. **Hitachi, Ltd. and Toyo Ink Manufacturing Co., Ltd.,** Japan Patent 85,87,337, 1985.
176. **Toyo Ink Manufacturing Co., Ltd. and Hitachi, Ltd.,** Japan Patent 85,86,551, 1985.
177. **Hitachi, Ltd. and Toyo Ink Manufacturing Co., Ltd.,** Japan Patent 85,87,336, 1985.
178. **Hitachi, Ltd. and Toyo Ink Manufacturing Co., Ltd.,** Japan Patent 85,87,335, 1985.
179. **Toshiba Corp,** Japan Patent 85,59,368, 1985.
180. **Oki Electric Industry Co., Ltd.,** Japan Patent 85,59,355, 1985.
181. **Hitachi, Ltd. and Toyo Ink Manufacturing Co.,Ltd.,** Japan Patent 85,87,334, 1985.
182. **Hitachi, Ltd.,** Japan Patent 85,26,344, 1985.
183. **Hitachi, Ltd. and Toyo Ink Manufacturing Co., Ltd.,** Japan Patent 85,87,338, 1985.

184. **Hitachi, Ltd. and Toyo Ink Manufacturing Co., Ltd.**, Japan Patent 85,87,332, 1985.
185. **Toshiba Corp.**, Japan Patent 85,70,454, 1985.
186. **Toshiba Corp.**, Japan Patent 85,73,540, 1985.
187. **Oki Electric Industry Co., Ltd.**, Japan Patent 85,59,354, 1985.
188. **Toyo Ink Manufacturing Co., Ltd.**, Japan Patent 85,50,538, 1985.
189. **Hitachi Chemical Co., Ltd.**, Japan Patent 85,52,856, 1985.
190. **Hitachi Koki Co., Ltd.**, Japan Patent 85,75,840, 1985.
191. **Hitachi, Ltd. and Toyo Ink Manufacturing Co., Ltd.**, Japan Patent 85,19,145, 1985.
192. **Nippon Telegraph and Telephone Public Corp.**, Japan Patent 85,49,342, 1985.
193. **Hosoya, A., Tsunoda, A., and Mori, Y. (Hitachi, Ltd.)**, Japan Patent 85,136,747, 1985.
194. **Enomoto, K. and Haino, K. (Mitsubishi Paper Mills, Ltd.)**, Japan Patent 85,135,952, 1985.
195. **Kito, S., Takegawa, I., Sakaguchi, Y., and Nakamura, K. (Fuji Xerox Co., Ltd.)**, Japan Patent 85,144,756, 1985.
196. **Matsura, K. and Oshima, K. (Toray Industries, Inc.)**, Japan Patent 85,146,249, 1985.
197. **Suzuki, A. and Shimada, M. (Ricoh Co., Ltd.)**, Japan Patent 85,189,760, 1985.
198. **Maeda, M. and Kajiura, S. (Toshiba Corp.)**, Japan Patent 85,189,756, 1985.
199. **Maeda, M., Mizushima, K., Sugiuchi, M., and Nakajima, Y. (Toshiba Corp.)**, Japan Patent 85,189,751, 1985.
200. **Nakajima, Y., Sugiuchi, M., Kajiura, S., Maeda, M., and Mizushima, K. (Toshiba Corp.)**, Japan Patent 85,189,753, 1985.
201. **Sugiuchi, M., Nakajima, Y., Kajiura, S., Maeda, M., and Mizushima, K. (Toshiba Corp.)**, Japan Patent 85,189,748, 1985.
202. **Ueda, H., Sakamoto, M., Fujikawa, Y., Tamura, S., and Mizukami, S. (Minolta Camera Co., Ltd. and Toyo Ink Manufacturing Co., Ltd.)**, Japan Patent 85,207,146, 1985.
203. **Tamura, S., Ehashi, S., Seta, T., Mizukami, S., and Oohashi, M. (Toyo Ink Manufacturing Co., Ltd.)**, Japan Patent 85,203,956, 1985.
204. **Sugiuchi, M., Nakajima, Y., Kajiura, S., Maeda, M., and Mizushima, K. (Toshiba Corp.)**, Japan Patent 85,189,755, 1985.
205. **Ueda, H., Tamura, S., and Seta, T. (Minolta Camera Co., Ltd. and Toyo Ink Manufacturing Co., Ltd.)**, Japan Patent 85,207,145, 1985.
206. **Sugiuchi, M., Nakajima, Y., Kajiura, S., Maeda, M., and Mizushima, K. (Toshiba Corp.)**, Japan Patent 85,201,347, 1985.
207. **Nakajima, Y., Sugiuchi, M., Kajiura, S., Maeda, M., and Mizushima, K. (Toshiba Corp.)**, Japan Patent 85,201,346, 1985.
208. **Koizumi, A., Kawamura, M., Watanabe, A., Matsuyama, S., Amada, H., and Kawamura, F. (Tomoegawa Paper Manufacturing Co., Ltd.)**, Japan Patent 85,189,749, 1985.
209. **Shimokihara, S., Yasui, M., and Ueno, S. (Sumitomo Chemical Co., Ltd.)**, Japan Patent 85,184,251, 1985.
210. **Nakajima, Y., Sugiuchi, M., Kajiura, S., Maeda, M., and Mizushima, K. (Toshiba Corp.)**, Japan Patent 85,201,348, 1985.
211. **Matsumoto, M. (Canon K. K.)**, Japan Patent 85,195,549, 1985.
212. **Tamura, S., Kashioka, M., Ohashi, M., Katsura, H., Ueda, H., and Nakamura, A. (Toyo Ink Manufacturing Co., Ltd. and Minolta Camera Co., Ltd.)**, Japan Patent 85,205,456, 1985.
213. **Kajiura, S., Maeda, M., Mizushima, K., Sugiuchi, M., and Nakajima, Y. (Toshiba Corp.)**, Japan Patent 85,165,654, 1985.
214. **Sugiuchi, M., Nakajima, Y., Kajiura, S., Maeda, M., and Mizushima, K. (Toshiba Corp.)**, Japan Patent 85,189,752, 1985.
215. **Yashiki, Y. (Canon K. K.)**, Japan Patent 85,189,747, 1985.
216. **Arita, T. (Canon K. K.)**, Japan Patent 85,184,258, 1985.
217. **Sugiuchi, M., Nakajima, Y., Mizushima K., Kajiura, S., and Maeda, M. (Toshiba Corp.)**, Japan Patent 85,198,552, 1985.
218. **Sugiuchi, M., Nakajima, Y., Kajiura, S., Maeda, M., and Mizushima, K. (Toshiba Corp.)**, Japan Patent 85,198,549, 1985.
219. **Tachiki, S. (Hitachi Chemical Co., Ltd.)**, Japan Patent 85,233,656, 1985.
220. **Nakajima, Y., Sugiuchi, M., Kajiura, S., Maeda, M., and Mizushima, K. (Toshiba Corp.)**, Japan Patent 85,198,547, 1985.
221. **Kajiura, S., Maeda, M., Mizushima, K., Sugiuchi, M., Nakajima, Y., Hayase, S., and Onishi, Y. (Toshiba Corp.)**, Japan Patent 85,165,655, 1985.
222. **Kawamura, F., Kamiyama, M., Watanabe, A., and Koizumi, A. (Tomoegawa Paper Manufacturing Co., Ltd.)**, Japan Patent 85,176,045, 1985.
223. **Dainippon Ink and Chemicals, Inc.**, Japan Patent 85,88,938, 1985.
224. **Hitachi Chemical Co., Ltd.**, Japan Patent 85,122,953, 1985.

225. **Ueda, H. (Minolta Camera Co., Ltd.)**, Japan Patent 85,256,148, 1985.
226. **Ueda, H. (Minolta Camera Co., Ltd.)**, Japan Patent 85,256,150, 1985.
227. **Umibe, K., Nishioka, Y., and Kato, M. (Oki Electric Industry Co., Ltd.)**, Japan Patent 85,260,054, 1985.
228. **Ehashi, S., Tamura, S., Suda, Y., Sakamoto, M., and Hikosaka, M. (Toyo Ink Manufacturing Co., Ltd.)**, Japan Patent 85,247,644, 1985.
229. **Kajiura, S. and Maeda, M. (Toshiba Corp.)**, Japan Patent 85,254,047, 1985.
230. **Ueda, H. (Minolta Camera Co., Ltd.)**, Japan Patent 85,256,149, 1985.
231. **Kawamura, F., Kawamura, M., Watanabe, A., Amada, H., and Koizumi, A. (Tomoegawa Paper Manufacturing Co., Ltd.)**, Japan Patent 85,168,150, 1985.
232. **Haino, K. and Enomoto, K. (Mitsubishi Paper Mills, Ltd.)**, Japan Patent 85,146,247, 1985.
233. **Haino, K., and Enomoto, K. (Mitsubishi Paper Mills, Ltd.)**, Japan Patent 85,146,246, 1985.
234. **Suzuki, A. and Shimada, M. (Ricoh Co., Ltd.)**, Japan Patent 85,175,056, 1985.
235. **Murakami, Y., Hisada, H., Niigae, R., and Kawatoko, S. (Matsushita Electric Industrial Co., Ltd.)**, Japan Patent 85,260,049, 1985.
236. **Murakami, Y., Hisada, H., Niigae, R., and Kawatoko, S. (Matsushita Electric Industrial Co., Ltd.)**, Japan Patent 85,260,048, 1985.
237. **Minolta Camera Co., Ltd.**, Japan Patent 85,130,745, 1985.
238. **Sakai, T. (Ricoh Co., Ltd.)**, Japan Patent 85,202,446, 1985.
239. **Sakai, T. (Ricoh Co., Ltd.)**, Japan Patent 85,202,447, 1985.
240. **Sasaki, M. (Ricoh Co., Ltd.)**, Japan Patent 85,175,052, 1985.
241. **Matsumoto, M., Yamashita, M., and Takahashi, H. (Canon K. K.)**, Japan Patent 85,178,456, 1985.
242. **Kobayashi, K., Akiyama, K., and Oishi, C. (Fuji Photo Film Co., Ltd.)**, Japan Patent 85,142,329, 1985.
243. **Tanaka, H., Takenochi, O., and Kawahara, T. (Dainippon Ink and Chemicals, Inc.)**, Japan Patent 85,191,265, 1985.
244. **Enomoto, K. and Haino, K. (Mitsubishi Paper Mills, Ltd.)**, Japan Patent 85,191,264, 1985.
245. **Tanaka, H., Takenochi, O., and Kawahara, T. (Dainippon Ink and Chemicals, Inc.)**, Japan Patent 85,177,354, 1985.
246. **Matsumoto, M. and Takahashi, H. (Canon K. K.)**, Japan Patent 85,189,758, 1985.
247. **Yamashita, M. (Canon K. K.)**, Japan Patent 85,208,761, 1985.
248. **Matsumoto, M. and Takahashi, H. (Canon K. K.)**, Japan Patent 85,189,757, 1985.
249. **Takasago Perfumery Co., Ltd.**, Japan Patent 85,130,744, 1985.
250. **Sato, S., Fukuma, T., Suzuki, M., and Kano, M. (Sony Corp.)**, Japan Patent 85,225,161, 1985.
251. **Anayama, H. (Canon K. K.)**, Japan Patent 85,225,856, 1985.
252. **Sato, S., Suzuki, S., Fukuma, T., and Kano, M. (Sony Corp.)**, Japan Patent 85,225,160, 1985.
253. **Anayama, H. and Ko, M. (Canon K. K.)**, Japan Patent 85,250,348, 1985.
254. **Matsumoto, M. and Takahashi, H. (Canon K. K.)**, Japan Patent 85,217,364, 1985.
255. **Yamaguchi, A. and Tsuruta, H. (Takasago Perfumery Co., Ltd.)**, Japan Patent 85,218,652, 1985.
256. **Hitachi Koki Co., Ltd.**, Japan Patent 85,104,953, 1985.
257. **Nakamura, C., Oe, K., and Muta T. (Dainippon Ink and Chemicals, Inc.)**, Japan Patent 85,191,244, 1985; **Nakamura, C. and Oe, K. (Dainippon Ink and Chemicals, Inc.)**, Japan Patent 85,203,630, 1985.
258. **Nakatani, K. and Komon, H. (Mita Industrial Co., Ltd.)**, Japan Patent 86,02,157, 1986.
259. **Matsumoto, M. (Canon K. K.)**, Japan Patent 85,195,547, 1985.
260. **Fujimura, N. (Canon K. K.)**, Japan Patent 85,218,655, 1985.
261. **Tachiki, S. (Hitachi Chemical Co., Ltd.)**, Japan Patent 85,233,655, 1985.
262. **Yoshihara, Y. (Canon K. K.)**, Japan Patent 85,225,851, 1985.
263. **Nakajima, Y., Sugiuchi, M., Kajiura, S., Maeda, M., and Mizushima, K. (Toshiba Corp.)**, Japan Patent 85,189,750, 1985.
264. **Kowamura, F., Kamiyama, M., Watanabe, A., and Koizuma, A. (Tomoegawa Paper Manufacturing Co., Ltd.)**, Japan Patent 85,164,749, 1985.
265. **Suzuki, Y., Fujiwara, S., Okuyama, T., and Uehara, T. (Toshiba Corp.)**, Japan Patent 85,138,562, 1985.
266. **Hitachi Chemical Co., Ltd.**, Japan Patent 85,120,362, 1985.
267. **Fujitsu Ltd.**, Japan Patent 85,131,544, 1985.
268. **Matsumoto, M., Yamashita, M., and Takahashi, H. (Canon K. K.)**, Japan Patent 86,73,960, 1986.
269. **Matsura, K., Oseto, H., and Oshima, K. (Toray Industries, Inc.)**, Japan Patent 86,20,952, 1986.
270. **Ueda, H. (Minolta Camera Co., Ltd.)**, Japan Patent 85,216,354, 1985.
271. **Ko, M., Okunuki, M., Mabuchi, M., Yoshihara, Y., and Egashira, Y. (Canon K. K.)**, Japan Patent 86,13,253, 1986.
272. **Matsura, K., Oseto, H., and Oshima, K. (Toray Industries, Inc.)**, Japan Patent 86,20,951, 1986.
273. **Ko, M., Yoshihata, Y., and Egashura, J. (Canon K. K.)**, Japan Patent 86,13,252, 1986.

274. **Shindo, S., Hosoi, T., and Niimoto, H. (Nippon Kayaku Co., Ltd.),** Japan Patent 86,03,149, 1986.
275. **Ko, M., Yoshihara, Y., and Egashira, Y. (Canon K. K.),** Japan Patent 86,13,254, 1986.
276. **Yamashita, M. (Canon K. K.),** Japan Patent 85,237,451, 1985.
277. **Sakama, K., Yamamoto, K., Sato, T., and Kojima, F. (Fuji Xerox Co., Ltd.),** Japan Patent 86,109,064, 1986.
278. **Sugiuchi, M., Mizushima, K., Nakajima, Y., Kajiura, S., and Maeda, M. (Toshiba Corp.),** Japan Patent 86,27,549, 1986.
279. **Hayashida, S. (Hitachi Chemical Co., Ltd.),** Japan Patent 85,237,454, 1985.
280. **Anayama, H. (Canon K. K.),** Japan Patent 85,227,264, 1985.
281. **Nogami, A., Shira, Y., and Kinoshita, A. (Konishiroku Photo Industry Co., Ltd.),** Japan Patent 86,80,157, 1986.
282. **Nakajima, Y., Sugiuchi, M., and Mizushima, K. (Toshiba Corp.),** Japan Patent 86,26,062, 1986; **Tachiki, S. (Hitachi Chemical Co., Ltd.),** Japan Patent 86,34,549, 1986.
283. **Ueda, H., Sakamoto, M., Fujikawa, Y., Tamura, S., and Mizushima, S. (Minolta Camera Co., Ltd. and Toyo Ink Manufacturing Co., Ltd.),** Japan Patent 85,207,147, 1985.
284. **Ueda, H. (Minolta Camera Co., Ltd.),** Japan Patent 86,35,452, 1986.
285. **Ueda, H. (Minolta Camera Co., Ltd.),** Japan Patent 86,07,840, 1986.
286. **Oaku, K., Nakano, H., and Aizawa, M. (Dainippon Ink and Chemicals, Inc.),** Japan Patent 86,45,249, 1986.
287. **Anayama, H. (Canon K. K.),** Japan Patent 85,260,046, 1985.
288. **Kajiura, S., Maeda, M., Mizushima, K., Sugiuchi, M., and Nakajima, Y. (Toshiba Corp.),** Japan Patent 85,201,345, 1985.
289. **Ueda, H. (Minolta Camera Co., Ltd.),** Japan Patent 86,84,651, 1986.
290. **Ueda, H. (Minolta Camera Co., Ltd.),** Japan Patent 86,77,054, 1986.
291. **Sugiuchi, M., Kajiura, S., Mizushima, K., Nakajima, Y., and Maeda, M. (Toshiba Corp.),** Japan Patent 86,25,150, 1986.
292. **Ueda, H. (Minolta Camera Co., Ltd.),** Japan Patent 86,80,158, 1986.
293. **Koizumi, A., Amada, H., Watanabe, A., Kawamura, M., Matsuyama, S., and Kawamura, F. (Tomoegawa Paper Manufacturing Co., Ltd.),** Japan Patent 85,195,548, 1985.
294. **Tachiki, S. and Sugimoto, Y. (Hitachi Chemical Co., Ltd.),** Japan Patent 86,34,547, 1986.
295. **Kitajima, R., Seto, M., Fukagai, T., Oshima, K., and Ishikawa, F. (Ricoh Co., Ltd.),** Japan Patent 86,23,169, 1986.
296. **Kitajima, R., Fukagai, T., Ishikawa, F., and Nishijima, H. (Ricoh Co., Ltd.),** Japan Patent 86,20,955, 1986.
297. **Nakajima, J., Sugiuchi, M., and Mizushima, K. (Toshiba Corp.),** Japan Patent 85,254,044, 1985.
298. **Okuyama, T., Suzuki, Y., Fujiwara, T., and Uehara, T. (Toshiba Corp.),** Japan Patent 86,06,657, 1986.
299. **Mino, H. and Kadota, K. (Sanyo-Kokusaku Pulp Co., Ltd.),** Japan Patent 86,99,151, 1986.
300. **Sugiuchi, M. and Nakajima, Y. (Toshiba Corp.),** Japan Patent 86,80,259, 1986.
301. **Hiratsuka, H., Arishima, K., and Nishi, T. (Nippon Telegraph and Telephone Public Corp.),** Japan Patent 86,67,866, 1986.
302. **Hasegawa, S. and Nagashima, A. (Fuji Photo Film Co., Ltd.),** Japan Patent 86,20,939, 1986.
303. **Kinoshita, S. and Yonehara, Y. (Dainippon Ink and Chemicals, Inc.),** Japan Patent 86,115,085, 1986.
304. **Oaku, K., Guen, C. K., Aizawa, M., and Nakano, H. (Dainippon Ink and Chemicals, Inc.),** Japan Patent 86,124,951, 1986.
305. **Kuroiwa, A., Nanba, N., and Kamijo, T. (TDK Corp.),** Japan Patent 86,156,129, 1986.
306. **Tamura, S. Ehashi, S., Seta, T., and Suda, Y. (Toyo Ink Manufacturing Co., Ltd.),** Japan Patent 86,181,867, 1986.
307. **Ehashi, S., Suda, Y., and Hikosaka, M. (Toyo Ink Manufacturing Co., Ltd.),** Japan Patent 86,189,556, 1986.
308. **Ueda, H., Yasutomi, H., and Kameda, H. (Minolta Camera Co., Ltd.),** Japan Patent 86,252,558, 1986.
309. **Kumon, A. (Matsushita Electric Industrial Co., Ltd.),** Japan Patent 86,209,458, 1986.
310. **Origasa, H., Watanuki, Y., and Kosho, N. (Fuji Electric Co., Ltd.),** Japan Patent 86,278,856, 1986.
311. **Kuroiwa, A., Nanba, N., and Kamijo, T. (TDK Corp.),** Japan Patent 86,198,242, 1986.
312. **Ehashi, S., Suda, Y., and Hikosaka, M. (Toyo Ink Manufacturing Co., Ltd.),** Japan Patent 86,166,551, 1986.
313. **Umehara, M. and Matsumoto, M. (Canon K. K.),** Japan Patent 86,175,646, 1986.
314. **Matsumoto, M. (Canon K. K.),** Japan Patent 86,210,363, 1986.
315. **Matsumoto, M. (Canon K. K.),** Japan Patent 86,210,362, 1986.
316. **Ricoh Co., Ltd.,** Japan Patent 83,142,343, 1983.
317. **Okunuki, M. and Matsumoto, M. (Canon K. K.),** Japan Patent 86,210,364, 1986.

318. **Guen, C. K., Fukawatase, M., Kawahara, T., and Aizawa, M. (Dainippon Ink and Chemicals, Inc.),** Japan Patent 86,188,543, 1986.
319. **Watanuki, Y. and Kosho, N. (Fuji Electric Co., Ltd.),** Japan Patent 86,129,649, 1986.
320. **Sugimoto, Y. and Tachiki, S. (Hitachi Chemical Co., Ltd.),** Japan Patent 86,132,954, 1986.
321. **Nakajima, Y. and Sugiuchi, M. (Toshiba Corp.),** Japan Patent 86,138,956, 1986.
322. **Tanaka, S. and Sumino, F. (Canon K. K.),** Japan Patent 86,186,964, 1986.
323. **Takegawa, I., Kito, S., Sakaguchi, Y., and Nakamura, K. (Fuji Xerox Co., Ltd.),** Japan Patent 86,94,057, 1986.
324. **Otsuka, S. (Mitsubishi Chemical Industries Co., Ltd.),** Japan Patent 86,134,767, 1986.
325. **Okuyama, T., Fujiwara, S., Suzuki, Y., and Uehara, T. (Toshiba Corp.),** Japan Patent 86,153,660, 1986.
326. **Kawakami, S., Takimoto, M., and Sawada, K. (Konishiroku Photo Industry Co., Ltd.),** Japan Patent 86,235,847, 1986.
327. **Takahashi, H. (Canon K. K.),** Japan Patent 86,144,657, 1986.
328. **Oaku, K., Nakano, H., Aizawa, M. (Dainippon Ink and Chemicals, Inc.),** Japan Patent 86,217,050, 1986.
329. **Oaku, K., Nakano, H., and Aizawa, M. (Dainippon Ink and Chemicals, Inc.),** Japan Patent 86,239,248, 1986.
330. **Uriyu, T. and Oshima, R. (Mita Industrial Co., Ltd.),** Japan Patent 86,130,950, 1986.
331. **Suzuki, Y., Fujimaki, Y., and Nomori, H. (Konishiroku Photo Industry Co., Ltd.),** Japan Patent 87,103,653, 1987.
332. **Horie, S., Nakano, J., and Sato, H. (Fuji Photo Film Co., Ltd.),** Japan Patent 87,123,469, 1987.
333. **Murayama, H. and Kawabata, T. (Ricoh Co., Ltd.),** Japan Patent 87,28,769, 1987.
334. **Kuriowa, A., Nanba, N., and Kamijo, T. (TDK Corp.),** Japan Patent 86,198,241, 1986.
335. **Takahashi, K. and Guen, C. K. (Dainippon Ink and Chemicals, Inc.),** Japan Patent 87,54,266, 1987.
336. **Suzuki, Y., Fujimaki, Y., and Nomori, H. (Konishiroku Photo Industry Co., Ltd.),** Japan Patent 87,121,460, 1987.
337. **Mori, Y. and Iwabuchi, T. (Asahi Chemical Industry Co., Ltd.),** Japan Patent 87,133,462, 1987.
338. **Origasa, H., Watanuki, Y., and Kosho, N. (Fuji Electric Co., Ltd.),** Japan Patent 87,141,563, 1987.
339. **Nakamura, M., Fukawatase, M., Takahashi, K., and Guen, C. K. (Dainippon Ink and Chemicals, Inc.),** Japan Patent 87,111,257, 1987.
340. **Seta, T., Mizukami, S., and Katsura, H. (Toyo Ink Manufacturing Co., Ltd.),** Japan Patent 87,100,766, 1987.
341. **Watanuki, N. and Kosho, N. (Fuji Electric Co., Ltd.),** Japan Patent 87,69,271, 1987.
342. **Suzuki, Y., Fujimaki, Y., and Nomori, H. (Konishiroku Photo Industry Co., Ltd.),** Japan Patent 87,103,652, 1987.
343. **Tamaki, K., Suzuki, Y., Kudo, K., Fujimaki, Y., and Nomori, H. (Konishiroku Photo Industry Co., Ltd.),** Japan Patent 87,115,167, 1987.
344. **Okamoto, M., Nakajima, Y., and Sugiuchi, M. (Toshiba Corp.),** Japan Patent 87,00,961, 1987.
345. **Guen, C. K., Aizawa, M., Nakano, H., and Oaku, K. (Dainippon Ink and Chemicals, Inc.),** Japan Patent 86,203,461, 1986.
346. **Kuramoto, A., Tanaka, E., Takimoto, A., and Akiyama, K. (Matsushita Electric Industrial Co., Ltd.),** Japan Patent 86,275,846, 1986.
347. **Kuramoto, A., Tanaka, E., Takimoto, A., Onomichi, K., and Aikiyama, K. (Matsushita Electric Industrial Co., Ltd.),** Japan Patent 87,19,861, 1987.
348. **Morimoto, A., Sano, Y., Nimura, E., Tomii, Y., and Maeda, T. (Mita Industrial Co., Ltd.),** Japan Patent 87,14,162, 1987.
349. **Suzuki, T., Murayama, T., Ono, H., Otsuka, S., and Rin, M. (Mitsubishi Chemical Industries Co., Ltd.,)** Japan Patent 87,67,094, 1987.
350. **Hatashi, S. (Mita Industrial Co., Ltd.),** Japan Patent 87,92,960, 1987.
351. **Suzuki, Y., Fujimaki, Y., and Nomori, H. (Konishiroku Photo Industry Co., Ltd.),** Japan Patent 87,47,054, 1987.
352. **Ueda, H., Otani, J., and Noda, T. (Minolta Camera Co., Ltd.),** Japan Patent 87,58,257, 1987.
353. **Kosho, N. and Watanuki, Y. (Fuji Electric Co., Ltd.),** Japan Patent 87,61,055, 1987.
354. **Suzuki, Y., Fujimaki, Y., and Nomori, H. (Konishiroku Photo Industry Co., Ltd.),** Japan Patent 87,103,650, 1987.
355. **Suzuki, Y., Fujimaki, Y., and Nomori, H. (Konishiroku Photo Industry Co., Ltd.),** Japan Patent 87,103,651, 1987.
356. **Murakami, Y., Niigae, R., and Takeda, H. (Matsushita Electric Industrial Co., Ltd.),** Japan Patent 87,100,758, 1987.
357. **Niigae, R., Murakami, Y., and Takeda, H. (Matsushita Electric Industrial Co., Ltd.),** Japan Patent 87,100,757, 1987.

358. **Murakami, Y., Niigae, R., and Takeda, H. (Matsushita Electric Industrial Co., Ltd.)**, Japan Patent 87,100,756, 1987.
359. **Niigae, R., Murakami, Y., and Takeda, H. (Matsushita Electric Industrial Co., Ltd.)**, Japan Patent 87,100,755, 1987.
360. **Tamaki, K., Suzuki, Y., Kudo, K., Fujimaki, Y., and Nomori, H. (Konishiroku Photo Industry Co., Ltd.)**, Japan Patent 87,79,470, 1987.
361. **Watanuki, Y. and Kosho, N. (Fuji Electric Co., Ltd.)**, Japan Patent 87,61,054, 1987.
362. **Suzuki, N., Goto, S., Maeda, Y., and Shimizu, S. (Konishiroku Photo Industry Co., Ltd.)**, Japan Patent 87,111,249, 1987.
363. **Suzuki, T., Ono, H., and Sudo, M. (Mitsubishi Chemical Industries Co., Ltd.)**, Japan Patent 87,119,547, 1987.
364. **Konishiroku Photo Industry Co., Ltd.**, Japan Patent 83,194,041, 1983.
365. **Konishiroku Photo Industry Co., Ltd.**, Japan Patent 82,210,345, 1982.
366. **Konishiroku Photo Industry Co., Ltd.**, Japan Patent 83,194,040, 1983.
367. **Suematsu, H., Yamamoto, A., Ikeda, T., and Imai, E. (Canon K. K.)**, Japan Patent 85,217,370, 1985.
368. **Suematsu, H., Ikeda, T., Yamamoto, A., and Imai, E. (Canon K. K.)**, Japan Patent 85,217,369, 1985.
369. **Nakamura, C., Oe, K., and Yugami, F. (Dainippon Ink and Chemicals, Inc.)**, Japan Patent 85,169,848, 1985.
370. **Shirai, Y., Nogami, A., and Kojima, Y. (Konishiroku Photo Industry Co., Ltd.)**, Japan Patent 86,36,759, 1986.
371. **Suematsu, H. (Canon K. K.)**, Japan Patent 86,130,961, 1986.
372. **Ueda, H. (Minolta Camera Co., Ltd.)**, Japan Patent 85,255,854, 1985.
373. **Toyo Ink Manufacturing Co., Ltd.**, Japan Patent 83,54,343, 1983.
374. **Canon K. K.**, Japan Patent 84,00,158, 1984.
375. **Oki Electric Industry Co., Ltd.**, Japan Patent 84,104,651, 1984.
376. **Tachiki, S. and Sugimoto, Y. (Hitachi Chemical Co., Ltd.)**, Japan Patent 86,217,052, 1986.
377. **Fujitsu Ltd.**, Japan Patent 83,192,043, 1983.
378. **Sumitomo Chemical Co., Ltd.**, Japan Patent 84,36,254, 1984.
379. **Fujitsu Ltd.**, Japan Patent 83,193,550, 1983.
380. **Minolta Camera Co., Ltd.**, Japan Patent 84,12,445, 1984.
381. **Minolta Camera Co., Ltd.**, Japan Patent 84,19,946, 1984.
382. **Matsushita Electric Industrial Co., Ltd.**, Japan Patent 82,212,463, 1982.
383. **Dainippon Ink and Chemicals, Inc.**, Japan Patent 84,62,862, 1984.
384. **Dainippon Ink and Chemicals, Inc.**, Japan Patent 84,62,861, 1984.
385. **Toyo Ink Manufacturing Co., Ltd.**, Japan Patent 84,116,755, 1984.
386. **Ricoh Co., Ltd.**, Japan Patent 84,75,258, 1984.
387. **Canon K. K.**, Japan Patent 84,204,839, 1984.
388. **Ricoh Co., Ltd.**, Japan Patent 84,75,257, 1984.
389. **Minolta Camera Co., Ltd.**, Japan Patent 84,142,581, 1984.
390. **Konishiroku Photo Industry Co., Ltd.**, Japan Patent 84,185,358, 1984.
391. **Konishoroku Photo Industry Co., Ltd.**, Japan Patent 84,185,357, 1984.
392. **Minolta Camera Co., Ltd.**, Japan Patent 84,185,352, 1984.
393. **Minolta Camera Co., Ltd.**, Japan Patent 84,204,052, 1984.
394. **Minolta Camera Co., Ltd.**, Japan Patent 84,218,475, 1984.
395. **Konishiroku Photo Industry Co., Ltd.**, Japan Patent 85,112,053, 1985.
396. **Ueda, H. (Minolta Camera Co., Ltd.)**, Japan Patent 85,207,142, 1985.
397. **Naganuma, T., Hirayama, S., and Kumagai, K. (Toppan Printing Co., Ltd.)**, Japan Patent 85,241,060, 1985.
398. **Matsumoto, M. (Canon K. K.)**, Japan Patent 85,177,347, 1985.
399. **Ito, K., Nakagawa, F., Ito, S., and Tooyama, H. (Oki Electric Industry Co., Ltd.)**, Japan Patent 86,148,468, 1986.
400. **Sugiuchi, M. and Nakajima, Y. (Toshiba Corp.)**, Japan Patent 86,148,452, 1986.
401. **Yasutomi, H. (Minolta Camera Co., Ltd.)**, Japan Patent 86,266,567, 1986.
402. **Copyer Co., Ltd. Canon K. K.**, Japan Patent 82,169,754, 1982.
403. **Ricoh Co., Ltd.**, Japan Patent 82,167,032, 1982.
404. **Hitachi, Ltd.**, Japan Patent 82,178,245, 1982.
405. **Canon K. K. Copyer Co., Ltd.**, Japan Patent 82,189,143, 1982.
406. **Canon K K.**, Japan Patent 82,192,961, 1982.
407. **Canon K. K.**, Japan Patent 82,155,550, 1982.
408. **Canon K. K.**, Japan Patent 82,158,645, 1982.
409. **Canon K. K.**, Japan Patent 82,200,043, 1982.
410. **Hitachi, Ltd.**, Japan Patent 83,87,557, 1983.

411. **Toppan Printing Co., Ltd.**, Japan Patent 84,95,548, 1984.
412. **Canon K. K.**, Japan Patent 83,111,948, 1983.
413. **Mita Industrial Co., Ltd.**, Japan Patent 84,02,050, 1984.
414. **Iijima, A. and Kita, N. (Konishoroku Photo Industry Co., Ltd.)**, Japan Patent 87,09,944, 1987.
415. **Fuji Photo Film Co., Ltd.**, Japan Patent 83,105,240, 1983.
416. **Ricoh Co., Ltd.**, Japan Patent 83,21,749, 1983.
417. **Konishiroku Photo Industry Co., Ltd.**, Japan Patent 84,170,862, 1984.
418. **Xerox Corp.**, Japan Patent 84,162,554, 1984.
419. **Toyo Ink Manufacturing Co., Ltd.**, Japan Patent 84,116,754, 1984.
420. **Toppan Printing Co., Ltd. and Toyo Ink Manufacturing Co., Ltd.**, Japan Patent 84,220,754, 1984.
421. **Toppan Printing Co., Ltd. and Toyo Ink Manufacturing Co., Ltd.**, Japan Patent 84,220,752, 1984.
422. **Toppan Printing Co., Ltd. and Toyo Ink Manufacturing Co., Ltd.**, Japan Patent 84,116,751, 1984.
423. **Matsushita Electric Industrial Co., Ltd.**, Japan Patent 84,109,062, 1984.
424. **Toyo Ink Manufacturing Co., Ltd. and Toppan Printing Co., Ltd.**, Japan Patent 84,116,753, 1984.
425. **Hitachi Koki Co., Ltd.**, Japan Patent 85,52,854, 1985.
426. **Ricoh Co., Ltd.**, Japan Patent 85,24,469, 1985.
427. **Hitachi, Ltd.**, Japan Patent 85,86,550, 1985.
428. **Sumitomo Chemical Co., Ltd.**, Japan Patent 85,55,346, 1985.
429. **Mita Industrial Co., Ltd.**, Japan Patent 85,108,860, 1985.
430. **Matsushita Electric Industrial Co., Ltd.**, Japan Patent 85,48,049, 1985.
431. **Oguchi, Y. and Katagiri, K. (Canon K. K.)**, Japan Patent 85,158,458, 1985.
432. **Shimada, F., Matsuzaki, M., and Uehara, M. (Konishiroku Photo Industry Co., Ltd.)**, Japan Patent 86,49,895, 1986.
433. **Yoshihara, Y., Okunuki, M., and Ko, M. (Canon K. K.)**, Japan Patent 85,260,045, 1985.
434. **Naganuma, T., Hirayama, S., and Kumagai, K. (Toppan Printing Co., Ltd.)**, Japan Patent 86,273,994, 1986.
435. **Takeda, N., Otani, Y., Kita, M., and Okajima, N. (Diafoil Co., Ltd.)**, Japan Patent 86,255,940, 1986.
436. **Kubo, T. and Hisamura, M. (Canon K. K.)**, Japan Patent 87,28,768, 1987.
437. **Rokutanzono, S., Sakon, H., and Set, M. (Ricoh Co., Ltd.)**, Japan Patent 87,01,282, 1987.
438. **Kakuta, A., Oka, H., Suzuki, S., Araya, K., Mori, Y., Morishita, H., Takano, S., Sawada, M., and Kumano, I. (Hitachi, Ltd. and Toyo Ink Manufacturing Co., Ltd.)**, Europe Patent Application 93,331, 1983.
439. **Sasaki, M. (Ricoh Co., Ltd.)**, German Patent 3,315,437, 1983.
440. **Horgan, A. M. (Xerox Corp.)**, U.S. Patent 4,415,639, 1983.
441. **Limburg, W. W., Pochan, J. M., and Beatty, C. L. (Xerox Corp.)**, Europe Patent Application 93,616, 1983.
442. **Everhardus, R. H. and Van Loom, G. J. E. (Oce-Nederland BV)**, Europe Patent Application 85,447, 1983.
443. **Kakuta, A., Oka, H., Suzuki, S., Araya, K., Mori, Y., and Morishita, H. (Hitachi, Ltd.)**, Europe Patent Application 92,255, 1983.
444. **Katagiri, K., Oguchi, Y., and Takasu, Y. (Canon K. K.)**, French Patent 2,527,797, 1983.
445. **Hiro, M. (Canon K. K.)**, German Patent 3,318,282, 1983.
446. **Matsumoto, M. and Yamashita, M.**, German Patent 3,329,054, 1984.
447. **Chu, J. C. and von Hoene, D. C. (Xerox Corp.)**, U.S. Patent 4,467,023, 1984.
448. **Sasaki, M. (Ricoh Co., Ltd.)**, German Patent 3,342,724, 1984.
449. **Kato, M., Nishioka, Y., and Kaifu, K. (Oki Electric Industry Co., Ltd.)**, German Patent 3,332,005, 1984.
450. **Ueda, H. and Kameda, H. (Minolta Camera Co., Ltd.)**, German Patent 3,405,817, 1984.
451. **Ong, B. S., Koeshkerian, B., and McAneney, T. B. (Xerox Corp.)**, U.S. Patent 4,474,865, 1984.
452. **Katagiri, K., Watanabe, K., Sakai, K., Ishikawa, S., and Kitahara, M. (Canon K. K.)**, U.S. Patent 4,487,824, 1984.
453. **Horgan, A. M. (Xerox Corp.)**, British Patent Application 2,141,249, 1984.
454. **Melnyk, A. R., Griffiths, C. H., and Brach, P. J. (Xerox Corp.)**, Europe Patent Application 123,461, 1984.
455. **Nakatani, K., Hanna, J., and Kokado, H. (Mita Industrial Co., Ltd.)**, Europe Patent Application 129,413, 1984.
456. **Kato, M., Nishioka, Y., and Kaifu, K. (Oki Electric Industry Co., Ltd.)**, German Patent 3,411,070, 1984.
457. **Tokoli, E. G., Horgan, A. M., and Carmichael, K. M. (Xerox Corp.)**, German Patent 3,420,305, 1985.
458. **Mammino, J., Sypula, D. S., Abramsohn, D. A., Abkowitz, M. A., and Scharfe, M. E. (Xerox Corp.)**, U.S. Patent 4,515,882, 1985.

459. **Doessel, K. F. and Lingnau, J. (Hoechst AG),** German Patent 3,329,441, 1985.
460. **Doessel, K. F. and Lingnau, J. (Hoechst AG),** German Patent 3,329,442, 1985.
461. **Tomida, Y., Hirai, Y., Haruta, M., Nishimura, Y., Matsuda, H., and Hamamoto, T. (Canon K. K.),** German Patent 3,444,354, 1985.
462. **Lin, J. W. P., Dudek, L. P., Limburg, W. W., and Schneider, E. J. (Xerox Corp.),** U.S. Patent 4,563,408, 1986.
463. **Branston, R. E. and Duff, J. M. (Xerox Corp.),** U.S. Patent 4,557,989, 1985.
464. **Mitsuhashi, Y. (Canon K. K.),** German Patent 3,508,379, 1985.
465. **Tanaka, S., Sumino, F., and Toma, H. (Canon K. K.),** German Patent 3,505,322, 1985.
466. **Schank, R. L. (Xerox Corp.),** U.S. Patent 4,565,760, 1986.
467. **Hor, A. M. and Loutfy, R. O. (Xerox Corp.),** U.S. Patent 4,555,463, 1985.
468. **Ong, B. S., Koeshkerian, B., Murti, D. K., and Mammino, J. (Xerox Corp.),** U.S. Patent 4,562,132, 1985.
469. **McAneney, T. B., Loutfy, R. O., Ong, B. S., and Popovic, Z. D. (Xerox Corp.),** U.S. Patent 4,559,287, 1985.
470. **Tanaka, S., Sumino, F., Kubo, K., Hisamura, M., Toma, H., and Fujimura, N. (Canon K. K.),** German Patent 3,504,370, 1985.
471. **Suzuki, T., Murayama, T., Ono, H., Otsuka, S., and Nozomi, M. (Mitsubishi Chemical Industries Co., Ltd.),** Europe Patent Application 180,930, 1986.
472. **Stolka, M., McGrane, K. M., and Facci, J. S. (Xerox Corp.),** U.S. Patent 4,588,666, 1986.
473. **Fujimura, N., Sakai, K., and Okunuki, M. (Canon K. K.),** German Patent 3,541,004, 1986.
474. **Lin, J. W. and Dudek, L. P. (Xerox Corp.),** U.S. Patent 4,584,253, 1986.
475. **Diener, G. and Keil, D. (VEB Filmfabrik Wolfen, Fotochemisches Kombinat),** East German Patent 234,089, 1986.
476. **Wiedemann, W. (Hoechst AG),** German Patent 3,502,689, 1986.
477. **Mino, H. and Monden, K. (Sanyo-Kokusaku Pulp Co., Ltd.),** German Patent 3,546,007, 1987.
478. **Nogami, S., Mori, Y., and Iwabuchi, T. (Asahi Chemical Industry Co., Ltd.),** Europe Patent Application 228,202, 1987.
479. **Suzuki, T., Murayama, T., and Aramaki, S. (Mitsubishi Chemical Industries Co., Ltd.),** Europe Patent Application 219,862, 1987.
480. **Hung, Y., Regan, M. T., and Staudenmayer, W. J. (Eastman Kodak Co.),** U.S. Patent 4,666,802, 1987.
481. **Suzuki, S., Kakuta, A., Mori, Y., and Morishita, H. (Hitachi, Ltd.),** U.S. Patent 4,435,492, 1984.
482. **Tamura, S., Seta, T., Takana, T., and Hikosaka, M. (Toyo Ink Manufacturing Co., Ltd.),** German Patent 3,311,485, 1983.
483. **Pan, F. Y., Morrison, I. D., and Teuscher, L. A. (Xerox Corp.),** Europe Patent Application 104,088, 1984.
484. **Yamamoto, H., Matsuda, H., Yubakami, K., and Takashima, Y. (Matsushita Electric Industrial Co., Ltd.),** PCT International Application 85,02,470, 1985.
485. **Tamura, S., Tanaka, T., Ohashi, M., Hikosaka, M., Seta, T., and Mizukami, S. (Toyo Ink Manufacturing Co., Ltd.),** French Patent 2,551,561, 1985.
486. **Inukai, Y., Naruse, Y., Nakayama, T., and Kaneko, N. (Fuji Photo Film Co., Ltd.),** Europe Patent Application 168,636, 1986.
487. **Endo, K., Kageyama, A., Katsuya, Y., and Mori, Y. (Hitachi Chemical Co., Ltd.),** Europe Patent Application 219,300, 1987.
488. **Pillai, P. K. C., Goel, T. C., Tripathi, A., and Tripathi, A. K.,** Electrophotographic characteristics of copper phthalocyanine (CuPh)-polyvinyl cinnamate (PVCn) binder layers, *J. Mater. Sci. Lett.*, 3(3), 207, 1984.
489. **Kitamura, T.,** Photoreceptor of phthalocyanine for electrophotography, *Nippon Shashin Gakkaishi,* 46(4), 283, 1983.
490. **Kitamura, T., Kumon, A., Shoda, K., and Imamura, S.,** Studies on metal-free phthalocyanine-polymer photoreceptor doped with electron acceptor molecule. IV. Measurement of thermally stimulated current, *Denshi Shashin Gakkaishi,* 21(2), 75, 1983.
491. **Yagishita, T., Ikegami, K., and Narusawa, T.,** Copper phthalocyanine-binder photoconductor sensitized with poly(*N*-vinylcarbazole), *Fujitsu Sci. Tech. J.*, 19(3), 341, 1983.
492. **Kosaihira, T., Suzuki, Y., Takahashi, Y., Yamamoto, Y., and Sakata, T.,** Measurement of reciprocity failure in laser exposured electrophotographic photoreceptors, *Denshi Shashin Gakkaishi,* 21(3), 183, 1983.
493. **Nakatani, K., Hanna, J., and Kokado, H.,** Electrophotographic properties of phthalocyanine-PVK [poly(*N*-vinylcarbazole)] multilayered photoreceptors, *Denshi Shashin Gakkaishi,* 23(1), 2, 1984.
494. **Kitamura, T., Kumon, A., and Imamura, S.,** Carrier transport in phthalocyanine-poly(N-vinylcarbazole) photoreceptor, *Denshi Shashin Gakkaishi,* 23(1), 18, 1984.

495. **Fridkin, V. M., Shlensky, A. L., Verchovskaya, K. A., Bilke, V. D., Markevich, N. N., Pietch, H., and Sudeow, M.**, Electrophotographic process based on the pyroelectric and photovoltaic effects in the ferroelectric polymer PVF_2 [poly(vinylidene fluoride)], *Denshi Shashin Gakkaishi,* 23(3), 193, 1984.
496. **Kishi, J., Inaba, Y., Takahashi, Y., and Sakata, T.**, Interactions of β-copper phthalocyanine pigment and binder resins and its electrophotographic properties, *Denshi Shashin Gakkaishi,* 23(3), 203, 1984.
497. **Kato, M., Nishioka, Y., Kaifu, K., Kawamura, K., and Ohno, S.**, Near infrared sensitive electrophotographic photoconductors using chloroindium chlorophthalocyanine, *Appl. Phys. Lett.*, 46(2), 196, 1985.
498. **Yagishita, T., Ikegami, K., Narusawa, T., and Okuyama, H.**, Photoconduction of copper phthalocyanine-binder photoconductor sensitized with poly-*N*-vinylcarbazole, *IEEE Trans. Ind. Appl.*, IA-20(6), 1642, 1984.
499. **Kakuta, A., Mori, Y., Takano, S., Sawada, M., and Shibuya, I.**, A new phthalocyanine photoreceptor for diode laser printers, *J. Imaging Technol.*, 11(1), 7, 1985.
500. **Pillai, P. K. C., Goel, T. C., Tripathi, A. K., and Sekar, R.**, Surface potential decay characteristics of corona charged CuPh-PMMA layers, *J. Mater. Sci. Lett.*, 4(9), 1131, 1985.
501. **Baumgaertner, A., Kersten, P., Petrikat, K., and Ellinger, M.**, Basic study on the use of organomettalic compounds as photoconductors in reprography, Forschungsber. *Bundesminist. Forsch. Technol., Technol. Forsch. Entwickl.*, BMFT-FB-T 84, 102 pp, 1984.
502. **Nakatani, K., Hanna, J., and Kokado, H.**, Sensitization of carrier generation layer by electron acceptors, *Denshi Shashkin Gakkaishi,* 24(1), 2, 1985.
503. **Nakatani, K., Hanna, J., and Kokado, H.**, Influence of space charges in carrier generation layer on electrophotographic characteristics of layered photoreceptors, *Denshi Shashin Gakkaishi,* 24(1), 8, 1985.
504. **Loutfy, R. O., Hor, A. M., Dipaola-Baranyi, G., and Hsiao, C. K.**, Electrophotographic photoreceptors incorporating aggregated phthalocyanines, *J. Imaging Sci.*, 29(3), 116, 1985.
505. **Loutfy, R. O., Hsiao, C. K., Hor, A. M., and DiPaola-Baranyi, G.**, Near infrared photoreceptor devices incorporating chlorindium phthalocyanine, *J. Imaging Sci.*, 29(4), 148, 1985.
506. **Arishima, K., Tate, A., Matsuura, T., and Okada, T.**, Organic photoreceptor for electrophotography highly sensitive to infrared light, *Kenkyu Jitsuyoka Hokoku-Denki Tsushin Kenkyusho,* 34(3), 537, 1985.
507. **Nguyen, C. K. and Aizawa, M.**, The use of magnesium phthalocyanine compound as electrophotographic receptor available for laser diode recording, *Nippon Kagaku Kaishi,* 3, 393, 1986.
508. **Verkhovskaya, K. A., Fridkin, V. M., Shlenskii, A. L., Ben'kova, L. F., Vlader, N. B., and Zolotova, V. I.**, Electrophotographic process based on pyroelectric and photovoltaic effects in polymeric ferroelectric PVDF, *Zh. Nauchn. Prikl. Fotogr. Kinematogr.*, 31(3), 176, 1986.
509. **Ohaku, K., Nakano, H., Kawara, T., Yokota, S., Takenouchi, O., and Aizawa, M.**, Titanyl phthalocyanine photoreceptor devices for solid state diode laser beam printers, *Denshi Shashin Gakkaishi,* 25(3), 258, 1986.
510. **Loutfy, R. O., Hor, A. M., and Rucklidge, A.**, Near infrared photoreceptor devices incorporating evaporated chloroindium phthalocyanine, *J. Imaging Sci.*, 31(1), 31, 1987.
511. **Shimada, A., Anzai, M., Kakuta, A., and Kawanishi, T.**, Electrophotographic properties of an organic photoconductor and a triboelectric magnetic toner for a new diode laser printer, *IEEE Trans. Ind. Appl.*, IA-23(5), 804, 1987.
512. **Kovacs, G. J. and Vincett, P. S.**, An instant, dry, updatable infrared film with high sensitivity and resolution, *J. Imaging Technol.*, 12(1), 17, 1986.
513. **Chan, A. Y. C.**, Some application orientated properties of xerographic copper phthalocyanine-acrylate films, *Mater. Lett.*, 4(5,6,7), 239, 1986.
514. **Kakuta, A., Takano, S., Enokida, T., Shibuya, I., and Mori, Y.**, New organic photoconductor for diode laser printers, *Denshi Shashin Gakkaishi,* 24(2), 102, 1985.

Chapter 42

ELECTROPHOTOGRAPHIC DEVELOPERS

A number of Japanese patents relate to the use of M phthalocyanines (pcs) and di hydrogen pcs as toners in electrophotographic developers. The emphasis is on Cu pc.

Cu tetracarboxy pc is a pigment in a liquid developer for electrostatography.[1] Cu pc is a coloring agent in electrostatographic liquid developers.[2,3] Cu pc is a colorant in an electrophotographic developer.[4] Cu pc is referenced in an electrostatographic developer.[5] Di hydrogen pc is incorporated in a liquid developer for electrophotography.[6,7] Cu pc is found in an electrostatographic liquid developer.[8] Cu pc is incorporated in a toner in an electrophotographic developer.[9-12] Cu pc is in a microencapsulated toner for an electrostatographic developer.[13] Di hydrogen pc is in a toner in electrostatographic developers.[14] Cu pc is in a toner in electrophotographic development.[15] Cu pc is a charge-controlling agent-colorant for electrophotographic developers.[16] Cu pc is in a toner for electrophotographic developers.[17,18]

Cu pc is in a microencapsulated toner in electrophotographic development.[19] Cu pc is in a toner for liquid electrophotographic developers.[20] Cu pc is in a liquid developer for electrophotography.[21] Cu pc is a toner for a liquid developer for electrophotography.[22] Cu pc is in a toner for an electrophotographic green developer.[23] Pcs are a coloring agent for electrostatographic developers.[24] Cu pc is in a toner for electrophotographic liquid developers.[25,26]

Cu pc is in an electrophotographic insulating magnetic toner related to electrophotographic development.[27] Cu pc is in a toner for electrostatographic jumping development.[28,29] Cu pc is used as a nonmagnetic electric insulating toner in electrostatographic development.[30] Cu pc is in a toner related to an electrostatographic developing method.[31] β-Cu pc is found in electrostatographic dye image developer particles.[32]

Toner cores are microencapsulated with di hydrogen pc in electrophotographic developers.[33] Microcapsule toners contain Cu pc in electrostatographic development.[34] Chlorosulfonated Cu pc is found in a toner for an electrophotographic developer.[35] Di hydrogen pc is an ingredient in a toner for jumping development of electrostatographic latent images.[36]

Cu pc is a component in a toner in electrophotographic development.[37-40] Di hydrogen pc is in an aqueous alkali solution in a developable photosensitive composition containing di hydrogen pc and a polyester for lithographic plate fabrication.[41]

Cu pc is in a toner in an electrostatographic developer.[42] Cu pc is a colorant in an electrostatographic color developer.[43] α-Cu pc is situated in a core particle to provide a carrier for electrophotographic developers.[44] Cu pc is in an insulating medium in liquid electrophotographic developers.[45] β-Cu pc is part of a toner in a developer for electrostatic image development.[46] Cu pc is in a toner in an electrostatographic liquid developer.[47] A developer for electrostatic latent image development contains Cu pc.[48]

Cu pc is a component in an electrostatographic developer.[49-51] An electrophotographic color liquid developer contains Cu pc.[52] Cu pc is a coloring agent for cyan color toners in color electrophotographic development.[53,54]

Cu pc is a toner in an electrostatographic color liquid developer.[55] A pigment including Cu pc, acid treated and mixed with resin, gives a blue resin powder with a triboelectric charge of -8.2 μC/g. An electrophotographic developer obtained by mixing the powder with colloidal silica and iron powder gives clear blue images.[56]

Pcs are also referenced in U.S. patents relating to developers.

Vanadyl pc is a component in a developer composition for electrophotographic photoreceptors.[57] Developer compositions relating to electrophorectic color toners contain arylsulfonimide and Cu 1-octadecanesulfonamide pcs.[58] Vanadyl pc and vanadyl nc (nc = naphthalocyanine) are components of toners in electrostatographic developers.[59]

Chrome-complexed dye developers relating to color diffusion transfer material contain Cu *N,N',N'',N'''*-tetrakis[2-(2,5-dihydroxyphenyl)-1-methylethyl]-pc-1,8,15,22-tetrasulfonamide.[60] A cyan dye developer contains this Cu pc complex for color diffusion transfer photographic films. Copolymer mordants are described for this system.[61]

Two British patents relate to developers. An electrophotographic toner contains Cu pc for a liquid developer.[62] Magnetic particles contain Cu pc for an electrophotographic developer.[63]

Several German patents also relate to developers. Cu pc is in an encapsulated toner in an electrostatographic suspension developer.[64] Cu pc is a component in an electrophotographic developer.[65] Cu pc is a component in a liquid electrophotographic developer[66] and in an electrostatographic liquid suspension developer.[67]

Di hydrogen pc is a component in an electrophotographic liquid developer.[68] Cu pc is a component in an electrostatographic suspension developer.[69] An electrostatographic toner containing di hydrogen pc relates to a development process with controlled toner packing density for improved image quality.[70]

Cu pc is a toner in an electrophotographic magnetic brush developer.[71] An electrophotographic developer contains toners with binder, C. I. solvent blue 70-type compound, and Cu pc.[72]

A di hydrogen electrophotographic toner is related to a process for toner development involving the use of a development zone.[73] A powdered developer toner capable of being fixed in a flash-fixable process includes Cu pc.[74] Cu pc is a coloring agent for electrophotographic liquid developers in a flash process.[75]

An international patent application describes Al pc OH with reduced green absorption as a toner in an electrophotographic developer.[76]

The physical chemical properties demanded of a liquid toner for electrostatically based liquid developers containing Cu pc are outlined in a journal article.[77]

REFERENCES

1. **Ricoh Co., Ltd.,** Japan Patent 83,178,365, 1983.
2. **Ricoh Co., Ltd.,** Japan Patent 82,186,759, 1982.
3. **Ricoh Co., Ltd.,** Japan Patent 82,186,760, 1982.
4. **Toshiba Corp.,** Japan Patent 84,37,554, 1984.
5. **Konishiroku Photo Industry Co., Ltd.,** Japan Patent 84,100,445, 1984.
6. **Dainippon Printing Co., Ltd.,** Japan Patent 84,87,463, 1984.
7. **Ricoh Co., Ltd.,** Japan Patent 83,107,550, 1983.
8. **Ricoh Co.,** Japan Patent 83,57,138, 1983.
9. **Canon K. K.,** Japan Patent 84,195,661, 1984.
10. **Canon K. K.,** Japan Patent 84,197,051, 1984.
11. **Canon K. K.,** Japan Patent 84,195,663, 1984.
12. **Ricoh Co., Ltd.,** Japan Patent 84,101,656, 1984.
13. **Canon K. K.,** Japan Patent 84,187,351, 1984.
14. **Canon K. K.,** Japan Patent 84,100,455, 1984.
15. **Canon K. K.,** Japan Patent 84,231,550, 1984.
16. **Canon K. K.,** Japan Patent 84,197,045, 1984.
17. **Canon K. K.,** Japan Patent 84,197,050, 1984.
18. **Toshiba Corp.,** Japan Patent 84,176,753, 1984.
19. **Canon K. K.,** Japan Patent 84,187,357, 1984.
20. **Ricoh Co., Ltd.,** Japan Patent 84,95,543, 1984.
21. **Ricoh Co., Ltd.,** Japan Patent 84,114,549, 1984.
22. **Ricoh Co., Ltd.,** Japan Patent 84,97,157, 1984.
23. **Canon K. K.,** Japan Patent 84,197,046, 1984.

24. **Ricoh Co. Ltd.**, Japan Patent 85,13,171, 1985.
25. **Ricoh Co., Ltd.**, Japan Patent 84,139,054, 1984.
26. **Ricoh Co., Ltd.**, Japan Patent 84,114,550, 1984.
27. **Canon K. K.**, Japan Patent 84,231,549, 1984.
28. **Canon K. K.**, Japan Patent 84,198,470, 1984.
29. **Canon K. K.**, Japan Patent 84,200,266, 1984.
30. **Canon K. K.**, Japan Patent 84,200,265, 1984.
31. **Canon K. K.**, Japan Patent 84,222,850, 1984.
32. **Matsushita Electric Industrial Co. Ltd.**, Japan Patent 85,118,850, 1985.
33. **Canon K. K.**, Japan Patent 85,33,567, 1985.
34. **Canon K. K.**, Japan Patent 85,26,354, 1985.
35. **Canon K. K.**, Japan Patent 85,10,259, 1985.
36. **Canon K. K.**, Japan Patent 85,87,346, 1985.
37. **Canon K. K.**, Japan Patent 84,187,355, 1984.
38. **Canon K. K.**, Japan Patent 85,103,360, 1985.
39. **Canon K. K.**, Japan Patent 85,117,257, 1985.
40. **Ricoh Co., Ltd.**, Japan Patent 85,108,865, 1985.
41. **Nakamura, C., Oe, K., Muta, T., Sasaki, T., and Nishio, Y. (Dainippon Ink and Chemicals, Inc.)**, Japan Patent 85,165,646, 1985.
42. **Shoji, H., Haneda, S., and Hiratsuka, S. (Konishiroku Photo Industry Co., Ltd.)**, Japan Patent 85,177,359, 1985.
43. **Tomura, S., Hosoya, M., Okuyama, T., and Uehara, T. (Toshiba Corp.)**, Japan Patent 86,80,262, 1986.
44. **Yamagishi, Y., Saruwatari, N., and Ko, K. (Fujitsu Ltd.)**, Japan Patent 85,214,371, 1985.
45. **Tsubishi, K. (Ricoh Co., Ltd.)**, Japan Patent 85,221,768, 1985.
46. **Mori, H., Takagi, S., and Hyozu, Y. (Canon K. K.)**, Japan Patent 86,35,457, 1986.
47. **Tsubushi, K. and Mori, K. (Ricoh Co., Ltd.)**, Japan Patent 86,156,264, 1986.
48. **Kawabata, T. and Murayama, H. (Ricoh Co., Ltd.)**, Japan Patent 86,210,366, 1986.
49. **Imai, E., Ikeda, T., Fukumoto, H., and Tanaka, K. (Canon K. K.)**, Japan Patent 86,128,256, 1986.
50. **Imai, E., Ikeda, T., Fukumoto, H., and Tanaka, K. (Canon K. K.)**, Japan Patent 86,128,255, 1986.
51. **Fukumoto, H., Tanaka, K., and Kawagishi, Y. (Canon K. K. and Orient Chemical Industries, Ltd.)**, Japan Patent 86,141,453, 1986.
52. **Kuramoto, S. (Ricoh Co., Ltd.)**, Japan Patent 87,143,065, 1987.
53. **Canon K. K.**, Japan Patent 84,26,757, 1984.
54. **Konishiroku Photo Industry Co., Ltd.**, Japan Patent 85,131,554, 1985.
55. **Kuramoto, S., Tsubushi, K., Mori, K., Emoto, S., Fukase, T., Sugiyama, Y., Takanashi, H., Matsuyama, A., and Nonogaki, T. (Ricoh Co., Ltd.)**, Japan Patent 87,14,658, 1987.
56. **Nakamura, M., Abe, Y., Hasegawa, M., Mori, A., Kanbara, Y., Tanaka, I., and Horiguchi, S. (Dainichiseika Color and Chemicals Manufacturing Co., Ltd.)**, Japan Patent 86,20,051, 1986.
57. **Lu, C. H. (Xerox Corp.)**, U.S. Patent 4,391,890, 1983.
58. **Gruber, R. J., Julien, P. C., Yourd, R. A. III (Xerox Corp.)**, U.S. Patent 4,464,452, 1984.
59. **Barbetta, A. J., Gruber, R. J., Isganitis, L. V., Law, K. Y., Lin, J. W., Koch, R. J., and Dunn, S. T. (Xerox Corp)**, U.S. Patent 4,539,284, 1985.
60. **King, P. F. and Stroud, S. G. (Polaroid Corp.)**, U.S. Patent 4,481,278, 1984.
61. **Bronstein-Bonte, I. Y. (Polaroid Corp.)**, U.S. Patent 4,563,411, 1986.
62. **Tsubuko, K. and Hashimoto, J. (Ricoh Co., Ltd.)**, British Patent Application 2,131,566, 1984.
63. **Imai, E., Takenouchi, M., Uchiyama, M., and Suematsu, K. (Canon K. K.)**, British Patent Application 2,149,525, 1985.
64. **Podszun, W., Simm, W., Gossens, I. J., and Sueling, C. (Agfa-Gevaert AG)**, German Patent 3,232,062, 1984.
65. **Suematsu, K. and Imai, E. (Canon K. K.)**, German Patent 3,413,833, 1984.
66. **Herrmann, H. and Schlosser, H. J. (Hoechst AG)**, German Patent 3,247,925, 1984.
67. **Podszun, W., Goossens, J., Sueling, C., Hoffarth, G., Richter, W., Uytterhoeven, H., and De Winter, W. (Agfa-Gevaert AG)**, German Patent 3,426,256, 1986.
68. **Okawara, M. and Tsubuko, K. (Ricoh Co., Ltd.)**, German Patent 3,531,132, 1986.
69. **Podszun, W., Goossens, J., Sueling, C., Uytterhoeven, H., De Winter, W., Hoffarth, G., and Richter, W. (Agfa-Gevaert AG)**, German Patent 3,426,198, 1986.
70. **Suematsu, K., Takenouchi, M., Urawa, M., and Ikeda, T. (Canon K. K.)**, German Patent 3,438,430, 1985.
71. **Mitsuhashi, Y., Takenouchi, M., Tajima, H., Uchiyama, M., Urawa, M., and Okado, K. (Canon K. K.)**, German Patent 3,438,271, 1985.
72. **Inoue, S., Asahina, Y., and Ochiai, H. (Ricoh Co., Ltd.)**, German Patent 3,520,290, 1985.

73. **Suematsu, K., Ikeda, T., Yamamoto, A., Goto, K., and Imai, E. (Canon K. K.)**, German Patent 3,515,191, 1985.
74. **Sawatari, N., Yamagishi, Y., Ebisu, K., Katagiri, Y., and Narusawa, T. (Fujitsu Ltd)**, German Patent 3,542,701, 1986.
75. **Tsubuko, K., Kuramoto, S., and Nagai, K.**, German Patent 3,619,078, 1986.
76. **Regan, M. T. and Wright, H. E. (Eastman Kodak Co.)**, PCT International Application, 85,00,440, 1985.
77. **Croucher, M., Drappel, S., Duff, J., Hamer, G., Lok, K., and Wong, R.**, Colloidal and transport properties of electrostatically based liquid developers, *Photogr. Sci. Eng.*, 28(3), 119, 1984.

INDEX

D

E

F

G

H

I

L

M

N

O

P

T

U

V

W

Z